AF841148

# Quality Management in Plastics Processing

# Quality Management in Plastics Processing

## Strategies, targets, techniques and tools

**Robin Kent**
**Tangram Technology Ltd**

ELSEVIER

Elsevier
Radarweg 29, PO Box 211, 1000 AE Amsterdam, Netherlands
The Boulevard, Langford Lane, Kidlington, Oxford OX5 1GB, United Kingdom
50 Hampshire Street, 5th Floor, Cambridge, MA 02139, United States

© 2016 Elsevier Ltd. All rights reserved

No part of this publication may be reproduced or transmitted in any form or by any means, electronic or mechanical, including photocopying, recording, or any information storage and retrieval system, without permission in writing from the publisher. Details on how to seek permission, further information about the Publisher's permissions policies and our arrangements with organizations such as the Copyright Clearance Center and the Copyright Licensing Agency, can be found at our website: www.elsevier.com/permissions.

This book and the individual contributions contained in it are protected under copyright by the Publisher (other than as may be noted herein).

**Notices**
Knowledge and best practice in this field are constantly changing. As new research and experience broaden our understanding, changes in research methods, professional practices, or medical treatment may become necessary.
Practitioners and researchers must always rely on their own experience and knowledge in evaluating and using any information, methods, compounds, or experiments described herein. In using such information or methods they should be mindful of their own safety and the safety of others, including parties for whom they have a professional responsibility.

To the fullest extent of the law, neither the Publisher nor the authors, contributors, or editors, assume any liability for any injury and/or damage to persons or property as a matter of products liability, negligence or otherwise, or from any use or operation of any methods, products, instructions, or ideas contained in the material herein.

**Library of Congress Cataloging-in-Publication Data**
A catalog record for this book is available from the Library of Congress

**Library Cataloguing-in-Publication Data**
A catalogue record for this book is available from the British Library

ISBN: 978-0-0-8102-082- 1

For Information on all Elsevier publications visit our website at https://www.elsevier.com

ELSEVIER
Book Aid International
Working together to grow libraries in developing countries
www.elsevier.com • www.bookaid.org

# Contents

# Preface

This Workbook is a companion to two previous books entitled respectively 'Cost Management in Plastics Processing' and 'Energy Management in Plastics Processing'. The first of these books dealt briefly with quality management but only as part of the general topic of cost management. Quality management is, however a vital topic, and deserves more attention than was possible within the constraints of a book on cost management.

Whilst the topic of quality management has been dealt with in many other books, I felt that the treatment in most of these fell short of what was actually needed by the industry. The books tend to fall into two broad categories:

- How to achieve Quality Systems Certification, i.e. 'ISO 9001 in 3 easy lessons' or similar.
- Statistical handbooks which use high-level statistics that leave most readers stranded after the first paragraph.

Neither of these types of books is written specifically for plastics processors and many plastics processors were left wondering where to start on the 'never-ending journey' towards quality management and improvement.

As with the previous books, this book provides a structured approach to the techniques of quality management and covers the main topics of relevance to plastics processors. It is a workbook for practical use and not an academic textbook. It does not cover all aspects of the selected topics but focuses on the key quality management and technical issues for each topic. Each topic is dealt with in a single 2-page spread and most can be read independently of each other - this is not a 'cover-to-cover' book. It should be easily understood and the actions recommended should be easily undertaken by most people in the plastics processing industry.

My own introduction to quality management was short, swift and brutal. In 1983, whilst listening to a presentation by a major motor manufacturer on quality the presenter stated that every supplier was expected to have a statistically trained quality professional on their staff. I turned to my boss and said 'Do we have one of those?' He looked at me and said emphatically 'Yes'. I was both naïve and confused and said 'Who?', whereupon he looked back at me and said 'You'.

I would like to thank my friends and colleagues Kevin Chandler and Iain Brown for their comments on the draft of this book. The errors are mine alone - Kevin and Iain were probably right!

Finally, I would like to dedicate this book to our children, Emma and Matthew, they are our two greatest achievements in life and we are intensely proud of them.

Robin Kent
Hitchin, 2016
rkent@tangram.co.uk
www.tangram.co.uk

**Note:** Additional resources are available at www.tangram.co.uk/quality.html

# Chapter 1

# Introduction to quality management

Quality management was one of the most discussed topics in manufacturing industry in the late 20th century. Many plastics processors embarked enthusiastically on the pursuit of the 'Holy Grail' of quality and many companies are still trying to make sense of the confusion surrounding quality. The results for many companies have been disappointing and they are still confused about what quality really means and how they achieve it.

In a sense, this was because the discussion was diverted by the mechanics of implementing formal quality systems such as ISO 9001, either in the mistaken belief that these would deliver quality or simply to conform to the requirements of suppliers or specifications. The implementation of statistical methods that could actually deliver quality through effective process control became secondary to the filling out of forms to satisfy the requirements of the systems.

Thirty years ago, quality management and statistical process control were the 'in things' and there was great interest and activity in the subjects but today little has really changed at the grass roots level.

At many of the plastics processors that I visit, quality management still consists of a few final inspectors and the mention of statistical process control still draws blank looks.

This really has to change if you are going to be a successful plastics processor in the future.

At many sites, the real basics of quality seem to have been missed and have been substituted with lots of forms but little real management information or involvement. In this book we will go back to basics to provide the essentials for plastics processors in an easily accessible format.

This chapter looks at quality management in a broad sense to put the problems in perspective and to allow actions to be effectively targeted. It is designed to provide the essentials of the management framework for quality management.

Readers should not be tempted to rush into the practical aspects of quality management without first understanding the management framework.

"Of course we believe in quality - look at all the inspectors we employ."

# 1.1 Where we are going

## The destination

One problem at many sites is that they know that they want to have quality but they really have no idea of the steps that they have to take to get there. They set off on the journey to quality with no real planning about where they want to go, why they want to get there (if anywhere), how they are going to go about it and how they will measure the progress they are making (if any).

Some sites set off by installing a 'quality system' (of any flavour) in the faint hope that this will somehow magically deliver quality, some install complex SPC systems, some increase the numbers of inspectors and others simply subject the quality manager to various arcane management tortures.

All of these partial and disjointed measures will inevitably fail. Some will fail more miserably than others, but all are eventually doomed (particularly subjecting the quality manager to management torture - as I am sure some of the readers will know from direct experience).

Even if a site has an operating and relatively successful quality system, there is always a need to improve quality and every site needs a plan (a road-map) to define where they are and where they want to be. Quality is often poorly defined (see Section 1.3) but still needs a road-map for the actions to be taken. The figure opposite shows some of the major areas, the processes and tools used, the benefits from using the processes and tools and the overall results of a properly functioning quality system.

## This is about the whole company

A major issue is that quality is still sometimes seen as a 'bolt-on' to the other operations of the company. It is seen as a 'good thing' for the good times but optional in the bad times. The reality is that 'quality' needs to permeate the complete operation of the company. It is not a 'production' issue, it is a management issue for the whole company.

Collis Potter Huntington (see sidebar) was determined that the Newport News shipyard would be known the world over for 'good ships' and this philosophy has been consistent since 1886. The company is still a major supplier of ships to the US Navy even when other shipyards around the world have failed - the focus on 'good ships' has been the ethos to drive the company forward. How many other companies could make such a statement and live by it? How many other companies have been around since 1886?

Focusing on profit alone may make a company profitable in the short-term but, as too many companies have found out, does not guarantee that the company will survive in the long-term. Focusing on quality will not only deliver 'good ships' but also sustainable long-term profits. This means that companies need to see quality not as a cost but as a benefit and to realise that improving quality can also improve profits in both the short and the long-term.

## The road-map

The road-map identifies the wide range of skills and activities that are necessary to achieve quality and even this is limited by space. The road-map shows the type of things that you will have to do to really achieve quality. It is not simply about systems or SPC or any other single thing. Specifically, it is not only about production. It is about the whole company and it is about a mind-set that says 'quality is our guiding principle' and that this principle can deliver both sustainable profits and continued employment.

The road-map covers all areas of a company's operations from people (and their training) through product design, sales/marketing and production. The implementation of an effective quality management system (see Chapter 4) impacts every aspect of a company's operations. This is not for the faint-hearted but the rewards are more than worth the effort.

- **Tip -** None of the actions in the road-map are ever completed. Get used to continual improvement.

"Alice: Would you tell me, please, which way I ought to go from here?

The Cheshire Cat: That depends a good deal on where you want to get to.

Alice: I don't much care where.

The Cheshire Cat: Then it doesn't much matter which way you go.

Alice: ...So long as I get somewhere.

The Cheshire Cat: Oh, you're sure to do that, if only you walk long enough."

**Lewis Carroll, Alice in Wonderland**

"We shall build good ships here; at a profit if we can, at a loss if we must, but always good ships."

**Collis Potter Huntington, Newport News Shipbuilding**

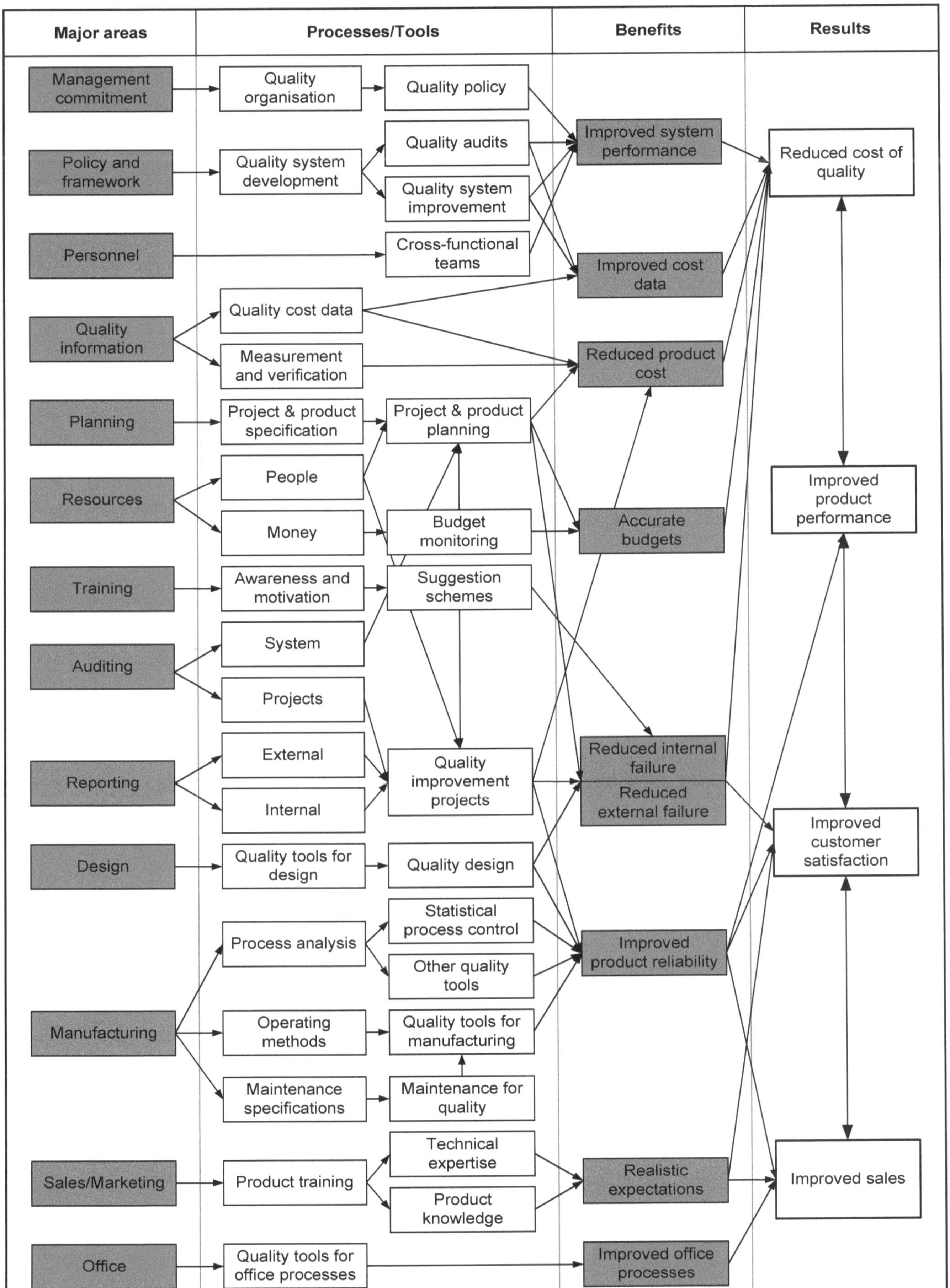


The quality road-map

Improved quality can deliver many benefits but it needs a road-map of the available processes and tools to get the best results.

# 1.2 The drivers for quality management

## What do we want?

In the last 30 years there has been a great deal of discussion about the benefits of quality and why we should want it. Everybody agrees that we want quality; however it is defined.

The only issues are how we get there and what happens along the way. The road-map (see Section 1.1) provides the directions and the tasks and should be used to map the overall process.

No manager at any company wants to force or allow his employees to produce poor-quality goods and, equally, very few employees of any company deliberately produce poor-quality goods. It is clearly in everybody's interests to produce quality products.

So what goes wrong along the way and why does it seem so hard to produce products that are of acceptable quality? What are the barriers and why does it seem all so hard?

One of the problems that many companies face is that they interpret quality as a single issue and attempt to improve quality by fixing a single variable. In fact, quality is affected by a range of variables and processes. All of these variables and processes must be aligned to produce a quality product or service. Even worse is the confusion about what quality management really is (see Section 1.5).

## What quality management delivers

Good quality management can deliver exceptional benefits to companies but even worse, poor quality management will quickly destroy a company's reputation and a company. This is as much about retaining what you have as about gaining something. Improved quality management will deliver:

### Better products

Better-quality products are an inevitable result of good quality management. This is what it is designed to do. However you define 'quality', a better-quality product is a better product.

### Improved customer satisfaction

Customers assume that the products they buy are of good quality. They do not set out to, or knowingly buy, poor-quality products. Therefore, they will be dissatisfied if the product they buy is of poor quality (and they will tell their friends). Customer satisfaction will improve if a product is of superior quality to what the customer expected.

### Improved sales

Improved customer satisfaction inevitably leads to increased sales as customers return, purchase more products and, better still, tell their friends. The most powerful and cheapest sales method of all is a recommendation from a satisfied customer.

### Greater market share

Satisfied customers will tell their friends and customers, they will return (and bring their friends) and will increase market share.

### Lower operating costs

Reducing the 'cost of quality' (see Chapter 3) can dramatically reduce operational

"The race for quality has no finish line - so technically it is more like a death march".

**Poster caption from www.despair.com.**

"Price is negotiable.

Quality is not."

**Frank Price**

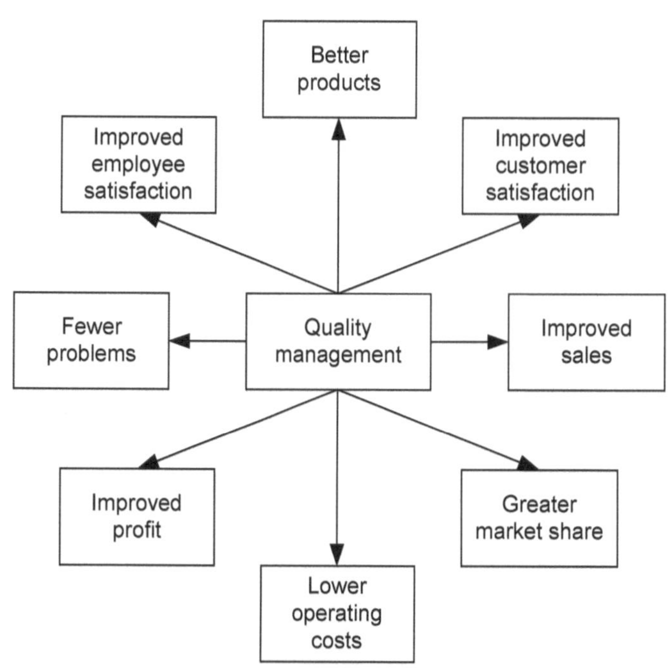


The benefits of quality management

The results of improved quality management produce benefits across a wide range of areas. Quality is a 'motherhood and apple pie' issue. Nobody wants bad quality at any time. You don't and your customers don't - so what are you doing about it?

costs by reducing scrap, reworked materials and external costs.

### Improved profit

Reduced operational costs will increase profits for any company even if sales remain the same.

### Fewer problems

Most of the time we spend at work is dealing with problems caused by poor-quality products or services. Eliminating or reducing quality problems leads to staff who are more productive.

### Improved employee satisfaction

Staff who have quality products to sell, who have fewer customer problems and who have the time to do productive work will be more motivated and more satisfied in their jobs.

## What do you need to deliver quality?

To deliver quality, plastics processors need to focus on the issues of:

### Organisation

Processors need to be organised correctly to deliver quality. Without the correct organisation, including top management support, achieving quality will be impossible.

### Systems

Systems (management and process) are an essential part of achieving quality and systems must not only be effective in quality terms but also cost-effective.

### Tools

The tools used to deliver quality are fundamental and independent of the industry sector. Many of the tools are based on statistics but most are easily understood and applied - do not be frightened by the statistics.

- **Tip** - Using the wrong tools will not deliver quality and will not be cost-effective.

### Culture

Culture is one of the hardest aspects of the issues because it is one of the many 'soft' issues that companies must deal with in operations. Cultural issues are one of the main reasons that quality programmes fail - the management cannot make the change from blaming the quality department for 'letting this stuff out the door' to asking the production department 'why did you make it wrong in the first place?'

If any one of these issues is not adequately dealt with then it is impossible to deliver a high-quality product. In this Workbook, we will deal with these issues specifically from the viewpoint of plastics processing.

**Is it really that important?**

Quality is what determines whether you will be here next year or not.

Quality products and services deliver benefits over a range of levels and the race for quality is truly never-ending.

**Who cares and what do they care about?**

- What are the customer's three main concerns?
- What are the company's three main concerns?
- What are the employee's three main concerns?

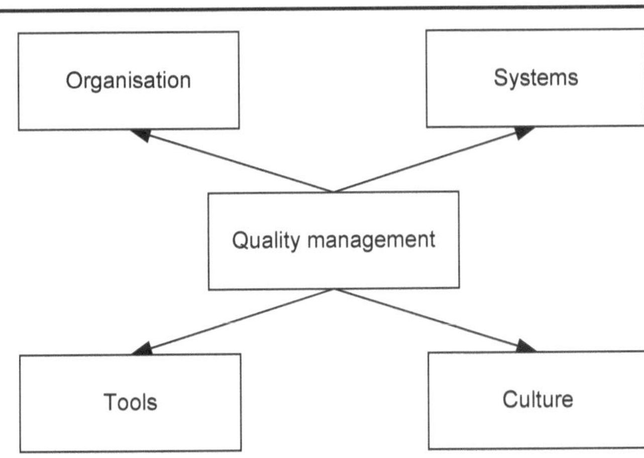


What do we need to deliver quality?

Quality is not delivered by meeting only one requirement. The only way to deliver quality is to meet all of the requirements. If any one of the requirements is missing then 'good' quality will not be delivered.

# 1.3 What is quality?

## Defining quality

Quality means many different things to different people and the apparently simple question 'What is quality?' is not always easy to answer.

When most people are asked to name a 'quality car' they will initially nominate an expensive car. This is due to the common confusion between 'expensive' and 'quality' and the linking of the two. When asked how they would rate an expensive car that always broke down then they will reconsider the question.

The reality is that an expensive car is not always a quality car. Price is a vital factor in assessing quality because we do not expect the same features or performance from a low-cost car. Our perception of quality is inevitably linked to the cost of the product and we always link quality and cost.

There are, in fact, a multitude of definitions of 'quality' (see top right) and a company can choose the one that best suits their needs. It is also important to realise that the definition of quality is not static, it is always transitory - the highest quality product of today can be an 'also-ran' product of tomorrow because of changes in competing products and prices. As the product and process environment changes so does the apparent quality of the product and process.

What was considered a 'quality' product in 1995 is definitely no longer a quality product in 2015.

- **Tip** - The customer's view of quality will change with time. Companies must be prepared to change too.

## Changing definitions

The initial definitions of quality were generally framed in terms of 'meeting specifications', 'meeting expectations' and 'customer satisfaction'. This was a challenge in itself for many companies at the beginning of the quality revolution.

Improvements in quality levels and changes in society's expectations have led to additions to the concept and definition of process and product quality. Current definitions look at 'exceeding' expectations and additional areas such as sustainability and other 'green' issues.

The definition of quality varies:

- What gives complete customer satisfaction.
- Fitness for purpose and value for money.
- Conformance to requirements.
- Conformance to specification.
- Value for money = Quality for price.
- 'On target with minimum variation' (Genichi Taguchi).
- 'Right First Time' (Frank Price).
- When the customer returns, not the product.
- When the product is what the customer wants it to be.
- The degree of excellence of something (Oxford Dictionary).
- When a product or service meets or exceeds its design specification.
- The whole package, not just what is in the box.

Quality Management

- Coordinated activities to direct and control an organisation with regard to quality (ISO 9000:2005).

Quality Assurance:

- Part of quality management focused on providing confidence that quality requirements will be fulfilled (ISO 9000:2005).
- All activities and functions concerned with the attainment of quality (BS 4778:1991).
- The activity of providing, to all concerned, the evidence needed to establish confidence that the quality function is being performed adequately (Juran).

Quality Control:

- Part of quality management focused on fulfilling quality requirements (ISO 9000:2005).
- The operational techniques and activities that sustain the product or service quality to specified requirements. It is also the use of such techniques and activities (BS 4778:1991).
- The process through which we measure actual quality performance, compare it with standards, and act on the difference (Juran).

Quality Improvement:

- Part of quality management focused on increasing the ability to fulfil quality requirements (ISO 9000:2005).

Quality System:

- The organisation and structure, responsibilities, activities, resources and events that together provide organised procedures and methods of implementation to ensure the capability of the organisation to meet quality requirements (BS 4778:1991).

Quality is no longer simply about what is in the box but considers the whole package that surrounds the product (and the box). Our concept of quality is constantly changing.

The definition of quality is now multi-dimensional and is based as much on perception as on reality. It varies with the person and with the 'brand' awareness (expectations), i.e. you may not be able to define it but you will know it when you feel it.

- **Tip** - For this Workbook we will focus on the 'hard' aspects of product and process quality and ignore the more metaphysical aspects of quality.

However we define quality internally, the only really important definition is that one used by the customer. They can define quality however they want and as the customer this is their right and they possess the ultimate sanction if the definitions vary.

- **Tip** - Be wary about trying to impose your definition of quality on the customer. They get to decide.

## Who is the customer?

Customers can be either external or internal. In terms of needing to provide quality goods and services, it doesn't matter much whether they are external or internal but external customers generally have a greater freedom of choice in their supplier.

Internal customers also need quality to carry out their functions effectively and one definition of the customer is 'the next person or process to use my output' - this can obviously be internal or external depending on where your operation is located in the supply chain. As an example, the output of the wages department is accurate wages and their customers are all internal, i.e. the staff.

- **Tip** - Think of your wages as calculated by the wages department. Is near enough good enough in your wages? Is ±10% OK? Is ±1% OK? Do you expect a quality output from the wages department to their internal customers? If we expect quality from our internal suppliers then why shouldn't our internal customers expect it?

Using this concept of the customer means that a supplier can also be a customer, e.g. the performance of the production area depends on accurate outputs from the purchasing department in terms of specifications, volume and order clarity.

- **Tip** - The fact that somebody is supplying you with goods or services does not mean that you can treat them badly or provide them with poor information and services.

## The customer's viewpoint

Despite the difficulty in defining 'quality', customers always look for quality in their suppliers. Many surveys have shown that quality is often more important than price in choosing between external suppliers. It is interesting to note that a 2013 survey showed the preferences to be:

- Quality
- Price
- Communication

There is a message there.

My first car was a 1966 Ford Anglia 105E which cost AUD$400. It ran perfectly for 5 years with the only expenses being 'consumables' as such as tyres, brakes and new exhaust.

A truly quality car for a young driver.

What is the difference between internal and external customers?

External customers have more choice.

Complaints are a chance to learn from our mistakes - do not dismiss them

'Our complaints are low so our quality must be good'

In terms of user complaints for washers and television sets only 1 out of 3 dissatisfied buyers complained to the manufacturer (although all complained to their friends)

Each appliance that fails soon after purchase results in adverse publicity that turns up to 10 potential customers against the product.

**H.E. Klein. Dun's Review and Modern industry, 1963.**

Technical Assistance Research Programs Inc. (TARP) published the results of a survey into consumer complaints. This found:

1. A manufacturer hears nothing from 96% of its dissatisfied customers. However, for every complaint received, there are, on average, another 26 customers who have a problem with the company's product or service but who do not voice their dissatisfaction. Of the 26 potential complaints, no less than 6 would be considered to be complaints of a 'serious' nature.

2. Once a consumer has complained, there is a higher probability that he will buy again from the company with which he has experienced problems than there is for a customer who has not complained and is not satisfied. This applies even when the complaint has not been resolved to the customer's satisfaction.

3. Of the consumers who complain, 70% will buy again from the supplier provided that the fault has been rectified. This figure increases to 95% for those consumers who feel that their complaint has been settled speedily.

4. On average, consumers who experience a problem tell no less than 9-10 other people. Every eighth person tells more than 20 other people about the problem which has caused, them annoyance or distress.

**TARP 1991**

# 1.4 A short history of quality management

## Since the stone age

Quality has been a manufacturing issue since the stone age when the manufacture of tools by 'flint knapping' resulted not only in large amounts of waste but also in sub-standard products that were rejected because the manufacturing process was not 'in-control' - not a lot has changed since then in many cases.

## The craft approach

The first attempts at quality were based on a craft approach where a single worker manufactured the whole object and objects were always unique. Production tooling was simple and flexible. In an attempt to control the quality of production, producers banded into guilds or companies. The guilds set 'standards', devised training schemes for apprentices and introduced certification schemes for the production quality. These 'hallmarks' for gold, silver and platinum are still used to certify material purity.

For the plastics industry, the first guild or company was the Worshipful Company of Horners which was first mentioned in 1284 - horn is a natural plastic and the Horners became associated with the plastics industry in 1943.

## The industrial revolution

The industrial revolution changed all manufacturing and led to the rise of the factory system where no single person was responsible for the complete process. Production was divided into small specialised tasks using dedicated and complex tooling and each specialised task was completed by an individual. The craftsmen became workers and became divorced from the final product. Quality was maintained by final inspection and scrapping or re-working of defective products.

## Scientific management

The early 1900's saw the introduction of 'scientific management', pioneered by Frederick Taylor. Taylor introduced 'scientific management' and the feared 'time and motion' studies to further reduce worker involvement in the product. The result was an increase in productivity but

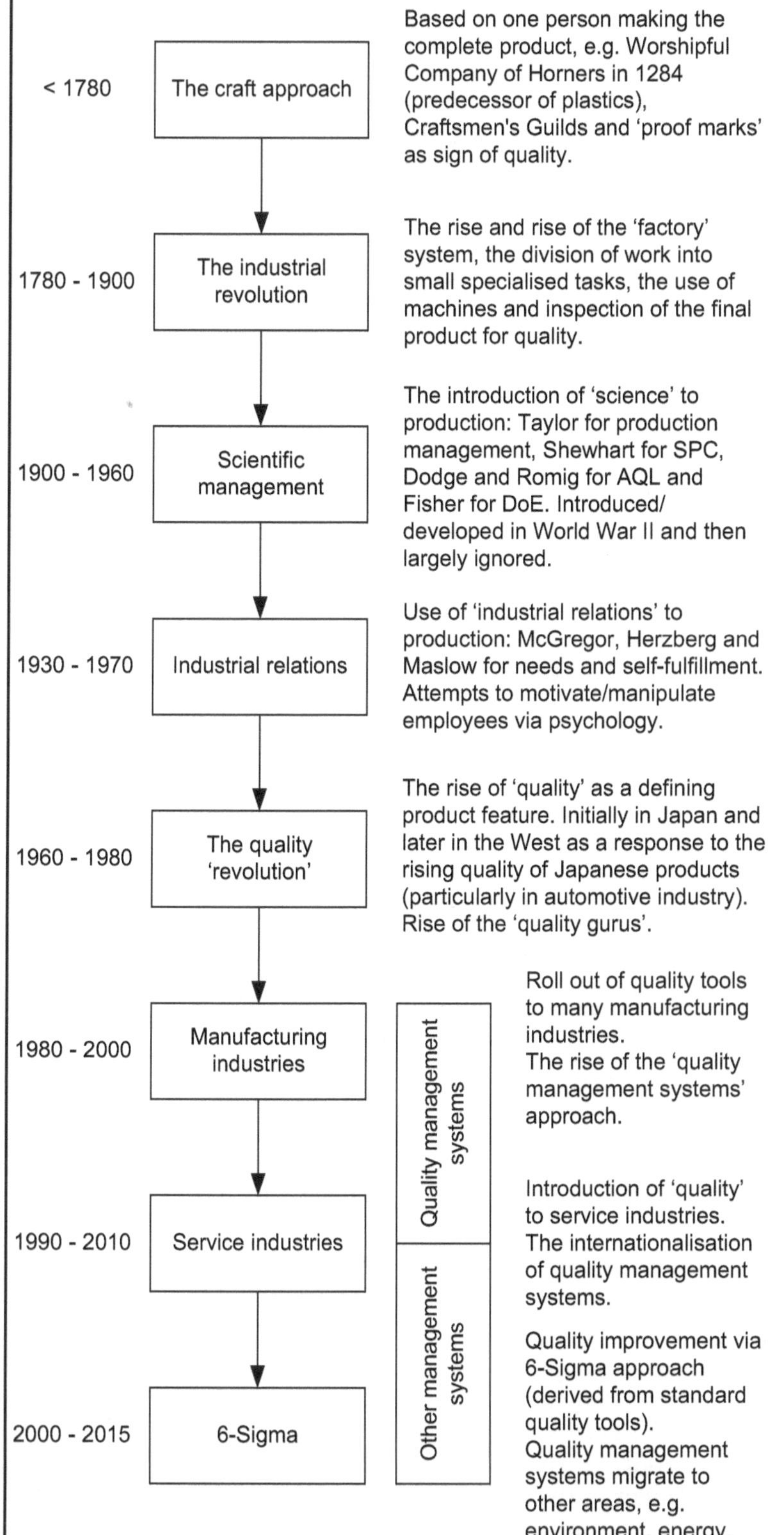


The history of quality management

Quality management has changed throughout history as new ideas, concepts and processes were developed. Quality has always been important but the methods of achieving it have changed dramatically.

often this was accompanied by a decrease in quality and the need for inspectors and quality departments.

The need to control processes to control product quality was first recognised by Shewhart at Bell Laboratories in the 1920's. Shewhart used process statistics to control the product and created the 'control chart' and much of the basis for Statistical Process Control (see Chapter 5). At the same time, Fisher was investigating the design of experiments and how to reduce the number of experiments to analyse crop experiments (see Section 8.15).

In the early 1930's Dodge and Romig (also at Bell Laboratories) developed the basic concepts of Acceptance Quality Levels (AQLs) such as acceptance sampling, including ideas such as consumer's risk, producer's risk, double sampling, lot tolerance percent defective and average outgoing quality limit (see Chapter 6).

During World War II, the need for consistency of production (millions of bullets must all fit millions of guns) led to efforts to improve production consistency across the world. Shewhart's SPC techniques and Dodge and Romig's AQL techniques were used to dramatically improve consistency (Mil-Std-105) but, at the end of the war, these were abandoned as companies went back to their 'superior' engineering approach.

## Industrial relations

In parallel with the scientific management movement, there were also significant developments in the field of industrial relations as psychologists tried to repair the damage done by the dehumanising effects of scientific management. McGregor, Herzberg and Maslow all worked to identify what motivated workers and what could be done to improve worker satisfaction and hence quality of output.

## The quality 'revolution'

The quality revolution in the West was driven by the earlier quality revolution in Japan. After the devastation of World War II, efforts to reconstruct Japanese industry included technical assistance and part of the assistance was with management. Deming and Juran both worked with Japanese industry to improve quality and developed a unique blend of technology and management techniques that revolutionised Japanese manufacturing. The Japanese did not only rely on Deming and Juran, Japanese workers such as Ishikawa, Kano, Taguchi and Shingo all contributed to developing not simply statistical techniques but many of the tools and techniques that are now considered essential to quality management.

### Manufacturing industries

The rapid rise in quality of Japanese products took the West by surprise and decimated whole sectors of Western industry, particularly the automotive and electronics industries.

The Western response was swift and the 1980's saw a rapid adoption of Japanese techniques (with variable results) in an effort to save the Western industrial base. The Total Quality Management (TQM) movement began in the USA and was championed by leaders such as Crosby who pioneered 'Right First Time' through his book 'Quality Is Free'. This movement spread throughout the world and the result was the 'quality revolution' of the 1980's and 1990's.

### Service industries

The TQM movement started in manufacturing but the techniques were just as applicable to service industries (see Chapter 11) and they were used to great effect in the late 1990's.

Unfortunately, TQM became seen as another management 'fad' (largely because management never really implemented it properly) and it has largely been replaced by the '6-Sigma' movement. This was developed at Motorola and famously implemented at General Electric. It uses many of the standard statistical tools (rebranded for a new generation) with a high reliance on reducing variation in the process.

## Quality management systems

Alongside the TQM development of tools and techniques there was a movement to institutionalise quality via the use of management systems. Initially developed in the UK in the 1970's (as BS 5750), the systems approach rapidly went international via the ISO 9000 suite of standards (see Chapter 4). Not content with institutionalising quality, the standards creators have migrated the same approach to other areas such as environment and energy.

Our concept of quality has changed with time.

The methods and techniques are important but it is our perception that makes the difference.

The development and transition of quality management into the service industries seems to have taken manufacturing industry by surprise.

Manufacturing has always regarded quality as something that was relevant only to the physical product.

Perhaps the service industries have something to teach us about quality in the services that are an essential part of what we do?

Your competitors can buy the same equipment, buy the same materials at the same prices and even employ the same type of people that you can.

The only way to really differentiate yourself is to be better in quality and people by investing in both of them.

# 1.5 Quality management and not simply quality control

## QA or QC?

'Quality management' is an umbrella term that covers all the areas of quality.

Most companies will be familiar with the terms quality assurance (QA) and quality control (QC) and carry out most of their activities in these areas. The third area of quality, i.e. quality improvement, is less often discussed or implemented with any enthusiasm.

The boundaries between these areas are shown on the right and more fully defined below.

## Quality assurance (QA)

This deals with the systems and processes that set the scene for quality. It is normally represented by a quality system to ISO 9001 and all of the support services that are necessary for this system to work.

Quality assurance does not guarantee a high quality product. It simply controls the management processes.

The early releases of BS 5750/ISO 9000 standards were written around 'quality' as a product property. As ISO 9001 expanded into service industries, e.g. solicitors, travel agents and government, the concept of quality also expanded. This new concept of quality goes beyond simple product properties and ISO 9001 is now written with a strong focus on total quality as perceived by the customer.

This aspect of quality management will largely be dealt with in Chapter 4 of this Workbook.

## Quality control (QC)

This deals with the methods of assessing the product (however defined) of the business. It is typically represented by control schedules for measurements, tests and sampling methods that assess the properties of the product or service to specifications (either internal or external). This aspect of quality management will largely be dealt with in Chapters 8 to 11 of this Workbook.

## Quality improvement (QI)

Quality improvement is an essential part of quality management but many companies actually do little to really improve quality. The early releases of BS 5750/ISO 9000 made no reference to improving the 'quality' of the product but this has now changed and a major part of ISO 9001 is now concerned with the continual improvement process via the Plan-Do-Check-Act (PDCA) cycle.

All of these components are important and it is important to realise that they are not the same thing. Having one does not imply that you have the other.

- **Tip -** How much money/time do you spend on QA, QC and QI? Is this a good balance? Should you be doing more QI?

"Cutting costs without improvement in quality is futile."

**Dr. W. E. Deming**

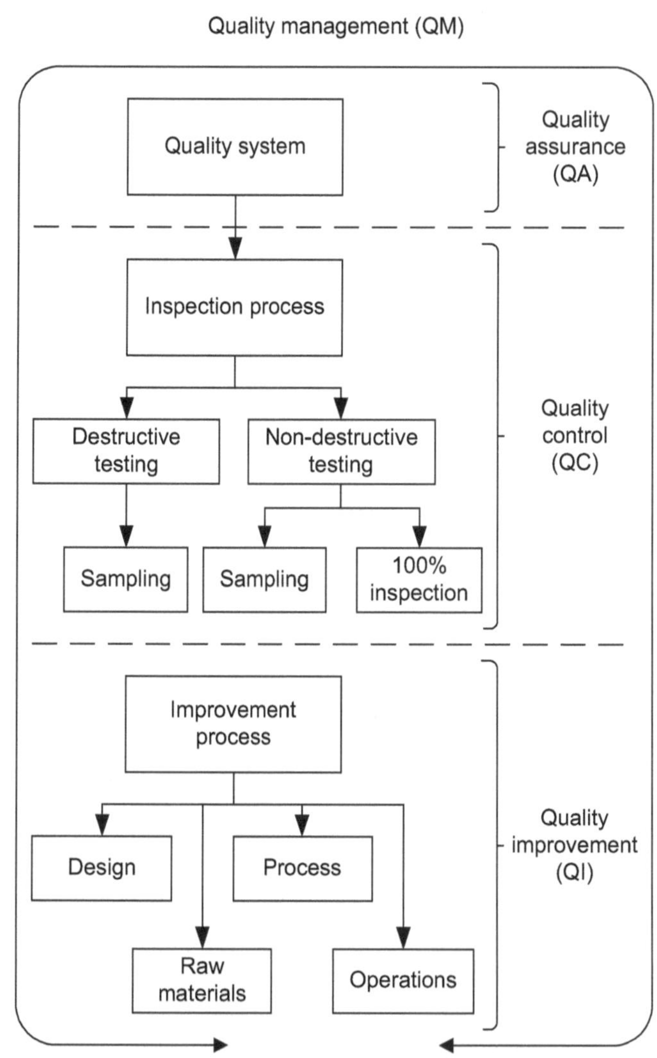


The components of quality management

Quality management has 3 components and these are all essential for real quality management. Many companies have good systems for Quality Assurance and Quality Control but real Quality Improvement is often neglected or missing.

## The quality building blocks

Quality is not simply a production concern but is the responsibility of everybody in the company. The first contact that the customer has is likely to be with the receptionist answering the phone. This initial contact can have a large effect on the quality relationship that you have with your customer. The difference between a welcoming and enthusiastic 'Good morning, how can I help?' and the gruff disinterested 'What do you want?' can alter the whole relationship. Quality is not only the responsibility of the production area or any one group or person.

Manufacturing the physical product faster, more efficiently and more effectively counts for nothing if the 'total product' produced is not of acceptable quality. This covers the total product and is not simply the physical product. The total product includes the whole range of services and contacts with the customer. The physical product itself is of great importance but it is not the complete product.

The drive for improved product quality, reduced variability, improved reliability and reduced cost really has little to do with public relations and marketing (although these are valuable spin-offs). It is about the simple issue of survival in a world where excellence is normal and not a guarantee of winning.

The building blocks for achieving true quality are:

### Management commitment

The management of the company must be committed to the concept of quality. This involves both leading the way and being the facilitators for quality improvement. The management must provide:

- The quality policy.
- The resources to get the job done.
- The organisation and framework for success.
- The training in the necessary skills.
- The delegated authority to allow the people to get the job done.

### Quality tools

These are the tools used for both QC and QI, e.g. Pareto analysis, Cause and Effect diagrams, flow charts and statistical process control (SPC). These tools are not only used for quality improvement but for all process improvement. Many companies do not use the tools effectively for either QC or QI and there is often much work to be done here in implementing and integrating the use of the tools into the complete company.

### Quality improvement

This is the discipline of never accepting that things are good enough, it is the search for never-ending quality improvement. It involves problem identification, problem solution, identifying causes and not symptoms and a continuous effort to reduce quality costs.

### Quality systems

These are the systems that make it all work and provide control, communication, continuity and confidence.

All of these building blocks must be present to succeed in the search for quality. As with the components of quality management, if any of the quality building blocks are not present then attempts to reduce costs through quality management will fail.

- **Tip** - How much money/time do you spend on the building blocks of quality? Are they all in place? Are they all effective?

**Does your quality system work?**

The lack of understanding of the essential quality building blocks is the reason that some companies complain about the lack of effectiveness of their QA system.

The presence of a quality system to ISO 9000 or any other standard will not automatically reduce costs or improve quality.

If the other building blocks are absent the quality system is seriously weakened from the start.

Quality is no longer a 'winner' it is only a ticket to the game.

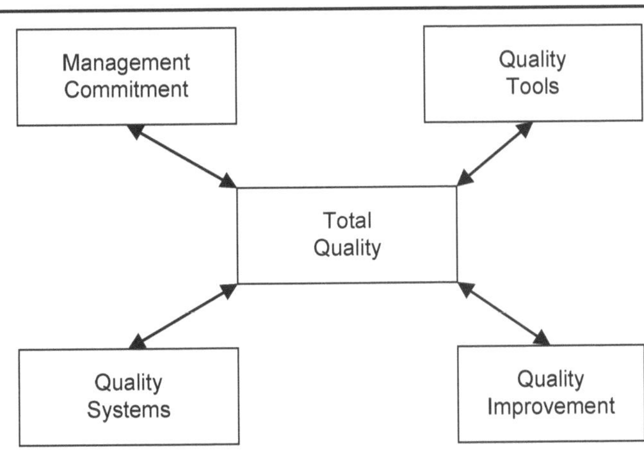


The quality building blocks

It is not enough to have some components present, all must be present for a quality programme to succeed. If any of the building blocks is missing then it will be impossible to deliver 'Total Quality' to the customer.

# 1.6 The importance of quality costs

## The terminology

In most books the terminology used is 'the cost of quality' or 'quality costs' and we will continue to use this for the sake of conformance with the existing literature. The problem with these terms is that they immediately imply that quality is a discretionary cost and that if you stop spending money on quality that your costs will decrease and profits increase. This also creates the impression that quality is a 'burden' on the company.

Nothing could be further from the truth - if you stop spending money on quality then profits will decrease and eventually the company will cease to exist.

At this stage we will look only at the size of the costs and the opportunities for improving the overall financial performance of a company. The cost of quality will be dealt with at greater length in Chapter 3.

## The magnitude of the costs

For the majority of plastics processing sites, the cost of poor quality is in the region of 20% of sales. This cost comes mainly from failure costs, i.e. where the product does not meet the requirements. This failure cost can be either an internal cost (if the product does not reach the external customer) or an external cost (if the product is dispatched to the external customer). Failure costs generally dwarf the other quality costs such as appraisal costs, i.e. inspection costs, and prevention costs, i.e. what we do to stop failure happening in the first place.

The typical 'real' costs are shown on the right and these costs will usually be significantly higher than the profit margin for a plastics processing company. Despite this, few sites are able to accurately (or even inaccurately) quantify and allocate their real quality costs. A basic understanding of the quality costs at any site will provide a real driving force for quality improvement. The exact costs for each site will depend on the operations but the 'real' costs will be approximately correct for the majority of sites.

- **Tip** - A simple walk around to look for reworked and scrapped material will often identify quality costs that the site has never really considered before.
- **Tip** - Look in the regrind areas, skips and scrap bins for a real education.

## The magnitude of the savings

The possible savings from good quality management are in the region of 50% of the current quality costs for most plastics processing sites. Savings of this magnitude represent up to 10% of sales and this is significantly higher than the profit margin for most plastics processors.

These savings can be delivered virtually irrespective of the industry sector or process used. The process used makes

Nobody wants to do it wrong and have to do it again but at many sites this is what happens day after day after day.

Quality costs are generally 5-25% of turnover.

| Actual | | Ideal | |
|---|---|---|---|
| Prevention | 5% | Prevention | 25% |
| Appraisal | 25% | Appraisal | 35% |
| Failure | 70% | Failure | 40% |
| Total cost of quality ≈ 20% of sales. | | Total cost of quality ≈ 10% of sales. | |

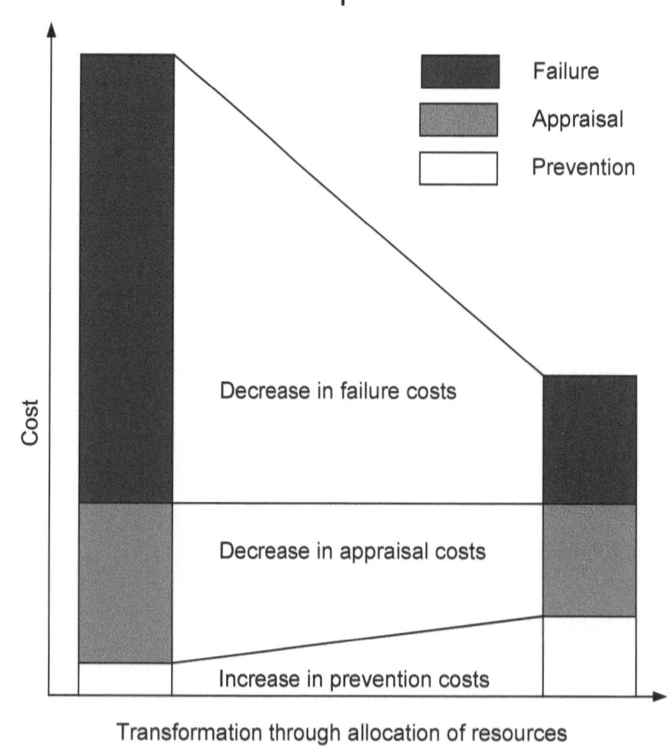


An approximate division of the cost of poor quality

In most cases, the total cost of quality is between 5-25% of sales. This easily exceeds the profits of most plastics processing companies and can be up to 5 times the profits. This can be transformed by re-allocating the resources.

little difference in the potential savings - it is management that makes the difference.

These savings are possible by changing resource allocation from dealing with the product after it has failed to preventing the product failing in the first place. Increasing resources in prevention may cost more money initially but greatly reduces later failure costs.

## The payback

The majority of the savings can be delivered through a balanced combination of no-cost, low-cost and investment (maintenance or capital) actions. The average payback for most investments in quality management is less than 12 months. This is true even when the payback is calculated using the costs for internal management efforts.

This type of payback makes investment in quality management extremely attractive from a purely financial point of view. Not many capital investment projects achieve a payback of less than 1 year and continue to deliver the benefits virtually indefinitely. Yet many sites fail to carry out attractive quality improvement projects because they have a faulty understanding of how quality can be improved. They continue to spend money on appraisal costs and accept the failure costs because these are regarded as inevitable. Capital investment proposals are still primarily presented based on direct labour reductions and rarely put forward based on reducing quality costs.

## Straight to the bottom line

In many cases, a site's quality costs are almost a discretionary cost, i.e. the site chooses to pay the cost of quality because it chooses not to do anything about it. Reducing the discretionary cost of quality at any site is directly geared to the profit of the site.

Any cost savings due to quality management translate directly to the bottom line and are shown as a direct increase in profit. Work in quality management is as valuable as work to reduce direct labour and in many cases the savings are more easily achieved because previous work to reduce direct labour has already removed the easy wins.

## Better than increased sales

When the cost of quality is reduced, it adds directly to the bottom line - something that increasing sales does not do. If the quality costs are 20% of the turnover and the profit margin is 5% then reducing quality costs by 25% (equivalent to 5% of the turnover) is the same as adding 100% to the profit. To achieve the same result through increased sales would require doubling the sales - not an easy task. What would your site do to gain new business that increased turnover and profit by 100% if you knew that:

- Your competitors couldn't stop you getting the business.
- The business was effectively guaranteed.
- The business required only internal effort.
- The business was low-risk and had a payback of less than 12 months.
- The business would continue into the future and probably increase in value.

Most sites are enthusiastic about gaining new business through increased sales. Yet when shown the opportunity to reduce quality costs and achieve the same results at the profit line, the same sites react with huge indifference. The saying is that 'turnover is vanity, profit is sanity' but unfortunately many sites still prefer the vanity of increased turnover to the sanity of increased profit.

Quality costs are the costs of not getting it right first time.

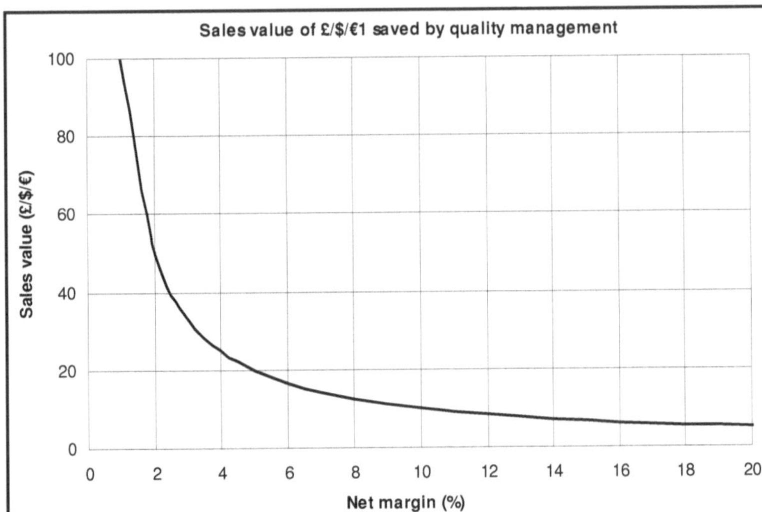


### Reducing quality costs is the equivalent of new sales

If the net margin is 10% then every £1 saved by quality management produces the same profit as £10 of new sales. If the net margin is 5% then every £1 saved is the equivalent of £20 of new sales. Savings due to quality management are easier to get than new sales and provide easy opportunities for increased profits.

# 1.7 Performance indicators

## How are you doing?

Performance indicators are a vital part of quality management and are essential for implementing and driving effective quality management. If performance indicators are not being prominently displayed, widely circulated, continually updated and widely discussed (at both the shop floor and board level) then attempts at real quality management will inevitably fail.

Performance indicators are needed to check that processes and systems are actually delivering improved quality and quantifying or measuring these results. The performance indicators chosen should not only provide information on the current status (relative to historical performance) but should also provide information on areas for potential improvement actions.

Indicators can be either internal or external, i.e. they can either reflect internal progress or external progress. Both types of indicators are valuable and necessary for a full understanding of quality management performance.

## Internal indicators

Internal indicators use measures which are only relevant internally, i.e. they are not valued by the customer. Internal indicators are typically:

- Provided by the accounting or production function.
- Real and quantifiable numbers, e.g. cost, material or time based.
- Focused on internal improvement, e.g. quality savings are quantifiable but only relevant internally.
- Can (and should) be integrated easily into the regular company reporting process.
- Can be updated monthly or weekly.

Cost of quality (see Section 1.6 and Chapter 3) is a typical internal performance indicator and is one of the best measures for driving improvement, provided the costs being reported can actually be influenced by the people who are given responsibility for them (see below). Other internal performance indicators include reject rate, scrap rate, machine downtime, etc.

## External indicators

External indicators use measures which are relevant from the customer's point of view. They are the 'voice of the customer'. This does not mean that they are less valid or that they should not be valued, in fact they can be the most valuable of all performance indicators. They provide essential feedback on performance as seen from outside the company. External indicators are typically:

- Provided by external customer surveys or other feedback, e.g. complaints data.
- Perception based rather than real and quantifiable numbers.
- **Tip** - If external customer feedback is numbers based then it usually comes as a surprise and is usually bad news. (Otherwise they wouldn't be recording it.)
- Focused on improvement of external relations with customers and suppliers.
- Are more difficult to integrate into the regular company reporting process. This does not mean that they should not be integrated, simply that it is more difficult.
- Are difficult to update monthly or weekly unless a sample of external customers is taken each month.

External indicators rely heavily on customer feedback and every company needs to establish some type of customer satisfaction scorecard for regular reporting.

- **Tip** - External indicators do not quantify savings and the results are often difficult to quantify.

### Customer scorecards

Setting up a customer scorecard involves:

- Defining the customer.
- Defining the main customer experience/ features. There should be no more than 7 features. Be lateral in thinking about what the customer wants.
- Setting a scale of 1-10, good to bad or other relevant measure.
- Trialing the initial scorecard with customers.
- Rolling the scorecard out to all customers.

1 complaint in 100 orders does not mean 99 satisfied customers!

It is deeper than that.

Try the 'phone-in' test.

Call your company and see how it feels to be a customer.

This would be a good idea for those companies who use telephone menu systems. These always seem to be designed so that you cannot ever get to where you want to go or speak to a real person.

They may sound a good idea and save money but have you ever asked your customers about them?

- Regularly convincing external customers to complete a scorecard.
- Feeding this information back into the performance indicator report.

Customer scorecards provide the essential external performance indicator and need regular refreshment as performance changes.

- **Tip** - At this stage we are only concerned with external customers although the same technique can be used for internal customers, i.e. the next person in the process who receives the output of your work.
- **Tip** - Web-based surveys can be used to reduce costs but getting customers to complete surveys can be difficult.

## Matching external indicators with internal perceptions

It is often revealing to match the external indicators with the internal perceptions of the same indicators. A quick way to do this is to use a 'snake chart' (see example on the right). This visually compares the internal view of performance with the external view of performance and can highlight areas where the company assessment of the achieved performance is very different to that of the customer. In the example on the right, the internal view of delivery reliability is greatly different to that of the customer - is there something to be learnt here?

- **Tip** - Assume the role of a customer and assess your own company giving a score and relevant comments.

## Problems with indicators

The basic reason for using performance indicators is to drive improvement but performance indicators need to be chosen with care. Focusing on the wrong performance indicator can actually drive performance in the wrong direction. Some of the problems with performance indicators are:

- Choosing performance indicators that people cannot influence. Unless it is possible to improve the indicator by action, then the indicator will be seen as irrelevant and the system will lose credibility. It is essential that the people responsible for the indicator can actually influence the result.
- Setting performance indicators that are statistically unstable and uncontrollable.
- Setting indicators that can be 'gamed' to improve the indicator but not affect real performance.

Performance indicators and some of the issues with these are dealt with in greater length in Section 3.9.

- **Tip** - Choose performance indicators with great care. Getting them wrong can be very costly.
- **Tip** - Beware of meeting targets for performance but still failing to meet internal or external expectations. Set the bar high!

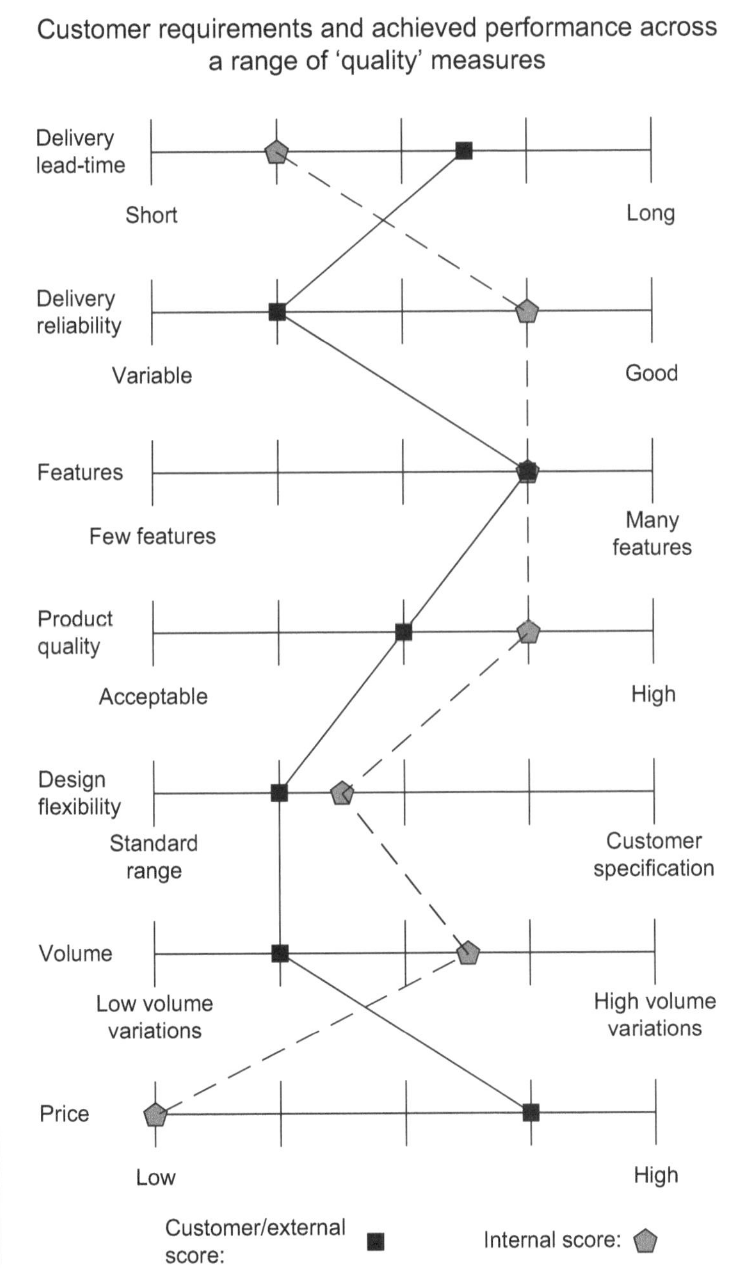


**A 'snake chart' to assess internal and external views**

A 'snake' can be used to compare the internal (company) and external (customer) views across a range of 'quality' measures. These measures do not always relate to the company vision of 'quality' but it is not the company vision that counts.

# 1.8 Quality management - where are you now?

## Where are you starting from?

Throughout this Workbook there are self-assessment charts designed to help assess the current position in a variety of areas.

The charts are easy to complete but we suggest copying the relevant pages before completing the forms.

Understanding the current situation provides the basis for an improvement strategy and many of the basic actions necessary for successful implementation.

## Completing the chart

Each chart has several columns covering various aspects of the main topic.

Read the descriptions in the column cells and select the cell that is closest to the current situation in your company.

It is unlikely that the description in the cell will fully describe a specific situation but choose the cell that has the most appropriate description. If in doubt score lower rather than higher. This will give a score ranging from 0 to 4, mark this at the base of the column.

After all the columns have been scored, transfer the scores to the radar chart for the relevant columns/axes. This gives a rapid visual assessment of the current situation for the specific topic.

If you don't know where you are starting out from then it is unlikely that you will end up where you want to get to!

**This is a team effort!**

Completing the chart on your own is not recommended. It is much better to either complete the chart as a group - you will be amazed at the divergence of opinions - or to get several people in the company to complete the chart separately and then to compare the results.

**Quality management**

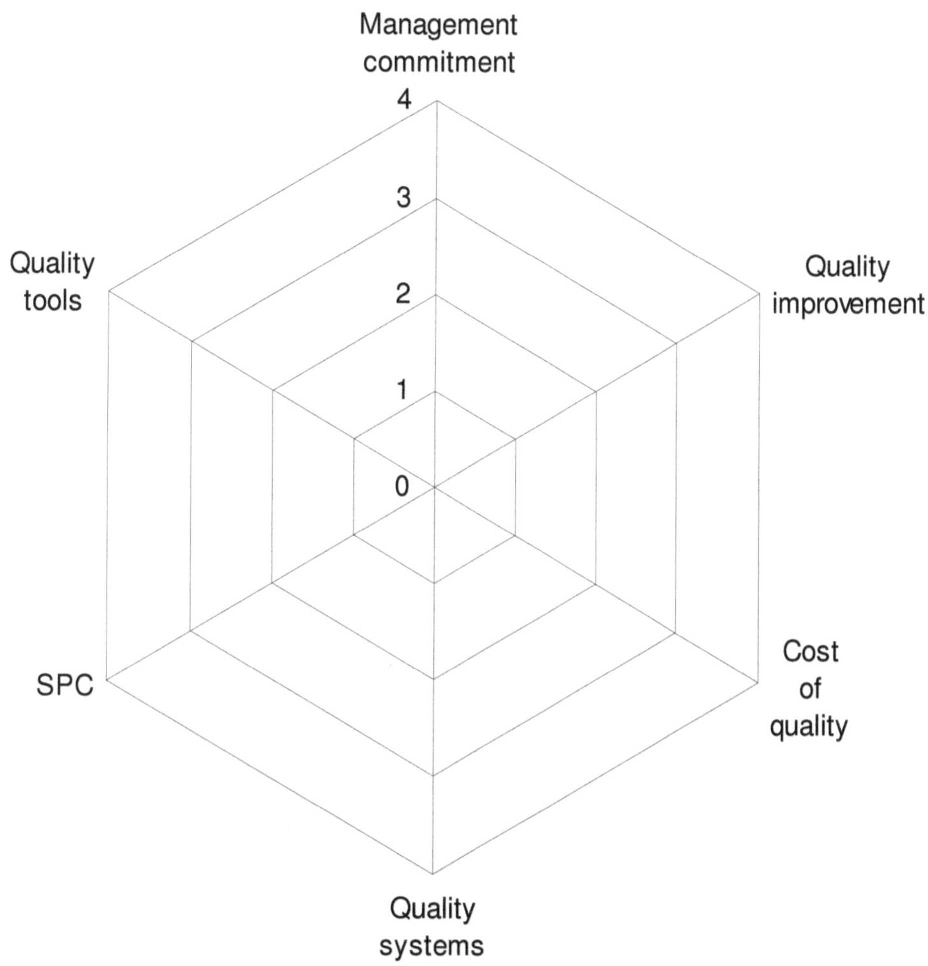


Use the scoring chart to assess where you are in quality management

The numbers from the scoring chart can be transferred to the radar chart for a quick visual assessment of where you are in terms of overall quality management.

**Assessing the results**

Ideally, a site would have balanced score with all columns/ axes in the same broad area. This is rare and in most cases, sites will show strengths in certain areas and weaknesses in others. The axes with low scores are the areas that the site needs to work on to improve the overall score.

**Download the software**

This and similar charts are available as a downloadable spreadsheet at www.tangram.co.uk/quality.

## Quality management

| Level | Management commitment | Quality improvement | Cost of quality | Quality systems | SPC | Quality tools |
|---|---|---|---|---|---|---|
| **4** | Management is totally committed. Quality is an integral part of business, all resources provided, staff trained & have delegated authority. | Quality improvement is a fundamental goal for market success. Improvement techniques used whether concerns present or not. | Full cost of quality known widely. Upper management see quality as a positive cash benefit. | Formal quality management system in place with full external verification of system. No major non-conformances found in last 2 years. | SPC used in all processes whether required by customer or not. SPC used for tooling & process equipment acceptance. | Full range of quality tools used to identify concerns, to determine root causes & to assess rectification actions. |
| **3** | Management has moderate commitment. Majority of requirements are in place but enforcement is sporadic. | Quality improvement is an important business goal. Improvement techniques only used when concerns are present & visible. | Full cost of quality known but knowledge is restricted. Upper management not aware of benefits of quality management. | Formal quality management system in place with full external verification of system. No major non-conformances found in last year. | SPC used on many processes but only if customer requirement. SPC sometimes used for tooling & process equipment acceptance. | Good knowledge & use of quality tools in analysis & problem solving. |
| **2** | Management has low commitment & only really involved when problems occur. Basic requirements are in place but not enforced. | Quality improvement is a minor goal. Improvement techniques sometimes used when concerns are present & visible. | Cost of prevention & appraisal known from accounts. Failure costs known vaguely but with no accuracy. | Formal quality management system in place with full external verification of system. Significant major non-conformances found in last year. | SPC used on some processes but only if customer requirement. SPC not used for tooling & process equipment acceptance. | Some knowledge of quality tools & often used for analysis. Problems often solved but key concerns remain unsolved & reappear. |
| **1** | Management not committed. Some aspects of quality are in place due to middle management dedication but few resources available. | Quality improvement is not seen as a goal. Improvement techniques not used even when concerns are present & visible. | Appraisal costs known from accounts but concept of prevention & failure costs unknown & no data available. | Formal quality management system in place but no external verification of system. | SPC known but not considered for use. | Little knowledge of quality tools & rarely used. When used they are not fully followed through to completion. Same concerns return time & again. |
| **0** | Management not committed. No quality policy, no resources, no training & no delegated authority. | Quality improvement is not seen as a goal. Getting the product out the door is the only goal. | Cost of quality unknown & all quality operations are seen as overhead for the business. | No formal quality management system in place. | SPC not considered or used. | No knowledge or use of quality tools. |
| **Score** | | | | | | |

# 1.9 Financial management - where are you now?

## Without money it won't happen

Quality management is the same as any other project or process - starve the process of the appropriate investment and it will fail. The only difference is that poor quality may well bring the whole company down with it. All projects, even nominally no-cost and low-cost projects need investment in staff time and much progress can be made in these areas.

Eventually, the process will require financial investment of some magnitude and this must be justified before progress can be made.

Quality management does not require preferential funding. Most quality management projects can easily meet the standard investment hurdles and analysis that are in place at most sites provided the relevant financial information is available to the site (see Chapter 3). The main concern is that quality management receives the appropriate level of funding for the benefits that it can deliver.

Financial aspects of quality are also covered in Chapter 3.

## Completing the chart

This chart is completed and assessed as for the previous chart.

**Financial management**

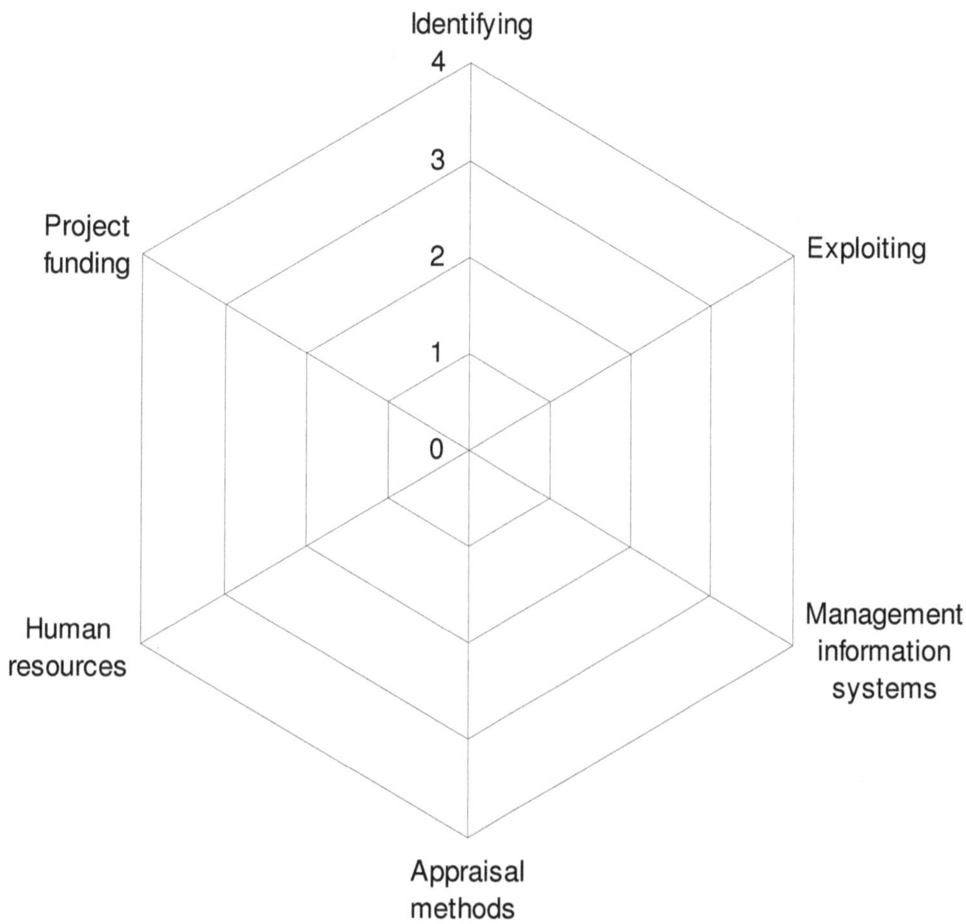


Use the scoring chart to assess where you are in financial management

The numbers from the scoring chart can be transferred to the radar chart for a quick visual assessment of where you are in the basics of financial management.

Areas of disagreement may show that there is something happening at the site that is not well known, i.e. the Finance function is actually keeping records but is not telling the Production function that they are doing so.

## Financial management

| Level | Identifying | Exploiting | Management information systems | Appraisal methods | Human resources | Project funding |
|---|---|---|---|---|---|---|
| 4 | Detailed cost of quality information regularly updated. List of quality improvement opportunities already costed & ready to proceed. | Formal requirement to identify the most cost-effective projects. Decisions made on the basis of total cost of quality. | Full MIS enabling identification of past savings & further opportunities for investment meeting organisation's financial parameters. | Full discounting methods using internal rate of return & ranking priority projects as part of an ongoing investment strategy. | Board take a proactive approach to long-term investment as part of a detailed quality policy in full support of the quality team. | Projects compete equally with other areas. Full account taken of indirect benefits, e.g. marketing opportunities. |
| 3 | Regular quality audits by experienced staff to identify quality improvement opportunities likely to yield largest savings. | Quality staff are required to comment on all projects. Quality improvement projects are sometimes considered in terms of total cost of quality. | Promising proposals are presented to decision-makers but insufficient information (e.g. sensitivity or risk analysis) results in delays or rejections. | Discounting methods using organisation's specified discount rates. | Quality manager working well with accounts/finance department to present well-argued cases to decision makers. | Projects compete for capital along with other business opportunities, but have to meet more stringent requirements for return on investment. |
| 2 | Sporadic quality audits used to identify quality improvement opportunities & potential savings but few records kept. | Quality staff notified of all proposals that affect quality. Proposals for quality improvements are at risk when capital is scarce. | Adequate management information available, but not in the correct format or easily accessed in support of quality improvement proposals. | Undiscounted appraisal methods, e.g. gross return on capital. | Occasional proposals to decision makers by quality managers with limited success & only marginal interest from decision makers. | Quality projects not formally considered for funding from capital budget, except when very short-term returns are evident. |
| 1 | Informal quality surveys conducted by staff with checklists to identify quality improvement opportunities. | Quality staff use informal contacts to identify projects where quality can be improved at marginal cost. | Insufficient information to demonstrate whether previous investment in quality improvement has been worthwhile. | Simple payback criteria are applied. No account taken of lifetime of the investment. | Responsibility unclear & those involved lack time, expertise & resources to identify projects & prepare proposals. | Funding only available from revenue on low-risk projects with paybacks of less than one year. |
| 0 | No mechanism or resources to identify quality improvement opportunities. | Quality improvement projects are rarely considered as being financially worthwhile. | Little or no information available to develop a case for funding. | No method used irrespective of the attractiveness of a project. | No-one in organisation promoting investment in quality improvement. | No funding available for quality projects. No funding in the past. |
| **Score** | | | | | | |

# 1.10 Technical management - where are you now?

## So where is the knowledge?

Quality management is often seen as 'the low man on the totem pole' and coming a very poor second to production volume.

Technical understanding of quality is often poor and it is considered to be an afterthought to the production process. Quality improvement requires a good technical knowledge of both quality issues and the processes used.

This is often absent both in the quality area and in the production area. The resulting contest between the Production Manager and the Quality Manager does not deliver quality - they need to be cooperating and not competing.

This chart provides an overview of the technical aspects of quality management from both the quality and the production aspects.

Even when the majority of the plant was not sourced with quality in mind there should be quality improvement plans in place to improve the process or to at least make the sales efforts align with the capability.

## Completing the chart

This chart is completed and assessed as for those presented previously.

Agreement by various people across the site shows consistency - a starting point for improvement.

**Technical management**

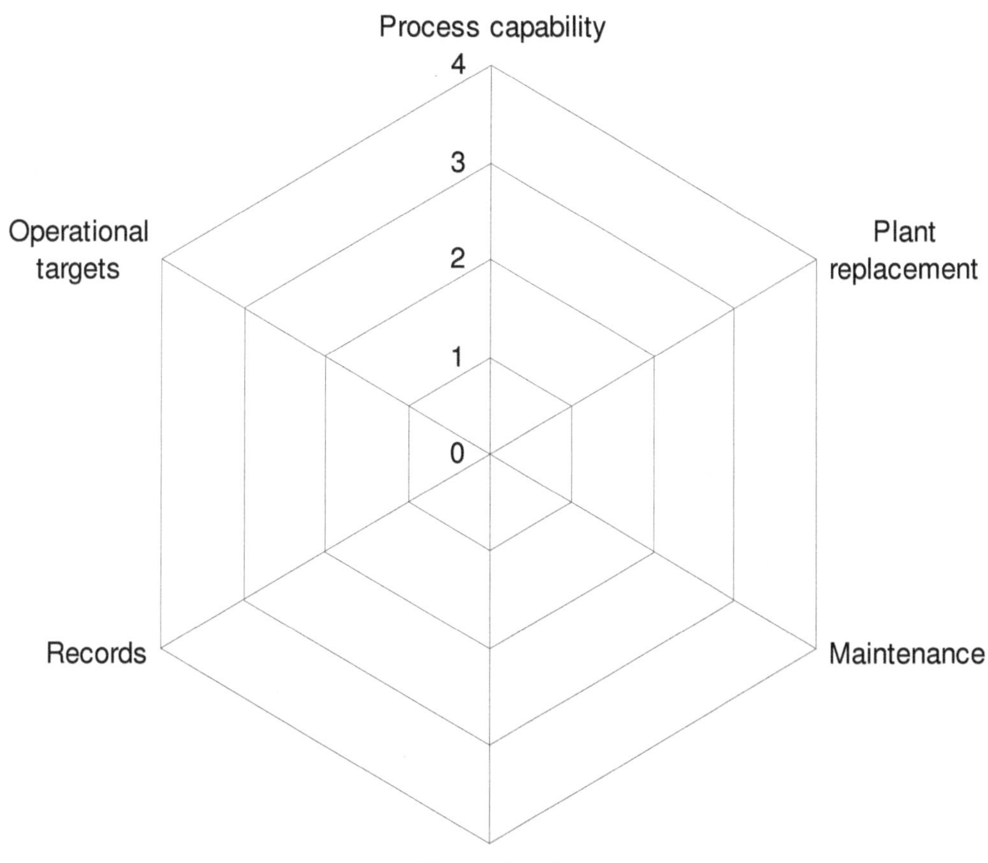


Use the scoring chart to assess where you are in technical management

The numbers from the scoring chart can be transferred to the radar chart for a quick visual assessment of where you are in the basics of technical management.

Operators often know how to improve quality by changing operational methods. Despite this, their knowledge is routinely ignored in favour of the less relevant knowledge of someone in an office.

An opportunity lost and a huge pity.

## Technical management

| Level | Process capability | Plant replacement | Maintenance | Operational knowledge | Records | Operational targets |
|---|---|---|---|---|---|---|
| 4 | Capability of all existing processes known & improvement plans in place. Process capability used as part of sales process. | Plant chosen is the most appropriate for application. Capability & consistency are major factors in selection. | Maintenance is based on needs, with condition appraisal used for all equipment affecting quality. Results acted upon. | Staff know how their actions affect quality & take positive steps to improve quality. Staff have targeted training in quality issues. | Detailed capability studies & control charts (or equivalent) available for all processes & products. Detailed schedules of quality processes & controls. | Targets set using historical performance with improvement plans in place for all products. Monitored for actual cost of quality. |
| 3 | Capability of major processes known but no improvement plans in place. Process capability not used as part of sales process. | Plant chosen is appropriate for application with capability considered. Capability is not a major factor in purchase decision. | Regular condition surveys carried out on equipment affecting quality. Action undertaken for most defects identified. | Staff are aware of how they affect quality & take good housekeeping measures to maintain quality. Training on a regular basis. | Some capability studies & control charts (or equivalent) for some processes & products. Some schedules of quality processes & controls. | Targets set using historical performance with no improvement plans in place. Monitored for actual cost of quality. |
| 2 | Capability of very few processes is known. No plans in place to determine capability & no improvement plans in place. | Plant selected to be fit for purpose. Little consideration of capability & assumed to be provided by supplier. | Condition surveys carried out regularly on all equipment affecting quality. Remedial work constrained by budgets. | Most good housekeeping practices are adhered to in an attempt to maintain quality. Occasional quality training received but primarily as part of operational training. | Few capability studies & control charts (or equivalent) available. Few schedules of quality processes & controls. | Some targets set but against unrealistic budgets. Little control via financial procedures to monitor the cost of quality. |
| 1 | Process capability not known for any processes. Process capability sometimes less than required for existing products. | Process capability & machine data on plant rarely obtained as part of selection process. Initial price is the major factor in decision. | Condition surveys carried out occasionally, prompted by plant failure or safety considerations. Remedial work only carried out on major defects. | Quality techniques are only adopted where they can be easily accommodated within traditional working practices. | No capability studies & control charts (or equivalent) for any processes or products. Minimal or poor-quality control records. | Few targets set. Those that are set are by default through budget-setting procedures that nobody understands. |
| 0 | Process capability not known for any processes. Process capability often less than required for existing products (with subsequent rejections & failures). | No consideration of process capability in equipment selection. | No regular surveys or maintenance carried out. | No consideration is given to quality during working operations. | No records available. | No targets set. 'We will make quality products.' |
| **Score** | | | | | | |

# 1.11 Awareness and information - where are you now?

## What is happening with quality?

As with any activity there is a need to both specify what people are going to do and to ensure that they are aware of their responsibilities.

One of the keys to any management task is to 'show results to get resources' and there is a need for clear reporting of successes in quality management to both get resources and to motivate the team.

Equally there is a need to provide staff with training and development opportunities. A training course on SPC may appear a luxury but if it saves real money by reducing defects and reducing variability then it is a good investment in both the staff and the company.

Quality management has developed considerably in the last 30 years and if the only training carried out is for an inspector on 'how to use a micrometer' then this is simply not enough. There is more at stake here.

Keeping staff well trained and up-to-date with the latest market developments will pay rewards in the short and long term.

## Completing the chart

This chart is completed and assessed as for those presented previously.

Low scores are not bad but simply show areas with improvement potential.

**Awareness and information**

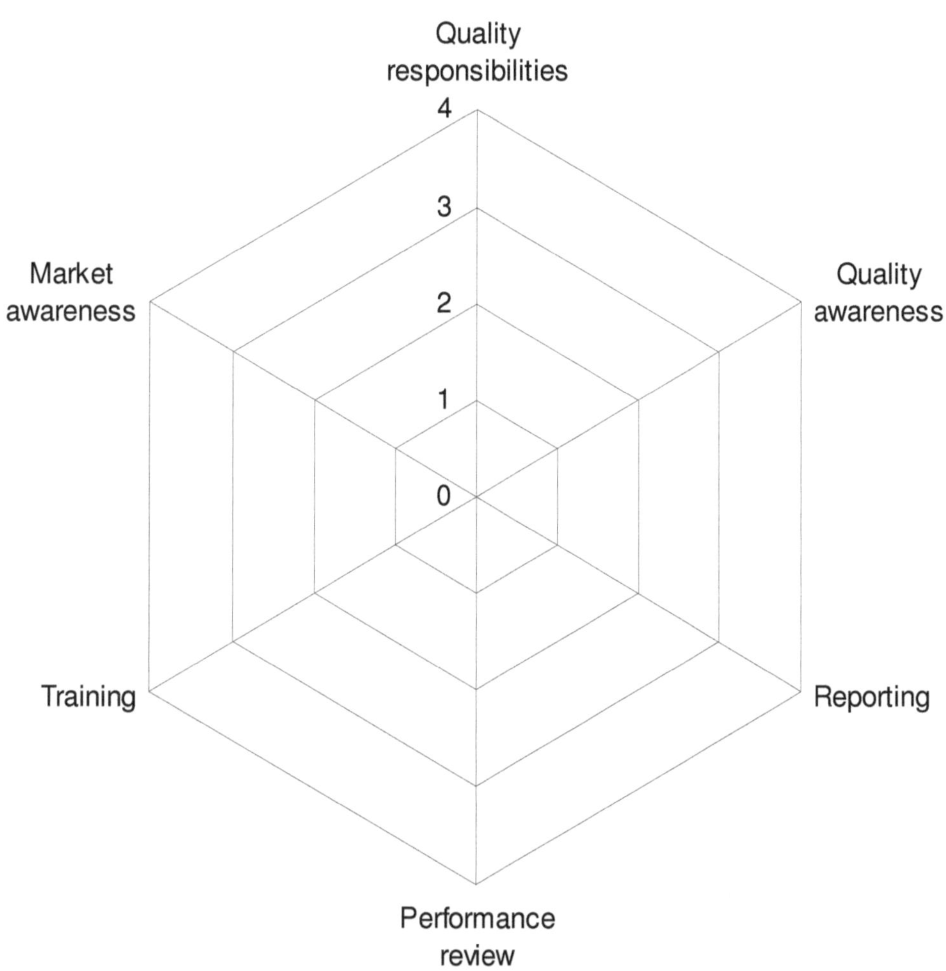


Use the scoring chart to assess where you are in awareness & information

The numbers from the scoring chart can be transferred to the radar chart for a quick visual assessment of where you are in the basics of awareness and information.

If you don't know that you are doing it wrong then how can you hope to improve?

What's worse: Training someone and risking they might leave, or NOT training someone and having them stay?

If you think training is expensive then consider the cost of ignorance.

| Awareness and information | | | | | | |
|---|---|---|---|---|---|---|
| Level | Quality responsibilities | Quality awareness | Reporting | Performance review | Training | Market awareness |
| 4 | Lists of responsibilities & their assignment exist & are comprehensive & regularly reviewed. All staff have responsibilities. | Quality performance regularly given to all staff. Full use made of publicity. All methods used to promote new measures for improving quality. | Wide reporting of current status compared with best practice, on regular basis & geared at a range of audiences. Full support to public statements. | Progress regularly reviewed. Performance compared against internal & external benchmarks. Ideas actively sought. | Training properly resourced for all staff. Active quality library. All staff have access to information on quality techniques & tools. | Company keeps up to date with all technical developments by monitoring of trade journals, literature & other sources on issues affecting quality. |
| 3 | Lists of responsibilities & their assignment exist for key quality staff & all departments. | Quality status presented to all staff at least annually. Occasional but widespread publicity to promote quality improvements. | Current status reports issued annually to shareholders & staff. Impartial reporting of performance to staff & departments on a regular basis. | Frequent quality reviews using performance & cost data. Analysis is regular, wide-ranging but ritualistic. | Continuous training & development for all staff. All staff are aware of & have access to a quality library. | Regular studies carried out on trade journals, literature & other sources to assess current developments impacting on quality. |
| 2 | Some staff & departments have written responsibilities. | Quality performance presented to staff on a regular basis. Occasional use of publicity to promote quality improvements. | Occasional issue of quality status reports. Concentrates on bad news. | Occasional technical quality reviews. Regular cost checks with exception reporting. Analysis of limited scope. | Staff development & training is a personal responsibility. Occasional initiatives to train staff in quality. | Trade journals, literature & other sources occasionally scanned for information on the latest developments in techniques & tools. |
| 1 | Unwritten set of responsibility assignments. | Quality performance occasionally reported & known to very few staff. Quality improvement measures are rarely promoted. | Reports only issued if prompted by a quality failure. Most reports will contain only bad news. | Quality review activity based simply on requirements of quality systems standards. | Few staff have knowledge of quality techniques & tools. Little training in quality for staff. | Trade journals, literature & other sources rarely studied for quality implications of production. |
| 0 | No evidence of assignment of quality tasks & duties. | No staff have explicit responsibilities or duties. | No reporting. | No monitoring activity to underpin review processes. | Staff have little knowledge of quality techniques & tools. No attempt to inform staff of techniques & benefits of quality improvements. | Quality is not a consideration when keeping up to date on products or technology. |
| Score | | | | | | |

# Key tips

- Quality is broader than simply 'product' quality. It is about the totality of the service that the company provides (including the product).
- The journey to 'quality' is not static - Quality is not a destination, it is a state of mind.
- Quality management needs a 'road map' to keep control of all the areas that need to be addressed.
- Quality management can deliver a multitude of benefits both inside and outside a company.
- Quality is not delivered simply by installing a quality system. It needs organisation, systems, tools and the right culture.
- Quality can be defined in many ways but the real definition is in the mind of the customer.
- Customers are always concerned with the quality of a product or service.
- The methods of achieving quality have changed with time and will continue to change. The one constant is the benefit it brings to a company.
- Quality management is not simply about quality control, it is about quality assurance, quality control and quality improvement.
- Without quality improvement the full benefits of quality management will not be achieved.
- Quality is not simply a production issues, it is an issue for the complete company.
- The building blocks for quality (management commitment, quality tools, quality systems and quality improvement) all need to be in place to achieve quality management.
- Quality costs are really the 'cost of poor quality'.
- The actual 'cost of quality' for most plastics processors is approximately 20% of sales. This is far higher than the profit margin for most processors.
- It is possible to reduce the actual 'cost of quality' to approximately 10% of sales and considerably increase profits for most plastics processors.
- Reducing the cost of quality is highly geared and the savings go straight to the bottom line.
- The payback time for projects to improve quality and reduce the cost of quality is very short.
- Reducing the cost of quality is better than increasing sales.
- Reducing the cost of quality can dramatically improve profits.
- Reducing the cost of quality is low-risk.
- Performance indicators can be either for internal or external. They are very different.
- Internal indicators are mainly data based and are only relevant for internal use.
- External indicators are mainly perception based and can be relevant for internal or external use.
- Understanding and matching external indicators to internal performance is critical.
- Performance indicators need to be chosen with care.
- Performance indicators need to be capable of being improved by the people responsible, statistically stable and relevant to the improvements needed.

# Chapter 2

# Organising for quality management

Quality management is the same as any other type of management and requires appropriate organisation to carry out effectively. This chapter reviews how a company can be organised to improve quality management and to manage the essential improvement process. This is not the same as discussing 'quality management systems', these are specified in ISO 9001 which sets out the international requirements for quality management systems and ISO 9001 is covered in depth in Chapter 4.

Instead, this chapter deals with the organisational aspects of quality and tries to answer the questions:

- How do you organise your company for effective quality management?
- How can the essential process of quality improvement be organised and stimulated?
- How do you find, stimulate and complete quality improvement projects?

Quality can only be monitored and improved within the organisational structure of a company and many companies get this fundamentally wrong. If this is in any doubt then ask yourself 'Who gets yelled at when quality falls or when we have a quality problem?'. If it is the Quality Manager then management fundamentally misunderstands whose job it is to produce quality products, i.e. the Production Manager - the Quality Manager is there to try to improve quality and keep the score, he cannot be held responsible for the actions of the Production Department if it makes faulty products. If this happens then clearly the organisational structure and concepts are wrong, the allocation of responsibilities is wrong and the company needs to revisit both. We look at how companies can get started in effective quality management and some of the organisational arrangements that are necessary and available for improving quality.

In the words of Juran, "Quality is improved project by project" and therefore we also look at the essentials of quality improvement projects, how these can be found, the organisation necessary to carry them out and how to deliver improved quality.

"The bitterness of poor quality lingers long after the sweetness of cheap price is forgotten."

**Anon**

# 2.1 Getting started

## Management commitment

The key element to making quality management work is top management commitment. Without this, then nothing else will work. Employee engagement and assistance is vital but without top management commitment then nothing else will work. If the top management is not committed then you might as well use this book to prop up a desk or hold a door open. Do not even try to start work until top management has made a firm and irrevocable commitment to quality management.

## The role of management

What does management need to provide to show commitment and how can they do it effectively? Some of the elements are:

- Quality policy - the management sets the tone of the process by producing the company quality policy. This is not an exercise in 'copying and pasting' a standard policy. It should be self-generated and really be the policy of the company. The policy should be generated by the top management and not by the quality department - let them think about it and express it in their own words.
- Allocation of resources - any process that is starved of resources will inevitably fail and quality management is no exception. Quality management, particularly the improvement process, needs resources. In time, these can be internally generated (see Chapter 3) but initially there will be a need for additional resources. Top management should be prepared for initial additional costs.
- Organisation and structure - many companies are not well organised for quality assurance and control (see Section 2.3). They fail to see the essential difference between the two and set the best organisation and structure in place.
- **Tip** - If the Quality Manager reports to the Production Director then it is the equivalent to expecting the fox to guard the henhouse.
- Quality improvement - whilst most companies have structures (good or bad) for quality assurance and control, few have any structure at all for quality improvement (see Section 2.5). This is simply poor planning for the future of the company and indicates a management weakness. Management needs to understand that quality improvement is a financial benefit to the company and an essential part of quality management.
- Motivation, awareness and teamwork - the top management sets the scene for quality management and the culture of the company. They need to encourage quality improvement, raise awareness of the real affects of poor quality and encourage teamwork both inside departments (quality circles) and across departmental boundaries (quality improvement teams). This needs motivation and awareness at all levels of the company (particularly for the top management).
- Creating the quality climate - above all, top management needs to create a quality climate where quality is seen as the issue that will decide if the company is in business next year or not. Poor profits can stop expansion and growth, poor quality will kill the company. Only top management can create this climate.

Managers are there to manage improvement.

The one question you have to answer to at the end of each year is - what have you improved?

You get what you measure.

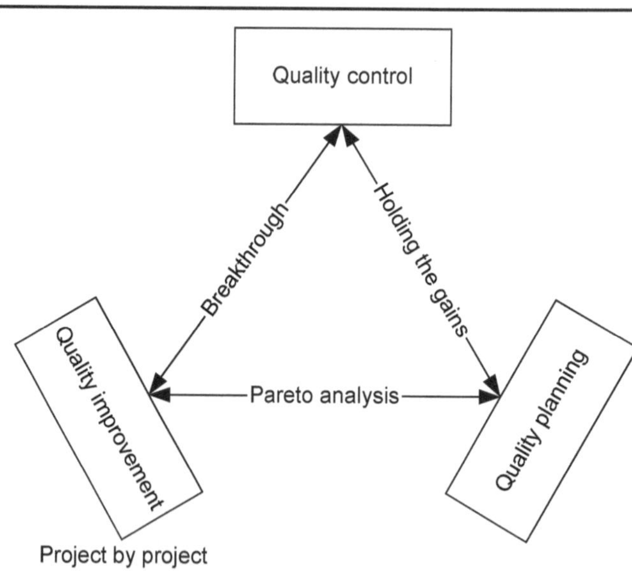


**Quality improvement is an essential part of quality management**

Quality control and planning are not enough. Quality improvement is needed to stay in business. Without continuous improvement a company is doomed to eventual failure.

## Systems and tools are not enough

Many managers think that because they have got ISO 9001 (see Chapter 4) and are making some use of SPC (see Chapter 5) that their work in quality management is complete. This completely ignores the essential part that quality improvement must take in quality management. Quality management is not complete without quality improvement.

- **Tip -** Many companies are more interested in the ISO 9001 logo than in real quality management. The 'Logoland' approach is not what quality management is about.

## The barriers to quality improvement

The road to quality improvement is not without barriers, the most common are:

- Non-acceptance of management responsibility - top management needs to realise that in most cases the causes of poor quality are the systems that they set up, not with the workers who operate them (see sidebar on the right).
- Separation of responsibility and power - if the Production Director has the power but the Quality Manager has the responsibility then all is lost.
- Lack of knowledge or awareness.
- Misunderstanding of quality costs - see Chapter 3.
- Fear of statistics - see Chapter 5.
- Lack of knowledge of problem causes.
- **Tip -** A sign saying 'Quality begins with you' often has the sub-text 'not with me' (written by the top management).

## Deming's points and diseases

One of the seminal publications on how to drive quality in a company was Deming's 14 points for management and his 5 deadly diseases. These are shown on the right and every company should review their performance against these points and diseases.

**Note:** For companies operating in the USA there were an additional 2 diseases:

- Excessive medical costs.
- Excessive costs of warranty from lawyers working on a contingency basis (no win - no fee).

These 2 diseases are not relevant in all parts of the world.

## The acid test

- When did quality management and improvement last make it onto the Board agenda as an important topic?

**Note 1:** Do not include when the Sales Director was complaining about how a quality issue affected customer relations and why are we paying the Quality Manager so much (whilst the Production Director looked on smugly).

**Note 2:** Do not include when the Financial Director was complaining about how much poor quality has impacted on the results this year when he turned down 4 quality improvement investments.

- Are quality costs and savings ever discussed at the Board meetings?
- Is quality a standing item on the Board agenda?
- Do they really care?

"85% of quality problems are caused by the system.

15% are caused by the workers.

Workers work *in* the system.

Management should work *on* the system."

**Dr. W. E. Deming**

For every action, there is an equal and opposite criticism.

### The 14 points

1. Consistency of purpose.
2. The new philosophy.
3. Cease mass inspection.
4. End 'lowest tender' contracts.
5. Constantly improve systems.
6. Institute training.
7. Institute leadership.
8. Drive out fear.
9. Break down barriers.
10. Eliminate exhortations.
11. Eliminate targets.
12. Permit pride of workmanship.
13. Encourage education.
14. Top management's commitment.

### The 5 deadly diseases

1. Lack of constancy of purpose.
2. Emphasis on short-term profits.
3. Evaluation by performance, merit rating or annual performance.
4. Mobility of management.
5. Running a company on visible figures alone.

### Deming's 14 points and 5 deadly diseases

At least some of Deming's 'points' and 'diseases' are present in every company. We know they are 'good' (or 'bad') but companies really cannot seem to stop themselves!

## 2.2 Who is responsible?

### Quality assurance

Function:

- Develop and maintain the quality system.

Quality management responsibilities:

- Ensure that all aspects of the company meet the quality system requirements.
- Prepare quality manuals or other documents for the quality system.
- Maintain records and documents of the quality system, e.g. procedures, works instructions.
- Ensure regular auditing (internal and external) of the quality system.
- Ensure management review of the quality system.

### Design/Research and Development

Function:

- Produce Product Design Specification (PDS) to document all product requirements.
- Translate customer requirements (internal or external) into final design and relevant drawings for internal or external use.

Quality management responsibilities:

- Chair Design Review meetings to produce PDS.
- Obtain sign-off from other departments for PDS.
- Produce production drawings, control drawings and other product drawings.
- Control drawing issue and release.
- Ensure that final design meets customer requirements (internal or external).
- Ensure that final design meets all relevant regulatory requirements.
- Ensure that the final design is capable of production (in conjunction with production engineering).

### Marketing/Sales

Function:

- Create positive perception of company in the market and take and process customer orders.

Quality management responsibilities:

- Assess market needs and demands.
- Present realistic product capabilities to the customer.
- Provide feedback from the customer on product improvements and competitors' products.
- Process customer complaints and resolution.

### Quality control

Function:

- Provide inspection and test system for purchased materials and manufactured products.

Quality management responsibilities:

- Take part in Design Review meetings to produce PDS and associated inspection and test system.
- Provide an inspection and test system (facilities and personnel) for:
  - Incoming materials and goods.
  - In-process materials.
  - Final product.
- Identify inspection and test requirements, procedures and locations.
- Recommend and validate preventive and corrective action.
- Provide information for cost of quality reporting.
- Provide information for process improvement actions.

Everybody is responsible

Every function in a company has a responsibility for quality. The quality assurance/quality control functions are only part of the chain of quality throughout the company.

### Production and process engineering

Function:

- Translate drawings and specifications into robust processes and procedures for production.

Quality management responsibilities:

- Assist production in any quality issues with tooling, machines or process.
- Control corrective action for production quality issues with tooling, machines or processes.
- Take part in Design Review meetings to produce PDS.
- Ensure that Design/Research and Development is provided with machine and process capability data to allow realistic tolerance setting.
- Identify materials requirements/specifications for purchasing/materials control.
- Lead process improvement actions for quality improvement.

### Purchasing/Materials control

Function:

- Control purchasing and movement of materials to achieve schedule compliance.

Quality management responsibilities:

- Translate customer orders into materials requirements.
- Identify and control suppliers.
- Ensure that all suppliers are capable of delivering material or product to specification.
- Ensure that all materials and product specifications sent to suppliers are complete and up-to-date.
- Issue purchase specifications and orders.
- Control production schedule to meet customer orders.
- Maintain inventories for schedule compliance.
- Control delivery, storage and issuing of materials to production and product to customer.

### Accounting

Function:

- Control all financial aspects of company to achieve profitable operations.

Quality management responsibilities:

- Manage company finances to ensure profitable operation.
- Produce reports, budgets and data for operational control and annual accounts for financial compliance.
- Provide clear, accurate and timely invoices and accounts for customers.
- Pay supplier invoices in clear, accurate and timely manner.
- Provide accounting information for cost of quality reporting.
- Provide accounting information for process improvement assessment.

### Production

Function:

- Produce products from agreed materials with agreed processes and procedures.

Quality management responsibilities:

- Produce product to specification.
- Maintain production schedule to meet customer requirements.
- Implement preventive and corrective action as required.
- Communicate with other functions on any problem affecting production or the product.
- Take part in Design Reviews to assess production capability.
- Manage production area quality improvement actions.

# 2.3 Organising for quality assurance and control

## The quality manager

The quality responsibilities outlined in Section 2.2 show that quality responsibilities run across the whole company but in this section we will only cover the quality assurance and quality control functions. A 'Quality Manager' is normally in charge of both of these functions although they are very different in what they do.

At most plastics processing sites, quality assurance generally has very few, if any, full-time staff, whereas quality control often has more staff (in the form of inspectors and laboratory and analysis staff ). This makes quality control more visible but not necessarily more important. The functions are equally important - simply different.

- **Tip -** Quality management is not only about inspection and inspectors whatever the staff numbers look like.

One of the most important points about organising for quality assurance and control is the separation of the quality and production functions. The quality function should not report to the production function because of the obvious conflict of interest between the two and it is normal for the quality function to report directly to the Managing Director or CEO. This has the advantage of giving quality a direct line to top management but has the disadvantage that the Quality Manager may fail to get the direction and support if the CEO is not committed to quality (see Section 2.1). It is still probably the best arrangement to ensure true and necessary independence of the quality function.

- **Tip -** The quality function should ideally be independent of the production function.

The responsibility of the production function is to produce products to the agreed specification. The quality function simply checks that this has been done. The quality function is responsible for providing the systems, expert knowledge and tools to ensure this.

- **Tip -** If you show a returned product to a manager and their first question is 'Which inspector let this out the door?' then you have already failed to get the organisation right. Start again.

## Quality assurance

This work is about management of the formal quality system, generally ISO 9001, and the duties are shown below. Most of these duties are specified as part of the system standard and after initial system development and approval there will only be maintenance involved in keeping the system up-to-date. The major work involves:

- Regular internal auditing of the system, procedures and specifications both for effectiveness and efficiency.
- The conduct of regular management reviews of the system for effectiveness and efficiency.
- Approving corrective and preventive action changes to the system. This does not mean that quality assurance carries out the corrective and preventive action. However, they must approve system or specification changes and document the process. The corrective or preventive action is carried out by the relevant department.
- Quality improvement action. Again, this does not mean that quality assurance is solely responsible for quality improvement but that they must approve system or specification changes and document the process as part of the quality management system (see Section 2.5).

The control of quality is a production function.

The quality control department should assist the other departments but should never be directly responsible for quality.

They are advisors, supporters and auditors.

Traditionally production has made the product and quality control has inspected it to remove products that do not meet specification.

This is costly, inefficient and doesn't work anyway.

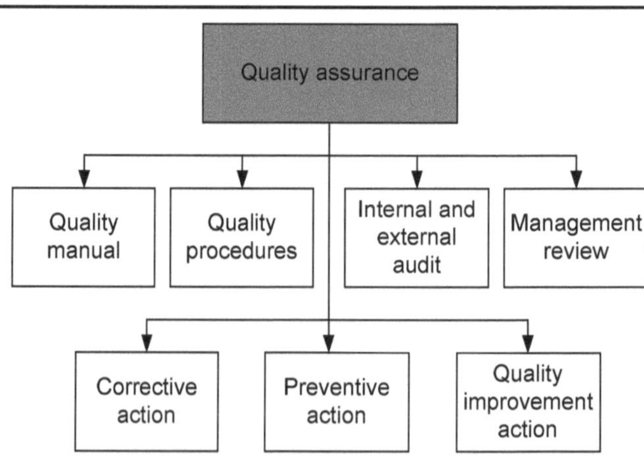


### Quality assurance work areas

Quality assurance is mainly concerned with the formal quality management system and the documentation of this system. This is very different from quality control but may not be a full-time job after the system is set up.

## Quality control

The quality control department is responsible for ensuring that all control procedures are carried out and the products are to the standard set by the customer. The main areas are:

### Inspection and test

In 'older' systems this involves regularly checking inwards goods and production and recording the results of these checks. It also involves controlling the status (HOLD, OK or REJECT) of inwards goods and production and issuing complaint reports on faulty products. At some sites the inspection area also recommends any necessary corrective action and is responsible for carrying it out, e.g. sorting.

In 'new' systems, the introduction of SPC (see Chapter 5) and other process controls and the devolution of responsibility for quality to production operators means that production has now accepted responsibility for quality production. In 'new' systems, the role of inspectors is to:

- Provide assistance in developing the necessary control systems and control plans.
- Provide assistance in developing the documentation, e.g. SPC charts.
- Checking the results of the operators (rather than checking the products).

This is a huge change in role for quality staff.

### Laboratory testing

Laboratory testing for most plastics processors is more relevant for inwards goods approval of plastics deliveries than for finished product. Quality control is responsible for carrying out, recording and filing the results of any tests specified in control or laboratory specifications.

### Measurement and standards

Measurement and test, whether at the machine or in the laboratory, needs accurate instruments and quality control is responsible for controlling the test and calibration of all measuring instruments and ensuring that calibration certificates are obtained and filed on the correct dates.

### Quality analysis

This is perhaps the most important area of quality control but is also the least visible. This area involves:

- Failure analysis of products.
- Recommending corrective action for any failed product (but not doing).
- Data analysis: This is the role of the quality engineer who:
  - Carries out capability studies for machines, tooling and processes (see Chapter 5) in conjunction with process engineering.
  - Controls and manages SPC data to allow the setting of achievable tolerances.
  - Trains all staff in SPC.

If your quality manager is titled 'Chief Inspector' then you have probably missed the point.

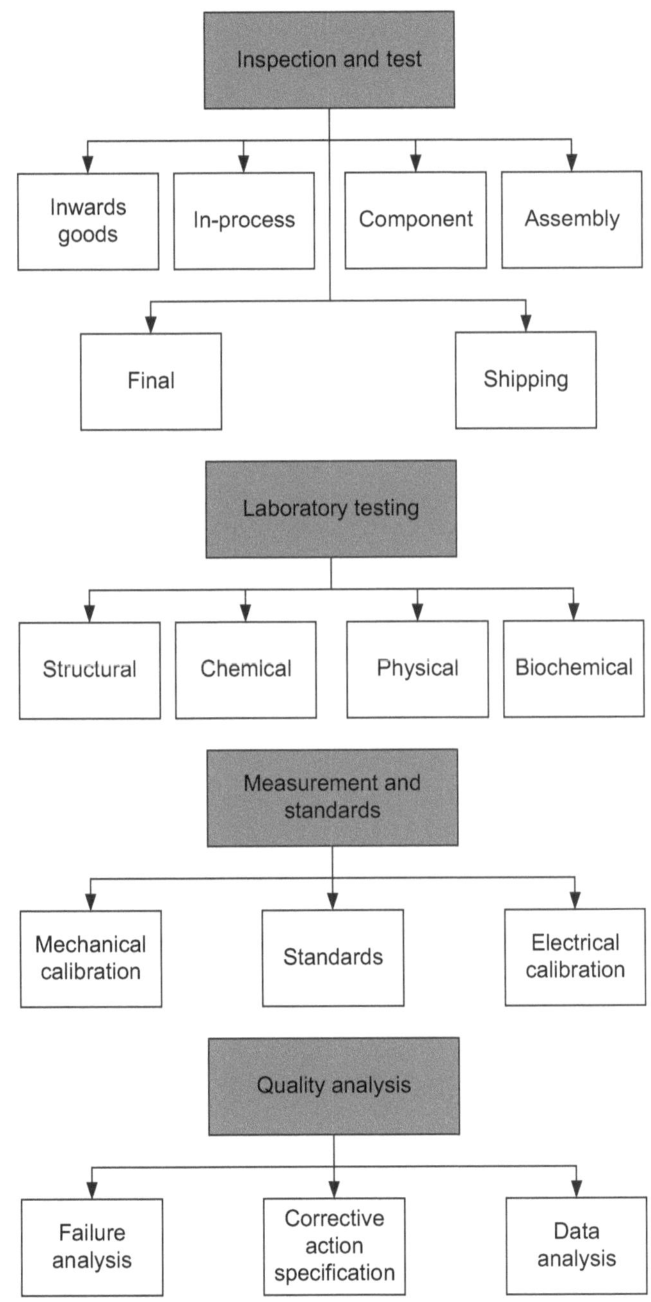


Quality control work areas.

Quality control is mainly concerned with the product. The role has changed greatly in the past few years, previously this was mainly 'inspection' but simple inspection does not work and the role is now based on supporting production to produce good products.

# 2.4 Quality assurance and control organisation - where are you now?

## Organisation counts

The organisation and management of the quality function and the wider implications of quality are the key to successful quality management.

Warning signs for poor organisation and management are:

- Quality is seen as being of low importance in the organisation.
- There is no clear chain of command.
- The quality function is regarded as subordinate to the production function.
- Quality systems are seen as solely the responsibility of the Quality Department with no wider responsibility throughout the organisation.
- Quality control is seen as the 'enemy' of rather than the partner of production.
- Process improvement is not undertaken even when there is clear market and customer pressure.
- There is a low statistical awareness throughout the organisation.

These are all signs that the organisation and management of quality is not effective and needs action.

## Completing the chart

This chart is completed and assessed as for the previous charts.

For the quality function to be effective, it must always have a separate chain of command to the production function.

Putting production in charge of quality is the same as getting the fox to look after the chickens.

**Quality assurance and control organisation**

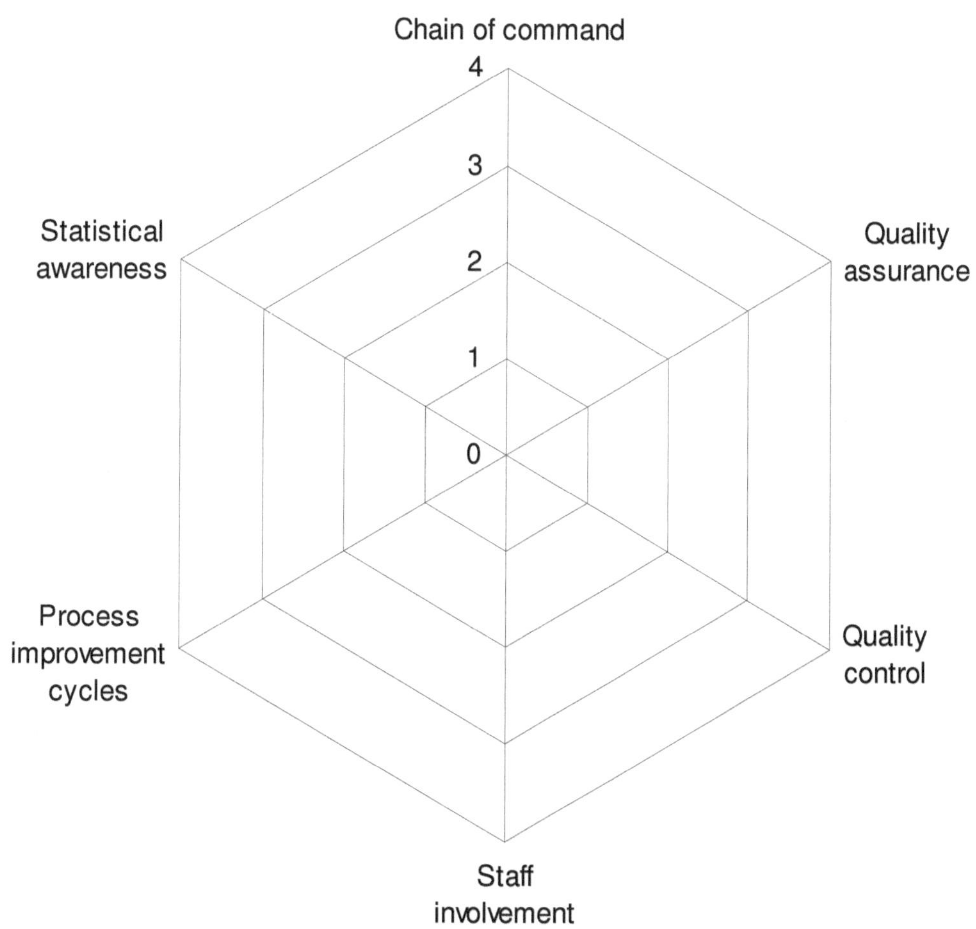


Use the scoring chart to assess where you are in quality organisation

The numbers from the scoring chart can be transferred to the radar chart for a quick visual assessment of where you are in terms of quality organisation.

Quality is not the enemy of production; it is an essential support function to check that production is performing to the required standards.

Quality is all about statistics and risk. Every site needs a good awareness of statistics to operate effective quality systems.

**Quality assurance and control organisation**

| Level | Chain of command | Quality assurance | Quality control | Staff involvement | Process improvement cycles | Statistical awareness |
|---|---|---|---|---|---|---|
| 4 | Clear & unambiguous chain of command for quality. Quality manager in place & reports to Chief Executive/ Managing Director. | Quality systems maintained by all staff as part of on-going process. Quality systems are clearly owned by all the staff. Quality Department acts as recording function only. | Quality control is seen as a partner of the production function. Production staff regularly seek guidance from inspectors on product quality. | High level of staff involvement in quality issues. All staff accept a high level of responsibility for quality performance of processes & products. | Process improvement cycles used to continuously improve products & processes without customer or market pressure. | High level of statistical awareness in all staff. Specialist statistical knowledge available to site. |
| 3 | Clear chain of command for quality. Quality manager in place & reports to Chief Executive/ Managing Director but often over-ruled by Production Manager. | Quality systems maintained by most staff as part of on-going process. Quality systems are clearly owned by most of the staff. | Quality control is seen as lower in importance than the production function. Production staff rarely seek guidance from inspectors on product quality. | Moderate level of staff involvement in quality issues. Some staff accept responsibility for quality performance of processes & products. | Process improvement cycles used to improve products & processes after customer or market pressure. | Moderate level of statistical awareness in some staff. No specialist statistical knowledge available to site. |
| 2 | Documented chain of command for quality. Quality Manager in place but does not report to the Chief Executive/ Managing Director. | Quality systems maintained by some staff as part of on-going process. Most staff see quality systems as being 'owned' by the Quality Department. | Quality control is seen as a 'servant' of the production function. Production staff never seek guidance from inspectors on product quality. | Low level of staff involvement in quality issues. Few staff accept any level responsibility for quality performance of processes & products. | Process improvement cycles sometimes used to improve products & processes after customer or market pressure. | Low level of statistical awareness in most staff. No specialist statistical knowledge available to site. |
| 1 | Documented chain of command for quality. Quality manager in place but reports to Production Manager. | Quality systems maintained by very few staff as part of on-going process. Other staff do not feel that quality systems are part of their duties. | Quality control is seen as the 'enemy' by production. Production staff sometimes try to get faulty products past inspectors & out the door. | Poor staff involvement in quality. Quality is the 'responsibility' of the Quality Department & not other staff. | Process improvement cycles rarely used to improve products & processes even after customer or market pressure. | Poor level of statistical awareness in most staff. No specialist statistical knowledge available to site. |
| 0 | Poor chain of command for quality. Quality Manager often criticised for poor quality of products & seen as being the only person responsible for quality. | Quality systems only maintained by Quality Department staff. Other staff do not feel that quality systems are part of their duties. | Quality control is seen as the 'enemy' by production. Production staff consistently try to get faulty products past inspectors & out the door. | No staff involvement in quality. Quality is the 'responsibility' of the Quality Department & not other staff. | No knowledge or use of process improvement cycles even when under customer or market pressure. | No level of statistical awareness in any staff. No specialist statistical knowledge available to site. |
| Score | | | | | | |

# 2.5 Organising for quality improvement

## This is where it gets serious

Quality improvement is an essential but often neglected part of quality management. Many companies focus simply on the quality system when quality improvement is the most important part of quality management. It is only through continuous improvement that any company can hope to stay competitive. The competition is getting better all the time and neglecting quality improvement is bound to be fatal.

- **Tip** - If you want to know if a company is serious about quality management then ask about their quality improvement activities. If there are none or if they cannot point to them very quickly then they are probably not all that serious. Leave and don't look back.
- **Tip** - Quality improvement is not only necessary but must be faster than the competition. Heading in the right direction is not enough - if you are not going fast enough then you will still be run over.

Quality management, and quality improvement in particular, is too important to be left to the quality department. The quality department cannot be expected to run this process alone. Everybody's job depends on quality so everybody must be involved.

## Customers/suppliers

A key issue for quality management is how we organise companies (see Section 2.7). The 'departmental' organisation model does not reflect the real 'process' orientation of actual companies and does not recognise that the 'tribes' of the company can become more important than the company itself, i.e. the people see themselves as part of the production department rather than as part of the company. The process model of a company accepts this and recognises the customers can be internal as well as external.

The old model of a company saw a set of suppliers sending raw materials to the company, the company doing 'work' (adding value) and then sending the product to the customer. The new model sees external suppliers sending raw materials to the company and then a series of supplier/customer relationships inside the company before the company sends the product to the final customer. These models are shown below and it is obvious that there will be many customer/supplier relationships in any company. Suppliers do not have to actually pass physical products to the customer to be considered a supplier, i.e. if the accounts department is considered then the product could be considered to be high-quality accurate reports.

This is a radical change and the concept of internal customer/supplier relationships is quite strange to many companies.

- **Tip** - I was pleasantly surprised when I asked an injection moulding operator who his customer was and his immediate unprompted reply was 'The assembly area'. Things are changing and the operators understand it too (maybe better in some cases).

The ultimate responsibility for a quality product lies with each and every one of the employees. A perfect system may not produce quality products because it is ignored or poorly applied but a less than perfect system may produce good-quality products because of the enthusiasm of the workforce.

The people make all the difference.

## Inside the department

The organisation of most companies is still based on the departmental/work area model and inside the department/work area is a natural place to look for quality improvements. The best mechanism for

You get to choose your opinions, you don't get to choose your facts.

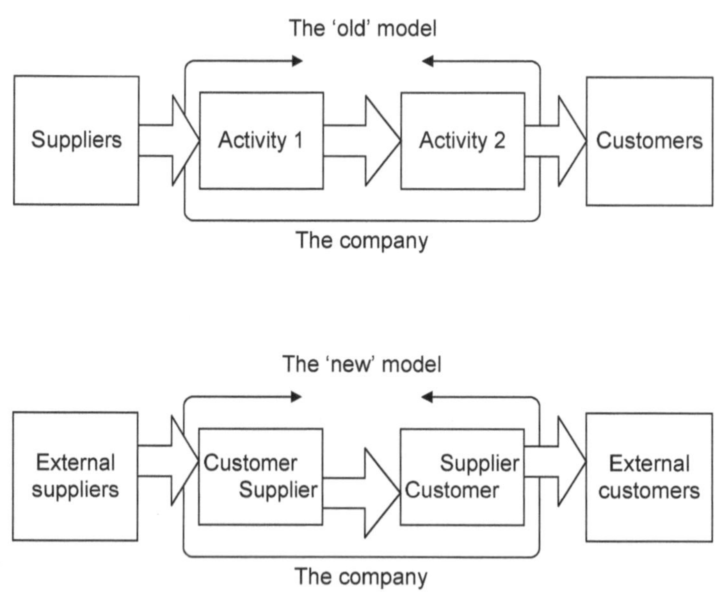


### Suppliers and customers

The 'old' model had a set of suppliers and a final customer. The 'new' model has 'internal' suppliers and 'internal' customers. The person you give the output of your process to is your customer and you are their supplier - you need to work together.

quality (and other) improvement inside the department/work area is the quality circle. The operation of quality circles is summarised below right and covered in detail in Section 2.6.

## Across the business

When business processes cut across departments then improving quality needs a cross-function/cross-departmental approach and quality improvement teams are the best approach for this. The operation of quality improvement teams is summarised below right and covered in detail in Sections 2.7 and 2.8.

## What are the questions?

Although quality circles and quality improvement teams are fundamentally different, they are both designed to improve company operations and need to ask (and answer) the same type of questions about what is happening in a company and how it can be improved.

These questions involve whether the company is doing the right things (effectiveness) and whether the company is doing things right (efficiency).

### Doing the right things (effectiveness)

The effectiveness questions are:

- What is the purpose of the department/process?
- Who are the suppliers of the department/process?
- Who are the customers of the department/process?
- What are the customer's requirements?
- Do we fully meet them and, if not, then where not and why not?
- **Tip -** All outputs from a department should be necessary and all necessary inputs should be on time.

### Doing things right (efficiency)

- What do we do that adds value to the customer/company?
- What do we do that does not add value to the customer/company?
- How can we reduce activities that do not add value?
- Can we eliminate them entirely?
- What are the customer's priority issues/problems?
- What are the root causes of these issues/problems?
- What can we do differently to solve these issues/problems?
- Have our actions eliminated the customer's issue/problem (to their satisfaction) and prevented it from happening again?
- **Tip -** Improving efficiency should only be sought after assessing effectiveness, i.e. don't get better at something you shouldn't be doing at all.

### Improvement

- What are the best opportunities/areas for improving both effectiveness and efficiency?
- What is our next project?

Quality improvement is not a single process, it is a continuous process to define and improve the value-adding activities of a company. It leads to a better appreciation of what the company does to earn money and improves the customer's perception of the company.

The goals of the business should be used to select quality improvement projects. Use a targeted approach, i.e. don't get great at a process if the rewards are not worth it.

Quality improvement should focus managers not on supervising the quality of their group's output but on working with their staff to improve the focus and direction of the process.

**T**ogether
**E**veryone
**A**nnoys
**M**e

The result of poor quality organisation.

| Area | Quality improvement teams | Quality circles |
|---|---|---|
| Project scope | Across departments | One department or work group |
| Project importance | Vital few | Trivial many |
| Membership scope | Across departments | One department or work group |
| Membership reason | Appointed | Voluntary |
| Timescale for team | Until end of project | Continuous |

**The differences between quality improvement teams and quality circles**

They are not the same thing and have very different functions. The major breakthroughs in quality will come with quality improvement teams but quality circles are good inside functions.

# 2.6 Quality circles

## An old one but still a good one

Quality circles were first devised and implemented in Japan in around 1962 by Kaoru Ishikawa (also creator of the cause and effect chart or Ishikawa diagram) and spread throughout Japan as part of their drive for quality improvement. They were later introduced into more than 50 countries but never really took root in Western culture for various reasons and are now largely unused in the West.

The basic concept of quality circles is that one of the key resources of a business is the people and that it is their experience and knowledge that allows a company to function and improve. Quality circles (although not always termed as such) are an effort to create systems, processes and procedures to capture this experience and knowledge to improve a company's performance.

When used effectively, quality circles can be an essential and vital part of continuous operational improvement in any company.

## What are the rules?

Quality circles have some general rules of operation and these are:

- Participation is voluntary by all employees. There is no requirement for any member of a work area to be a member of the area quality circle.
- There is no pay increase (or decrease) associated with being part of a quality circle or not. However, there is often other public recognition of the success of the quality circle.
- Membership is between 4 and 10 people from the same work area with a shared background in the process.
- The work area may be any area and quality circles are not limited to the production areas, i.e. quality circles can operate in sales, administration or any work area.
- Meetings are held regularly and may be during work time or outside work time depending on the company and arrangements.
- The focus is finding solutions to selected problems in the work area. A quality circle is not simply a 'talking shop'.
- The selected problems are identified, investigated and solved by the quality circle. The problems are not designated or selected by management.
- Quality circles present their findings to management for approval to implement but then implement and monitor the effectiveness of the proposed solution.
- There is no discussion of pay, employment conditions or salary grades in quality circles.
- Circle members (and other relevant staff) are trained in team working, problem-solving techniques (see Chapter 7) and potentially in presentation techniques.

Quality circles are a way of capturing the creative and innovative power of the work force.

Don't let them hang up their brain with their coat!

## Implementation

Quality circles have been implemented in all types of work areas in companies of all sizes. The type of work area and size of the company do not appear to be relevant to the success of quality circles.

Successful implementation of quality circles depends on several key factors, these are:

Quality circles can capture invaluable knowledge about the process.

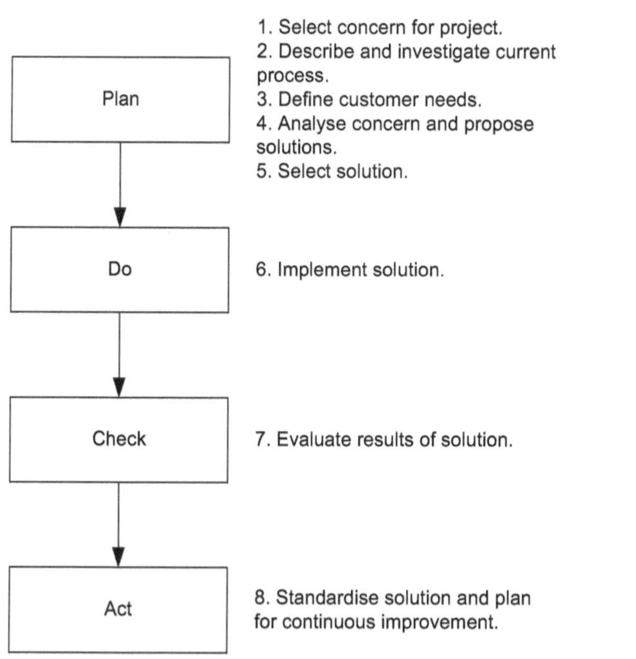


### Quality circles fit into the PDCA cycle

The PDCA cycle (see Section 4.3) easily fits in with the idea of quality circles. PDCA provides a good model for logical problem solving and prepares quality circles for their future transformation.

- Top management support is needed before quality circles are considered. The top managers must actively support the programme and have a willingness to allow employees to contribute to the company irrespective of their level. The top managers must also have the patience to wait for the results which can sometimes take up to 12 months to realise.
- In addition to top management, operational management support is needed to help quality circles implement their chosen solutions.
- A willingness to implement the quality circle projects even if there is no direct or immediate financial benefit. As long as the projects recommended by quality circles do not cost a lot of money then they should be implemented.
- Good employee relations in the company. Quality circles are not the solution to poor employee relations but can improve them when they are already good.
- Good information and widespread consultation before implementation is started. If this is not carried out then the programme may be treated with suspicion by the workforce. Unions may need to be approached with some care but, in the UK, unions have previously supported quality circles in a variety of industries as a method of improving workforce engagement.
- A steering or management committee to oversee the implementation. This should include quality specialists and trade union representatives (if applicable).
- A management structure that will implement the programme (not the projects) and report to the steering committee.
- A 'facilitator' who will be a coach, advisor and expert for the quality circle, the steering committee and management. The facilitator may be part-time (internal or external). This role will involve training quality circles in problem-solving techniques, attending the meetings, monitoring progress and applying to management for any funds necessary for implementation.
- A 'quality circle leader', this will normally be the team leader in the work area. There may be a need for additional training for the team leader in how to run quality circles because their operation is much more 'egalitarian' than standard management of a work area.

## Results

Quality circles show good financial results although this is not a specific aim of the process and is not the driving force for most programmes. Many companies do not even record the financial benefits of quality circles as they feel that this does not help the process. Where the financial results are recorded then they show that the savings are generally twice the costs and can be up to 10 times the costs.

- Start small.
- Start with winners.
- Do not measure success by cost savings.
- Use outside help to get started.

## The future

Quality circles have decreased and almost disappeared in the West and this is perhaps a great shame - they are still relevant in terms of improving worker participation and capturing the knowledge of the workforce. The real problem with quality circles is that they operate only inside departments whereas the main issues and problems are actually interdepartmental in nature and only inter-departmental teams can solve the problems (see Sections 2.7 and 2.8).

Despite this, quality circles are still in use in Japan and many other Asian countries where they fit more closely with the culture.

Quality circles are not a substitute or replacement for suggestion schemes - they focus on issues that are important to the workforce.

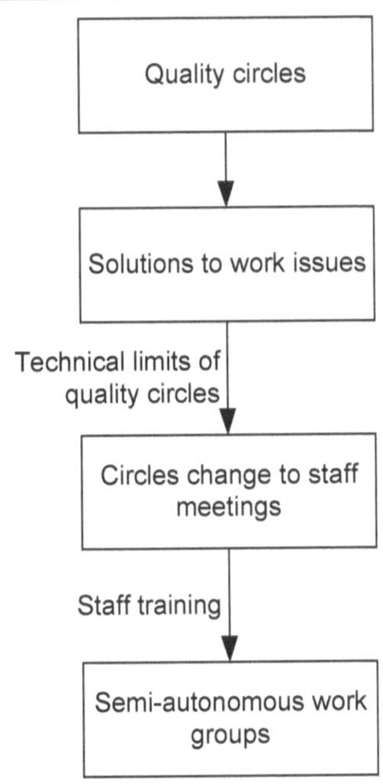


### Quality circles can (and should) change with time

Quality circles are part of a journey and a process. They should first solve the available and visible issues in their area but they will eventually solve most of the important and available issues. This is the time to train them more and change their focus.

# 2.7 Quality improvement teams - 1

## Interfaces

Quality management is not a simple quality department function, it is not a design function and it is not a production function. It is a company-wide process that transcends the pure departmental barriers. In fact, if quality improvement projects do not cross-departmental barriers then it is unlikely that they will be truly effective in improving quality.

Some quality improvement projects can be run and completed in a single department (in the rare case that the project affects only that department). This is rarely a difficult process because departments are controlled by a single person but this will only happen if the process owner recognises the need for improvement. The real difficulties (and opportunities) are in the interfaces between departments and 'managing the interfaces' is a key barrier to effective quality improvement.

## Organisation

Quality improvement is not always assisted by the company organisation and in many cases, the organisation structure actually hinders the process.

The typical company organisation chart is a pyramid with the Managing Director at the top, department or function heads reporting to the Managing Director, at the bottom is a large flat base of all the other employees.

This type of organisation makes control easy but is based on largely obsolete management practices. These use a model of activities and processes that comes from the distant past. The question that needs to be asked is: Why is the control of companies organised vertically when all projects and processes operate horizontally across the organisation?

This type of organisation leads to the classic 'over the wall' approach that is shown on the right. Each department operates as a distinct entity or tribe with little consideration of the other departments. After their work is complete the project is thrown 'over the wall' to the other department for further independent work. The result is products or processes that fail to deliver what the customer wanted to the quality level that the

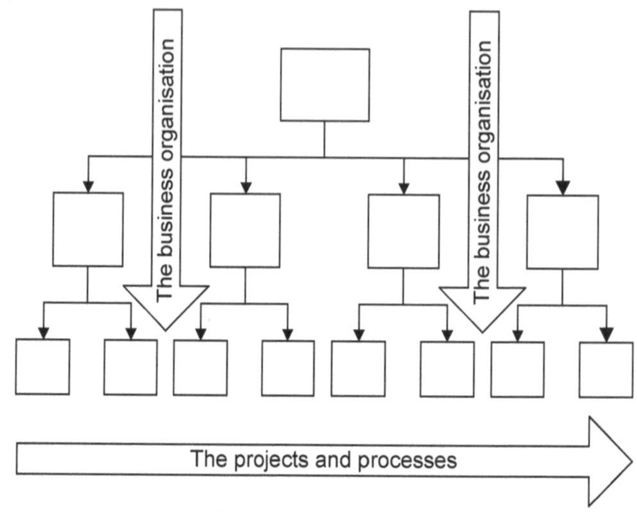


### How companies are organised

Companies are organised vertically along functional lines where people carrying out the same function are grouped together.

### How projects and processes are organised

Projects and processes are organised horizontally and flow across the functional boundaries. At every interface there are translation problems and misunderstandings.

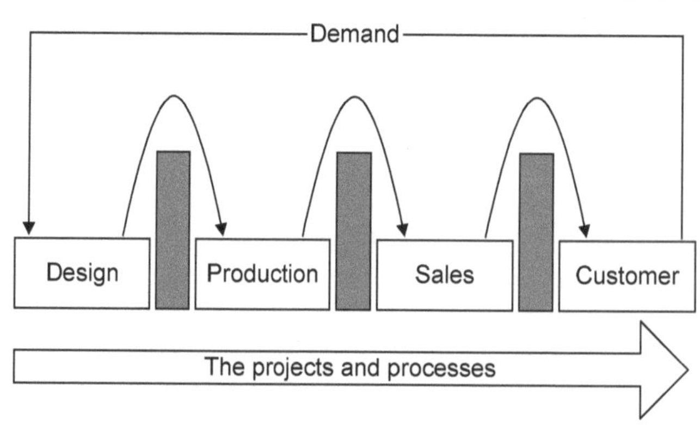


### The 'over the wall' approach to quality

Quality is important to the complete company but is often not seen as important to the individual departments (they have other agendas). The result is 'over the wall' working:

- An inadequately defined demand is thrown over the walls to the design department by sales who didn't understand the needs.
- The design department design a product with little consideration for manufacturability and then throw it over the wall to production.
- Production make the product to a poor specification and then throw it over the wall to sales.
- Sales try to sell an inadequate product and fail.
- The whole company fails.

customer wanted. This leads to the failure of products, processes and eventually companies.

Even when departments consider the next department as 'their customer' this type of behaviour often continues because of a failure to understand the needs of the next department.

Team-based working cuts across the traditional activity-based company divisions to follow processes and projects from start to finish. This method is applied in quality improvement projects by the formation of project teams to deliver the quality improvement projects.

The 'team' approach to quality is shown on the right and this uses the benefits and experience of all the team members to design and produce products and processes that are both high quality and robust.

- **Tip** - Consider co-opting a customer onto the quality improvement team, especially if they have complained about the product or process. Their external view of the product or process can be very illuminating and provide essential information. It also makes them feel part of the quality improvement process and improves their loyalty to the company.
- **Tip** - The process owner should be the project team leader.

For quality improvement to be truly rewarding, companies must start to use a team-based approach to reshape the fundamental methods of doing business.

## Products and processes

Team-based quality improvement projects should be considered not simply for the production process but also for all of the other customer-facing processes. Quality is not simply about the product, it is also about the service.

- **Tip** - How many times do customers complain about invoices or accounts? Why not improve the quality of the process and why not get the customer involved at the same time?

### Process measures

The process of quality improvement begins by understanding that either a quality concern or an opportunity for quality improvement exists. This requires some on-going measure of the process that can be monitored to detect if the process is failing. All major processes in a company should have some measure associated with them to provide feedback on the performance.

- **Tip** - Using SPC and attributes data (see Section 5.19) can provide excellent information on process performance.

## Non-value activities

Improving the performance of activities that do not add value to the company is a worthless exercise. Quality improvement can include removing these activities; it is no use being efficient at something that adds no value.

- **Tip** - Identifying and removing non-value activities is a valid quality improvement project.

"If you have a stable system, there is no use specifying a goal. You will get whatever the system will deliver. If you do not have a stable system then there is no point in setting a goal, because you have absolutely no way of telling what the system will produce."

**Dr. W. E. Deming**

When selecting quality improvement teams try to get a balance of personalities as well as roles. Personality types matter in teams.

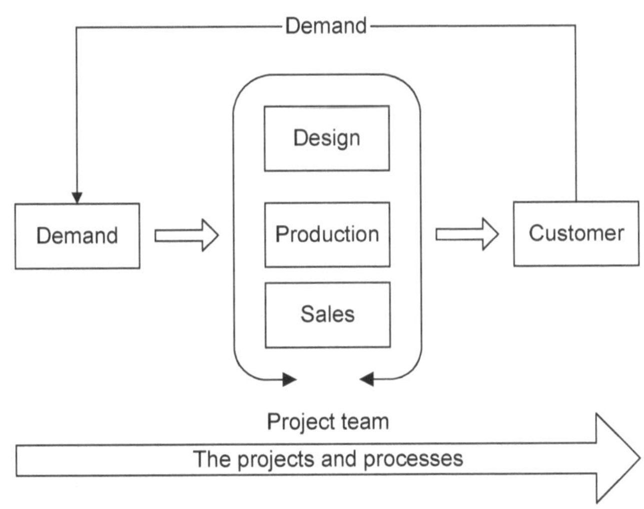


### The 'team' approach to quality

Quality products can only be delivered using a team approach where the department agendas are aligned to produce quality products. The result is 'team' working:

- The customer demand is clarified by the complete team.
- The design is reviewed for manufacturability before being finalised. Any production issues are resolved during design.
- Production have been part of the process and have a clear idea of the specification before production.
- Sales understand the product and succeed in selling it.
- The whole company wins.

# 2.8 Quality improvement teams - 2

## Starting the process

Quality improvement teams can deliver remarkable projects and results but the process needs careful management. The prerequisites are:

### Management commitment

Top management must set the scene for quality improvement (see Section 2.1). Whilst not necessarily a member of the team, each team needs a designated member of top management to be responsible for ensuring that the team gets the co-operation, resources and assistance it needs to function - this is also a very visible display of top management commitment.

### Identify the issue - initial scan

Management must specify the issue for improvement. At this level the specification can be quite broad, i.e. reduce scrap levels. The issue can cut across the whole company but it must be a real issue which can be quantified with real data.

- **Tip** - Problems can relate to productivity, profit, growth or any other area of process quality that management wants to improve.

### Identify the problem - refine the problem definition

After the initial problem selection it is often necessary to refine the problem statement to create a workable project. Refining the problem needs data on the specifics of the problem, i.e. is the scrap only at start-up, is the scrap only with certain products, is the scrap only with certain colours. These data can be used with the standard problem solving tools to home in on an achievable but valuable project (see Chapter 7).

- **Tip** - Fully defining the problem may not be possible at this stage due to lack of data and may need to wait until the team starts work (see below).
- **Tip** - Do not select the largest problem for the first project. Start small with a pilot project to learn the mechanics of the process and prove its worth.
- **Tip** - Projects set by management must have defined objectives and outcomes that can be measured.

### Select the team

The process improvement team is selected by management and participation is not voluntary. Team membership is between 4 and 10 people with experience of the process but from different departments and backgrounds.

- The team should always include at least one operator of the process being considered to provide the essential reality check.
- **Tip** - Process operators always know what actually happens, as opposed to process engineers who know what should happen. The team needs to know about the former and not the latter!
- The team should always include a customer of the process (internal or external) to assess proposed improvements.
- The team should also include at least one member who is familiar with the standard problem-solving tools (see

"All quality improvement of a lasting nature occurs as the result of a project."

**J.M Juran**

In God we trust - everybody else brings data.

Statistics do not solve problems, people do. Statistics are simply a tool people can use.

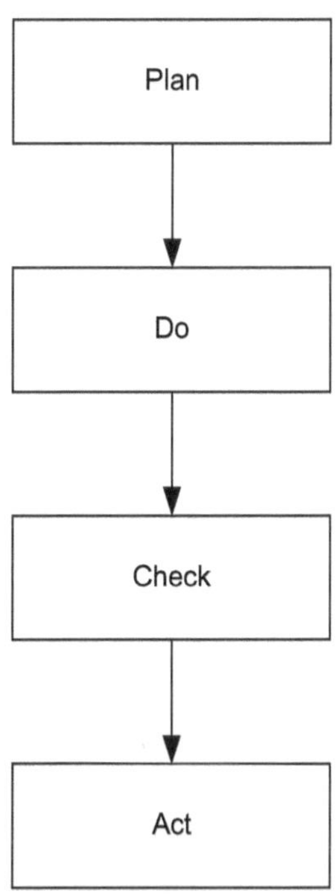


1. Select process for project.
2. Describe and investigate current process.
3. Define improvement needed.
4. Use quality tools to analyse concern and propose solutions.
5. Select solution.
6. Implement solution.
7. Evaluate results of solution. If process improved then go to 8. If no process improvement then go back to 4.
8. Standardise solution and plan for continuous improvement.
9. Complete and team disbanded.

### Quality improvement teams fit into the PDCA cycle

The PDCA cycle (see Section 4.3) easily fits in with the idea of quality improvement teams. PDCA provides a good model for logical problem solving and should be part of the quality improvement teams method of operation.

Chapter 7). This will give the whole team experience in using the tools.

### Managing the team

Every team will need a Team Leader to manage the project and the Team Leader will need experience in team management but does not need to be senior in the company. The Team Leader will record action plans, project reports and be responsible for meeting the objectives.

## Running the process

Quality improvement teams have some general rules of operation and these are:

- Even after the problem is refined by management, the team should start by rigorously defining the problem. This is to check that the management set project is achievable. The results of this work should be agreed with management as the fully defined problem.
- The team should define the problem and the variables using the standard problem-solving techniques such as flow charts, cause and effect (see Chapter 7).
- The problem-solving tools will almost inevitably highlight the need for more data and these data will need to be collected, e.g. is the start-up scrap only high with certain tools, is the scrap only high with certain product or colours.
- **Tip -** Data collection/research is often one of the longest phases of thc project but must be done correctly. Bad or incomplete data can ruin a project. Take your time to get good defensible and well-defined data.
- The team should use the collected data with the cause and effect diagram, Pareto analysis and other tools to propose the corrective action.
- The team should present the corrective action to management for evaluation and approval to implement.
- The team should implement the corrective action and gather data to validate the improvement.
- If there is no improvement, or if the improvement does not meet the defined objectives, then the data gathered can be fed back into the process to redefine the solution.
- If the improvement meets the defined objectives the team is disbanded.
- Management finds another problem and starts another quality improvement team.

## Types of improvement project

Quality improvement teams can work on two types of project, the incremental or innovative. The differences are:

### Incremental

This type of project is evolutionary, it takes an existing process and, using the existing knowledge base, makes it better. This is the type of project that makes an existing machine or process go faster or operate more accurately. This is the classic quality improvement project. Progress may be slow but will be continuous.

### Innovation

The type of project is revolutionary, it introduces a new and much better process using a new knowledge base. This is the type of process that introduces an entirely new type of machine, combines processes to reduce complexity or in some other way fundamentally changes the rules. This is the classic research and development-based quality improvement project. Progress, when it comes, will be large and a step change in performance.

The quality improvement process should aim to stop constant 'fire-fighting'.

It should aim to 'drain the swamp rather than continue to wrestle alligators'.

Effect
to
Causes
to
Corrective action

| Item | Incremental improvement (evolution) | Innovation improvement (revolution) |
|---|---|---|
| Project focus | Existing departments | New science or technology |
| Project targets | Broad range covering all aspects of company | Limited range focusing on specific technique |
| Knowledge base | Existing knowledge base | New knowledge base |
| Investment needed | Low | High |
| Rate of progress | Small, slow but continuous | Large and a leap forward |
| Team | Wide, deep and inclusive | Narrow, shallow and exclusive |
| Model | Group works as team for progress | Small individual efforts are important |

**Quality improvements can be incremental or innovation based**

Quality improvement teams can either work on incremental improvement or on innovation improvement. These different types of improvement need different approaches to the team.

# 2.9 Quality improvement organisation - where are you now?

## Across the company

Quality management is rarely a single department activity and most often cuts across the traditional boundaries. The process needs to be organised with this in mind to achieve the gains that are necessary.

Cross-functional processes rarely have formal process owners who have responsibility for achievement of the process quality measures. These processes often fail or under perform because of the lack of ownership and lack of a driver for success.

Companies need to define their key business processes (preferably in terms of how the customer views the activity), assign owners for each business process and task the owners with setting up the appropriate measures and quality improvement plans.

Particular effort needs to be spent on identifying non-value activities and eliminating these. Projects that eliminate non-value activities have a great return on effort.

## Completing the chart

This chart is completed and assessed as for the previous charts.

Focus on the customer is a key point.

- Find out which activities benefit the customer and try to do more of these.
- Find out which of your activities do not benefit the customer and do less of these.
- Find out which of your activities actively hinder the customer and eliminate these activities.
- Find the 'sales prevention officers' (they are in every company) and then find them another job, preferably with a competitor.

**Quality improvement organisation**

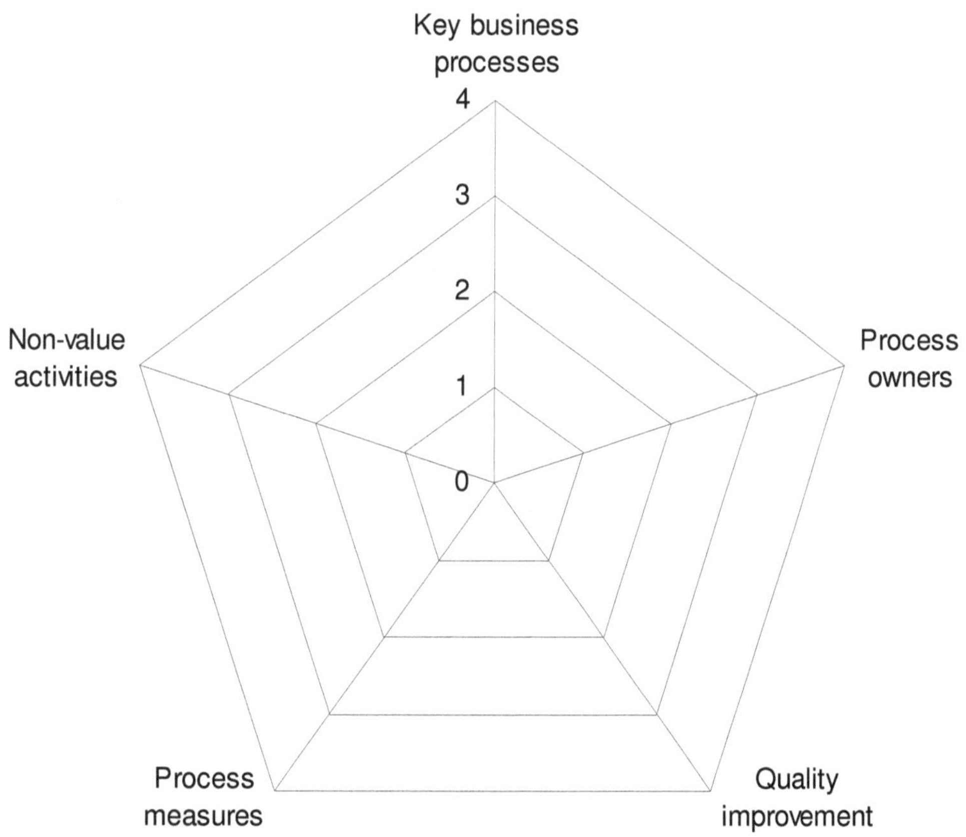


Use the chart to assess where you are in quality improvement organisation

The numbers from the self-assessment should be transferred to the radar chart for a quick visual guide to where you are in the basics of quality improvement organisation.

I once worked for a company which was permanently in 'fire-fighting' mode. We would have a moulding problem, everybody would yell at the Quality Manager, we would diagnose the cause and find the solution, write the reports, management would decide that it was too expensive to implement the solution and the problem would go away.

Everything would go quiet for a few months and then the problem would re-appear.

Thank heaven for 'cut and paste' - at least we didn't have to write the reports again.

## Quality improvement organisation

| Level | Key business processes | Process owners | Quality improvement | Process measures | Non-value activities |
|---|---|---|---|---|---|
| 4 | All key processes well defined, documented, resourced & optimised. Electronic documentation links to other processes, status, version & approval well controlled. | Process owners for all key business processes clearly & unambiguously defined with responsibilities & authority also clearly defined. | Formal quality improvement plans in place for all key business processes. Plans are regularly reviewed for relevance. | Process quality measures in place for all key business processes. Measures regularly monitored, reported & show continuous quality improvement. Prompt action taken on adverse results. | Non-value-adding activities minimised. Continuous review of processes to prevent accumulation of non-value-adding activities. |
| 3 | Good understanding of most key processes but inadequately documented & resourced. Automated processes filed on company Intranet. | More than half of key business processes have defined process owner but owner primarily has responsibility without power. | Formal quality improvement plans in place for some key business processes but no regular review for progress or relevance. | Process quality measures in place for most key business processes. Measures regularly monitored & reported but do not show quality improvement & little action is taken. | Non-value activities identified by process flow charts but not totally eliminated. Some gains still to be made in activity reduction. |
| 2 | Good understanding of some key processes but generally inadequately documented & resourced. Processes in common file format on company Intranet. | Less than half of key business processes have defined process owner. | Informal quality improvement plans in place for most key business processes but no regular review for progress or relevance. | Process quality measures in place for some key business processes. Those in place are rarely monitored & reported. No action taken on adverse measures. | Non-value activities identified across the company but many areas remain to be investigated. Considerable gains still to be made in elimination of activities. |
| 1 | Poor understanding of key processes & of how they work. When created processes are in common file format on personal computers only. | Few key business processes have defined process owner. | Informal quality improvement plans in place for some key business processes but no regular review for progress or relevance. | Few process quality measures in place for key business processes. Those in place are not monitored & reported. | Sporadic attempt to minimise non-value activities (primarily in production areas). No consistent approach to process analysis & activity reduction. |
| 0 | Little or no understanding of what the key processes are or how they work. When created processes are paper-based. | No key business processes have defined process owner. | No quality improvement plans in place. | No process quality measures in place for any key business process. | No attempt to identify non-value-adding activities in company. Substantial non-value activities seen. |
| Score | | | | | |

# 2.10 Quality improvement projects

## Project-based

Quality improvement is about changing products and processes and project management is at the heart of any process that involves change in a business. This means that any quality improvements will be delivered via successful projects.

## Project selection

A precursor to a successful project is good selection of the project to be undertaken. Most companies will already have a range of potential quality improvement projects but if they are in any doubt then look at the big costs and the projects will be obvious. Choosing from a range of competing projects is always difficult but it is always best to choose a limited number of projects and succeed rather than to start many projects and never to complete any.

Project selection is best based on a simple 2x2 'effort-reward' matrix as shown in the diagram at right. Projects can be ranked quickly based on:

- The reward: The estimated effect of the quality improvement.
- The effort: The ease of implementation.

Projects with a high reward and low effort are preferred and these will be in the A-segment of the matrix. These should be the first projects attempted. Projects in the D-segment of the matrix are realistically never going to start.

When faced with a competing group of projects with comparable 'effort-reward' indices then the project with the shortest time to completion should be chosen.

## Resource bottlenecks

Every company has resource bottlenecks, these can be with regard to staff - there are always more things to do than there are time and people or with regard to finance - there are always more demands on capital than there is money in the bank. Whatever the particular resource bottleneck, follow these simple rules to get the best results:

- Set a limit on the number of quality improvement projects that are allowed to be active at any time. Do not open any new projects unless one of the current projects is closed or suspended. Make no exceptions to this otherwise staff will not know where to focus efforts or spending.
- Never start projects that you cannot finish no matter how attractive they may appear in terms of time or return. An unfinished project is a total waste of time.
- If the resource bottleneck is finance then insist that the quality improvement programme is totally self-funding.

## Project planning

There are three approaches to project planning:

- No planning, i.e. We'll do it!
- Simple planning methods.
- Complex computer-based planning methods.

Project planning is essential for successful projects but simple planning methods are far preferred. Many quality improvement projects will be small in scale in both investment and time and the planning process should match this.

One excellent method is to use top-down

Quality improvement in all about changing things.

This means selecting and running projects successfully.

This needs an effective project system.

There are no points awarded for starting projects, there are only points awarded for successfully completed projects.

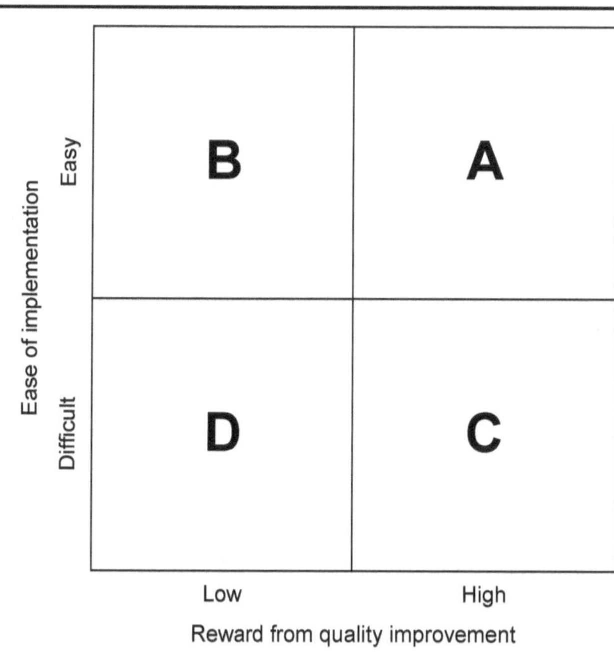


**Project selection for quality improvement**

It is essential to decide the order of implementation before starting any quality improvement process.

Rank potential projects according to the effect of the improvement and the ease of implementation. Go for the greatest improvements and easiest implementations first.

planning using Post-it notes. This method is described in the box on the right and, despite the apparent simplicity, is a very powerful and flexible method for small project planning. The method encourages an open approach to planning where the whole process is visible; this is in contrast to the computer-based approaches where the project plan is controlled by the operator of the software.

Whichever method of project planning is chosen, every project plan must have the following elements:

### Aims and objectives

These are the clearly stated aims and objectives of the project. These must be measurable and achievable to allow later performance assessment.

### Milestones

These are dates (from the actual start of the project not when it was planned to start) that show when particular tasks are to be completed. Milestones enable assessment of project plan time results.

### Budget

All projects should have an initial allocated budget.

### Assessment

Projects must be assessed after the aims and objectives have been completed or when the project manager decides that no more progress can be made. Assessment allows a review of the achievements against the previously agreed aims.

**Note:** For quality improvement projects the assessment should not only consider the financial aspects such as return on investment but also other non-financial benefits of the project, e.g. reputation, increased customer satisfaction, reduced downstream losses.

### Closure

All projects should be formally closed after the assessment phase.

## Project management

### Project teams

- Project teams need a leader or 'project champion'.
- Delegate control and accountability to the project manager.
- Project teams can make decisions without fear of being over-ruled later.
- Project teams are free to innovate.
- Project teams are assessed on their results.

### Meetings

- Project teams meet regularly.

### Schedules

- Check progress (and slippage) against agreed time and financial targets.

### Communication

- Project teams report on progress via an agreed and pro-active reporting and communication plan.
- One of the best methods is to communicate on 1 page and the One-Page Project Manager[1] is an excellent tool for both planning and reporting.

**Home truths of project management**

- What you don't know hurts you.
- Any project can be estimated for cost accurately - after it is finished.
- Nothing is impossible to the person who doesn't have to do it.
- What is not on paper has not been said.
- If you can keep your head while all around you are losing theirs then you haven't understood the plan.

•1. Campbell, C.A. 2006. '*The One-Page Project Manager*'. Wiley.

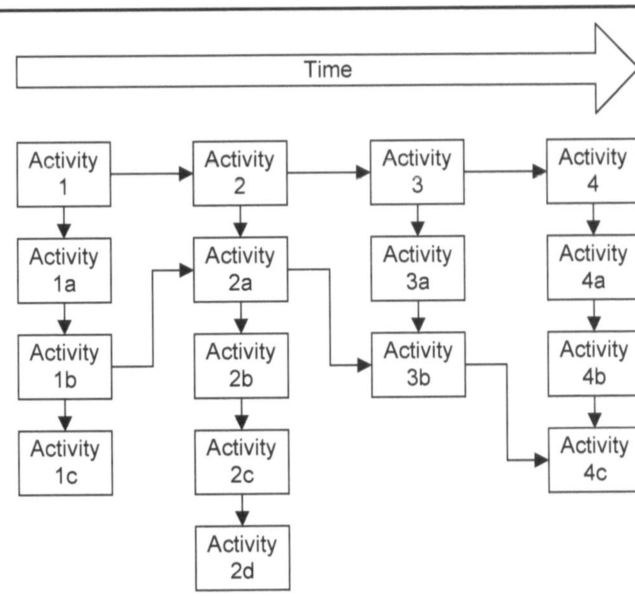


**Project planning using Post-it notes**

Top-down project planning for quality improvement can be carried out using Post-it notes and a flip-chart.

Write each task and the time to be allocated to it on a note. Move the notes around and group them according to the major tasks. Combine or divide tasks as the process continues.

Leave the chart in view and add, subtract or move the notes around as the plan develops. Finally, move the notes to overlap activities that can be done at the same time and reduce the total time taken for the project (simultaneous engineering). The critical path is easily seen from the sum of the individual tasks.

# 2.11 Quality improvement projects - where are you now?

## Quality improvement projects

Choosing between quality improvement projects has always been a concern and difficulty. There are always too many projects competing for too few resources.

Companies need to rapidly assess the potential gains and difficulty of implementing any potential project before rushing into a complex project that has a relatively low quality improvement potential.

Project selection is a key to quality improvement.

After projects have been selected then an effective project management system is an essential to actually delivering the project and achieving the potential gains.

Cross-functional teams are an invaluable tool for quality improvement due to the organisation of most companies.

## Completing the chart

This chart is completed and assessed as for the previous charts.

Select your projects wisely - go for the big bucks and easy projects.

**Quality improvement projects**

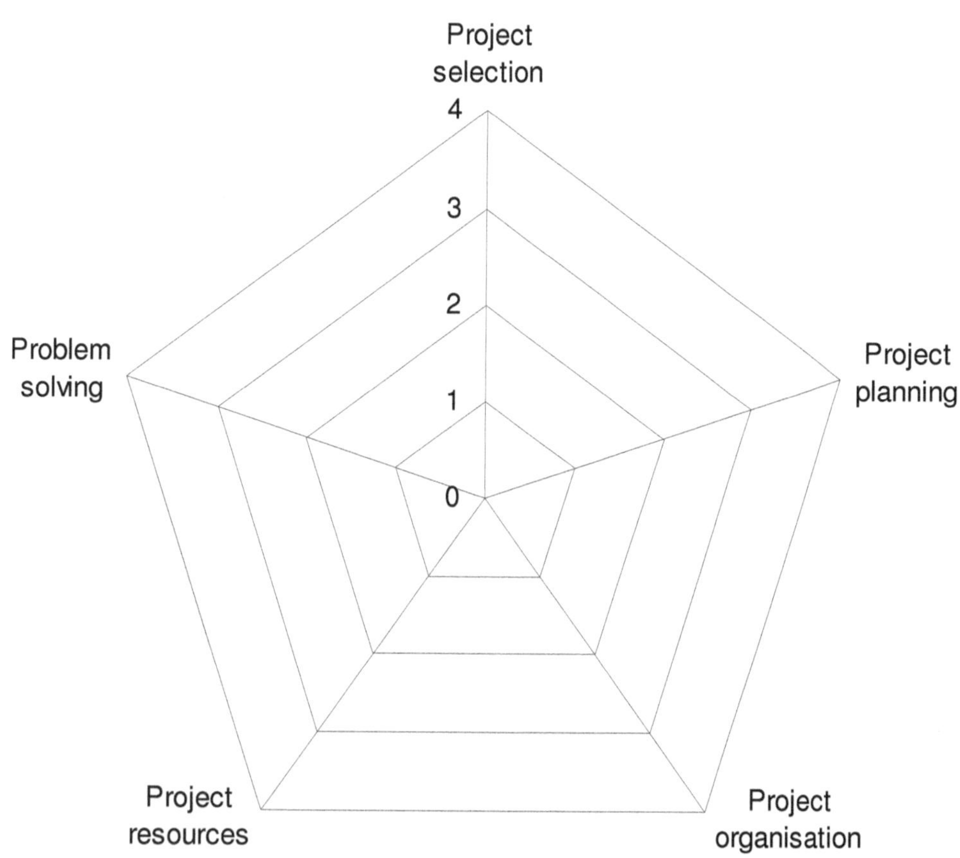


Use the chart to assess where you are in quality improvement projects

The numbers from the self-assessment should be transferred to the radar chart for a quick visual guide to where you are in the basics of quality improvement projects.

**Why quality management projects have problems delivering savings**

- Over-enthusiastic estimates or wrong forecasts.
- Financial systems are not able to calculate the savings (accurately or even inaccurately).
- Production and recording systems are adjusted to the losses of poor quality.

## Quality improvement projects

| Level | Project selection | Project planning | Project organisation | Project resources | Problem solving |
|---|---|---|---|---|---|
| 4 | All relevant quality improvement opportunities identified & prioritised for action. | Formal project definition & project plan necessary for any project. Progress is regularly reported & post-project assessment is carried out. | Excellent project management system used in all cases. Projects have clearly defined management & cost/benefits. | Project resources defined & allocated before project start. Projects are rarely delayed due to resource constraints. | Firmly embedded culture of improvement & problem solving through planning, action & review. Root causes identified & resolved. |
| 3 | Most available quality improvement opportunities identified but not prioritised for action. | Formal project planning carried out for all projects but control, reporting & assessment are variable. Failed projects are sometimes hidden & no lessons learnt. | Good project management system but use is variable. Good integration across departments but many projects have poor cost/benefit definition. | Project resources defined but not allocated at project start. | Problem solving is largely reactive with focus on solving root causes. Solutions developed but not always fully implemented. |
| 2 | Some quality improvement opportunities identified but no real planning process. | Project planning carried out for most projects but control, reporting & assessment is poor or rarely carried out. Failed projects are often hidden & no lessons learnt. | Project management system available but not used. Some integration of projects across departments & poor cost/benefit definition. | Project resources poorly defined at project start. | Problem solving is largely reactive; solutions are developed but rarely fully implemented. Focus on dealing with urgent effects & not on solving root causes. |
| 1 | Few quality improvement opportunities identified via unplanned process. | Cursory & undocumented project planning but no formal project planning or monitoring. Projects can become dormant & remain unfinished. | No project management system. Some integration of departments for projects that clearly cross departmental boundaries. | Project resources rarely considered at project start. | Problem solving is purely reactive & focused on dealing with urgent effects & not on solving the root cause. |
| 0 | Significant quality improvement opportunities ignored due to 'urgent' daily pressures. | No effective project planning. Actions are ad hoc & driven by events. Action is seen as more important than planning. | No project management system. Every project is 'different'. Projects are run by departments with little input from other departments. | Projects often started without adequate resources (due to poor planning) or starved of resources during project. Urgency is rated more highly than strategic importance. | Problems are ignored until they go away. |
| **Score** | | | | | |

# Key tips

- Top management commitment is essential for effective quality management.
- Management needs to provide:
  - Quality policy.
  - Resources.
  - Organisation and structure.
  - Quality improvement processes.
  - Motivation, awareness and teamwork.
  - The quality climate.
- Having a quality system and using the quality tools is not enough - quality improvement must be a continuous part of the process.
- There are many barriers to quality management but the biggest is when management does not accept responsibility.
- Every function in the company has a responsibility for quality. The QA/QC functions are only part of the chain.
- QA is responsible for the management and operation of the formal quality system, e.g. ISO 9001.
- QC is responsible for ensuring that all the agreed control procedures (checking, testing, etc.) are carried out.
- QC is not simply about inspection.
- The control of quality is a production function. QC is there to assist them.
- Quality improvement is an essential part of quality management but is often neglected.
- Quality improvement can be either inside a department, e.g. quality circles, or across the departments, e.g. quality improvement teams.
- Quality is about doing the right things (effectiveness) as much as it is about doing things right (efficiency) but improvement has to be part of the process.
- Quality improvement (inside the department or across departments) should begin with small projects to allow staff to gain experience with problem-solving tools.
- Quality circles work only within the departmental boundaries and are entirely voluntary.
- Quality circles are now neglected but can provide a good mechanism for improving quality and workforce engagement across a range of issues.
- Quality circles do not discuss pay, employment conditions or salary grades.
- Quality circles need assistance in problem solving in the early stages.
- Quality circles need good management to succeed.
- Quality improvement teams work across departmental boundaries and need top management support to succeed.
- Quality improvement teams can follow products or processes that cross the traditional departmental organisation.
- Quality improvement teams can look for:
  - Product improvements.
  - Process improvements (internal and external).
  - Non-value activity reduction.
- Quality improvement teams can deliver excellent results.
- Quality improvement teams are data-driven and need access to the problem-solving tools.
- Quality improvement work can be either incremental (evolutionary) or innovation (revolutionary).
- Quality improvement projects need good project selection, project planning and project management.

Senior Managers at all levels and with various backgrounds make the decisions and then expect/demand that the workforce live with the consequences of those decisions.

Setters and operators rarely get a say in the machines, plant, tools etc and yet then have to make quality products from it (and are disciplined if they fail).

Sounds fair, doesn't it?

# Chapter 3

# Targeting and controlling the cost of poor quality

It is possible (but unwise) to regard the cost of quality management as being additional to the other costs of the business. The cost of implementing quality management is then seen as an additional cost that does not add value to the business.

Quality management may not be a value-adding activity, such as processing or assembly, but the main function of quality management is to prevent the loss of value to the business.

Section 1.6 presented a brief introduction to 'quality costs' and used the Prevention, Appraisal and Failure (PAF) model to show the approximate size of the 'cost of quality'. The PAF model is the oldest model but it is only one of several different models available and these are all discussed in this chapter.

As with anything, the terminology matters and, in reality, the 'cost of quality' is actually the 'cost of poor quality'.

The concept of the 'cost of quality' was first developed and promoted by Armand Feigenbaum during the 1940's whilst working at General Electric in the USA (Feigenbaum was one of the greats of quality management and was also responsible for the development of the concept of 'Total Quality Control').

The cost of quality is therefore an old concept but it seems to come as a surprise to many in the plastics processing industry. Few companies collect and analyse the true cost of poor quality on a regular basis (<10%).

It is not uncommon to be unable to find any cost of poor quality information in even the best-run companies. This is an opportunity missed and every company needs relevant and accurate information on the cost of poor quality. This chapter gives recommendations for the necessary actions to set up and maintain a cost of poor quality system and how to effectively report these costs to get the maximum impact.

Collecting these costs is relatively easy for plastics processing, it is simply a matter of wanting to do it, getting the accountants on board and setting up the reporting system. Regularly collected and analysed cost of quality information will provide companies with an incentive to improve quality and will always provide quality staff with a strong argument for investment and support to reduce these costs. Reporting and analysis will also highlight areas for additional work and profitable projects.

Even companies who think that they have relatively low quality costs will often be surprised at the magnitude and distribution of the cost of poor quality and how it pervades the whole organisation. This is not an opportunity to be missed and is really at the heart of quality management - this is an opportunity to reduce costs as well as improve the company's reputation in the market place.

"Quality costs are the costs of not getting it right first time."

# 3.1 Quality costs - overview

## The fundamental measurement

There are many potential measurements that can be made to measure progress in quality. Measurements can be from internal or external sources (see Section 1.6) but will generally fall into one of four areas:

- Cost of quality data, i.e. the total cost of poor quality in financial terms.
- Process quality data, i.e. defect levels as measured from the process.
- Process improvement data, i.e. activities taking place to improve quality.
- Customer perception and expectation data, i.e. how well is the company meeting the customer's needs and expectations?

Financial data are a very strong driver for quality management and will be very similar whatever the process being used. The other types of data will vary with the specific company operations.

This chapter will therefore focus on the financial cost of quality measurements and how these can be quantified, treated and reported. Data from the other areas will be covered briefly in the latter parts of the chapter.

Section 1.6 introduced the importance of quality costs, discussed the magnitude of the cost of quality (really the cost of poor quality) and introduced the basics of quality costs using the PAF (Prevention, Appraisal and Failure) model. The PAF model is not the only model and in this chapter we will also review other models for quantifying the cost of quality.

The important thing to realise is that when the cost of quality is in the region of 20% of sales then quality is not an abstract theory but a vital tool for cost management in any company.

## The quality iceberg

The total cost of quality to a company is often demonstrated using the 'iceberg' model. Management will focus initially on the 'visible' cost of quality. As with an iceberg, this is the smallest part. The 'invisible' cost of quality is far higher but management often treats these costs as the normal 'cost of doing business' and ignores them. Many of these costs happen every day and are never questioned or noticed. In many cases, top management is unaware of the true cost of quality and often unaware of the concept itself.

Poor quality costs you money, good quality saves you money...

### How big is the 'cost of poor quality'?

- Feigenbaum (quoted by Deming) estimated the cost of poor quality to be between 15 and 40% of sales.
- Juran estimated the cost of poor quality to be between 20 and 40% of sales.
- Crosby estimated the cost of poor quality to be between 15 and 20% of sales.
- Our experience shows the cost of quality for most plastics processors in the region of 20% of sales.

Most companies are shocked to find that their actual quality costs are far higher than simply the cost of the Quality Control Department and that they have never monitored, recorded or targeted these costs.

The cost of poor quality is a very real opportunity for profit improvement!

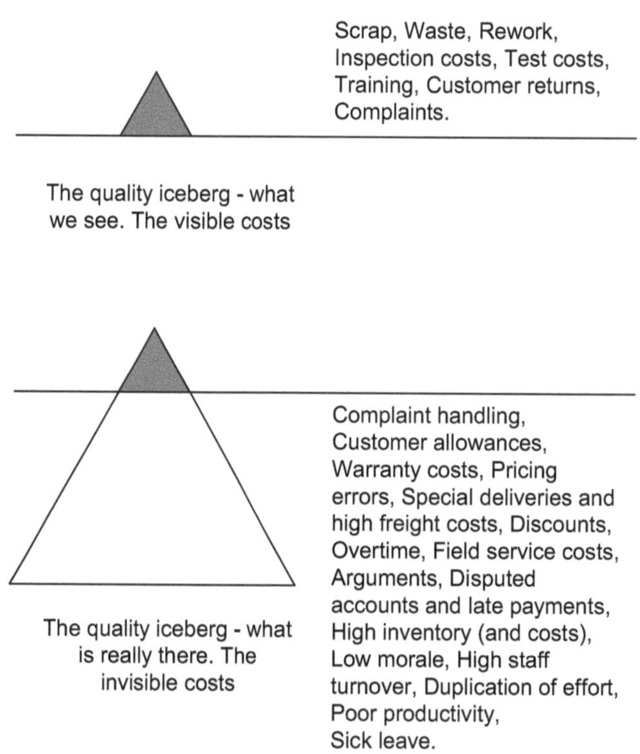


### The quality iceberg

The visible cost of quality is not the reality. As with an iceberg, most of the costs are hidden and invisible. The visible costs are often those tackled by companies new to quality costing but they are often not the major costs - they are just more visible.

- **Tip** - Why don't we demand the same financial control for quality that we expect for purchasing, materials use, payroll, etc? They are all costs to the business.

## Why do we need quality cost data?

Quality cost data allows quality and all the related activities to be expressed in the common language of management - money. Quality cost data allow companies to:

- Treat quality the same as any other business parameter, e.g. sales, marketing or overheads.
- Report progress on quality management in the same standard monthly reporting format as used for other management controls.
- Identify areas of high or abnormal expenditure.
- Identify areas for potential savings.
- Compare products, processes, areas and sites in a consistent manner.
- See variations or anomalies in cost allocation.
- Treat investment in quality management as a business decision that can be made using standard business tools.
- See the importance of quality as a business critical issue.
- Budget for quality as a business cost.

Ultimately, however, the main reason for using quality cost data is that introducing measurement is one of the first steps in control.

Quality cost data can be a major driver in the quest for improved quality and when used properly it can be even more effective than customer pressure. It shows the real benefits of quality management in the effect on the bottom line.

- **Tip** - Always talk about quality in the effect on the profits of the company. Too many quality people at the front end of the business get bogged down in the detail. They forget that the main reason a company exists is to make money. Talk money and the management will listen.

## How do we get quality cost data?

The best source of quality cost data is the accounting function. This allows solid and validated business data to be used and removes any emotion from the data.

Involving the accounting function also provides a company with a consistent method for assessment and quality can be seen as a company-wide issue and not simply a 'production' or 'quality department' issue.

- **Tip** - We are all in the business of quality management whatever your title.

"Defects are not free. Somebody makes them, and gets paid for making them."

**Dr. W. E. Deming**

Quality management can be the most highly geared cost management action that a company can ever make.

### Accountants can be your friends

Many years ago we decided to collect the COQ numbers. We were fortunate in having not only a good management accountant but also a good set of management accounts. Sitting with the accountant I simply went through the cost codes for each department and allocated these into the PAF categories for each department. The accountant then ran the report from the system and suddenly, as if by magic, we had a complete COQ for the site.

The reaction from the accountant was one of horror as he realised the total COQ was nearly 30% of our costs.

His simple questions were:

- Why weren't we taught to look for this in the accounting courses?
- Why did it take an engineer to show this to me in my accounts?

Equally, we now had the accounts function on board for the process of reducing the COQ at the site. They knew the numbers and they worked with them to reduce the overall cost of quality.

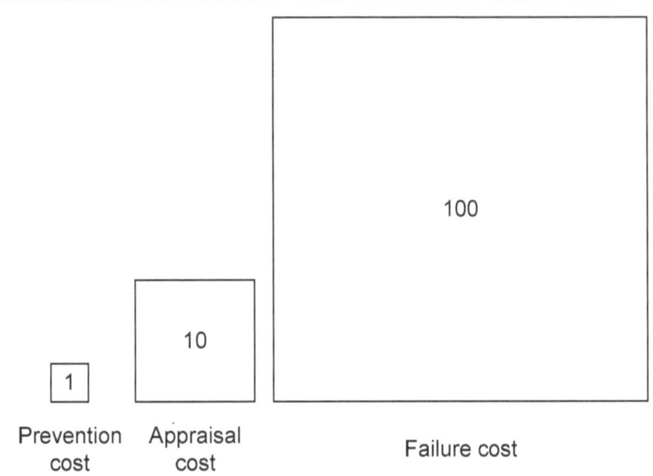


### 1:10:100 - the essential truth

The 1:10:100 model shows that it costs £1 to prevent a defect being produced, £10 to find the defect after it has been produced and £100 to fix the defect if it makes it to the customer. Prevention is always cheaper than failure

# 3.2 Quality costs - the models

## Choose your model

As with any method of cost collection and assessment there are several methods for defining and collecting the cost of quality.

For cost of quality there are four basic models and many variations of these. Whilst these vary in detail, they all seek to quantify the financial impact of poor quality on a company and to use this as a driver for quality improvement.

We will cover all of these briefly in this section and then review the main types in more detail in Sections 3.3 to 3.5.

## PAF models

The prevention, appraisal and failure (PAF) model that was used in Section 1.6 is the oldest and is the traditional model for assessing quality costs. It is also one of the easiest to apply to a traditional set of company accounts for easy cost collection and reporting.

### Feigenbaum

Armand Feigenbaum originally developed the PAF type of model to divide the cost of quality into four separate areas, i.e. prevention, appraisal and internal failure and external failure, although later models combine internal and external failure into one category of 'failure'.

Feigenbaum originally identified two major categories and several sub-categories:

- Cost of control
  - Prevention costs.
  - Appraisal (or inspection) costs.
- Cost of failure of control
  - Internal failure costs.
  - External failure costs.

### Juran

Joseph Juran was a pioneer in the cost of quality (as in many other areas of quality management) and used the same major categories but with different names:

- Cost of good quality
  - Prevention costs.
  - Appraisal or (inspection) costs.
- Cost of poor quality
  - Internal failure costs.
  - External failure costs.

### Crosby

Phillip Crosby defined quality as 'conformance to specification'. This led him to use the same major categories but again with different names:

- Price/Cost of conformance
  - Prevention costs.
  - Appraisal or (inspection) costs.
- Price/Cost of non-conformance
  - Internal failure costs.
  - External failure costs.

As a rule of thumb: If you can identify the 'cost of good quality' then the 'cost of poor quality' will be about 3 times this cost.

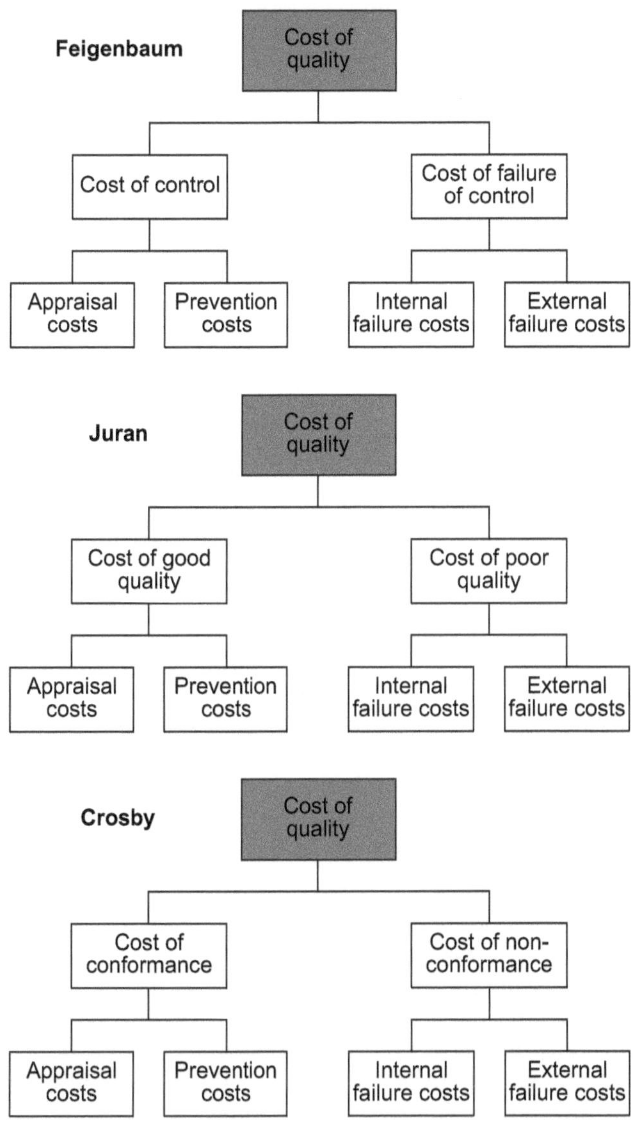


### Spot the difference?

Many of the major models use the same PAF categories but use different terms for the things. The differences are relatively minor and appear to be based on a desire to differentiate their approach. Are they trying to confuse us?

## Full PAF model

The 'full' PAF model with none of the top-level divisions for quality costs is perhaps best described in the relevant British Standard (BS 6143-2:1990 'Guide to the economics of quality. Prevention, appraisal and failure model'). This provides excellent guidance on the type of costs that should be included in the PAF categories.

## Opportunity cost model

This is an extension of the PAF model to include the importance of opportunity and intangible costs in the model. The traditional PAF model only deals with the tangible costs and intangible costs such as profits not earned because of lost customers are not included in the cost of quality. The opportunity cost model extends the PAF model to use the major categories:

- Price/Cost of conformance (POC)
  - Prevention costs.
  - Appraisal or (inspection) costs.
- Price/Cost of non-conformance (PONC)
  - Internal failure costs.
  - External failure costs.
- Cost of lost opportunity
  - Under-utilisation of installed capacity.
  - Inefficient resource utilisation.
  - Inadequate material handling.
  - Poor delivery of service.

## Process cost model

The focus of ISO 9001 and other quality approaches on 'processes' (see Chapter 4) has led to the development of the process cost model. This is very different to the traditional PAF model. It uses some similar terms, e.g. 'cost of conformance' and 'cost of non-conformance' but they mean very different things in this model (see Section 3.4).

In this model, the 'cost of conformance' is not the sum of the prevention and appraisal costs but is the process cost of producing products right first time by a process. The 'cost of non-conformance' is not the sum of the internal and external failure costs but the costs that can be allocated to the process not being carried out to the required standard.

This model is best described in the relevant British Standard (BS 6143-1:1992 'Guide to the economics of quality. Process cost model'). This provides excellent guidance on how to develop process models and find the costs for a given process.

## ABC model

Activity-based costing (ABC) systems are very different to traditional costing systems and can be used to identify areas of waste and non-value adding activities such as poor quality (see Section 3.5)[1].

ABC has broader implications than simply cost of quality measurement. ABC is used to change the method of cost collection and to drive Activity-Based Management (ABM)[2]. ABC and ABM are methods for achieving world-class manufacturing as a general philosophy and not simply reducing the cost of quality.

## Which one to use?

Each of the models has positives and negatives:

- If you simply want to get cost of quality data then my own preference is for the PAF type of model because of the easy integration with existing costing and accounting systems.
- If you are looking at processes and how to improve these then the process cost model offers an excellent method of assessing how efficient the process is and locating potential areas for improvement.
- If you regard cost of quality as simply part of the move to world-class manufacturing then ABC/ABM provides a set of tools that have benefits across the whole company and not only in quality.

Take your choice but do use one of them.

- 1.Kent, R.J. 2012. '*Cost management in plastics processing*', Section 1.6. Plastics Information Direct.
- 2. ibid, Section 1.7.

Reducing the cost of quality is a key area in cost management because of the large financial implications but more importantly, quality will be the one thing that decides if a company survives or not.

- The PAF model is based on activities.
- The process cost model is based on processes.
- The ABC model is based on activities for cost assignment and based on processes for the process view.

Whichever method is used, most companies are shocked to find that their true quality costs are far higher than simply the cost of their Quality Control Department.

Even rough figures will show that the cost of quality is a significant proportion of turnover and far higher than the profits.

# 3.3 Quality costs - PAF model

## The traditional model

The variations of the traditional PAF model all divide the cost of quality into three separate areas, i.e. prevention, appraisal and failure (PAF). In addition, most of the models divide the cost of failure into internal and external failure because these are very different in their magnitude and their accuracy, i.e. internal failure costs can often be assessed reasonably accurately whereas some of the external failure costs, especially the opportunity type of cost, can be very difficult to assess and collect.

As shown in Section 1.6, in most companies only 5% of costs are in prevention and 95% of quality costs are in failure and appraisal. Increasing the money spent on prevention and appraisal will save money in failure costs and change the ratios of the costs as well as reduce the overall costs. Re-assigning the costs and efforts to prevention can reduce the overall cost of quality from 20% of sales to 10% of sales and reduce the overall cost of quality by up to 50%.

## Getting and using the information

Section 3.2 refers to BS 6143-2:1990 and this standard gives guidance on how to identify the relevant data, how to report the data so that they actually become information and some broad guidelines on what to do with the data.

It does not matter which subset of the PAF model you use, e.g. Feigenbaum, Juran, Deming or Crosby, the basic data collection is the same. The differences are mainly in the terms you use.

In the initial stages of data gathering the main work will be carried out by the quality and accounting functions as they decide on the cost categories to include in the assessment, validate the data being collected and establish a consistent reporting format (see Section 3.10).

In the second stage, the main work will be by project teams (see Sections 2.7 and 2.8), the quality function and accounting who will use the reporting to drive improvement.

Reporting the information is only the start of the long-term process to transfer costs from the appraisal and failure categories to the prevention category and to reduce the overall magnitude of the costs. The cost of quality report and other non-financial numbers provide a focus for these activities (see Section 3.9). Standard tools such as Pareto analysis, 'cause and effect' charts and statistical process control can then be used to prevent defects and reallocate resources from failure to prevention and appraisal.

In the last stage (although the second stage is actually 'never-ending'), the reporting will be mainly used by the accounting function to hold the gains of improved quality management.

- **Tip** - Transferring costs from failure to prevention and appraisal is NOT cost free. In the initial stages there will be an increase in costs as prevention and appraisal costs rise before the failure costs decrease.
- **Tip** - Do not forget that prevention and appraisal costs should also be targeted for reduction.

Get some rough figures first - do not seek great accuracy at the early stages but simply try to find out where you are spending money on poor quality.

These areas can be targeted for rapid improvement using the standard tools.

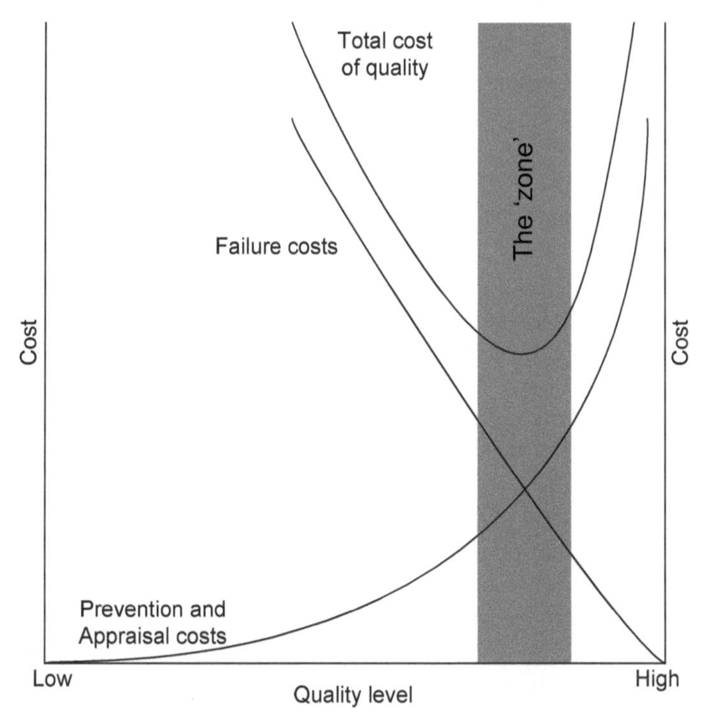


The classical model of minimised cost of quality

As prevention and appraisal costs rise the failure costs will reduce but after a minimum cost then the prevention and appraisal costs will rise. The objective is to minimise the total cost of quality by balancing the costs and being in the 'zone'.

| Prevention costs | Appraisal costs | Failure costs |
|---|---|---|
| ■Pre-contract meetings.<br>■Contract review.<br>■Design review.<br>■Realistic design tolerances.<br>■Quality plans and planning.<br>■Supplier selection and approval.<br>■Process design.<br>■Planning and scheduling.<br>■Planned maintenance.<br>■Housekeeping.<br>■Packaging.<br>■Statistical process control.<br>■Special investigations into failure costs.<br>■Quality reports.<br>■Quality improvement.<br>■Quality system auditing (internal and external).<br>■Quality system approval costs.<br>■Training personnel in quality. | ■Sales target monitoring.<br>■Customer satisfaction surveys.<br>■Supplier monitoring.<br>■Inwards goods inspection.<br>■Design verification.<br>■Stock control.<br>■ In-process inspection.<br>■Final inspection.<br>■Storage facilities for 'hold' products.<br>■Purchase of test equipment, gauges and measuring devices.<br>■Control and calibration of test equipment, gauges and measuring devices.<br>■General quality control overheads.<br>■Inspectors' wages.<br>■Training assessment.<br>■Personnel appraisal. | **Internal**<br>■Scrap.<br>■Labour overhead.<br>■Rework.<br>■Re-inspection.<br>■Sorting of product.<br>■Extra operations.<br>■Selective production and assembly.<br>■Downtime.<br>■Scrap material.<br>■Losses on sale of substandard items.<br>■Late or replacement deliveries.<br>**External**<br>■Customer complaint servicing.<br>■Customer returns.<br>■Credit notes.<br>■Discounted sales.<br>■After sales service (under guarantee or warranty).<br>■Fault investigations.<br>■Fault rectification.<br>■Interest on unpaid invoices.<br>■Product recall costs (transport, paperwork etc.).<br>■Litigation costs.<br>■Poor press/PR.<br>■Lost sales and customer accounts through bad reputation and production delays.<br>■Lost sales opportunities.<br>**Internal failure costs for previous 12 months:**<br>**External failure costs for previous 12 months:** |
| **Prevention costs for previous 12 months:** | **Appraisal costs for previous 12 months:** | **Total failure costs for previous 12 months:** |

**Total cost of quality for previous 12 months:**

## Quality costs - the PAF divisions

Most of the data for the PAF division of the cost of quality can come directly from a standard set of accounting codes. This makes the PAF cost approach very easy to integrate with standard company reporting processes and data. The codes should be checked so that the data are appropriate and accurate.

# 3.4 Quality costs - process cost model

## Integrating cost of quality and process improvement

The PAF model is the older (but probably easier to implement) method of costing quality but the process cost model is thought to be better for modern process-based management systems and finding where the cost of quality actually is in a process.

The management systems approach for quality (see Chapter 4) uses a process model and this model can also be used for cost of quality assessment. This approach means that the same model is used for both the management system and the financial data and the process model integrates these. The process cost model is also much better integrated with the Total Quality Management (TQM) approach to quality.

Section 3.2 refers to BS 6143-1:1992 and this standard gives guidance on how to identify the inputs and outputs of a process, several excellent examples of completed process models and some examples of the costs associated with the process model approach.

## Modelling the process

The basic process model shown on the upper right can be applied to any manufacturing or service operation that has inputs and outputs. The model defines a process as an activity that transforms inputs (materials or data) into outputs (materials or data) using resources (factors that are not transformed into an output). A process also has controls that define, regulate and influence the process.

A key difference between this model and the PAF model is in the definitions of the 'cost of conformance' and 'cost of non-conformance':

- In the PAF model the cost of conformance is the total of the prevention and appraisal activities and the 'cost of non-conformance' is effectively the cost of failure (internal and external).
- In the process cost model the 'cost of conformance' is the total cost of the process if it is operating at 100% efficiency and the 'cost of non-conformance' is the cost of all inputs and

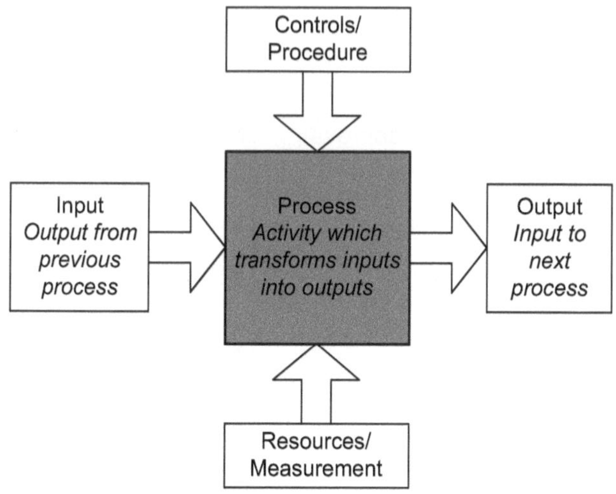


The basic process model

A process converts inputs (outputs from the previous process) into outputs (inputs to the next process).

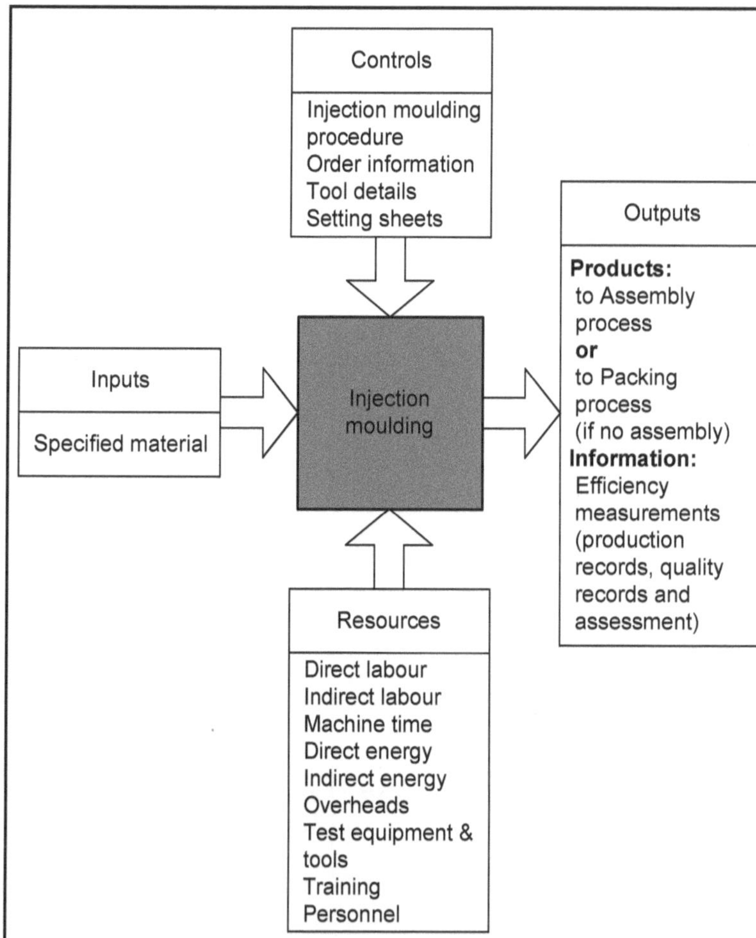


A process model for injection moulding

Most of the basic plastics forming processes are quite easy to map in the process model. The inputs and the outputs are limited and relatively easy to define.

resources that are due to the process operating at less than 100% efficiency.

These are very important differences in terminology and you cannot mix and match the terms in the models.

- **Tip** - The process cost model assesses efficiency (how well are you carrying out the process) but does not assess effectiveness (should you be carrying out the process at all). That is another issue.

An example process model for injection moulding is shown on the lower left and this gives the details of most of the important inputs, outputs, resources and controls.

## The process

Assessing the costs is a 4-step process:

### Define the processes

The first step is to identify the main processes (see Section 4.4), clarify the boundaries of the processes and the owners of the processes. The process owners are then responsible for identifying the process inputs, outputs, resources and controls for their processes.

- **Tip** - Every process must have a 'process owner' who is responsible for the cost of quality in the process.
- **Tip** - One advantage of the process cost model is that it can be used not only for production processes but also for service processes.

### Define the costs

Every process will have key activities and a cost of conformance and non-conformance for each activity. These can generally be grouped under the 'cause and effect' headings of (see Section 7.4):

- Men
- Machines
- Materials
- Methods
- Environment

Identify the cost of conformance for the activity (the cost of producing at 100% efficiency). This is the minimum cost for operating the process.

- **Tip** - One advantage of the process cost model is that it can reveal 'hidden' costs, i.e. if a process has traditionally had a yield of 90%, i.e. 10% scrap, then this will be written into the accounting system and only process yields of less than 90% will show as part of the cost of quality. The process model explicitly shows this type of process inefficiency and reveals areas for improvement.

Identify the cost of non-conformance for the activity (the cost of not producing at 100% efficiency and all the related costs). These are non-essential process costs. If a process is not producing at 100% efficiency then there is an immediate cost of non-conformance.

Some activities will be 100% cost of conformance, e.g. cost of test equipment and others will be 100% cost of non-conformance, e.g. scrap and machine breakdown costs. Other activities will have a mix of cost of conformance and cost of non-conformance.

- **Tip** - Every process needs process cost measurements.

### Set up a cost report

The process cost model needs a well developed cost report (see Section 3.10) with all the costs and their source data recorded.

### Start improving the process

Separating the costs of conformance and non-conformance highlights:

- High conformance costs, indicating a need for process re-design to reduce costs. This is the responsibility of the process owner.
- High non-conformance costs, indicating a need to reduce failure costs and to improve process yield. This is the responsibility of both the process owner and the process operators.

## Does it work?

The process cost model is a relatively recent introduction (it is only 30 years old). Some claim that it is easier to measure and quantify costs using the process cost model but it takes more time to initially set up the models and sometimes does not integrate as well with traditional accounting systems. As a result of these issues, it is used even less than the PAF model and has made limited impact in industry.

As with PAF, get some rough figures first and trial this method on a single process. Do not seek great accuracy at the early stages but simply try to see if the process cost model works for your company.

The integration of quality costs and the economics of these is attempted in ISO 10014:2006 'Quality management - Guidelines for realizing financial and economic benefits' but this gives no explicit support for either the PAF or the process cost model.

# 3.5 Quality costs - ABC model

## ABC for accurate costing

Activity-Based Costing (ABC) was originally developed to assign indirect overhead costs (or transactions) more accurately to products. ABC says that activities drive costs and it concentrates on the 'cost drivers' for the product/process. A product's cost is not just related to the variable costs but also to the overheads necessary to get and produce the particular job. This requires more analysis but gives clearer understanding of what a product really costs.

ABC was designed to improve the relevance and value of management accounting and to account for changes in manufacturing. The fundamental analysis necessary for ABC reveals the basic processes taking place in a company and the true cost of each process. This previously unavailable information provided a focus for cost reduction and leads on to even more important techniques such as Activity-Based Management (ABM).

As a result of this ABC and ABM are much more compatible with the ideas of World Class Manufacturing (WCM) than the more traditional management accounting methods. This makes ABC and ABM much more suitable for calculating the cost of quality in a WCM environment where 'waste' of any type is targeted for reduction and the cost of quality is a clearly identifiable 'waste'.

## Any 'waste' is a cost of quality

ABC provides knowledge about the fundamental processes (and the associated costs) and clearly shows where these are. Traditional accounting methods hide the opportunities for cost reduction by the way that the numbers were presented.

The information from the ABC process is used to develop Activity-Based Management (ABM) where all waste activities are targeted for reduction.

The two example cost reports on these pages show the traditional classification-based cost report (including the inevitable variance report line) and the ABM report.

The traditional report contains a re-statement of the budget and performance against that budget stated in terms of variances. Most managers will have seen dozens of these useless reports and been required to write long reports explaining the incomprehensible. This 'old' report gives no indication of value-adding processes or areas to target for cost reduction (except for 'everything').

The ABM report shows us that the value-adding activities ('produce mouldings' and 'assemble product') cost £245,000 or 60% of the total. All the other activities add cost (£167,000), but no value, and amount to 40% of the total costs. The areas for quality improvement are clearly seen and can be targeted for improvement.

Transaction costs offer the key to an understanding of the real costs.

Moulding Department - Cost Report
Classification based

| Allocation area | Cost (£) | Variance (%) |
|---|---|---|
| Salaries - Managers | 20,000 | -5 |
| Wages - Labour | 80,000 | +5 |
| Expenses | 10,000 | +12 |
| Factory space | 100,000 | +1 |
| Quality control | 15,000 | -12 |
| Packaging | 25,000 | +4 |
| Tool room | 30,000 | +6 |
| Security | 10,000 | -4 |
| Telephone | 1,000 | +6 |
| Training | 1,000 | -6 |
| Utilities | 60,000 | +10 |
| Insurance | 5,000 | +4 |
| Miscellaneous | 30,000 | -7 |
| Depreciation of machinery | 25,000 | -2 |
| TOTAL | 412,000 | -2 |

The traditional cost report showing the classifications and the variances

What do you do to get better and to improve performance? What value do the variances add? Do they help you to get better?

If a company is focused on 'waste' and cost of quality reduction then ABM gives far more valuable information than traditional management accounting. The cost of quality information is there but the overall rationale is cost reduction on a larger scale through waste reduction.

It is totally useless being efficient at something that you shouldn't be doing!

## Implementation

Implementing ABM starts by looking at and assessing the activities taking place in each department or process group. Each activity is assessed for time taken, total cost and the amount of value added by the activity. This allows the preparation of an ABM cost report similar to that shown in the example on the right. The value added by each activity can be clearly defined to provide targets for quality improvement.

For a longer treatment of ABC and ABM see:

- 1. Kent, R.J. 2012. '*Cost management in plastics processing*', Section 1.6. Plastics Information Direct.
- 2. ibid, Section 1.7.
- 3. Kaplan, R.S. and Johnson, H.T. 1991. '*Relevance Lost - The Rise and Fall of Management Accounting*'. Harvard Business Review Press.

## The benefits

ABM highlights the cost of all non-value-adding activities, including the cost of quality, and points the way to effective cost reductions. In the example, the ABM-based report shows the wasted costs associated with quality. The example report shows only the prevention, appraisal and internal failure costs of quality (to use the PAF terms). Nevertheless, the external failure cost of quality will appear in another cost report related to another process. This type of report acts as an incentive to reduce the time and cost associated with poor processes and their output.

## The downside of ABC/ABM

Despite the benefits, ABC/ABM is not a simple cost of quality system but a company-wide approach to accounting. Problems have been found with:

- The amount of work involved in accurately collecting the information on activities and drivers. This is especially true if the collection process is treated as ongoing.
- The overall complexity of the method.
- The continued focus on financial measurements and control of the process in the accounts department where it was still not 'owned' by the line managers.

**Moulding Department - Cost Report**
**Activity based**

| Allocation area | Cost (£) | Value added |
|---|---|---|
| Produce mouldings | 185,000 | 55,000 |
| Transport product | 20,000 | 0 |
| Tool changes & set-up | 35,000 | 0 |
| Store product | 22,000 | 0 |
| Count product | 8,000 | 0 |
| Inspect product | 22,000 | 0 |
| Segregate scrap | 9,000 | 0 |
| Recycle & rework materials | 8,000 | 0 |
| Assemble product | 60,000 | 30,000 |
| Repair and maintain machines | 37,000 | 0 |
| Sales of scrap | -2,000 | 0 |
| Rescheduling | 8,000 | 0 |
| TOTAL | 412,000 | 85,000 |

The Activity-Based Management cost report showing the costs of the activities

The areas for improvement are obvious and can be targeted for improvement by the local management.

# 3.6 Quality costs in plastics processing

## Not unique but special

Most of the cost of quality models will easily work for all the plastics processing methods but some care needs to be taken in the treatment of scrap and reprocessed material.

In many industries if the output of the process is defective then it is possible to rework the product to bring it up to specification. This is not generally the case in plastics processing and reworking of products is often minimal. Start-up scrap and defective products are almost always reground and the material is used again. Depending on the process, this reground material can be used internally or disposed of externally as scrap sales.

Internal use involves either:

- Material being regranulated and fed directly back into the process (as with tops and tails in extrusion blow moulding and in injection moulding of many caps and closures).
- Material being taken off-line, regranulated and fed back into the process at a later stage at a set rate (as with larger injection mouldings and most extrusion products).
- Material being taken off-line, reground and fed back into another process at a later stage (as with the use of lower-grade material for specific products).

External use involves either:

- Material being reground for volume reduction and sold to external firms for reprocessing and sale to other companies (or sometimes back to the original processor).
- Scrap products being sold to external firms for regrinding and sale to other companies (or sometimes back to the original processor).

This ease of reprocessing and the number of routes available make plastics special (but not unique) and we need to know how to deal with scrap and reprocessing.

## How do we treat scrap?

### Start-up scrap

Too often start-up scrap is not considered in the accounts and is set at a fixed level or percentage. Only if start-up scrap exceeds this fixed percentage will an alarm be generated. Start-up scrap (and all the other associated costs of set-up and start-up) should be included in the cost of quality system. Only by understanding the size of these costs will a company see the benefits of reducing these costs and improve the quality of operations.

- **Tip** - The cost of quality report should assume no allowable levels of start-up scrap (effectively an allowable level of 0%) and report all scrap generated at the purchase cost of the material.
- **Tip** - It is impossible to reduce start-up scrap to zero. The important thing is to not accept a set level and to only react if this level is exceeded. Reducing start-up scrap not only reduces the amount of material used but also gets processes on-line quicker.

Don't forget that in most cases it is the process that is at fault and not the operator.

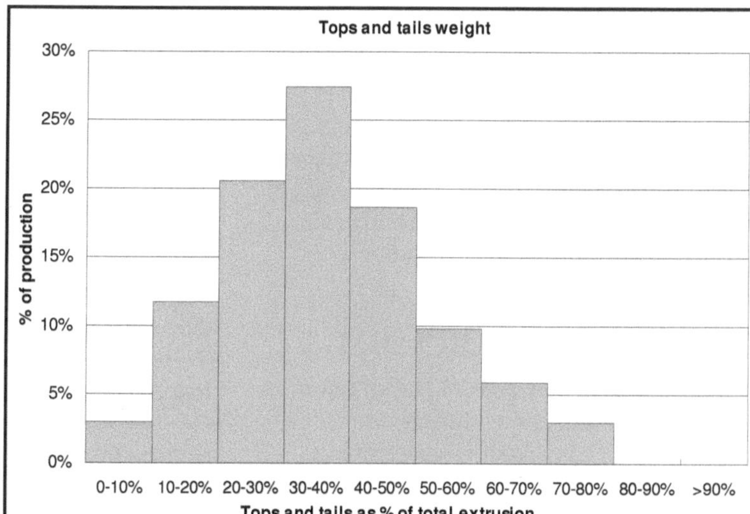


### Scrap in extrusion blow moulding - tops and tails

Tops and tails are an inevitable part of the extrusion blow moulding process. The real problem is that they are rarely controlled to an adequate degree. Companies think that because all the material is recycled back into the process that the scrap level is effectively zero.

This ignores all the costs of processing the tops and tails, i.e. the energy cost of operating the extruder, the conveyors for the tops and tails, the regranulators, the blending equipment and all the other reprocessing equipment. It also ignores the process efficiency losses.

Any material that is processed and reground costs money.

The UK industry average for tops and tails is approximately 35% and this is a significant number. It needs to be accounted for in the cost of quality.

## General (running) scrap

As with start-up scrap, general (running) scrap is often set as an allowable percentage of the amount of material processed and little investigation carried out unless the actual value exceeds the 'allowable' scrap level. The cost of quality report should assume no allowable levels of scrap (effectively an allowable level of 0%) and report all scrap generated.

Scrap is generated as a result of an inefficient process (see Section 3.4) and reporting the cost of quality needs to identify this cost. Often the allowable percentage was set some time ago and never revisited for improvement. Any level of scrap is a cost to the company compared to 100% efficient production.

- **Tip** - The cost of quality report should assume no allowable levels of running scrap (effectively an allowable level of 0%) and report all scrap generated at the purchase cost of the material.

## Re-processing

All reprocessing operations for scrap (whether for internal or external use) is a cost of quality compared to 100% efficient production and needs to be included in the cost of quality report.

Although the material is not lost to the system, reprocessing produces other losses such as machine time losses, labour time losses, energy use losses and opportunity losses and these all need to be considered as part of the cost of not producing at 100% efficient production.

Typical reprocessing costs also include:

- The capital costs of all the equipment needed for regranulation, e.g. conveyors, sorters, regranulators and storage equipment.
- The establishment costs needed for regranulation, i.e. space associated with regranulation if this is charged in the accounts.
- The operating costs of all the equipment needed for regranulation, e.g. labour, energy, administration.
- The sorting, segregation and transport costs necessary for regranulation and for treatment of the regranulated material, e.g. labour and equipment costs.
- The capital costs of all the equipment needed to insert the material in the processing system again, e.g. blenders, silos and conveying systems.
- The operating costs of all the equipment needed to insert the material in the processing system again, e.g. labour, energy, administration.

These are not insignificant costs that arise purely because of system inefficiencies and are legitimate costs of quality, i.e. if the system was 100% efficient then they would not be there.

## Scrap sales?

When material is sold externally, it is often tempting to include the receipts from the sales as a credit to the cost of quality account. In strict accounting terms, this is a receipt and needs to be included in the accounts as such but not in the cost of quality account!

If the original raw material was purchased at £800/tonne, scrap was generated and then the scrap was sold (either with volume reduction or not) at £400/tonne then in cost of quality terms the company has lost £400/tonne. This is not a positive sign!

We do not include material purchases in the cost of quality report and it is therefore illogical to make a claim for material sales that were generated purely as a result of less than 100% efficiency.

- **Tip** - Do not deduct any income from sales of scrap from the cost of quality calculations. You should not have produced it in the first place so it is a real cost.

**What is your cost of scrap?**

I once asked this question of a moulder and was told there was no scrap because it was all reground and used again.

It took me some time to convince them that there was a real cost here and that it was worthwhile measuring (and reducing)!

Some years ago we collected these costs for an in-house moulding company. We used the standard management accounts and set up a report based on the allocation codes. After the numbers were complete and the magnitude of the cost was revealed there was no longer any problem with investment approval for quality management and the 'cost of quality' report became a regular management report at Board level.

# 3.7 Starting out

## What do we want this for?

The real reason for calculating the cost of quality is to provide a concrete financial justification for quality improvements and to identify areas where the cost of quality can be reduced quickly and easily. This is not simply a measurement system, it is a monitoring and targeting system.

At this stage we will concentrate on a measurement system based on the PAF model. This is not because it is better but simply because it is more common and more easily integrated into a standard set of company accounts.

## What can we achieve?

As noted in Section 3.1, the typical cost of quality for most plastics processors is in the region of 20% of sales. Upper management will initially be doubtful about this value (they always think it is much less) but the real numbers can only be quantified by a cost of quality report.

A good quality management programme, backed up by good reporting, can reduce this to 10% of sales within 2 years. For most plastics processors, where the net profit will be in the region of 5%, reducing the cost of quality will have a dramatic effect on profitability.

The cost of quality is a real number that can be reduced. However, as the easy projects are completed, it will eventually be more difficult to find new projects and savings - this is where cost of quality reporting can provide an essential guide to areas for attention and improvement.

## Before starting

There are several question that need to be answered before spending any time on generating a cost of quality report. These are:

- Is the management committed to using the cost of quality information to improve the business and improve profits?
- Will cost of quality information fit within the existing framework for operational information and reporting? i.e. will cost of quality information either duplicate existing reporting or conflict with existing reporting?

### Let's do this quickly!

The best way to get the cost of quality programme running is to use basic and easy to obtain information to quickly identify some easy targets.

'Fast starts' are one of the best methods to validate a cost of quality programme. These use a tactical approach to show the rough order of magnitude of the problem to a company.

Some rules for 'fast starts' are:

- Start out with a single easily defined area as a pilot project.
- Start with the data that are easiest to collect.
- Where possible, use data that are 'approved' by the accounting function.
- Use the existing accounting cost codes as a starter for most of the data.
- Use existing time code classifications as a starter if cost code classifications are not available.
- If data are not available then use 'unit cost' approximations, i.e. if you don't know the cost of all the product returns then estimate the cost of a single return and multiply this 'unit cost' by the number of returns to give an estimate of the total cost.
- Where possible, use data that are 'standard' for the company.
- Get good data on the largest costs first rather than worry about the smaller costs. Remember Pareto!
- Use sensible estimates when no data are available. Accept that estimates may be wrong and do not aim for perfection on the first pass. Use your judgement, be consistent and record any and all assumptions (they will be criticised).
- Do not waste time on costs that are known to be small, either ignore them or use an estimate.
- Get accounting to sign-off on estimates as being reasonable or get a second opinion (preferably from the process owner).
- Develop a report format for the 'fast start' that can be expanded to cover the whole company.
- Trial the report format with the process owners.
- Allocate costs to the PAF categories and to each area.
- Identify the largest costs and allocate responsibility for the large costs.
- Look at failure costs first and set a target of 50% reduction.
- Assess the results:
    - Is the report understandable by the users?
    - Does the report provide guidance on where savings can be made?
    - Are the metrics appropriate.

Only after a successful pilot project has demonstrated good information should the project go live across the whole company.

- Will cost of quality information focus only on the easily defined costs (i.e. pure accounting data and numbers) or is there a need for less easily defined costs (i.e. customer feedback and reputation costs)?
- Will cost of quality information, however generated, be used throughout the business, without exception?
- How often is cost of quality information to be prepared? i.e. is this a once-off assessment or is this a regular report that will be prepared weekly or monthly?

These questions need answering to ensure that the work involved in generating the cost of quality information is both valued and valuable.

## Project team

Setting up a cost of quality information system is a project that needs management and control (see Section 2.7). It is strongly recommended that a small project team be set up to start the work and to assess the results. This team must include the accounting function in order to be useful. Get the accountants involved early so that their input and data are used to the best effect.

## What are the problems?

Fast starts are an opportunity to trial a cost of quality system and to reduce system problems before the system is rolled out to the complete company.

Getting good cost of quality data is not always easy and starting out will always be difficult. There are several main reasons for this:

- Data can be hard to obtain because quality costs often cut across normal accounting boundaries. This is particularly true for overhead costs where these are allocated by an accounting formula. These need to be treated carefully to ensure that realistic allocations are made.
- For some processes the cost of quality is included in the expected yield of the process, i.e. the cost of quality is often implicitly included in accounting model of the process. The process model approach can often clarify this (see Section 3.4).
- There is a lack of transparency in what is actually being recorded in the standard accounting cost codes where these are being used as part of the data. This is a particular problem if staff are inconsistent in recording data because of varying interpretations about what the cost of quality really is.
- There is a lack of understanding about what cost of quality reporting is designed to do. It is not simply about reporting, it is about action.

Use 'fast starts' as a tactic to show the benefits of a longer-term strategic system to the sceptics in the company.

## What can we expect to see happen?

Even using a good, well-designed fast start that generates good information that leads to action the cost of quality will not initially decrease. The reason for this is that prevention actions immediately cost money to implement but failure cost savings will take time to work their way through the system, i.e. today's failure costs relate to products made some time ago. It is therefore normal for the total cost of quality to rise in the early stages of implementing quality management. A typical result is shown below and illustrates the initial rise in quality costs as the transition from failure to prevention is made.

Study the data before developing the system.

'Fast starts' are designed to show what can be done with cost of quality information.

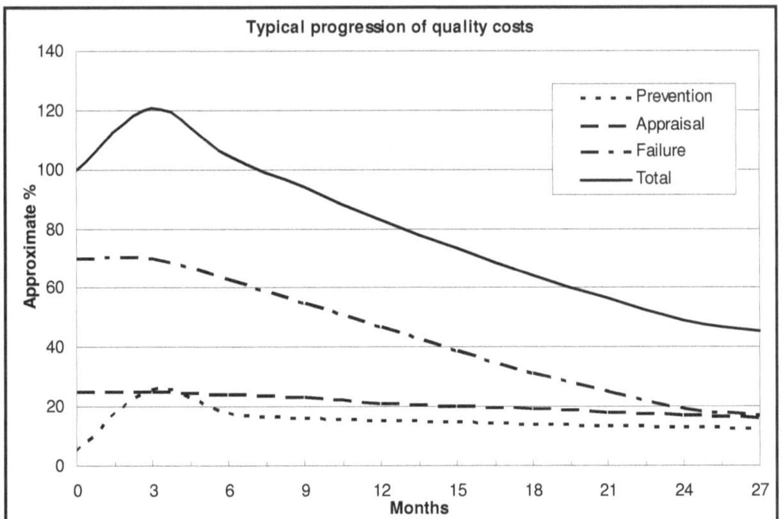


### Quality costs will initially increase (ideal)

The transition from failure to prevention is not cost free. Initially the costs will increase as prevention costs increase before failure costs decrease. Most companies will see an initial increase in costs - quality might ultimately be free but not in the initial stages.

# 3.8 Integrating quality costs into the accounts

## Accounting - a key function

The potential cost of quality savings from simple actions are known to be large but when companies attempt to reduce their cost of quality the achieved reductions are often far smaller than those known to be possible. One of the major reasons is that quality management is seen as a quality or production department issue that is not linked to the financials of the company. This means that quality activities are not seen as rewarding enough for the other staff to get involved.

Nothing could be further from the truth and some of the most effective quality management efforts can come directly from the accounting function. So why is it that the accounting function pays little attention to quality and the positive benefits of cost of quality reductions?

## The magnitude of the costs

The cost of quality in plastics processing is in the region of 20% of sales but surprisingly few accountants are aware of this and even fewer try to account for it. Where did we go wrong?

Quality improvement is an internal low-risk activity that can raise profits substantially and should certainly attract attention in any company. Most companies do not hesitate to spend money on trying to increase sales but fail to see the benefits in spending money on increasing profits by decreasing the cost of quality.

The efforts and risks of improving profits via quality management are entirely internal and within the control of the company but the effects are wider than this. Increasing profits through the conventional external approach of increasing sales means increased sales effort, increasing production capacity and increased risk. Increasing profits from cost of quality reductions is low risk, generally only needs low investment and involves mainly internal efforts.

## Data collection

Measuring the cost of quality involves data collection but there is no point in spending more money to collect data than can be saved by the useful application of those data. It is always necessary to assess the cost-benefit balance for data collection. It is also important to recognise that good-quality data are not necessarily the same thing as highly accurate data. The need is for enough relevant information to promote action - collect only the minimum amount of data necessary to produce the relevant information.

Much of the core data is probably being collected already for other cost-control measures - in most cases it is simply a matter of integrating the available data into an accessible report and converting it into information to be used for action.

- **Tip** - Use the available data, they do not cost extra money and are already there.

## Assigning the costs

Cost reductions will only happen if somebody is made directly responsible and held accountable for them. After the data have been converted into information it is necessary to assign the costs to create ownership.

When monthly accounts are prepared for any business the operational costs are assigned to people who are responsible for the performance of that section of the operation. The cost of quality information should be treated similarly in the monthly accounts package.

- **Tip** - Always assign costs to a process and the process owner to get action.

Accountants can add value to a business by using their analytical skills to improve quality.

Aim to integrate the collection, analysis and reporting of quality-related costs into the accounting system and optimise quality costs.

As with any measurement system, the cost of implementing the measurement system should be less than ≈ 1% of the savings made by using the measurement system.

- Total cost of quality (£).
- Total cost of quality as % of added value.
- Total cost of quality as % of sales.
- Total cost of quality as % of profit.
- % Failure costs = Failure costs/total cost of quality.
- % Appraisal costs = Appraisal costs/total cost of quality
- % Prevention costs = Prevention costs/total cost of quality
- % Return of investment for quality = Cost of quality savings/ Implementation cost.

### The top-level indicators: What to report?

In addition to the area-specific information (see Section 3.5), some top-level indicators are needed to allow assessment of progress. Some of the more common top-line indicators are shown - but feel free to use your own.

## Standard measures

The top-level financial indicators are shown in the box (on the left) but many other measures can be used depending on the audience. Some of these are financial and some are more operational in nature.

### Financial numbers:

- Cost of materials.
- Cost of prevention labour.
- Cost of appraisal labour.
- Cost of defects/100 good products.
- Cost of late deliveries.
- Return on quality = increase in profits/ cost of quality management programme.
- Quality rate = input material - (defects + start-up scrap + rework)/input material).
- Process quality = (Available time - rework time)/Available time.
- First time yield = % of product with no rework.

### Non-financial numbers:

- Supplier performance.
- Inventory accuracy.
- Number of products (line items) delivered on schedule.
- Value of products delivered on schedule.
- Machine down times.
- Set up times.
- Material availability.
- Direct labour productivity.
- Indirect labour productivity.
- Number of complaints.

The measures chosen should be selected by each company to meet their operational needs.

- **Tip** - All of these measures come from hard internal data. It may also be necessary to include some non-standard non-accounting data (see Section 1.7) to reflect the customer's needs and assessment, e.g. customer satisfaction surveys. This is always subjective and the accounts function may need assistance in the collection and formatting of these data.

## Measuring to manage

These ideas allow quality management to be integrated into the accounts as with any other cost element and accounting for quality management can be treated in the same way as other items in the accounts systems. The overall aim is to achieve cost-effective quality management.

Integration of quality reporting into the accounting function allows quality to take its place on the management agenda as a part of the normal management process.

In spite of this, the majority of companies have no active cost of quality reporting system. They consider that it is already reported (but under a variety of headings), is not central to their core business and are unaware of the potential for improvement or the substantial returns that can be made through investments in quality management. This needs to change to improve profitability.

Data are not the same thing as information. Data are simply a collection of numbers whereas information provides the basis for management action.

Sites can be awash with data and still have little real information.

## Monitoring & targeting (M&T)

Gathering data will not, in itself, provide results. The potential benefits of quality management cannot be achieved by collecting large amounts of data or preparing lengthy reports. Data are meaningless without careful analysis, and reports are useless if they are not targeted at people with the authority and the will to act.

Quality costs are not always due to bad production but can also be created by initial specifications or customer expectations that are unrealistic with respect to production capabilities.

**Integrating the cost of quality into the accounts:**

| Do: | Don't: |
|---|---|
| ■ Understand what you are trying to do with the data. | ■ Start alone - always get the accountants involved at the start. |
| ■ Remember this is about the money. | ■ Expect the accountants to start the process. |
| ■ Remember this is about business not philosophy. | ■ Expect that accounts data will always give you what you will need. |
| ■ Start small. Try a couple of 'fast starts' with easy data and with accounting support. | ■ Be too ambitious - this can take some time to get right. |
| ■ Try to only use data approved by the accountants. | ■ Focus exclusively on the known costs - the unknowns can often be larger. |
| ■ Use standard data whenever possible. | ■ Worry too much about PAF categories and how a cost fits into them - a single cost can sometimes be fit into two or three categories. |
| ■ Only use costs that are relevant to quality. | ■ Worry too much about the small costs - pick them up later. |
| ■ Start with failure costs (they are probably the biggest). | ■ Use data that are suspect in any way. |
| ■ Go for the big costs first and try to get better data. | |
| ■ Get the easy costs (however big) next. | |
| ■ Try to capture some of the staff costs and the indirect costs. | |

# 3.9 Setting targets

## The basics

After the cost of quality is integrated into the accounts and suitable measures (financial and non-financial) are defined then it is necessary to set some targets for improvement. Setting achievement targets is necessary for action but effective targets must be set to provide a real incentive for action.

The characteristics of effective target measures are[1]:

- They should be clearly communicated to the employees: 'I will succeed if you tell me what to do and where you want me to improve'.
- They should be largely non-financial and used by all employees: 'Tell me in terms I can understand'.
- They should reflect the performance required at the location: 'Give me something that is relevant'.
- They should vary with time to reflect changing organisation goals: 'I've got that right so let's move on to the next priority'.
- They should be simple and easy to use: 'I understand that'.
- They should be fast to give quick response and feedback: 'Is today OK?'
- They should aim to teach rather than monitor and produce improvement rather than simple historic reporting: 'OK, now I see where I can get better'.

When setting targets it is essential that the target is set relative to the audience that is expected to achieve the target (see Section 3.10).

Above all, targets should be challenging but achievable to provide a real incentive to improve but not be so difficult that the people required to deliver them become disenchanted with the targeting system. Targets therefore need to be derived from data that are relevant to the site.

- **Tip -** Every company needs to set their own targets based on their internal and market needs. This can be from external competitor benchmarking, internal financial or non-financial data or from external market/customer data.
- **Tip -** Be aware that not all targets will be financially based.

## Zero defects?

The 'classical' model of quality costs (see Section 3.3) implies a minimum cost of quality at something less than zero defects and an infinite cost of quality at zero defects. This is based on the assumption that prevention and appraisal costs are nearly infinite at zero defects. In fact, although prevention and appraisal costs rise significantly as zero defects is approached, they do not rise to infinity. It is possible to achieve nearly zero defects without an infinite cost of quality but most companies will not achieve zero defects due to diminishing returns on investment.

- **Tip -** Failure costs are minimised when the company can find no more profitable projects for reduction.
- **Tip -** Appraisal costs are minimised when failure costs are minimised and there are no more profitable projects for reduction.
- **Tip -** Prevention costs are minimised when most of the cost of quality is in quality improvement.

Aim for a 50% decrease in the overall total cost of quality within 18 months - challenging but achievable.

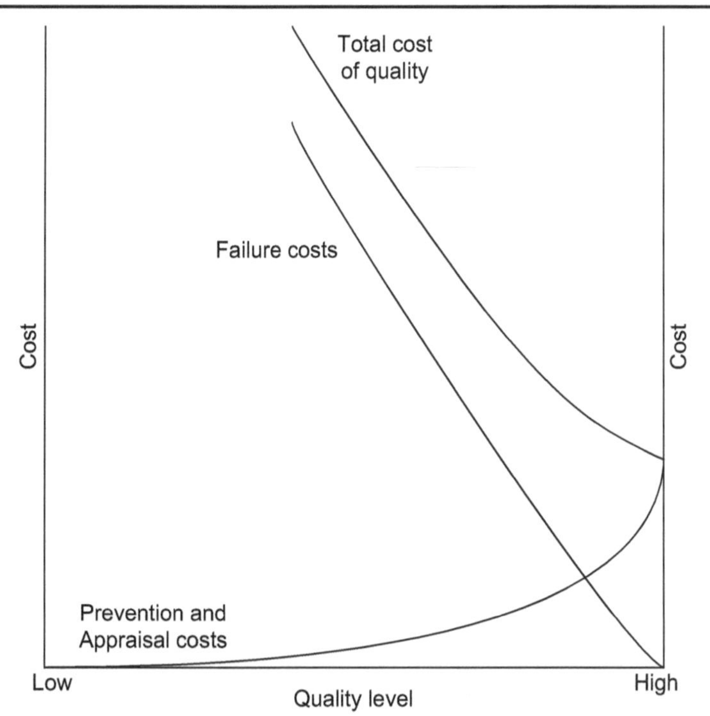


### The modern model of minimised cost of quality

The modern search for zero defects argues that prevention and appraisal costs do not rise to infinity as defects are reduced and there is no 'zone' (see Section 3.3). Failure costs can be reduced to nearly zero and this is where the minimum cost of quality is.

## Common and special causes

As Chapter 5 will make clear, variations will occur as a result of 'common' and 'special' causes. The actual progress in cost of quality reduction will not be the smooth ideal curve shown in Section 3.7 but will be more variable due to both common and special causes. Common causes are the natural variation of the system but special causes are significant events that will cause short-term variations - these will often cause targets to be missed.

Special causes can often be traced back to introduction of poor material or other system disturbances. These need to be identified, rectified/removed and systems put in place to prevent reoccurrence (see Section 4.16).

## Holding the gains

As the cost of quality decreases there is a need to pay attention to holding the gains. It is common for companies to make good initial progress but then to slip back as a result of failing to hold the gains. Improvements to systems and operations are either reversible or irreversible and these need to be separated:

### Reversible improvements

These are improvements to working methods that can be forgotten, ignored or reversed as management looses focus on quality. These are the improvements that really need work to hold the gains.

- **Tip** - Regular process auditing will help to hold the gains for reversible improvements. This is similar to, but not the same as, ISO 9001 auditing (see Section 4.17) and should be part of the normal work of the company.

### Irreversible improvements

These improvements are the result of investments in new equipment or processes that will prevent reoccurrence of the concern. After the investment is made, these improvements will generally be stable but it is still recommended that auditing is carried out to ensure that the gains made are those that were promised.

- **Tip** - Auditing of irreversible improvements should still be carried out to validate the investment and then regularly but less frequently than for reversible improvements. This is particularly necessary after investment in machines to ensure that they really deliver quality improvements.

## When targets start to fail

If the rate of improvement in the cost of quality starts to slow then look for the following:

- The business processes are out of control and need reviewing and/or simplification.
- The cost of quality indicators are not analysed adequately, i.e. trends are missed.
- The cost of quality information is not related to other business information and trends.
- Changes made in one area to decrease costs are having an effect in another area.
- The management team is not involved in the process and only gets involved when there is a crisis (generally to shout at somebody).
- The management team has no new ideas on cost of quality reduction despite the information in the report.
- The bulk of the company employees are not engaged with the direction the company is moving in and do not agree with the company priorities.

It is tempting to try to reduce the prevention and appraisal costs because these are seen as direct costs that can easily be cut. The problem with this is that it will lead to a short-term reduction in the cost of quality that will be followed in the mid-term but much higher internal failure costs and in the long-term by very much higher external failure costs.

The first place you look is not always the best place for savings.

Do not be tempted to respond to falling sales (as a result of poor quality) by reducing the efforts to improve quality.

•1. Maskell, B. 2009. '*Making the numbers count*'. Productivity Press.

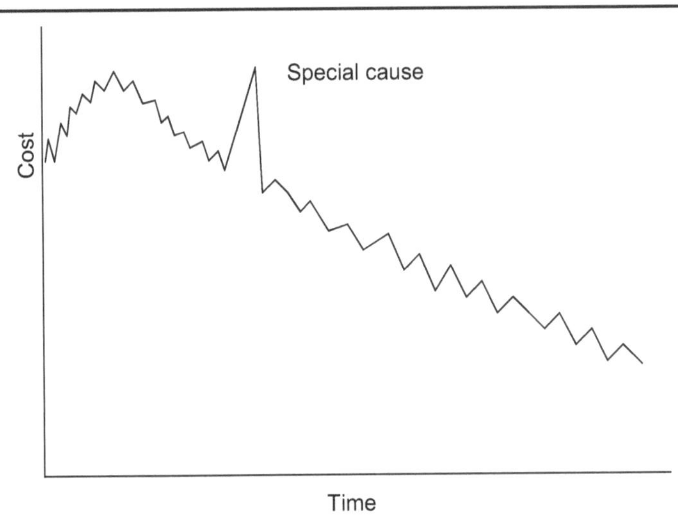


### Special causes and holding the gains

The graph shown in Section 3.7 is an idealised graph. In reality, there will be special causes that will increase the cost of quality for a short time. Setting targets is also about holding the gains that are made during the transition to prevention.

# 3.10 Reporting quality costs

## The management action cycle

Accounting, measuring and allocating the cost of quality does not automatically reduce the cost of quality. Cost of quality reductions only result from actions. As with any information system, there is not only a need to set targets and to report the costs but even more importantly to use these reports to identify and stimulate action. The management action cycle (right) provides a template for successful management.

If any action is to be successful the system should:

- Have the full support and commitment of the senior management team.
- Support the existing lines of authority and responsibility.
- Use analysis appropriate to the process to ensure adequate control.
- Involve the people it serves, accommodating both their views and desires to offer solutions and have their performance judged on fair criteria.

## Monitoring and reporting

After targets for improvement have been agreed and set, it is necessary to report on progress to ensure that effective action has taken place. Any cost of quality accounting report should support the targets set and meet the following requirements:

- The report should reach the manager who controls the relevant process (the process owner).
- The manager should be able to understand what the report means to them.
- There should be a minimum of extraneous information.
- There should be some means of ensuring that action is taken when it is needed.
- The reports should be integrated as far as possible into the existing management information systems to make cost of quality accounting a part of the normal operations of the company.

It is more important that the information reporting system does what the company needs rather than does many things that it does not require. In most cases, simple accounting reports offer a method of starting out with very basic (and validated) numbers and then developing the report as the needs grow.

The type of report produced should match the needs of the audience to the information in the report:

- Annual reports should provide top-level performance information to show progress over the period.

- Talk in £ to upper management.
- Talk in both £ and operational numbers to middle managers.
- Talk in operational numbers to other staff and operators.

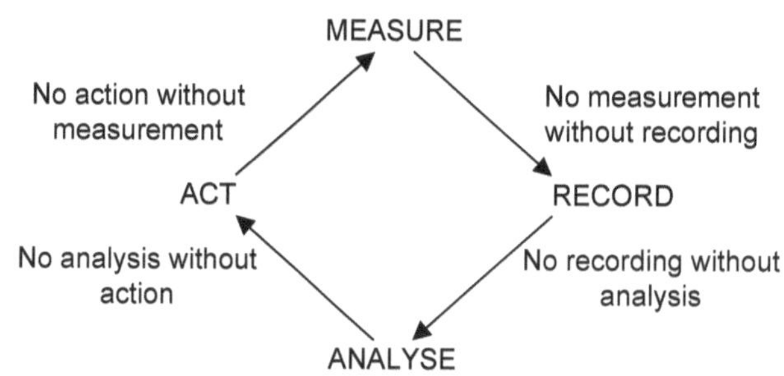


### The management action cycle

Analysing data to provide information will not generate action unless there is a formal system for both reporting and taking action.

- We will take no measurements without recording the results.
- We will not record results without analysing them.
- We will not analyse results without acting on them.
- We will not act without measuring the results of our actions.

### Distributing the information

| | Annual report | Monthly report | Weekly report | Key indicators | Exception report |
|---|---|---|---|---|---|
| Managing Director | ✓ | ✓ | | ✓ | |
| Finance Director | ✓ | ✓ | | ✓ | |
| Department Heads | ✓ | ✓ | ✓ | ✓ | ✓ |
| Purchasing | ✓ | | | ✓ | |
| Supervisors | ✓ | | ✓ | ✓ | ✓ |
| Workforce | ✓ | | | ✓ | |

### What information do recipients need?

The same report is not needed for all the staff at the site. Reports need to be tailored to the needs of the audience and not everybody needs all the information. Be selective about distributing information otherwise people will turn off.

- Monthly reports should match the format for conventional financial reporting and be incorporated into these.
- Weekly reports should be more detailed and include operational target information to provide rapid feedback on operational issues.
- Key indicator reports should be designed for general consumption and produced as colour charts suitable for posting on notice boards.
- Exception reports should be produced for Departmental Managers on areas where targets have not been met or for identification of areas for improvement.

Distribution of reports is always a key area. Not all levels of staff require the same information and the reporting should vary to reflect the needs of the audience.

## Sustaining the system

The greatest threat to cost of quality accounting is where the system operates but no one takes any notice, there are four reasons for this:

- Lack of authority: where the system does not have senior management support.
- Lack of ownership: where the staff are servants of the system rather than vice versa.
- Lack of accountability: when there is no clear linkage between responsibility and authority.
- Lack of resolution: where the system highlights concerns that do not really exist or does not highlight concerns that do exist.

Do not fall into any of these traps.

## Involve the accountants

The accounting function can provide a range of tools and experience to target and control the cost of quality. This is part of their function and they cannot disclaim responsibility. Cost of quality information can be easily integrated into the conventional accounts package and the magnitude of the costs makes the accounting function negligent if they are not involved in this area.

### Measure

- ■Be prepared to change the measurements with time. As one concern is solved/reduced then move on to the next most important concern and change the measurements.
- ■Do not be afraid to include 'special' measurements to drill down into specific issues that need solutions.
- ■Use units that are familiar to the people responsible for the measurement.
- ■Check that the people responsible can actually influence the measurement before measuring. If they cannot control it then they will be frustrated and the system will not drive action.

### Record

- ■Only collect data that will be used.
- ■If data comes from accounting then check that the recording is accurate.
- ■If data do not come from accounting then use simple straightforward forms.
- ■Perform simple checks to test data quality and validity.

### Analyse

- ■Analyse the data for each time period, as they are available and do not wait too long.
- ■Use graphs and visual displays to highlight important information and progress. A well-designed 'dashboard' communicates information quickly.
- ■Keep analysis simple to meet the needs of the company and to show areas for improvement.
- ■Integrate the process with existing systems such as the accounts package.
- ■Issue reports in a timely fashion.
- ■Do not overload people with unnecessary information.
- ■When setting standards ensure that any derived relationship is reasonable and defensible.

### Act

- ■Discuss reports informally with the relevant staff after each time period.
- ■Agree future plans for each period.
- ■Encourage discussion about future cost of quality savings.
- ■Include quality in other management information discussions (e.g. team briefings).
- ■Motivate to encourage action rather than reaction; it is better if the production area is ahead of the game.
- ■Do not be afraid of fundamental changes to the process.

### Control actions from the management action cycle

Control actions should be consistent and follow the management action cycle format to be most effective. Simple easy-to-read reports are most effective at generating improvement action. When in doubt, leave it out - they can always ask.

# 3.11 Capital expenditure

## Changing the rules

Many people think that quality management is about simply increasing the number of inspectors to sort the bad from the good. Cost of quality reporting will soon show that this is not the case but it is not the only use of cost of quality information.

The cost of quality reporting will show that true gains in quality will only come from real improvements in prevention and appraisal and not simply more inspectors. This aspect of quality management is about using cost-effective improvements and technology to permanently reduce a company's cost of quality.

Whilst many quality management improvements are relatively low-cost and fall more properly into the management or expenditure budgets, there are often projects that need capital investment and it is worth investigating how quality improvement projects are treated in terms of capital investment.

## Capital expenditure - how to get it wrong

Capital expenditure is 'something on which a business spends money in order to earn more money'. This definition includes all the time and money we have tied up in the business. These investments are what we have put into the business or have at risk should the business fail. This is a broad definition of investment but it does reflect reality.

Capital expenditure is often unconsciously divided into two areas:

- Business development: This is exciting and gets the most attention, e.g. 'New toys for the production area'.
- Business maintenance: This is routine and gets ignored unless it is really large numbers: 'We already have a toy that does that'.

This division is shown on the right with the business maintenance activities further sub-divided into essential and discretionary expenditure. Expenditure on quality management is typically seen as being a discretionary business maintenance expenditure (until it is too late). This is the lowest priority in the ranking of business expenditure and the necessary investment often falls off the bottom of the list when the available money or management time runs out.

Despite this perception, the business maintenance activities are often those that generate immediate cash for the least effort. Investment in business maintenance, mainly cost management, is proven to be a valuable source of internally generated investment funds, i.e. initial investment in business maintenance generates funds for later business development.

Purchasing capital equipment to improve quality can permanently change the cost of quality for a company.

## Financial appraisal

The conventional accounting treatment of capital expenditure is largely based on methods such as Net Present Value (NPV), Return on Investment (ROI) and Discounted Cash Flow (DCF). These methods enable ranking of competing investment opportunities and choosing between competing projects - they do not decide whether to invest in a specific project or not, that is the management function.

Unfortunately, when looking at quality improvement projects there is often no 'income' associated with the project. The accounts department keeps records of expenditure but without a cost of quality system, it is not set up to keep accounts of money saved - especially as the cost of quality is often unknown. This makes it difficult to justify expenditure on quality improvements because the cost reductions

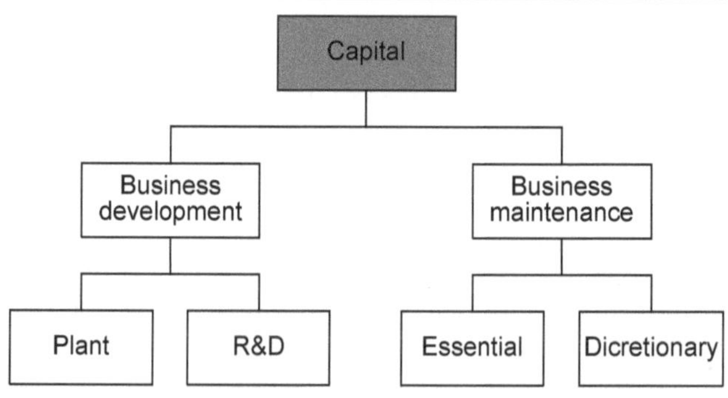


**Expenditure on quality management is seen as discretionary business maintenance expenditure**

This category is generally seen as the lowest priority in the ranking of business expenditure and it can be difficult to show an 'income' from business maintenance projects.

are not capable of being recorded as being due to the investment.

Fortunately, the cost of quality system can be used to not only identify the time of changes to the cost of quality performance but also the magnitude of changes.

Quality managers can use the cost of quality reporting system to maintain a 'capital return budget' that identifies improvements, costs and effective income generated by the investments. This can be a valuable tool in justifying continued investment in quality management.

## Investment decisions

Cost of quality information needs to be used carefully in investment decisions.

It is possible that capital investment and/or improvements in one PAF category will simply transfer costs to another PAF category. If cost of quality data are not used effectively then investment in a new machine may reduce internal failure costs (scrap) but increase other costs in other areas such as set-up, maintenance and inspection. When this happens, the cost of quality will simply be moved from one category to another with no net decrease. Any capital investment project must be analysed for the overall impact in the cost of quality.

Capital investment always involves the choice between competing products. It is easy to take the soft option and go for the lowest capital cost equipment without considering the total life cycle cost of the equipment. This cycle cost includes not only the operating costs, e.g. energy, but also the effect on the cost of quality. Efficient, reliable machines can reduce the cost of quality significantly but this must be justified and this can only be done if the cost of quality information is available.

In the long-term, and that is what capital expenditure is all about, it is cheaper to specify and purchase high-specification machines at the start than to pay continuously for the poor cost of quality throughout the life of the equipment. The initial purchase cost should not dominate the decision-making process. The 'whole life' cost of the equipment (initial cost + operating costs) is the important cost for any plastics processor who wants to continue operating in the long term.

Purchasing high-quality capital equipment is the simplest and easiest way to permanently improve a company's cost of quality.

- **Tip** - Make 'cost of quality' part of the capital expenditure approval process. If there is no assessment of the effect on the cost of quality, then the capital expenditure should not be approved.
- **Tip** - Ask suppliers for proof of the ability of their products to improve the cost of quality (see Section 5.22 on how to assess machine/process capability). If the process capability is not improved then why are you signing up to increase the cost of quality in the future?
- **Tip** - Be prepared to pay slightly more in initial purchase cost for long-term cost of quality savings.
- **Tip** - Look for projects where the rules can be changed to the company's advantage and make cost of quality savings automatic.
- **Tip** - Beware of simple payback calculations which do not include an assessment of the cost of quality.
- **Tip** - Don't forget energy costs!

As with energy, the cost of quality for a cheap machine during the machine's life will probably exceed the initial purchase cost.

You get what you pay for.

High-specification machines and controls may cost more at the start, but they cost less in the long-term.

"If you need a new machine and don't buy it then you will eventually find out that you paid for it anyway."

**Henry Ford**

# 3.12 Quality costs - where are you now?

## What are the costs?

Good quality cost data are one of the most powerful drivers possible for quality improvement.

These data must be of high quality to justify investments and expenditure. This means that cost of quality data, as with all types of data, should:

- Be collected from good sources.
- Be collected regularly.
- Be analysed correctly.
- Be reported to the appropriate staff members for action.

When managers and shareholders initially see the cost of poor quality they will be shocked but provided the data are of high quality there will be no reason to doubt the truth of the numbers generated.

Good financial data provides an undeniable reason for quality improvements.

## Completing the chart

This chart is completed and assessed as for those presented previously.

Cost of quality data are a potent driver for change but make sure that they are good-quality data.

Do not lose the battle with poor-quality data.

**Quality costs**

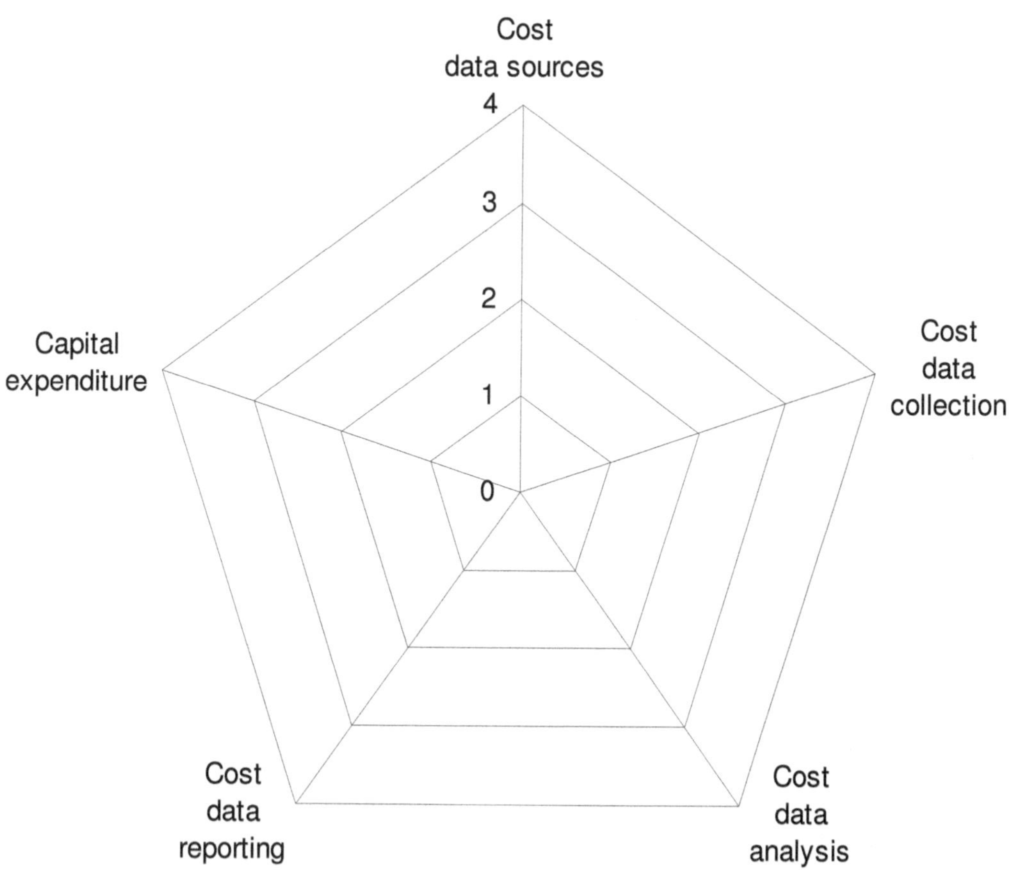


Use the scoring chart to assess where you are in quality costs

The numbers from the scoring chart can be transferred to the radar chart for a quick visual assessment of where you are in terms of quality costs.

As with most things:

- Report clearly.
- Report regularly.

Start to collect the data and put a value on the cost of poor quality to the site.

Finance wins the day so make them your friends.

## Cost of quality

| Level | Cost data sources | Cost data collection | Cost data analysis | Cost data reporting | Capital expenditure |
|---|---|---|---|---|---|
| 4 | Excellent data sources updated automatically from machines & processes. All data sources calibrated & reliable. Quality data recorded by shift. | Data collected in accordance with a written plan. Data collated & combined with relevant trading & business data. | Data analysed with production costs, volumes & other factors. | Concise reports for managers to allow technical & financial data to be used. Data normalised for comparison. Impact of uncertainties defined. | Projects compete equally with other areas. Full account taken of indirect benefits, e.g. marketing opportunities. |
| 3 | Manual data sources updated regularly. All data sources calibrated & reliable. Quality data recorded by shift or operation. | Data collected regularly. Data collated & combined with relevant trading & business data. | Data analysed with production costs, volumes & other factors. | Concise reports for managers to allow technical & financial data to be used, with deviations from budget & comparisons with previous period. | Projects compete for capital along with other business opportunities, but have to meet more stringent requirements for return on investment. |
| 2 | Manual data sources updated sporadically. Most data sources calibrated & reliable. Quality data recorded weekly. | Some data collected. Data collated & combined with relevant trading & business data recording. | Data analysed with production costs, volumes & other factors rarely & only as required. | Reports prepared for managers giving both technical & financial data, deviations from budget & comparisons with previous period. | Quality projects not formally considered for funding from capital budget, except when very short-term returns are evident. |
| 1 | Manual data sources updated sporadically. Few data sources calibrated & reliable. Quality data rarely recorded. | Occasional data collected when thought to be adverse. Occasionally collated & combined with relevant trading & business data recording. | Data analysed with production costs, volumes & other factors only in response to adverse trends. | Reports prepared & provided to managers incorporating both technical & financial data for the period. | Funding only available from revenue on low-risk projects with paybacks of less than one year. |
| 0 | No measurements taken & no records kept. | No information collected. | No data analysis prepared. | No management reports prepared. | No funding available for quality projects. No funding in the past. |
| Score | | | | | |

# 3.13 Targeting and controlling - where are you now?

## The initial steps in targeting and controlling

As with the score charts shown earlier, this is a self-assessment exercise to allow a site to benchmark their status. This will set the scene for planning improvements in the area of targeting and controlling.

Targeting and controlling is primarily concerned with having operational systems in place to get and use the cost of quality data to set challenging but achievable targets for the whole site. Preferably these targets will be reported and will directly affect the pay package of a large proportion of the staff - if their salary is linked to performance then they will be motivated to improve quality.

The preceding sections have shown how to obtain and treat the data to provide real information on the quality performance of the site and how to use this information to both set targets and to assess quality performance.

## Completing the chart

This chart is completed and assessed as for those presented previously.

Targeting the costs is the first step and controlling the costs is the second step. Information is the key to both steps.

**Targeting and controlling**

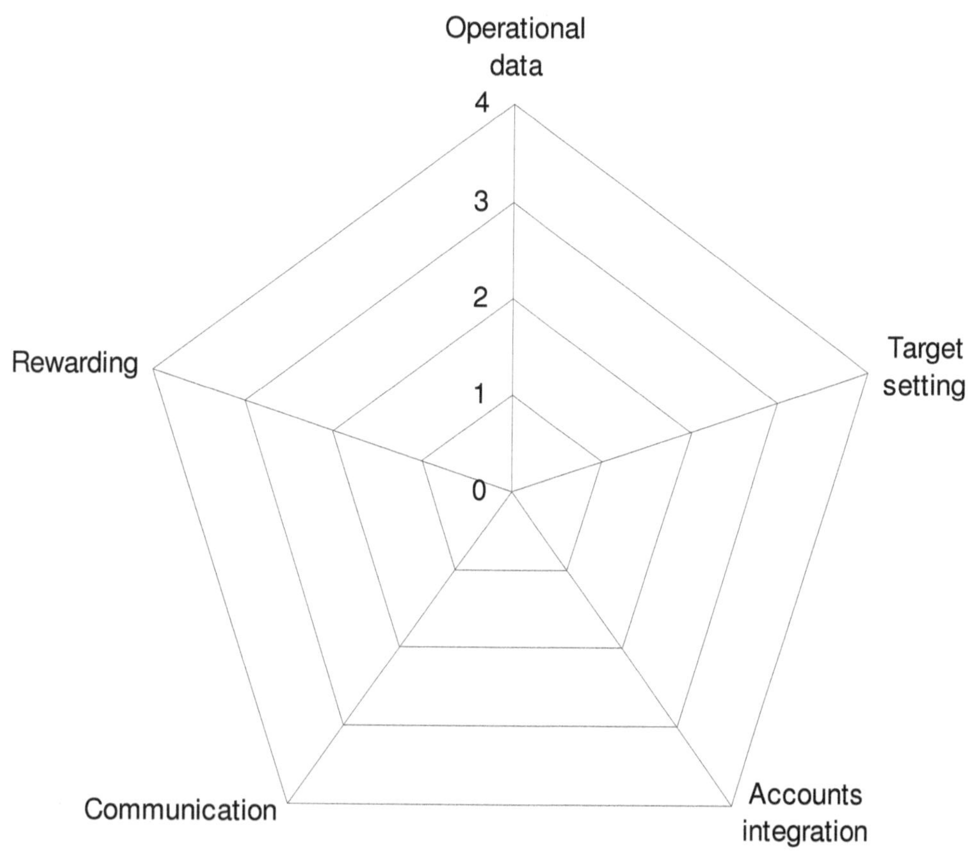


Use the scoring chart to assess where you are in targeting and controlling

The numbers from the scoring chart can be transferred to the radar chart for a quick visual assessment of where you are in terms of targeting and controlling.

Data sources can be direct from machines, scrap figures, accounting data on returns , accounting data on lost material or any other data used to make up the cost of quality report.

High scores in this chart indicate that a site has most of the relevant information to progress to selecting profitable quality improvement projects and measuring the profitability of these projects.

| Targeting and controlling | | | | | |
|---|---|---|---|---|---|
| **Level** | **Operational data** | **Target setting** | **Accounts integration** | **Communication** | **Rewarding** |
| **4** | Regular cost of quality data promptly obtained & analysed. Management & operational information provided promptly & in relevant detail. | Targets set & based on reducing targets to encourage continuous reduction of cost of quality. Achievement of targets monitored via reporting. | Targets & achievements integrated into accounts package for regular reporting on target achievement. | Regular targets clearly communicated to all relevant managers & staff. | Cost of quality targets & performance form part of pay package for all staff. |
| **3** | All cost of quality data obtained is up to date & analysed to provide management information in adequate detail. | Targets set based on arbitrary target with some assessment of achievement. Comparison possible with previous periods but no targets. | Targets & achievements partially integrated into accounts package for occasional reporting on target achievement. | Regular targets communicated to only a select few managers. No communication with staff. | Cost of quality targets & performance form part of pay package only for upper managers. |
| **2** | Provision of budget cost of quality data adjusted for changes in base data (e.g. production volume). | Targets set based on arbitrary target with no assessment of achievement. Some comparison possible with previous periods but no targets. | Targets manually calculated & achievements partially integrated into accounts package for occasional reporting on target achievement. | Targets sometimes communicated to some managers. No communication with staff. | Cost of quality targets & performance form part of pay package for some managers such as Quality Manager. |
| **1** | Provision of approximate budget cost of quality data. | Broad statement of intent to reduce cost of quality but target is poorly defined. | No integration with accounts package but manual calculations occasionally made for accounts data. | Targets rarely communicated to any managers. No communication with staff. | Cost of quality targets & performance do not form part of pay package for any staff except Quality Manager. |
| **0** | No cost of quality data available. | No targets set. | No integration with accounts package as no targets set. | No communication as no targets set. | No reward for performance possible as no targets set. |
| **Score** | | | | | |

# Key tips

- Implementing cost-effective quality management is not an additional cost to the business - it will decrease costs overall.
- Cost of quality information provides an essential financial driver for quality management and quality improvement.
- The total cost of quality is not simply what is 'seen'. There are a multitude of other 'unseen' costs of poor quality.
- The cost of quality can be as high as 20% of sales.
- We need cost of quality information to find savings, justify quality improvements and budget for quality.
- Accountants are an essential ally in managing cost of quality information.
- Accountants will need help in calculating the cost of quality.
- There are a variety of models for cost of quality information:
  - Prevention, Appraisal and Failure (PAF) cost model.
  - Process cost model.
  - Activity-Based Costing (ABC) model.
- The PAF cost model is the most widely used model and can be integrated well with many of the existing accounts data.
- The PAF model comes in several different variants but most are basically the same.
- The process cost model is not widely used but is best suited for integration with modern quality systems such as ISO 9001.
- The definitions in the process cost model are very different to those in the PAF model - do not get them confused!
- The ABC cost model can be very good for overall assessment of business process costs and reducing these. This model provides clear guidance on where to start looking for savings but can be difficult to implement and does not easily incorporate non-financial measures.
- For plastics processors, scrap treatment needs careful consideration and analysis.
- Never deduct income from scrap sales from the cost of quality calculations.
- Use 'fast starts' to trial a cost of quality system in a single area.
- Start small, start with a limited area, set up a project team and get some experience.
- Look for the largest cost (usually failure costs) and target this first.
- Integrate the cost of quality into the regular accounts for the most effective cost of quality system.
- Use the available data where possible.
- Assign costs to people - it is the only way to get action.
- Use a mix of financial and non-financial information to get the best results.
- Set targets that are relevant to the people who have to achieve them.
- Use non-financial targets for non-financial people.
- Targets may need to change with time to reflect changing situations.
- Audit all system and process improvements to hold the gains.
- The reporting should vary with the needs of the people who get the report.
- Report regularly, widely and concisely.
- Capital expenditure requests should always include an assessment of the effect on the cost of quality.
- Machine and process suppliers must provide information on the capability of their machine or process and this must be used to reduce the cost of quality.

# Chapter 4

# Quality management systems and standards

Quality management standards and systems have been a worldwide phenomenon over the past 30 years. In the UK, this began with the introduction of BS 5750 in 1979 and on the international stage with the introduction of the ISO 9000 series of standards in 1987 when the ISO series largely followed the BS 5750 series in requirements and layout.

These ISO standards are now used in virtually every country in the world and total worldwide ISO 9001 registrations were over 1.1 million in 2014. This worldwide movement towards formal quality management systems (QMS) is accelerating and moving from the established economies of the West to the rapidly developing economies of the East.

When these standards were first developed they were primarily designed for manufacturing industries, i.e. they were largely based on the concept of inspection and physical product control. However, the use of the standards for non-manufacturing industries has now led them to evolve into what are now termed 'management system standards'.

This was particularly true of the substantial revisions that took place in 2000 where the ISO 9001 standard was revised to take a more generic approach based on business processes. The 2008 revision to ISO 9001 introduced few new requirements but cemented the process approach and ISO 9004 was updated in 2009 to act as guidance for companies who wanted quality improvement for internal reasons rather than simply meeting external requirements.

The September 2015 revision to ISO 9001 introduced some new and updated concepts, e.g. the concept of risk-based thinking to support the process approach, changed some of the terminology and reduced some requirements by placing less emphasis on documents and referring to 'documented information'. The major change for this revision was the adoption of a consistent high-level structure (Annex SL) for all ISO management systems standards. This means that all ISO management systems standards, e.g. quality, environment, energy, etc. will share a common structure, identical core text and common terms and definitions.

Nobody can deny that ISO 9001 has been successful in terms of market penetration, revenues generated for standards-setting bodies, revenues generated for certification bodies and revenues generated for consultants advising companies seeking registration. What is less certain is the benefit to the companies who have spent large amounts of money on consultants (myself included) and on achieving registration.

A quality management system (however defined) is a necessity to achieve quality management. It provides the quality assurance aspect of quality management and is the framework for much of the other work in quality management. This chapter considers ISO 9001:2015 quality management systems, and removes some to the mystique surrounding such systems and shows how effective systems can be developed and implemented to reduce costs.

# 4.1 What do you want your QMS to deliver?

## Define what you want

Installing a QMS is a major project in most companies and, like any project, it needs to be managed and controlled. An essential part of any project, before starting, is defining the objectives of the project.

However, many companies embark on installing a QMS with little clear idea of what they want the QMS to deliver. They are convinced that installing a QMS will deliver the elusive 'quality' factor that has previously been lacking and that this will save the company. In many cases, this is driven by the expressed customer demand of 'If you don't comply then we won't buy' - where the customer thinks that forcing every supplier to have a QMS will also magically produce good-quality raw materials and improve their own quality.

The problem with this approach is that quality is:

- Often poorly defined in terms of the business needs (see Section 1.3).
- Often poorly defined in any way that can be quantified or assessed.
- Often defined only from an internal company view (people, processes and products) rather than from the customer view (outcomes, benefits and results). This is a common failing and is compounded by a lack of initial information on the customer perspective and a lack of follow-up to see if it has changed or improved. Quality may improve from the internal view but this is of little benefit if the customer view is that quality has not changed or has gotten worse.

Quality can be defined in terms of conformance to specification or one of the other methods given in Section 1.3 but the relevant definition should be one that everybody in the company can agree with (whatever that may be) and that also includes the customer's perspective.

Having established a consistent internal view of what 'quality' actually means the next step is to determine how this can be measured or assessed from both the internal and external viewpoints. This is a key issue but is often not an easy task.

Defining quality in quantifiable terms is necessary because if you have no clear definition of quality then it will be impossible to know if you have achieved it and if the installation of the QMS has been successful or not.

The internal definition must be translated into the customer view so that the delivery of quality (in whatever form) has an impact not only internally but also externally.

- **Tip -** Many companies fail to do any of this. Is it any wonder that they find that their QMS is not successful and that it does not deliver what they imagined that it might deliver either internally or externally?
- **Tip -** What does your customer want in terms of quality and how can you deliver it to delight them? These are the vital questions and ISO 9001 requires explicit answers.
- **Tip -** If you don't have a clear view of what the QMS should deliver then the QMS will simply generate paper and no effective improvement in quality.

## Set the objectives of the QMS

After quality is defined from both the internal and external views, the objectives

The biggest driver of formal quality management systems?

**"If you don't comply then we won't buy."**

Any target, and objectives are just targets by another name, must be SMART:

**S**pecific

**M**easurable

**A**chievable

**R**elevant

**T**ime-limited

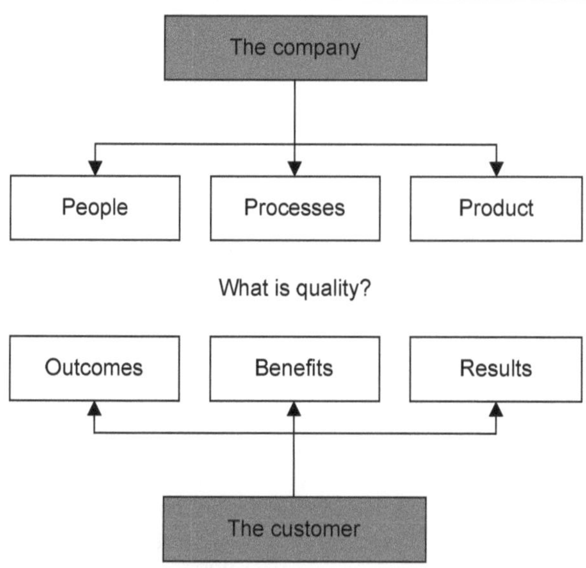


Quality can be viewed from the company or the customer's perspective - they are different

The typical company views quality in terms of people, processes and products but the customer has a completely different perspective of outcomes, benefits and results.

of the QMS should be defined. Too many companies install a QMS in the mistaken belief that it will automatically deliver quality but without specified objectives, this is unlikely. If objectives are not defined before the QMS is installed then it is unlikely that the QMS will achieve them!

- **Tip** - ISO 9001 requires a QMS to have measurable quality objectives but this is not the same thing as deciding the objectives of installing a QMS. The ISO 9001 objectives are further in the future than the project objectives.
- **Tip** - Customer-related objectives should be measured using their metrics and not yours.

## Define the cost

This is the cost of the project (installing the QMS) and not the cost of quality (see Chapter 3). It should include the staff time (a major element that is usually ignored), the cost of consultants (internal or external) and the cost of any certification bodies.

The cost of installing a QMS is sometimes very high even when the processes are relatively well defined.

## Define the savings and benefits

A good QMS can deliver both quantifiable and unquantifiable benefits.

If a CoQ system (see Chapter 3) is in place before installing a QMS then this can be used to quantify the internal and external benefits of installing the QMS. A good QMS should result in reduced failure costs but you need to be able to measure them first.

Unquantifiable benefits are more difficult to assess and some of these are:

- Improved internal workflow from process improvement.
- Improved customer relations from better process management.
- **Tip** - A good CoQ system can act as a driver for installing a QMS if you think that a QMS will reduce the CoQ.

## Assess if the QMS has delivered

After the QMS is installed and operating, it should be assessed to see if it is cost-effectively delivering the objectives. At this stage the project objectives can be updated and transferred to form the objectives of the overall QMS.

### The QMS - a $C^3I$ view

Understanding what a QMS is and what it delivers is essential to have a realistic expectation of what a QMS can deliver. Many people have an unrealistic expectation and think that simply installing a QMS will somehow produce 'quality'.

**What is it?**

A QMS is a set of procedures and processes that are consistently followed to make sure that the product is what the customer wanted, what was promised and conforms to the specification.

**What is it designed to do?**

The QMS is designed to make sure that all processes are carried out in a repeatable and predictable manner by everybody in the company. The processes are designed to make sure that there are checks and balances to protect the organisation and the quality of the product. The QMS is the roadmap for how things are done in the company.

The QMS functions can be summarised by the term $C^3I$, which stands for Command, Control, Communication and Intelligence.

**Command**

The command function is the way of specifying how things are done. The QMS sets out, in plain words, and often for the first time, the processes used. Until a process is defined and consistent, it is impossible to either control or improve it. Installing a QMS often results in changes to existing processes when it is realised that they are not suitable for consistent operation or do not produce the required outputs. This is especially true when processes were established in the early stages of a company and have then been modified and changed until they no longer reflect what the company is either doing or wants to do.

In the early stages of installing a QMS, the command function is the most important function.

**Control**

The control function is the way of making sure that all the staff in the company follow the defined and documented processes all the time. The control function means that the staff carry out defined actions at defined times, record things, sign things, check each other's work and use other methods to ensure that everybody is following the defined and documented processes.

**Communication**

The communication function is the method of telling other people that something has been done and the results of this. It is not enough to inspect a product and know that it is bad; other people must be told so that they do not send it to the customer anyway.

Communication involves filling out forms and sending them or other information to other people and marking the status of products so that other people know what to do with them.

**Intelligence**

The intelligence function is the process of reviewing what can be done to get better in the future. It involves reviewing what has been done, how it was done, whether it was effective and, most importantly, whether it can be done better/easier/more effectively in the future.

# 4.2 Does a QMS deliver quality?

## Do they deliver?

One question that is largely absent from any of the discussions about ISO 9001 is 'Does an ISO 9001 actually deliver quality?' In fact, in some circles, even asking this question is treated as heresy.

The simple fact is that ISO 9001 has little of nothing to do with absolute product quality, it is concerned with the management system used and how this is used. This is one of the reasons that the new high-level Annex SL 'management systems' model can be used for management systems as diverse as quality management, environmental management and energy management.

ISO 9001:

- Provides a general framework for management of systems: It tells you how to manage and document systems but this presumes that the top-down ISO 9001 model of 'command and control' is the best model for all companies.
- Provides a framework for documented information: It tells you how to handle and control documented information.
- Provides a framework for process definition and improvement: It concentrates on processes and the inputs and outputs of the process. These can then be linked to the ISO 9001 requirements in a matrix fashion so that a single ISO 9001 requirement might be met in various process descriptions. The process approach provides the basis for process improvement if used appropriately.
- Must be made specific to company activities: The original ISO 9001 standard was developed for manufacturing companies but has now been extended to include any type of organisation from lawyers through to parks. This makes the requirements very generic and they must be adapted to the specific organisation to be relevant.
- Does not specify product quality standards or control methods: The standards leave the actual product standards and the methods used to control quality to the discretion of the company. ISO 9001 is all about management of the system and the data, not the actual results. ISO 9001 does not specify the use of SPC (see Chapter 5) although the use of statistics is generally recommended.
- Leaves the company to control costs, plan communications and ensure that the system is cost-effective: As long as the QMS meets the requirements of ISO 9001, and they are very flexible, the company still has to run the system.

**Note:** A product cannot be certified to ISO 9001 and it should never be suggested that this is the case. The certification is for the management system used to produce the product, not the product itself.

Supporters of ISO 9001 claim that implementing the standard:

- Makes a company more effective.
- Reduces waste (as measured by CoQ).
- Increases productivity.
- Increases profit.
- Increases employee motivation.
- Increases customer satisfaction.
- Reduces external customer audits.
- Provides a marketing tool.

The evidence for these claims is variable and all of the benefits will be achieved only if the company is already heading in the right direction, in which case ISO 9001 may well be a distraction from achieving real quality improvement.

Many companies are driven to installing ISO 9001 by a demand from their customers to achieve certification, the 'if you don't comply then we won't buy' driver, others are driven by a need to have ISO 9001 certification 'because all the competitors have it'. Some even install ISO 9001 without fully understanding what it is that they are trying to improve (see Section 1.3). In these companies, the introduction of ISO 9001 can become an exercise in 'window-dressing' where current and failing systems are simply documented and engraved in stone. For these companies and, when a company is only interested in the certificate on the wall, the benefits of ISO 9001 are likely to be minimal.

However, in developing companies, the introduction of ISO 9001 can force a re-evaluation of the current systems. This can be a powerful catalyst for change and

Throughout this chapter we will simply refer to 'ISO 9001'.

In all cases, unless otherwise noted, we will be referring to ISO 9001:2015.

Quality management systems should not be regarded as a method of gaining logos or as an abstract theory but as a vital tool in the management of the company.

'You can produce shit to ISO 9001 as long as you always do it the same way and complete the paperwork'.

**Anon**

improvement with the standard acting as a reference for 'best practice' and providing guidelines for the development of internal systems.

### Do they deliver?

The answer depends on what you want them to deliver and your motivation for installing ISO 9001.

## Criticisms of ISO 9001

ISO 9001 has suffered from a good deal of criticism over the years but not much of this makes it into the press. The reasons for this are obvious, i.e. there is a huge amount of money and prestige tied up in the ISO 9001 certification industry. The criticism falls under several main areas, criticism of the process, criticism of elements of ISO 9001, criticism of the fundamentals of ISO 9001:

### Criticism of the ISO 9001 process

This focuses on the process of certification and concerns mainly the time, energy and money needed to become certified to ISO 9001. There is a whole industry that has grown up around ISO 9001 certification where a lot of people have made a lot of money installing ISO 9001 and have a stake in extending the process.

### Criticism of elements of ISO 9001

The process focus of ISO 9001 does not assess the quality of the process or whether the correct values are being measured, only that the specified process is being followed. These may be the wrong parameters but if you meet the process then this is good for ISO 9001. ISO 9001 does not look at the actual technical solution to the problem but simply at the process and assumes that this will deliver a robust technical solution.

Poorly designed systems can easily lead to products which are poorly designed even when the processes function as designed.

### Criticism of the basis of ISO 9001

ISO 9001 is essentially about top-down control rather than worker empowerment. This can lead to a workplace that is crowded with paper and a preference to following the procedure rather than actually improving the quality of the product. Some critics claim that ISO 9001 stifles improvement, although this has been improved since the 2000 revision.

Other critics claim that the focus on processes/procedures reduces fundamental understanding and improvement of the product because it is more interested in specification and control than in actual quality products.

One of the major reasons for the ISO 9001: 2000 revision was the perception (and reality) that many certified companies were not using their QMS to actually improve quality in any way but were simply concerned with certification. The effect of this change of emphasis is still being felt throughout the ISO 9001 industry and many companies (although they may have updated their certification) have not made the full transition to the process-based approach.

All of these criticisms are valid for those companies who fail to understand ISO 9001 and who use it as an end in itself rather than as a tool for management.

## The signs of a failing QMS

Installing ISO 9001 does not guarantee that the system will improve quality (despite the best efforts of legions of auditors) and there are several warning signs to look out for, these are:

- More time is spent talking about the system than about the actual product. When the focus is more on the system than the product then the system is getting in the way.
- More time is spent talking about the system than the customer experience. When the focus is more on the system than the customer then the system is definitely getting in the way. Customers are not interested in systems, they are interested in products.
- Systems that are never changed because they are difficult and tedious to modify, i.e. systems that are engraved in stone because 'ISO requires this'. Stagnant systems will result in stagnant quality and this is not what ISO 9001 is about.
- Systems that dominate and stifle innovation in the company.
- Systems where the upper management has no idea about how they operate and is not involved. Upper management commitment is essential to success of a QMS and if they are not involved then the system is doomed.

Any of these signs shows that the ISO 9001 system is getting in the way rather than assisting in managing the company. If you see any of these then get very worried, very quickly.

In most cases, installing ISO 9001 is an act of faith. The evidence that ISO 9001 improves quality is mostly produced by those who have the most to gain, e.g. standards bodies, consultants and assessors.

Despite this, there are still a lot of data out there that indicates ISO 9001 has many positive benefits.

Certification and accreditation are not the same thing.

A company is certified, registered or approved to ISO 9001 by a 'Certified Body'. These will generally be 'for profit' companies and there will generally be many in a given country.

A 'Certified Body' will be accredited to issue ISO 9001 certification by the national 'Accreditation Body'. These will generally be national bodies and will be backed by the government. There will normally only be one 'Accreditation Body' in each country and they agree to accept each other's judgements and accreditations.

Plastics processors would be certified as complying with ISO 9001.

# 4.3 The 7 quality management principles and Plan-Do-Check-Act

## The 7 quality management principles

In the 2000 revision, ISO 9001 was based on 8 quality management principles but in 2015 this was trimmed to 7 quality management principles and these apply equally well to all types of management.

The 7 principles of quality management align the standard with the philosophy of the TQM movement. The 7 principles are:

### 1. Customer focus

Quality is more about meeting and exceeding the customer's expectations than it is about production but many companies take a purely internal view of quality, i.e., they only consider the product aspects of quality and do not consider the other aspects of the business that affect the customer's perception of quality.

Only by understanding the customer's present and future needs can a company manage these needs and expectations and deliver product to the required quality.

### 2. Leadership

Leadership is about setting a common purpose and direction for the company so that the whole company is focused on meeting the customer's expectations.

Good leadership ensures good communication at all levels of the company, motivates staff to achieve the company goals and communicates the customer's needs within the company.

### 3. Engagement of people

People at all levels of the company need to be competent, empowered and engaged to deliver their best performance and to work towards the company's specified goals.

People who are respected and involved will deliver better performance, be more motivated, be more creative, accept more responsibility and, most importantly, deliver better-quality products and services to the customer.

### 4. Process approach

Any business is a collection of processes and good management consists of effectively managing these processes, particularly the interfaces between processes and functions.

The process-based approach not only improves quality but can also streamline the business, lower costs and improve consistency of the outputs.

Processes combine to form systems and understanding what these systems deliver allows the processes to be integrated, aligned, optimised and to be made more consistent as well as allowing ineffective or redundant processes to be identified, improved or eliminated.

### 5. Improvement

Improvement is necessary to stay competitive and should be the constant objective of any company, not simply in quality management but in all aspects of the company's operations.

Embedding an improvement philosophy delivers performance advantages, flexibility in operations, allows continual improvement of products and services and allows a company to react positively to changes in internal and external conditions.

### 6. Evidence-based approach to decisions

Many business decisions, not only decisions related to quality, are emotion-based rather than evidence-based. Effective decisions need a proper analysis

The difference between involved and committed: When you have ham and eggs for breakfast, the chicken is involved but the pig is committed.

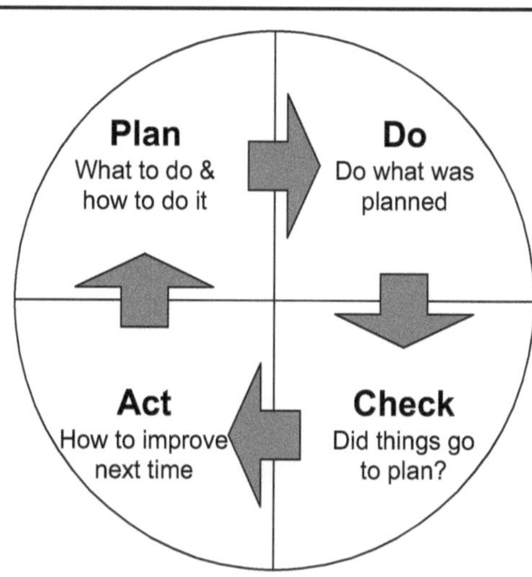


The standard PDCA cycle

The standard PDCA cycle provides a robust method for improvement of virtually any process in virtually any industry provided that the upper management provides motivation for improvement.

of data and the conversion of these data into useful information for decision making.

Evidence-based decisions can be reviewed, challenged and changed to meet changes in the situation. Decisions based on emotion will remain fixed whatever else changes.

### 7. Relationship management

Every company is not only a supplier but also a customer and managing relationships is a key to success. Partnerships and good relationships, throughout the supplier chain and with other interested parties, are an essential part of running an effective company. Good relationship management is mutually beneficial; it can deliver improved products and profits to suppliers and customers (see Chapter 9) as well as improved relationships with other interested parties such as investors, employees and the general society.

## Designing the QMS

Designing a QMS based on the 7 principles, and particularly on the process approach, allows a QMS to deliver real value rather than simply conforming to the various clauses and items of the ISO 9001 specification. The 7 quality management principles are reflected throughout ISO 9001.

The change to a process approach in the ISO 9001:2000 revision was much more far-reaching than was appreciated by many companies. Many companies took their previous system and bolted the 2000 revisions on top of the existing system. This missed many of the benefits of the 2000 revision and, 16 years later, many companies have still not realised this potential for improvement. The alignment of the TQM and ISO 9001 approaches was one of the major benefits but this remains unrecognised in many companies.

## The Plan-Do-Check-Act cycle

The Plan-Do-Check-Act (PDCA) cycle is one of the fundamentals for process improvement and is also a fundamental for ISO 9001. The PDCA cycle is shown on the left as a stand-alone cycle and on the right overlaid on the process-based model of ISO 9001.

PDCA is sometimes called the 'Deming cycle' (although Deming is said to have referred to it as the 'Shewhart cycle' due to Shewhart's work in this area) and can be used to improve virtually any process in virtually any industry. The cycle is easy to use, provides a model for improvement and is analogous to the conventional 'scientific method'.

For an excellent outline of the 7 quality management principles and actions, see www.iso.org/iso/pub100080.pdf.

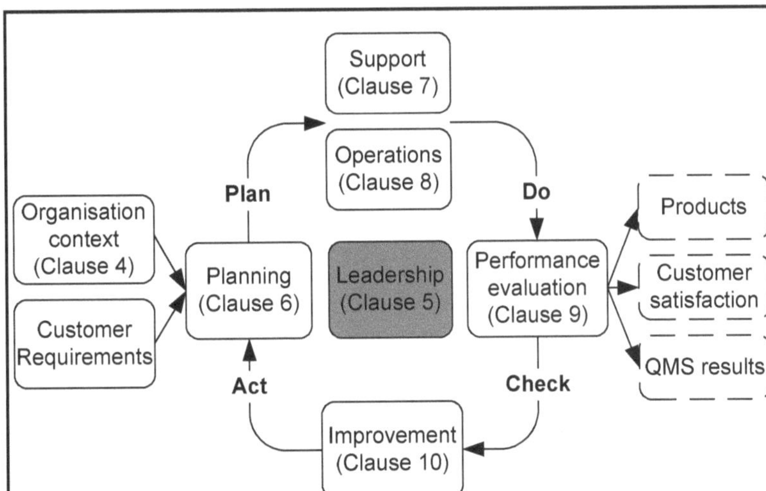


Model of a process-based QMS

The process-based QMS used by ISO 9001 showing how the PDCA cycle aligns with an ISO 9001 QMS.

**After: ISO 9001:2015**

# 4.4 The process approach - an overview

## Process approach

The development of the process approach in ISO 9001 has continued in the ISO 9001:2015 and this has now been extended to incorporate the newer concept of risk-based thinking (see Section 4.19).

The process approach is a natural and logical approach to implement in most sites despite the initial apparent complexity. It focuses on the processes taking place at the site rather than simply on meeting the requirements of the standard. The process approach is 'the systematic management of processes and their interactions to achieve intended results' (ISO/TC 176/SC 2/N1290).

## What is a process?

Processes are related activities that transform 'inputs' (which can be anything from data to actual physical products and almost always involves other resources such as labour) into 'outputs' (which can be products or services). The way that processes are carried out is set down in 'procedures', which ensure consistent processes, and the process can be monitored and measured for both the effectiveness of the process (is the process doing what it was designed to do?) and the efficiency of the process (is the process using the minimum amount of resources to achieve the task?).

Processes can be carried out within one functional unit or department or may spread to other functional units, departments or even sites within an organisation.

## The organisation and not the standard

The process approach can be contrasted with the earlier procedure approach where the main focus was on checking that the documented procedures met the requirements of ISO 9001 and where many of the procedures existed in isolation from the real activities of the site, i.e. procedures were written to comply with the clauses of the standard rather than with the actual operations of the company.

In the procedure approach, each clause of the standard could be related directly to a specific procedure and vice versa.

**Process:**

A set of interrelated or interacting activities which transforms inputs into outputs.
The inputs to the process are generally the outputs of another process.
The outputs of a process are generally the inputs to another process.

**Product:**

The result of a process. A product can be a service, some data, an object, processed material or a combination of these.

**Procedure (Controls):**

The specified method of carrying out a process. This may be documented or undocumented.

**ISO 9000:2005**

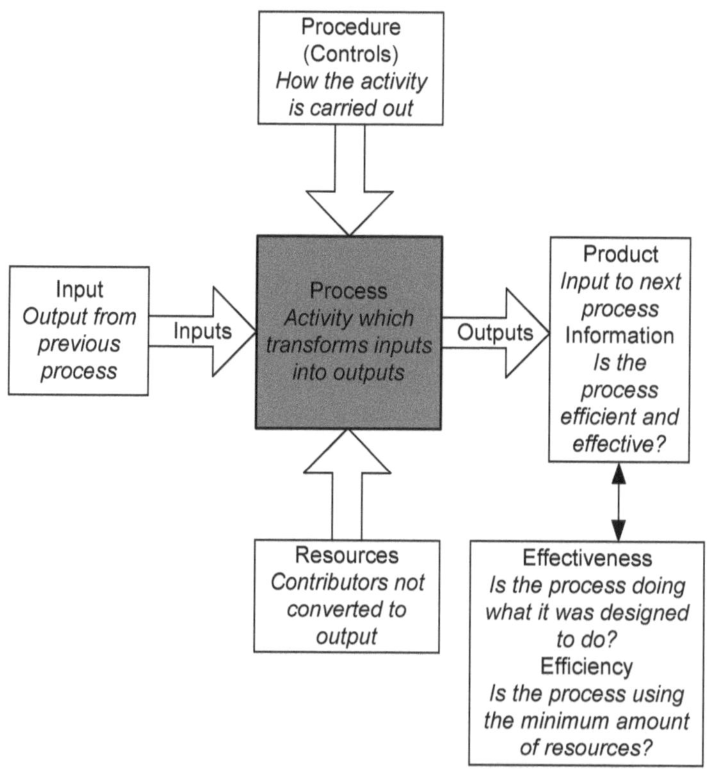


### A process is very different to a procedure

A process does something to the inputs to create the outputs and can be measured for both effectiveness and efficiency. A procedure simply tells you how to carry out a process so that it is carried out the same every time. In the process approach, the procedure is simply one of the overall set of controls that define what the process will achieve. Drawings, standards and other documented information are also part of the controls.

A process can also be measured for both efficiency and effectiveness to check that the process is using the minimum amount of resources and is doing what it was designed to do. These measurements are part of the output of the process.

The process approach looks at the actual processes used by the company and relates these to the requirements of the ISO 9001 standard. The processes are documented by procedures and other controls that may not directly relate to any specific clause of the standard but do directly relate to the operation of the process.

This means that the documented information, and the eventual audit, focuses on the processes rather than on direct compliance with any specific clause of the standard. There is no longer a need for any direct relationship between the procedures and each clause of the standard, i.e. compliance with a specific clause of the standard can be demonstrated through multiple processes and procedures.

Whilst the process approach is more realistic and beneficial for companies, it can present some difficulties in documenting the system and in proving compliance with the standard.

## Multiple processes

Processes never exist in isolation and in most companies the processes are highly interlinked, i.e. the output of one process is almost always the input of at least one other process and sometimes the input of several other processes. This is shown on the right where a company is shown as a complex web of interlinked processes.

## Mapping and defining the processes

The first step in the process approach is understanding the range of processes in the company and mapping these both as individual processes and as part of the web of processes that define how the company operates. This web of processes is what achieves the strategic and operational objectives of the company.

This is often a difficult exercise that demands a higher degree of understanding of the company operations than simply preparing a procedure. It is rare that any one person has all the knowledge and information to completely map the processes in any company.

- **Tip -** This is a team effort and needs a team approach.
- **Tip -** The initial global process map is an essential tool and is necessary to comply with the requirements of ISO 9001 so this is not wasted effort.
- **Tip -** The initial outline process map serves as a guide and checklist for the complete documentation process.

After the basic processes have been identified, it is necessary to start to define and separate the processes. This is often made easier by considering the limits of each process and defining the outputs of the process and how they are used in the subsequent processes.

- **Tip -** If the output of a process goes to more than one other place it is likely that this is a 'process boundary'.

## Risk-based thinking

The concept of risk-based thinking is not new, it has always been implicit in ISO 9001 but ISO 9001:2015 makes risk-based thinking explicit (see Section 4.19). Essentially, risk-based thinking ensures that risk (to the company or any stakeholder) is considered at all stages and makes preventive action part of the overall strategic and operational planning of the company.

- **Tip -** Risk must be considered as processes are developed and documented.

The focus is on the effectiveness and efficiency of the processes rather than on slavish 'box ticking' of the clauses of the ISO 9001 standard.

Effectiveness and efficiency measurements can be used to drive improvement.

Companies must identify and manage the processes that make up their business.

Make sure that it is processes which are identified and mapped and not simply activities (which are very different in ISO 9001 terms).

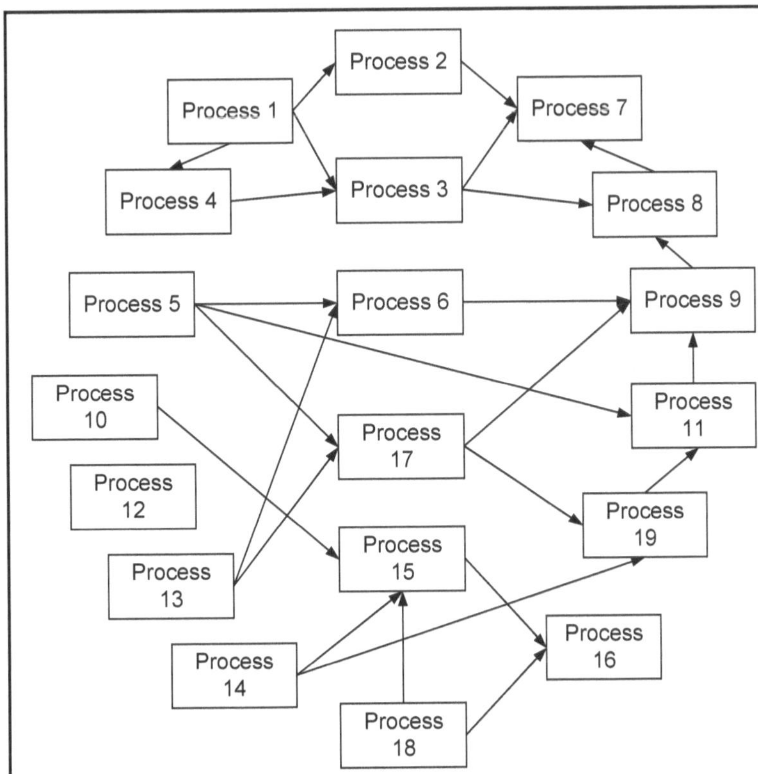


**The company as a collection of processes**

The processes at a typical company will cover all the operations of the company and will be interlinked in a web of processes. The output of one process will typically be the input of at least one other process.

# 4.5 Manufacturing processes - a global view

## What are the processes?

Mapping and defining the processes in the company is the first step in the process approach. This is not always an easy task but one of the easiest ways to do this is to first map the workflow at the global level. A typical map of the workflow in a plastics processing company is shown below. This map includes most of the processes in a company but does not include all of them. Other processes will be present in a company, e.g. accounting processes, and these need to be included on the initial map so that it represents the complete range of processes carried out.

- **Tip -** It is important to note that the actual processing of plastic is only one of the many processes.

Do not forget that the processes should be focused on the customer.

For some processes, the customer will be external but for others the customer will be internal.

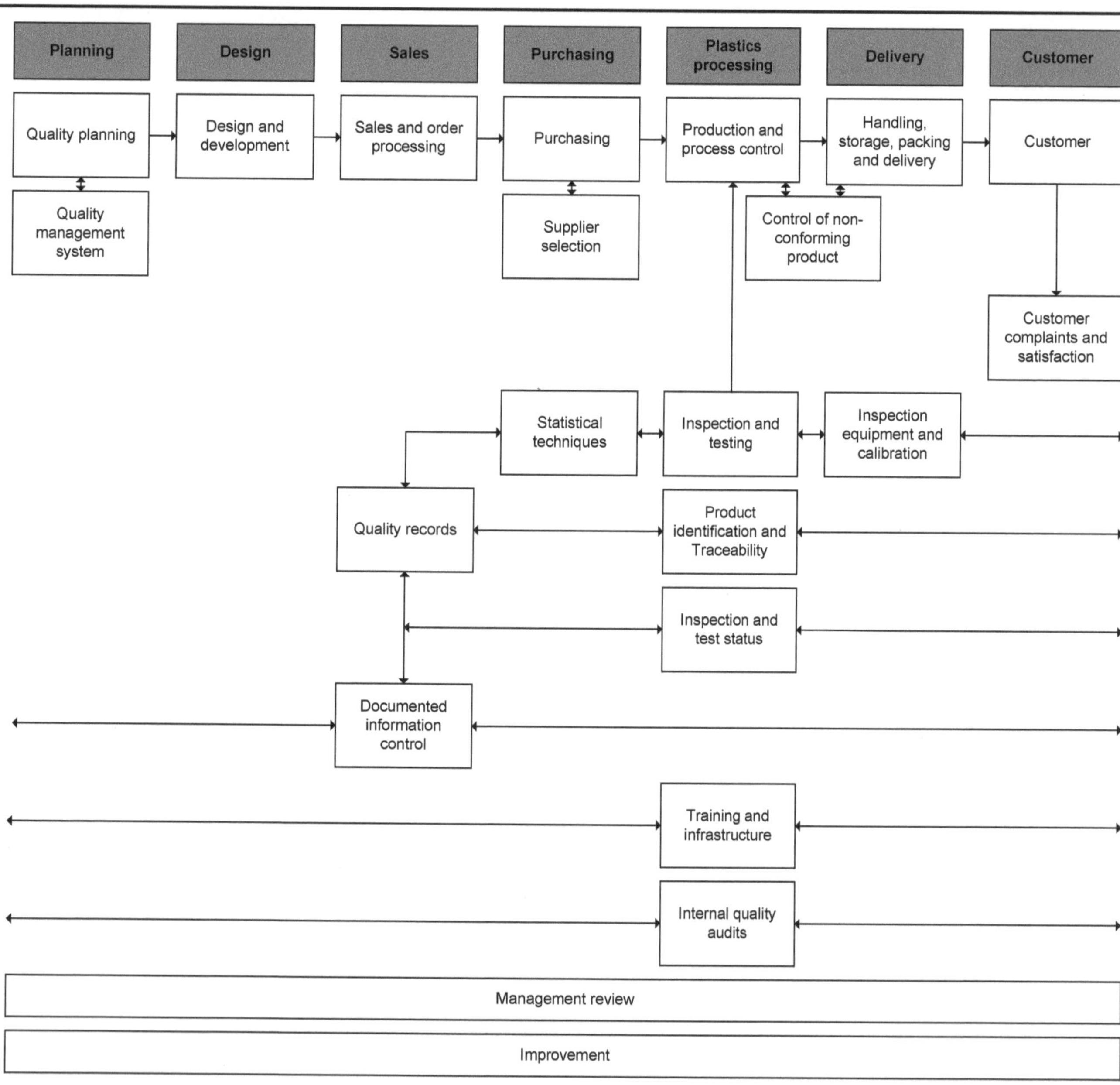


Workflow at a typical plastics processing company - the global process map

Mapping the workflow in a company will help to identify the processes. As paper and products move through the site then processes are involved. These may not be recognised as processes, they may be called 'standard operating procedures, or 'how we do things here' but they are processes.

**Note:** Some processes are not shown on this diagram for ease of presentation.

ISO 9001 is not simply about the plastics processing or manufacturing operations but is about how the company as a whole provides the customer with quality products and services.

The global process map shows how the individual processes are joined together by inputs and outputs to form the complete management system. This is the model for the complete management system of the company - not simply the quality management system but the complete and integrated management system.

## Creating a global process map

A global process map can be initially difficult to produce because of the range and number of processes involved.

As a starting point, the table on the right lists typical processes in a plastics processing company. These are grouped by operational area (or operational manager) for clarity and should allow any company to produce a list of the main processes. Create a global process map by:

- Identifying the main processes in the company from the list (add or delete processes for the specific company).
- List the main processes on Post-It Notes.
- Move these around on a blank wall or a very large piece of paper marking the main inputs and outputs with tape, pencil or other connections until the inputs and outputs linking the processes are correct.
- Use the wall or paper to produce the final global process map.
- This may be very large and need editing but remember you are trying to map the processes for a complete company here.
- **Tip** - Do not attempt to produce a global map on a computer until you have most of the inputs and outputs marked on the wall - you will simply be wasting your time.
- **Tip** - The global map can become very complicated so list only the main inputs and outputs to keep the information to a useful minimum.
- **Tip** - Do not forget 'external' processes and their relevant inputs and outputs.

The global process map provides the basis for documenting the individual processes and their inputs/outputs. However, each main process will need further work to fill in the inputs/outputs, procedures and process measurements (see Section 4.6).

By operational area:

**Management processes**

- Design and development.
- Marketing.
- Sales development.
- Sales and order processing.
- Production planning.
- Production control.
- Supplier selection.
- Purchasing.
- Invoicing.
- Accounting processes.
- Customer complaints and satisfaction.
- Improvement.

**Operations processes**

- Goods-in.
- Stores and stock control.
- Production (moulding, extrusion, blow moulding).
- Assembly.
- Handling, storage, packing and delivery.

**QA processes**

- Quality management system.
- Documented information control.
- Quality records.
- Quality planning.
- Corrective and preventive action.
- Internal audit.
- Management review.
- Training.

**QC processes**

- Goods-in and goods-in inspection (raw materials testing).
- In-process inspection (quality and quantity).
- Final inspection.
- Inspection equipment and calibration.

### Some typical processes for plastics processing

Not all these processes will be present in every company but most of them will be there. If they are not recognised as discrete processes then they may be combined with other processes, i.e. the goods-in process may be combined with goods-in inspection.

# 4.6 Manufacturing processes - a detailed view

## Global to detailed

The global process map will rapidly become very complex as the interactions between the main processes are revealed. A section of the global process map for injection moulding is shown on the upper right and this shows just some of the processes that link to the main injection moulding process, even this section of the chart has had some of the process and links left out for clarity.

After completing the global process map and gaining an overview of the company operations it is recommended that each of the individual processes be charted as a separate process on a separate sheet as shown on the lower right.

This starts to complete the process information for the individual process, injection moulding in this case, and lists the main inputs and outputs of the process as well as the measurements for efficiency and effectiveness that will be made during and after the process.

- **Tip** - Don't worry if you don't get it completely right at the start. Aim to get the main inputs and outputs for each process down onto the chart first. The details can be filled in later.

The detailed input/output chart for each process shows the flow of products, resources, services and information both into and out of the selected process. This flow of inputs/outputs is the integrated quality management system that ISO 9001 requires.

- **Tip** - It is also recommended to start to map the risks and opportunities for each process and the opportunities for improvement to the process and the QMS, you will need this information to conform to Clause 4.4 of the standard (see Section 4.8).

## Process maps to quality system information?

ISO 9001 requires the maintenance of 'documented information' on the operation of processes (see Section 4.17). The global and detailed process maps are an excellent way of communicating this information and will also show conformance with Clause 4.4 of ISO 9001. Process maps are shorter and clearer than a 'text' approach,

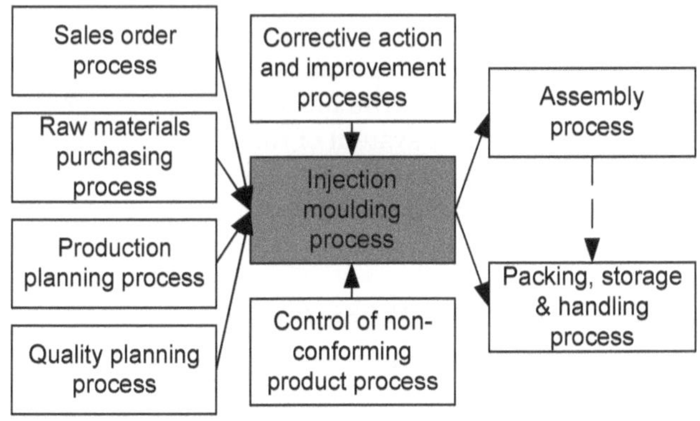


Section of global process map showing injection moulding process links

The injection moulding process is linked to other processes such as the sales order process and the raw materials purchasing process. Some processes and links are not shown for clarity.

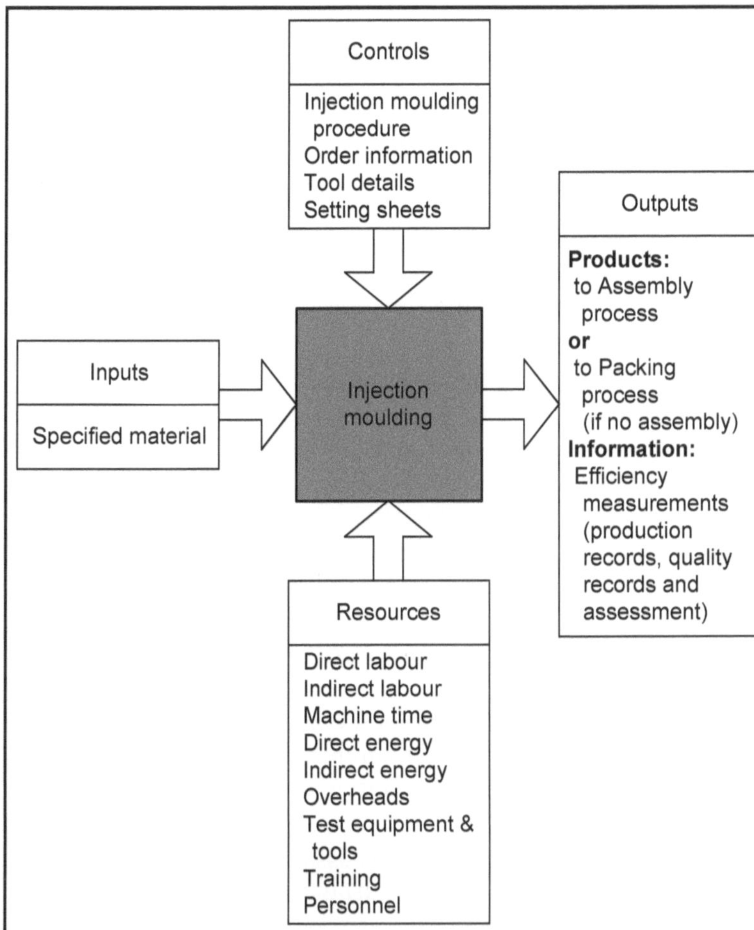


Detailed inputs and outputs of injection moulding

The detailed map of the injection moulding process shows the main inputs and outputs of the process, the measurements to be made of the process efficiency and effectiveness and the links to other processes.

are easier to update and are more relevant to the site.

- **Tip** - The global and detailed process maps are the basis for the QMS and establish the criteria for processes.

## Process maps to controls

After the main process maps have been completed they can be used to develop the related control procedures. Procedures can again be in 'text' form or in 'flow chart' form as shown on the right. Flow charts are clearer, easier to understand, easier to 'debug' and highlight all the process inputs/outputs required by ISO 9001. Decisions and responsibilities that are 'hidden' in text descriptions become very visible in flow charts and have greater clarity.

- **Tip** - Flow charts are not only acceptable to document ISO 9001 procedures but are actually better at communicating the flow of the process and the relationship between the various processes.
- **Tip** - Many sites insist on producing procedures for everything they do and clogging the system up with information. ISO 9001 has very few mandatory documents (not necessarily procedures).

  Any other procedures are documented at the discretion of the company. Less is better in documenting the system because if you say you do it then you have to do it all the time.

## Process maps - completing the picture

Process maps (global and detailed) and flow charts for the procedures offer a simple and elegant solution to the documentation requirements of ISO 9001.

- **Tip** - We have documented a complete system using flow charts and reduced the size of documents by over 50% whilst increasing clarity and revealing several errors in the 'text' procedures.

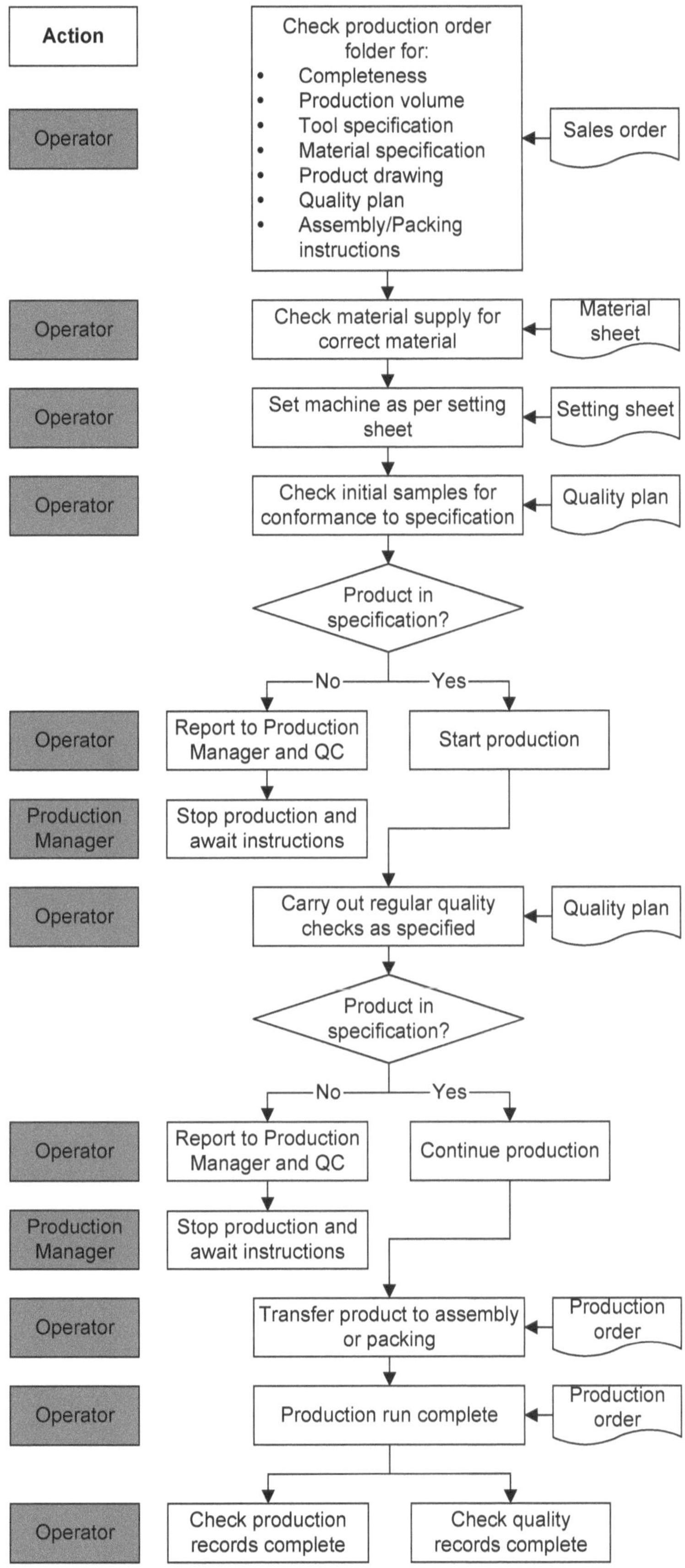


### Outline procedure for injection moulding

A simple flow chart of the injection moulding procedure allows process consistency and completeness to be quickly checked. Flow charts show what happens more clearly than written documents and are equally acceptable to ISO 9001 auditors.

# 4.7 An overview of ISO 9001

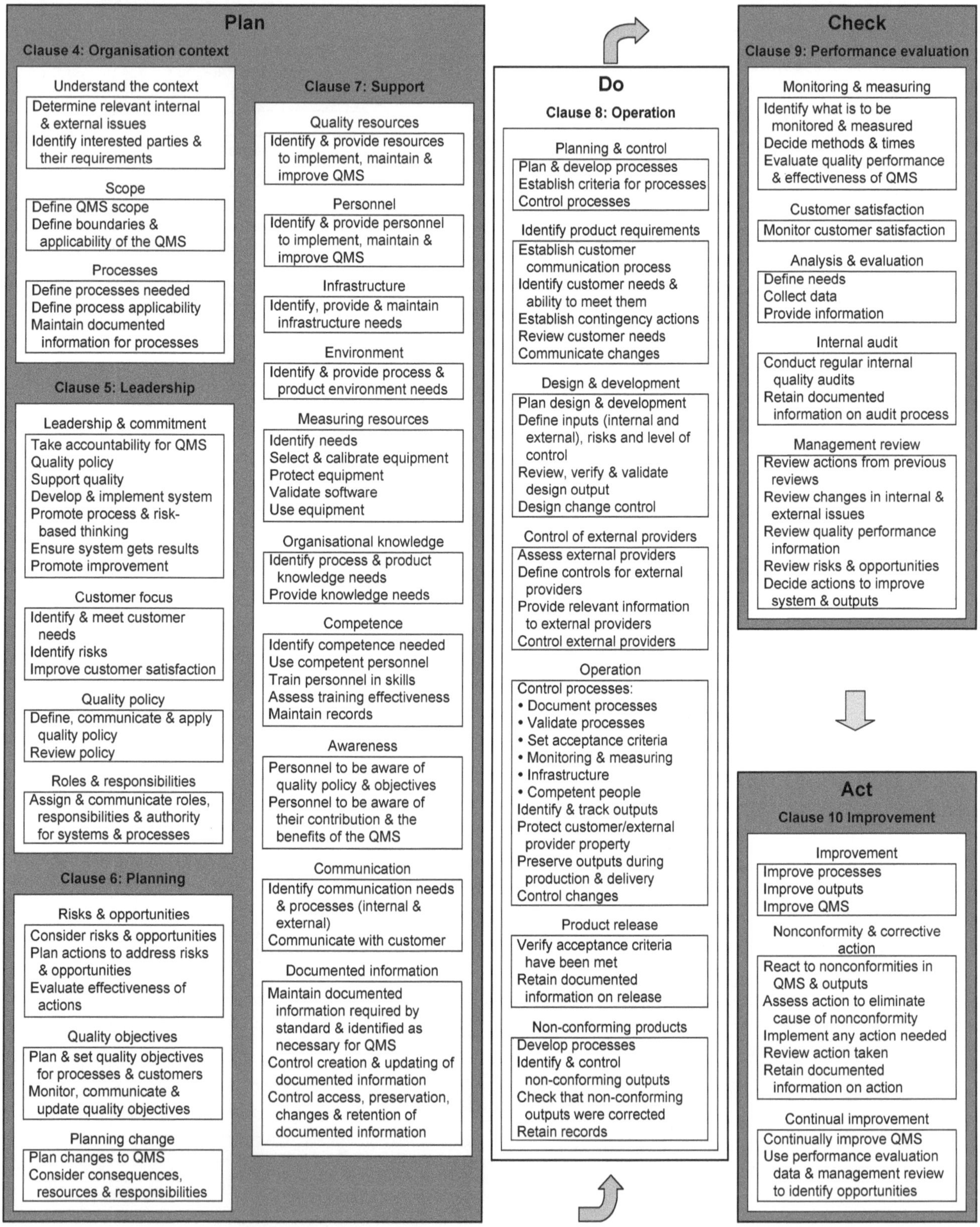


## An overview of the ISO 9001 requirements

ISO 9001 was initially a daunting document because the language was unfamiliar to many in the plastics industry. This has been improved in ISO 9001:2015 to make it easier to understand. The requirements are really quite simple and common sense in most cases. This diagram shows the main activities that must be carried out to conform to the standard. Most successful companies already carry out most of these operations.

## Introduction

ISO 9001:2015 conforms to the Annex SL high-level structure for 'management systems standards'. This gives more clarity on how the standard is laid out and will bring more consistency to these standards. The new common structure is:

- 1 - Scope
- 2 - Normative references
- 3 - Terms and definitions
- 4 - Context of the organisation
- 5 - Leadership
- 6 - Planning
- 7 - Support
- 8 - Operation
- 9 - Performance evaluation
- 10 - Improvement

## The requirements

Sections 4.8 to 4.16 give more detail on the clauses of ISO 9001 but the diagram on the left gives an outline of what is required.

The majority of these requirements will be met by any company that has successfully been in business for more than 3-5 years. The quality system will be there - it simply may not be recognised as such. Every company needs a management system and ISO 9001 simply formalises this and drives the creation of documented information to validate the system.

## Tracking processes to ISO 9001

The process approach can make it difficult for a single process to fully comply with the requirements of a specific ISO 9001 clause. Most of the typical processes from Section 4.5 can be easily related to the main clauses of ISO 9001. This is done on the right to show how company processes can be related to ISO 9001 requirements

**Note:** This grouping is by the primary process element. Some processes may cover more than one clause of ISO 9001.

- **Tip -** To check that all the requirements have been met, and to make it easy for auditors, it is strongly recommended that processes (and their procedures) are tracked to the ISO 9001 clauses using a spreadsheet. This should contain a list of the ISO 9001 clauses in the rows and the company processes in the columns. Simply indicating which process covers which clauses of the standard will make compliance checking easier (and your auditor will thank you for it as well).

By ISO 9001 clause:

**4 - Context of the organisation**

- ■Quality management system.

**5 - Leadership**

**6 - Planning**

- ■Quality planning.

**7 - Support**

- ■Training.
- ■Inspection equipment and calibration.
- ■Calibration.
- ■Quality records.
- ■Documented information control.

**8 - Operation**

- ■Marketing.
- ■Sales development.
- ■Sales and order processing.
- ■Invoicing.
- ■Accounting processes.
- ■Design and development.
- ■Production planning.
- ■Supplier selection.
- ■Purchasing.
- ■Goods-in and goods-in inspection (raw materials testing).
- ■Stores and stock control.
- ■Production control.
- ■Production (moulding, extrusion, blow moulding).
- ■Assembly.
- ■In-process inspection (quality and quantity).
- ■Final inspection.
- ■Handling, storage, packing and delivery.
- ■Corrective and preventive action.

**9 - Performance evaluation**

- ■Customer complaints and satisfaction.
- ■Internal audit.
- ■Management review.

**10 - Improvement**

- ■Improvement.

Typical processes for plastics processing arranged by ISO 9001 clause

These are the same 'operational' processes listed in Section 4.5 but in this case they are listed according to the relevant clause of ISO 9001:2015.

# 4.8 Organisation context requirements of ISO 9001

## Across the system

Clause 4 of ISO 9001 covers the 'context of the organisation', 'understanding the needs and expectations of interested parties', the scope of the system and the management system itself. The first two of these are new requirements that have been introduced to make the standard conform to the Annex SL structure. This structure will be eventually be the same across all the management systems standards and make them more consistent.

This clause refers to aspects of ISO 9001 that are spread throughout the whole QMS and are system-wide in their effects and coverage.

Clause 4 is therefore an introduction to the requirements of the standard and sets out the basics for the QMS.

**Note 1:** Sections 4.9 to 4.16 discuss the main points of ISO 9001. Sites should obtain a full copy of the standard for the exact wording of the standard. Copies of the standard can be obtained from your country standards body.

**Note 2:** In this chapter the clause numbers of ISO 9001 will be used to enable cross-referencing to the full standard, i.e. if we refer to 'Clause' then we are referring to the relevant clause in ISO 9001.

### 4.1 Understanding the organisation and its context

This is a new requirement that requires the company to set out why it is there (the purpose) and the internal and external issues that can affect the company's performance, purpose or strategic direction. Issues that should be considered are those that are relevant to the purpose and strategic direction.

The rationale behind this clause is that these issues will have an effect on the planning of the QMS, the ability of the company to deliver the QMS and can also be used as an input to frame the development of the QMS.

- **Tip** - A good way to understand and document the internal and external issues is to carry out simple SWOT (Strengths, Weaknesses, Opportunities and Threats) and PEST (Political, Economic, Social and Technological) analyses. These are a standard 2 x 2 matrix (see Section 7.11) but completed for different factors.
- **Tip** - Internal issues can relate to the values or culture of the company.

### 4.2 Understanding the needs and expectations of interested parties

This is also a new requirement and requires a company to consider the needs and expectations of 'interested parties'. These parties are those that are 'relevant' to the QMS, e.g. customers, regulators etc, as defined by the company. The company is required to monitor and review information about relevant 'interested' parties and their relevant requirements.

- **Tip** - The company should define who is relevant in terms of the QMS. There is no need to extend this beyond their relevance to the QMS.
- **Tip** - List all the stakeholders, e.g. shareholders or owners, employees, customers, suppliers etc. and rank them according to their interest and influence on the QMS. This can eliminate from consideration a lot of stakeholders who

**Transition to the new standard:**

The issue of ISO 9001:2015 in September 2015 started a 3-year transition period, i.e. companies have 3 years to transition to the new standard.

Certifications to ISO 9001: 2008 will no longer be valid after September 2018.

There is plenty of time to make the transition.

The new structure should not cause a company to renumber existing documents at once but eventually this will need to happen for consistency.

If a company uses software to generate numbers then this may cause a problem.

**Clause 4: Organisation context**

- Understand the context
  - Determine relevant internal & external issues
  - Identify interested parties & their requirements
- Scope
  - Define QMS scope
  - Define boundaries & applicability of the QMS
- Processes
  - Define processes needed
  - Define process applicability
  - Maintain documented information for processes

ISO 9001 organisation context requirements

These requirements cover the whole system. They deal with how the company operates and what affects it, the scope of the QMS and the processes through which quality is managed and maintained.

are not really relevant to the QMS.

- **Tip** - There is no new requirement to ensure that products and services meet these needs and expectations (other than those already present in ISO 9001:2008).

## 4.3 Determining the scope of the quality management system

This is where the company decides on what the QMS covers and lists the products and services covered by the QMS. The scope is decided by considering Clause 4.1, Clause 4.2 and the products and services of the company.

Previously it was possible for a company to exclude almost any requirement from the scope of the QMS but this is no longer the case. A requirement of the standard can only be excluded if it:

- is not applicable.
- cannot be applied.
- does not affect the product quality of products and service stated in the scope.
- **Tip** - If a clause of the standard is excluded on the basis that it cannot be applied then this exclusion should not affect the ability of the company to ensure conformity of the products and services listed in the scope.

Every exclusion must be justified in terms of Clause 4.1, Clause 4.2 and the products and services of the company. Every exclusion must also be documented.

- **Tip** - The scope of the QMS (the list of products and services covered) must be maintained as 'documented information' (see Section 4.17).

## 4.4 Quality management system and its processes

A company is required to set up, run and continually improve their QMS and the processes needed to run the system.

This means defining and mapping the company processes using methods similar to those described in Sections 4.4 to 4.6. A company will need to define:

- The global processes needed for the company (see Section 4.4).
- The overall process flow and interaction in the company (see Section 4.4).
- The individual process inputs and outputs (see Section 4.6).
- The measurements needed to check that individual processes are operating effectively and efficiently.
- The resources needed for each process.
- The responsibilities and authority for each process.

Much of this information was required in previous issues of ISO 9001 but the standard now requires consideration of:

- Opportunities for improvement of the process and the QMS (previously this only considered the process itself). This is part of the increased focus on improvement (see Section 4.16).
- The risks and opportunities and the planning and implementation needed to address these risks and opportunities. This is part of the increased focus on risk (see Section 4.19).
- **Tip** - Try to develop a standard format for process description (see Section 4.6) and use this to describe every process, i.e. keep it consistent across the company.
- **Tip** - The determination of the processes and their related information must be maintained as 'documented information' (see Section 4.17).

## The quality manual (gone but not forgotten)

ISO 9001 no longer requires a 'quality manual' but the requirements that a quality manual fulfilled are still in the standard, i.e. there are still requirements for 'documented information' (see Section 4.17) that must be maintained but the method is not specified. This lack of a requirement for a quality manual does not forbid a company from having one and there will still be a need to have some sort of documented information to show auditors and other interested parties:

- What the company does and the internal and external context.
- The quality policy.
- The quality objectives.
- The scope of the QMS.
- Any exclusions from the QMS (and why).
- The main processes and how and they link together.
- **Tip** - These all need to be maintained as 'documented information'. Why not keep them all together in a company quality brochure (the 'new' quality manual)?

The new 'expansion' of ISO 9001 to be more applicable to services and non-manufacturing companies means that it talks about 'products and services' rather than simply 'products'.

ISO 9001 now requires more thinking about what the company does and why it does it (the strategy) rather than simply looking at inspection and other procedures. This is now a real 'management system' standard but only time will tell if it is better.

The QMS should work for you and not the other way around.

# 4.9 Leadership requirements of ISO 9001

## Managing the process

Clause 5 of ISO 9001 covers the leadership requirements of the standard - where 'leadership and commitment' have replaced the previous concept of 'management commitment'. This change requires far greater leadership involvement in the QMS and the responsibilities of top management have been extended. They are required to do more and be responsible for more - just as it should be.

## 5.1 Leadership and commitment

### 5.1.2 Leadership and commitment for the QMS

Top management is required to show leadership and commitment by:

- Taking accountability for the effectiveness of the QMS.
- Setting the quality policy and objectives and making sure that these are compatible with the overall company context and strategic direction.
- Ensuring that the QMS requirements are integrated into the business processes, i.e. they are part of the 'way you do business' and not bolted on top as an afterthought.
- Promoting the process approach and risk-based thinking.
- Providing the necessary resources (including people) for the QMS.
- Communicating the importance and benefits of quality and the QMS.
- Ensuring that the QMS gets the desired results.
- Promoting improvement.
- Helping other management roles demonstrate leadership.

This is a much more 'active' requirement than before where the role of top management could be very 'passive'. The previous requirement was to 'provide evidence of commitment' but there is now a requirement to 'demonstrate leadership and commitment'. This is a bigger requirement.

It was always part of the management responsibility to establish a quality policy but ISO 9001 now requires that this is related to the context of the company (see Section 4.8).

- **Tip** - There is no requirement for documented information on leadership and commitment. Ideally, this will be shown by actions, i.e. the standard requires the 'demonstration' of leadership and commitment. This is important and it would be useful to record demonstrable actions in the Management Review as proof of commitment.

The quality policy commits the company to both meeting quality requirements and improvement.

It is a key document for checking the performance of the QMS.

### 5.1.2 Customer focus

Focusing on the customer to find their current and future needs and expectations is an essential part of the standard and of a successful business.

Customer needs and satisfaction are also covered in Clauses 8.2.1 and 9.1.2 of the standard although these elements of the standard deal more with actual product requirements and measuring customer

Customer focus is at the heart of ISO 9001 and should be part of every process.

**Clause 5: Leadership**

Leadership & commitment
- Take accountability for QMS
- Quality policy
- Support quality
- Develop & implement system
- Promote process & risk-based thinking
- Ensure system gets results
- Promote improvement

Customer focus
- Identify & meet customer needs
- Identify risks
- Improve customer satisfaction

Quality policy
- Define, communicate & apply quality policy
- Review policy

Roles & responsibilities
- Assign & communicate roles, responsibilities & authority for systems & processes

**ISO 9001 leadership requirements**

The leadership requirements deal with the responsibilities of the company's top management for implementing leadership as opposed to management. They set the scene for the overall QMS and define the necessary top management leadership.

satisfaction (see Sections 4.13 and 4.15 of this Workbook respectively). This clause of the standard is concerned with the strategic and management aspects of satisfying a customer's needs and expectations. Companies are expected to identify not only the customer's requirements but also any applicable statutory and regulatory requirements and to ensure that these are determined and met.

Customer focus also requires that:

- Top management assess and address any risks and opportunities concerning product conformity.
- Top management ensure that the focus on enhancing customer satisfaction is maintained.

This clause has been expanded to put customer focus and risk-based thinking at the heart of the standard.

- **Tip** - Customer focus is not simply about the product, it should include all of the services, such as sales order processing and invoicing, i.e. any process that interacts with or affects the customer.
- **Tip** - How this is done, and proven, is left to the company but it would be useful to record demonstrable actions in the Management Review as proof of customer focus.

## 5.2 Quality policy

Top management is responsible for establishing, implementing and maintaining a company quality policy. The quality policy must:

- Be relevant to the company, i.e. aligned to the purpose and context of the company.
- Provide the framework to set the quality objectives (see Clause 6.2 and Section 4.10 of this Workbook).
- Make a commitment to satisfying applicable requirements.
- Make a commitment to continual improvement.

The policy must be relevant to both the company and to the customer, i.e. a totally inward-facing policy does little to improve customer satisfaction.

The standard also requires that the quality policy is maintained as 'documented information' (see Section 4.17), be communicated, understood and applied and be available to appropriate interested parties.

- **Tip** - Do not be tempted to simply take a 'standard' quality policy and use it for your company. The policy must be related to the purpose and context of the company.
- **Tip** - ISO 9001 now has an explicit requirement that the quality policy is applied within the organisation - it is no longer enough to simply 'ensure communication and understanding'.
- **Tip** - The quality policy should be regularly reviewed for continued relevance to the context of the company.
- **Tip** - It would be useful to record how the quality policy is applied and any reviews of the policy in the Management Review as proof of compliance.

## 5.3 Organisational roles, responsibilities and authorities

Previously, ISO 9001 required a Quality Management Representative (QMR) who was responsible for the management of the QMS and who had both the responsibility and the authority for the QMS. In the current ISO 9001, top management is responsible for ensuring that responsibilities and authorities for relevant roles are assigned, communicated and understood. This means that top management can retain the responsibility for management of the QMS or delegate responsibility and authority across different roles.

In either case, top management must assign responsibility and authority for:

- Ensuring that QMS conforms to the requirements of ISO 9001.
- Ensuring that processes deliver the desired outputs.
- Reporting on the performance of the QMS and opportunities for improvement.
- Promoting customer focus throughout the company.
- Ensuring the integrity of the QMS when changes are made to the system.
- **Tip** - The addition of the requirement that responsibilities and authorities are 'understood' by the people that they are assigned to is new. Where companies have formal job descriptions then it may be necessary to amend these to communicate quality responsibilities.

"Be sincere, be brief, be seated."

**Franklin D. Roosevelt**

Advice given by FDR on public speaking.

Quality policies should follow FDR's advice, i.e. a short relevant quality policy is far better than a long policy copied from another company.

ISO 9001 no longer requires a Quality Management Representative (QMR) but you can still use one to comply with Clause 5.3.

If you feel that the existing arrangements are suitable and meet the intent of the standard then this is acceptable. Top management is simply assigning responsibility and authority for managing the QMS to a single person. As long as the responsibility and authority is assigned, communicated and understood then having a QMR is still acceptable.

The Quality Manager's job is safe after all!

# 4.10 Planning requirements of ISO 9001

## Planning to succeed

Clause 6 of ISO 9001 deals with planning for quality and particularly with planning for risks and opportunities and the needs of interested parties (see Section 4.8), setting the quality objectives and planning for changes to the QMS.

Planning is an essential part of the PDCA cycle and is therefore an essential part of ISO 9001. Do not be tempted to treat this clause lightly (even though it is quite short), this is an essential part of the standard.

- **Tip** - The consideration of risks and opportunities is part of the emphasis on risk-based thinking in ISO 9001 and must be addressed (see Section 4.19).

## 6.1 Actions to address risks and opportunities

The planning process starts with defining the context of the company (see Clause 4.1 in Section 4.8) and companies need to think about:

- The internal and external factors affecting the company.
- The SWOT or PEST factors that affect the company.

It is then possible for the company to start to look at the risks and opportunities that need to be addressed to deal with potential changes in the internal or external factors. The risks and opportunities should be assessed in terms of:

- How the QMS can continue to achieve the desired results when dealing with uncertainty.
- How risks (undesired effects) can be prevented or reduced.
- How opportunities (desirable effects) can be enhanced.
- How improvement can be achieved.

After this assessment, the company should start to plan:

- Actions to minimise the risks - these could be:
  - Elimination.
  - Reduction.
  - Avoidance.
  - Reducing the consequences.
  - Sharing the risk.
- Actions to maximise the opportunities - these could be:
  - Introducing new practices.
  - Introducing new products.
  - Entering new markets.
  - Developing new customers.
- How to integrate these actions into the QMS - addressing risks and opportunities can mean changes to processes and any process changes will need to be incorporated in the overall process map (see Clause 4.4).
- How to evaluate the effectiveness of actions taken - this is the 'check' element of the PDCA cycle and any actions taken should be evaluated for effectiveness.
- **Tip** - ISO 9001 states that the actions taken should be proportionate to the potential impact of the risks and opportunities on products and services. This means that if the risk or opportunity is minimal then the action taken can also be minimal. Do not try to take action on everything - choose your battles.
- **Tip** - There is no requirement for documented information on this clause but it would be useful to record the

Opportunity is sometimes said to be a positive risk but opportunity is simply a set of circumstances that make it possible to do something.

Opportunities simply present different risks.

"If you fail to plan, you are planning to fail."

**Benjamin Franklin**

**Clause 6: Planning**

Risks & opportunities

Consider risks & opportunities
Plan actions to address risks & opportunities
Evaluate effectiveness of actions

Quality objectives

Plan & set quality objectives for processes & customers
Monitor, communicate & update quality objectives

Planning change

Plan changes to QMS
Consider consequences, resources & responsibilities

### ISO 9001 planning requirements

Planning deals with understanding the risks and opportunities as well as setting the quality objectives and planning for changes to the QMS. A system that is not planned from the start will be hard to implement and not achieve the objectives.

planning process as proof of compliance. This could take the form of meeting minutes, notes etc.

## 6.2 Quality objectives and planning to achieve them

Setting quality objectives and monitoring their achievement (or not) is the only way of assessing the effectiveness of the QMS and the achievement of the aims of the quality policy.

Companies need to set objectives for all relevant functions, levels and processes in the company and these must

- Be consistent with the quality policy.
- Be measurable.
- Take the applicable requirements into account.
- Be relevant to product conformity or customer satisfaction.
- Be monitored.
- Be communicated.
- Be kept up to date.

Setting the quality objectives is not enough in itself, the company needs to plan how to achieve the objectives. This should follow the standard planning process:

- What should be done.
- What is needed to get it done.
- Who will do it.
- When it will be done by.
- How it will be evaluated.
- **Tip** - The quality objectives must be maintained as 'documented information' (see Section 4.17).
- **Tip** - Rudyard Kipling said it best in The Elephant's Child:

  'I keep six honest serving-men
  (They taught me all I knew);
  Their names are What and Why and When
  And How and Where and Who'.
- **Tip** - Objectives should always be SMART (see Section 4.1).
- **Tip** - Maintenance of documented information on the quality objectives can be done via the older concept of a quality manual (see Section 4.8).
- **Tip** - Planning the quality objectives is not the same as setting the objectives for installing the QMS (see Section 4.1).

## 6.3 Planning of changes

No QMS can be expected to remain static and changes are part of the growth of the QMS and in the context of the company. When the inevitable changes to the QMS are made then they must be planned so that the integrity of the QMS is maintained. This planning should consider:

- The purpose and consequences of any proposed change.
- The resources required for any proposed change.
- Changes to the responsibilities and authorities as a result of any proposed change.
- **Tip** - Plans should also be made for improvement of the QMS.
- **Tip** - Change is also covered in other clauses of the standard, e.g. Clause 7.1.6, Clause 8.1, Clause 8.5.6. These should also be considered when addressing this clause.

Do not forget to make the objectives SMART (see Section 4.1).

Quality objectives should not be static. They need to be changed as the business changes.

This is part of improvement and can be used as evidence for this.

# 4.11 Support requirements of ISO 9001 - 1

## Providing the resources

Clause 7 of ISO 9001 deals with reviewing and providing resources to support the implementation and improvement of the QMS. This covers all types of support, i.e. people, infrastructure, process environment, measuring equipment and information for the QMS and the processes to operate.

Any system that is inadequately supported is bound to fail and this clause is designed to ensure that the QMS is adequately supported and resourced.

## 7.1 Resources

### 7.1.1 General

Adequate resources should be provided to ensure effective operation and continual improvement of the QMS. These resources can come from internal resources or, if required, from external providers (suppliers).

### 7.1.2 People

ISO 9001 has significant requirements for competence, training, skills and record keeping. These are mainly dealt with in Clauses 7.2 and 7.3 (see Section 4.12). This clause requires that the company determines and provides the people necessary for the effective implementation of the QMS and the operation and control of the processes that make up the company operations.

- **Tip** - This requirement is not limited to the 'quality department' staff but is relevant to anybody whose activities, either directly or indirectly, affects the quality of products or services.

The essential point is that the company has staff who can run the QMS and the processes.

### 7.1.3 Infrastructure

Companies should assess what infrastructure is needed, provide it and then maintain it so that processes can achieve conformity with requirements.

- **Tip** - Infrastructure is not only the buildings (what is typically associated with the word 'infrastructure') but also the process equipment, software and support services. It includes anything that is associated with the product or service and can even include transport services.

This assessment of needs should also be regularly reviewed and recorded.

- **Tip** - This may be a formal record or can be the records of condition assessments, etc. but there should be some linkage to the QMS.

### 7.1.4 Environment for processes

Companies should assess the environment needed, provide it and then maintain it so that processes can achieve conformity with requirements. This is about the conditions under which the product is produced or service is carried out to ensure quality. It may include issues such as hygiene (for packaging or food products), clean rooms (for medical products) or health and safety for staff. These all need to be appropriate to meet the requirements.

- **Tip** - Environment in this context can include not just the physical factors but also the wider social and psychological environment.

### 7.1.5 Monitoring and measuring

#### 7.1.5.1 General

Monitoring and measuring is an essential part of providing evidence that products meet the requirements and companies are required to determine and provide resources:

- That are suitable for the monitoring and measurement being carried out.
- Maintain the resources.

#### 7.1.5.2 Measurement traceability

If traceability is required or considered necessary by the company then measuring instruments should be calibrated or verified (or both) either regularly or before use. Calibration and verification should be traceable to international or national standards but if these do not exist then the basis of calibration or verification should be retained as documented information.

A company is also required to identify and protect measurement instruments to ensure that results are valid.

- **Tip** - If it is found that measurement instruments are out of calibration then companies need to evaluate the validity of any previous measurements.

Providing the right resources is an essential part of delivering a quality product or service.

The treatment of organisation knowledge as a resource is a new requirement in ISO 9001. This may need some thought for companies to get it right.

Having documented information does NOT mean that the organisation's knowledge is properly recorded. Companies should use the PDCA cycle to review their recording of knowledge and improve their practice in this area.

When considering organisational knowledge resources, focus on the knowledge needed for operation of the QMS and achieving product conformity.

### 7.1.6 Organisational knowledge

This is a new requirement and aims to drive companies to capture and preserve knowledge as they do any other resource.

Companies are required to determine the knowledge that is required to operate processes and to ensure conformity of products, to maintain this knowledge and to make it available as necessary. It is also required that companies consider any changes that make acquiring or accessing additional knowledge necessary.

Organisational knowledge can be:

- Specific to the company and their products.
- Gained from experience (the way we do things).
- Used and shared in the company.

Organisational knowledge will normally come from one of two sources:

- Internal, e.g. intellectual property, company experience with products and processes or general undocumented company information.
- External, e.g. standards, conferences, industry discussions or customers.
- **Tip** - This clause does not require documentation to be maintained to prove compliance and it is up to the company to demonstrate compliance. This could be from process maps, procedures, work instructions, customer manuals, a company library or other sources.
- **Tip** - Knowledge can be formal, i.e. it can, and often is, documented and written down, or tacit, i.e. it is not often documented or written down. Tacit knowledge is mainly experience based, i.e. this works for us. Companies should consider both types of knowledge with regard to this clause.

**Note:** This clause of the standard is continued in Section 4.12.

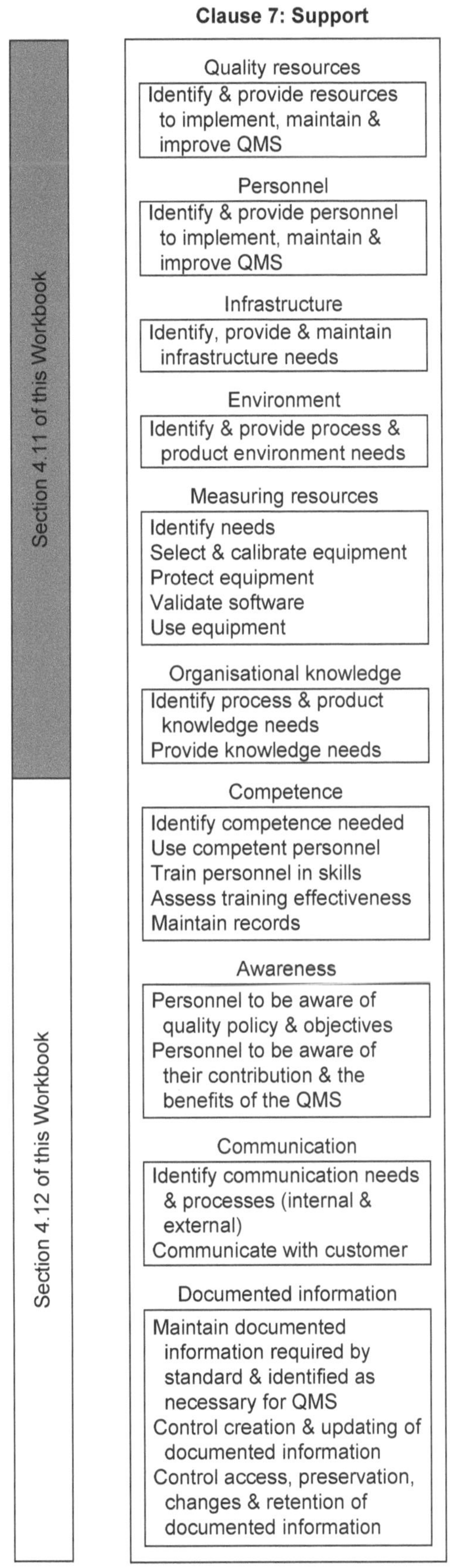


ISO 9001 support requirements

A QMS must be adequately supported with resources to function effectively. The requirements of this clause deal with all aspects of supporting the QMS to ensure successful installation and operation.

# 4.12 Support requirements of ISO 9001 - 2

## 7.2 Competence

Clause 7.1.1 requires that the company has staff who can run the QMS and the processes. People are at the heart of the QMS and competent people are needed to deliver quality products and services.

- **Tip** - This requirement is not limited to the 'quality department' staff but is relevant to anybody whose activities, either directly or indirectly, affects the quality of products or services.

Companies should first assess the competencies (skills) needed to operate processes so that they deliver the required outputs and then assess the people to identify any gaps in skills.

- **Tip** - A simple spreadsheet can be useful for this, i.e. what competencies are needed for each process, who has them and who is competent for each process. This can be used to provide evidence of conformance and a systematic approach.
- **Tip** - 'Competence' is determined by the company. It is up to the company to define what is needed.
- **Tip** - Experience can count towards competence and should not be discounted: there is no sense in training somebody in something they have been doing perfectly well for 20 years. It is an insult to them and a waste of time - but do record their experience as part of their competence.

Any competency gaps identified should be filled by training, coaching or self-learning.

- **Tip** - The standard requires competence but does not specify how this is to be achieved, i.e. training does not have to be external but can be internal (delivered by the company or by supervisors/ managers) or by self-learning.
- **Tip** - Training can be as long or as short as the company decides is necessary to be competent but do retain records for evidence that training has taken place.

The essential point is that the staff competency meets the needs of the QMS and delivers the required outputs.

- **Tip** - Companies should retain records to show that staff are competent. This can be recorded on the spreadsheet used to record any gaps in competencies.

After the company has taken action to ensure competency (training or other action), the action should be evaluated to see if it has been effective in achieving the necessary competence.

- **Tip** - The standard requires that documented information (records) are retained as evidence of competence .
- **Tip** - This requirement will usually be audited as part of the QMS process audit, i.e. an auditor will not be looking at this in isolation. They will look at the processes and try to see if the people are 'competent' to run their process.

## 7.3 Awareness

ISO 9001 requires that everybody carrying out work under an organisation's control should be aware of:

- The quality policy.
- The quality objectives.
- Their part in the effectiveness of the QMS (and the benefits of quality improvement).
- What happens if the QMS requirements are not met.
- **Tip** - This clause includes not just people working directly for the company but for everybody under the organisation's control, i.e. agency staff, contractors etc.
- **Tip** - Do not be surprised if an auditor randomly asks somebody about the quality policy and the implications for them. It can be very embarrassing if they don't have a clue!

## 7.4 Communication

The company is responsible for ensuring that communication processes relevant to the QMS (both internally and externally) are established within the company and that this communication takes place. The 'six honest serving men' (see Section 4.10) should all be covered in the communication process.

- **Tip** - There is no requirement for documented information for this clause but it could be considered and documented as a process.
- **Tip** - Customer communication is specifically covered in greater detail in Clause 8.2.1 (see Section 4.13).

A 'documented procedure' is very different to a 'record' although they are all 'documented information' in ISO 9001.

A 'documented procedure' tells you how to do something and is before the event.

A 'record' shows that something has been done and is after the event.

In ISO 9001 the term 'documented information' is used generally for what would previously have been either documented procedures or records.

The big 'hint' is now in the terms 'maintain' and 'retain' - documented procedures would be maintained and records would be retained.

Use the ISO 9001 document control process from the very start in producing the documents. It will save time later.

## 7.5 Documented information

### 7.5.1 General

ISO 9001 has various requirements for documented information and companies will also have their own requirements for documented information. Previously, ISO 9001 differentiated between 'documented procedures' and 'records' but these are now referred to generally as 'documented information' (see sidebar on the left).

- **Tip** - Many companies 'over-document' ISO 9001 and create work for themselves. Minimise the amount of documentation to increase the ease of use, improve understanding and reduce problems with auditors. Do the minimum needed to comply with the standard and run the system.

### 7.5.2 Creating and updating

Documented information should be:

- Identified with a title, the date, the current version and the status.
- Reviewed and approved before use.
- **Tip** - There is no requirement for the format of documented information, e.g. it can be printed, stored on local or cloud computers, etc. It must be 'appropriate' but the final decision is up to the company (see Section 4.17).

### 7.5.3 Control of documented information

Documented information should be controlled so that it is:

- Available and suitable for use when and where required.
- Adequately protected.

Documented information controls are needed to control:

- How documented information is distributed, accessed, retrieved and used.
- How documented information is preserved to retain usability.
- How changes are managed, including version control and unintended or unauthorised changes.
- How documented information is retained and/or disposed of.
- **Tip** - ISO 9001 no longer requires controls to prevent obsolete documents from being used but this is covered under 'version control'.
- **Tip** - External documents should be controlled as well as internal documents, e.g. standards should be controlled to ensure that the correct issue is used.

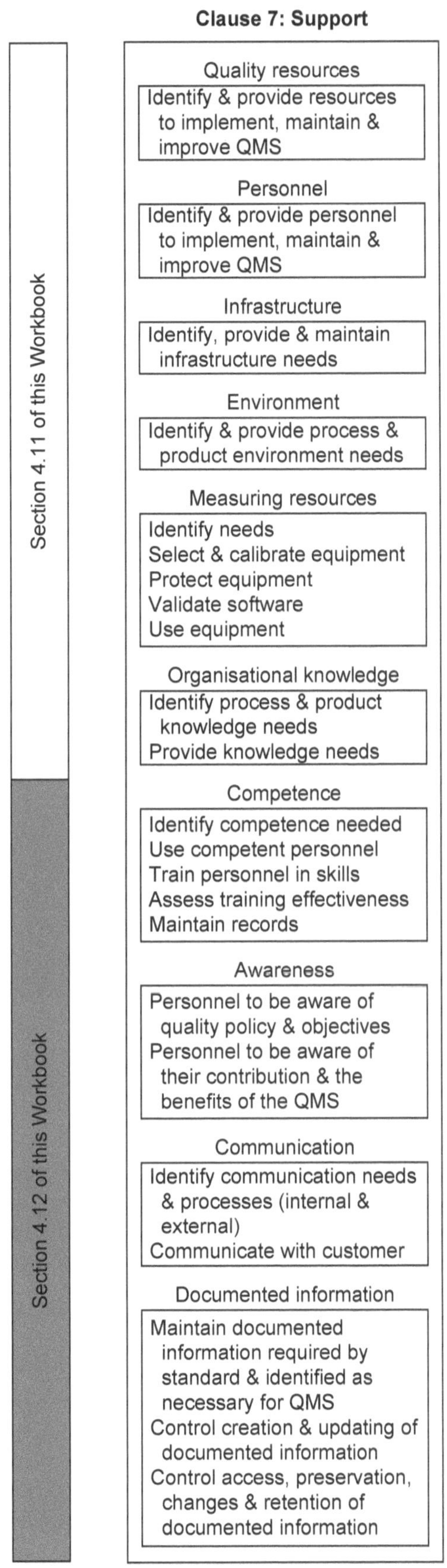


ISO 9001 support requirements

A QMS must be adequately supported with resources to function effectively. The requirements of this clause deal with all aspects of supporting the QMS to ensure successful installation and operation.

# 4.13 Operation requirements of ISO 9001 - 1

## Getting the product right

Operation is the collection of processes that go into designing and making the product. It is primarily about the production process but also includes requirements for design and other aspects of the larger process of taking the customer's needs and converting these into a product or service. This covers all areas, e.g. sales, design and development, purchasing and production/service activities.

**Note:** Any exclusions must be justified in the system scope (see Section 4.8).

## 8.1 Operational planning and control

Production processes are inevitably complex and planning the processes is an essential part of getting the product correct. ISO 9001 requires planning to establish the product requirements, processes (and their criteria), documents and resources necessary for the product to conform to the customer's requirements.

The outputs of the planning process should be suitable and consistent with the company's requirements.

- **Tip** - Records need to be retained to show that planning has taken place.
- **Tip** - If you make simple products then do not make the planning process too complex - it has to be suitable for you, not for everybody.
- **Tip** - A 'quality plan' is an invaluable asset here and quality plans are covered in Section 4.18 of this Workbook.

## 8.2 Product and service requirements

### 8.2.1 Customer communication

Identifying and reviewing the customer's needs is an internal process but the results of these reviews and other information needs to be communicated to the customer. Customer communication covers areas such as:

- Product and service information.
- Handling of enquiries, contracts or sales orders and changes to sales orders.
- Handling of complaints and customer feedback.
- Handling of customer property.

### 8.2.2 Product and service requirements

Before agreeing to supply the product, it is necessary to understand and identify the customer's needs - if you cannot supply to the customer requirements then it is best not to supply.

- **Tip** - If sales are via a catalogue or standard product line then reviewing the sales literature can cover this element, i.e. it is not necessary to identify and review every customer order.
- **Tip** - These are not simply the customer needs, they may be from regulatory or other bodies (or even internal needs).

### 8.2.3 Requirements review

After the customer's needs have been identified, it is necessary to review these to check that the company can provide the products as required by the customer. This should check that the product meets the customer's requirements.

- **Tip** - This review should also cover statutory and standards requirements.
- **Tip** - Retain a record of the requirements review and any order changes.
- **Tip** - The review process can be minimised for repeat orders.

### 8.2.4 Changes to requirements

If any changes are made to the requirements then the documented information should be changed and the relevant people notified.

## 8.3 Design and development

### 8.3.1 General

Where design and development is carried out then it needs to be controlled to produce a design that is capable of being produced to the requirements, i.e. it is a process.

### 8.3.2 Planning

Design and development needs planning and ISO 9001 requires a plan to control the design and development process and the interactions with other functions. This plan needs to be retained to show conformance to the standard.

ISO 9001 now refers to 'products and services' to reflect the use of the standard by services companies. As this Workbook is concerned with plastics processing, we will refer to products but in all cases this covers any associated services as well as the physical product.

There is an explicit requirement that a company be able to substantiate claims for products and services offered to customers.

The design and development process should not present any concerns to good companies. You are probably doing all this anyway.

You simply need to have documented information and, in this case, maintain this to be able to prove that the process has been followed.

## 8.3.3 Inputs

The inputs to the design and development process are:

- Product performance requirements.
- Product legal and regulatory requirements.
- Other requirements:
  - Resource needs.
  - Consequences of failure.
  - Customer control.

These requirements influence the process and companies are required to define the inputs (requirements) for the design and development process, to keep a record of these inputs and to review the inputs.

- **Tip** - The design and development inputs are effectively the Product Design Specification (see Section 8.2).

## 8.3.4 Controls

Any design produced should be subject to review, verification and validation to check that the design meets the specified requirements and the results of this review, verification and validation process must be recorded.

- **Tip** - Design review, design verification and design validation are different but can be carried out together or separately as decided by the company.
- **Tip** - Documented information on this process must be retained.

## 8.3.5 Outputs

The process outputs must:

- Meet the input requirements.
- Be adequate for subsequent processes.
- Have monitoring and measuring requirements and acceptance criteria.
- Be fit and safe for the intended use.
- **Tip** - Documented information on this process must be retained.

## 8.3.6 Changes

Changes during design and development are inevitable and these changes must be controlled. Any changes made should be identified, reviewed, controlled and authorised before implementation.

- **Tip** - This process must include an assessment of any risks and actions to reduce these (see Section 4.19).
- **Tip** - Documented information on this process must be retained.

**Note:** This clause of the standard is continued in Section 4.14.

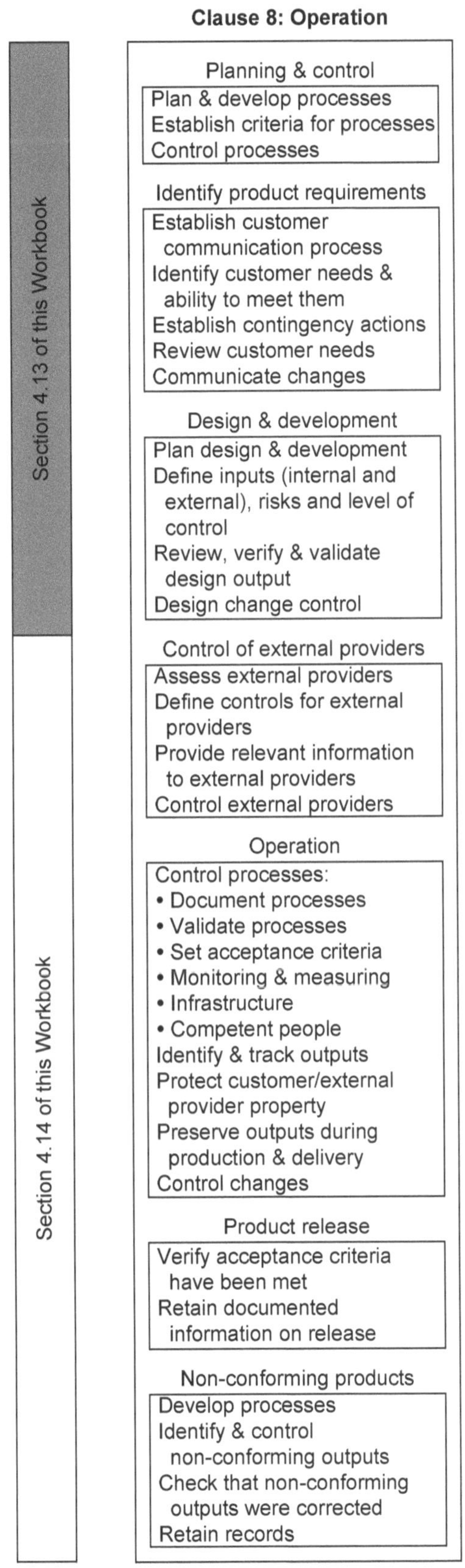


### ISO 9001 operation requirements

Operation is the process of taking the customer needs and translating these into a product or service. The requirements deal with all aspects of this process from ordering through to delivery and also include the design process (where applicable).

# 4.14 Operation requirements of ISO 9001 - 2

## Producing the product

After the product is designed then the actual production can start and this is part of the operation requirements.

## 8.4 Externally provided products and services

### 8.4.1 General

Getting the right raw materials and services is the first step in producing quality products and this clause covers the controls necessary to achieve this. Control of purchasing requires that products be purchased from approved providers (suppliers) and meet specified purchase requirements.

Companies should set appropriate criteria for supplier approval and evaluate/select suppliers for their ability to supply products that meet the company's needs.

- **Tip** - Records of the supplier selection process should be retained.
- **Tip** - Purchasing should be controlled for any product or service that affects the customer. This may include services such as transport of the final product to the customer.

### 8.4.2 Types and extent of control

The controls applied after the product or service is delivered need to be established so that it is possible to check that it meets the requirements and does not affect the ability of the company to meet the customer requirements.

- **Tip** - You do not need to have the same level of control for all suppliers. You can have different controls for suppliers of non-production materials, e.g. office equipment, to the controls for suppliers of production materials, e.g. plastic raw materials.

### 8.4.3 Information

Purchases and their controls should be specified to describe the relevant requirements to the supplier.

- **Tip** - Specifications may refer to the supplier's catalogues or sales literature.
- **Tip** - This does not prevent changes to the specification after discussion with the supplier but any change to the specification should be recorded.

## 8.5 Product and service provision

In most companies this is the largest section of the documentation but the requirements are quite simple, i.e. companies should manage their processes and operations. If they are not doing this adequately then it is unlikely that they will be in business for very long.

### 8.5.1 Control

A company must have effective methods of planning and controlling their processes. Documented information is to be retained on:

- The process and product characteristics.
- The activities and results to be achieved (think 'work instructions').
- Infrastructure.
- Monitoring and measuring.
- People.
- Release, delivery and post-delivery activities.

If the product cannot be verified until after it is delivered then use 'process validation' (as opposed to 'product verification'). This is not often the case in plastics processing but could include the sterilisation of medical products which is difficult to check after the process.

### 8.5.2 Identification and traceability

Knowing what, and where, products are is an essential part of production control and not just quality control. This requires methods to establish the identity of process outputs (products) and the monitoring and measurement status of the products during the production process.

If product traceability is required by the customer then the company should provide adequate traceability controls.

- **Tip** - Documented information on this process must be retained.

### 8.5.3 Externally supplied property

Companies should identify, verify and protect any property provided by customers or external providers, e.g. product for incorporation into the final product.

- **Tip** - This can include intellectual

ISO 9001 now talks about 'external providers' instead of 'suppliers' to reflect the more service-oriented nature of the standard. For plastics processors, in most cases you can think of 'suppliers' and you will get to the same place.

It also talks about 'process outputs' rather than 'products' but again for most plastics processors you can think of 'products'.

Procedures and documents should be available which specify:

- What is checked
- How it is checked
- When it is checked
- Who carries out the check
- What is to be done if the process is not to the standard
- What records are kept and where they are kept.

property or confidential information.

### 8.5.4 Preservation

Getting the production processes right is wasted if the product is damaged during the production or delivery processes.

Companies should control processes to prevent damage to the product and to keep it in good condition.

### 8.5.5 Post-delivery activities

Companies also need to control all post-delivery activities, e.g. warranties, maintenance, recycling etc.

- **Tip** - Be sure to consider any risks associated with the product's use.

### 8.5.6 Change control

Any changes during production must be identified, reviewed, controlled and authorised before implementation.

- **Tip** - This should include an assessment of any risks and actions to reduce these.
- **Tip** - Documented information on this process must be retained.

## 8.6 Product and service release

Verification requirements for production should be set so that it is possible to check that the product meets the requirements. Release to the customer should not take place until these requirements have been met unless otherwise approved by a relevant authority or, as applicable, the customer.

- **Tip** - Documented information should be retained to provide traceability to the person authorising the release.

## 8.7 Control of nonconforming outputs

If products do not conform to the specified requirements, they are nonconforming and companies must establish and document a nonconforming products process. This should specify actions to:

- Correct the nonconforming product and then re-verify conformance to the original requirements.
- Inform the customer.
- Obtain customer authorisation to use the product, discontinue production or supply under a concession.
- **Tip** - Records should be retained of any product nonconformities that describe:
  - The nonconformity.
  - The actions taken.
  - The authority for these actions.

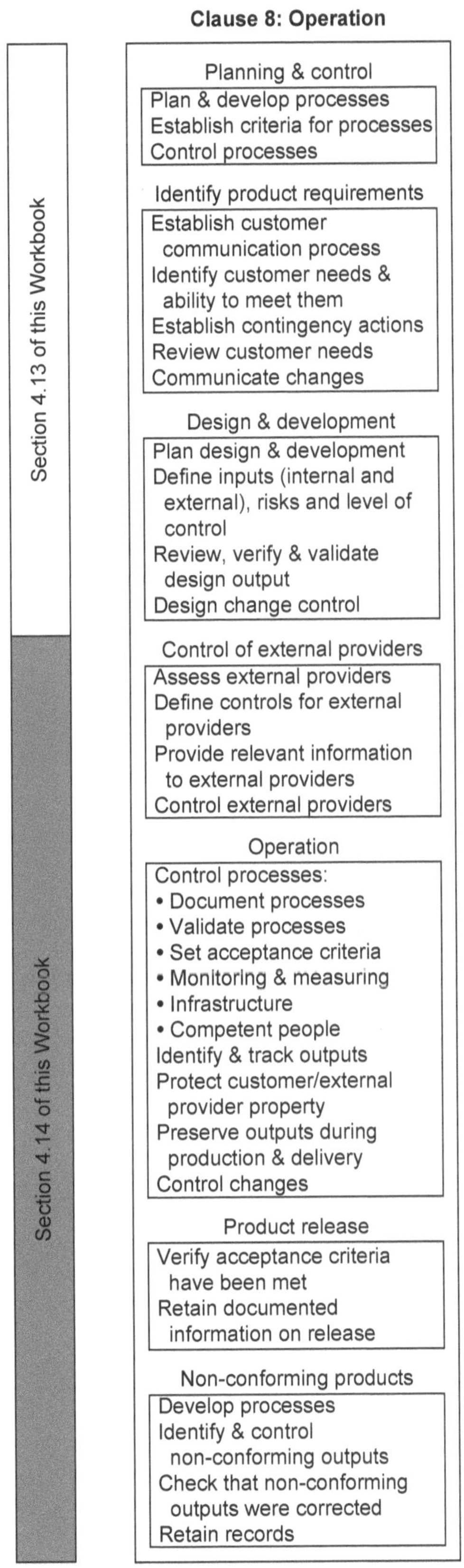


ISO 9001 operation requirements

After design and development is complete, production takes place using controlled raw materials and controlled production processes. Verified products are released to the customer and any nonconforming products are controlled and dealt with.

# 4.15 Performance evaluation requirements of ISO 9001

## Getting the information

The standards requirements do not stop after the product has been delivered to the customer and this section of the standard deals with finding out if the processes delivered the desired results, if the customer was satisfied and just as importantly if the overall QMS is functioning correctly.

## 9.1 Monitoring, measurement, analysis and evaluation

### 9.1.1 General

Processes need to be monitored and measured to check that they are operating as they were designed to operate and that they are effective in their operation. Monitoring and measuring are needed to be able to show that a product or service meets all the relevant requirements (customer, internal, legal and regulatory). Companies need to identify and plan:

- What needs to be monitored and measured.
- The methods for monitoring, measuring, analysis and evaluation.
- When monitoring and measurement will take place, i.e. where in each process and how often.
- How the monitoring and measuring will be made.
- What will happen to the results of the monitoring and measuring, i.e. how the results will be analysed and evaluated.

This clause also requires that companies evaluate quality performance and the effectiveness of the QMS.

- **Tip** - Documented information needs to be retained to show that this has been carried out.

### 9.1.2 Customer satisfaction

Customer satisfaction and perception are key measures in assessing if the QMS is operating effectively and if the company is meeting the customer's requirements. The standard therefore requires monitoring of customer views and opinions of the products and services and customer satisfaction. The method chosen is left to the discretion of the company.

- **Tip** - This does not mean that companies have to send out 'customer satisfaction surveys' to customers. This is only one method of assessing customer satisfaction and the method chosen is up to the company. Other methods could include market-share data, warranty claims or other reports. The important point is that data are obtained and evaluated.
- **Tip** - Previously this clause of ISO 9001 was focused on meeting customer requirements. It is now much more focused on the customer's perception of the company's products and services. This is very different.
- **Tip** - Documented information needs to be retained to show that customer satisfaction is monitored and evaluated.

Complaints can provide a method of assessing customer satisfaction but this is a dangerous assessment method.

**Clause 9: Performance evaluation**

Monitoring & measuring
- Identify what is to be monitored & measured
- Decide methods & times
- Evaluate quality performance & effectiveness of QMS

Customer satisfaction
- Monitor customer satisfaction

Analysis & evaluation
- Define needs
- Collect data
- Provide information

Internal audit
- Conduct regular internal quality audits
- Retain documented information on audit process

Management review
- Review actions from previous reviews
- Review changes in internal & external issues
- Review quality performance information
- Review risks & opportunities
- Decide actions to improve system & outputs

### ISO 9001 performance evaluation requirements

Performance evaluation is needed to assess if the QMS and processes are delivering what they were designed to deliver. The information from this process is the foundation for the improvement requirements of the standard.

### 9.1.3 Analysis and evaluation

After the monitoring and measurement process has generated data then it is necessary to analyse and evaluate these data. Evaluation should cover:

- Product and service conformity to requirements.
- Customer satisfaction.
- QMS performance and effectiveness.
- The success of planning.
- Process performance.
- Supplier performance.
- Opportunities for improvement.

The outputs of the analysis and evaluation can be used for improvement (see Clause 10 on improvement in Section 4.16) but should also be used as input for the management review (see below).

- **Tip** - Quality management provides tools for analysis and evaluation (see Chapter 7) and these tools are ideal for rigorous analysis and evaluation. There is no requirement in ISO 9001 to use these statistical tools and the standard simply says that these methods may be used.

## 9.2 Internal audit

Internal audit is an assessment of how the QMS is performing and if things are being done according to plan. Companies are required to conduct internal audits at planned intervals to check that the QMS meets the requirements of ISO 9001 and that the QMS is effectively implemented and maintained (see Section 4.20 for more detail on internal auditing).

Companies are required to plan and implement an audit programme to check that these requirements are met. The audit plan should:

- Specify the audit frequency, methods, responsibilities and reporting arrangements.
- The criteria and scope for each audit.
- How auditors are selected and how audits are conducted to give impartial and objective results.
- How audit results are recorded and treated, i.e. what action is to be taken.
- **Tip** - Internal quality audits check that you are 'doing what you say you do'.
- **Tip** - You should aim to internally audit each process/procedure at least once per year as a minimum but important processes (those that can have a large effect on the conformity of the product) may be audited more often. A schedule should be set up for system audits and adhered to.
- **Tip** - ISO 19011:2011 provides guidelines on auditing management systems.
- **Tip** - Documented information needs to be retained to show that auditing has taken place and that the results have been implemented.

## 9.3 Management review

### 9.3.1 General

Management review is an essential process and is a review of the complete system to consider possible improvements. Companies are required to carry out regular management reviews to consider the effectiveness and adequacy of the current QMS and potential improvement opportunities. The review should assess the need to make changes and keep a record of the actions to be taken.

### 9.3.2 Management review inputs

The input information for the management review can be:

- Previous management reviews.
- Changes in internal and external issues that are relevant to the QMS.
- Internal and external audit reports.
- Customer satisfaction and other feedback.
- Corrective actions taken.
- Opportunities for improvements to the QMS.
- Other information felt to be relevant to the management of the QMS.

### 9.3.3 Management review outputs

The output from the management review should consist of:

- Improvement opportunities for the QMS, processes and customer satisfaction.
- The need for changes to the QMS (including resource needs).
- **Tip** - Management reviews can also decide to simplify or remove processes. This is a valuable but often neglected function of the management review. Use the review to simplify and streamline the QMS.
- **Tip** - Documented information, e.g. management review minutes, needs to be retained to show that a management review has taken place and that the results have been implemented.

Data are not the same thing as information. Data are simply a collection of numbers whereas information is structured to provide the basis for management action.

Companies can be awash with data and still have little real information.

The management review acts as:

- A formal review of the system status and progress towards the objectives.
- A method to change the system as the company's needs change.

Whilst ISO 9001 does not specify the frequency of management reviews, it is good practice that they should take place at no less than 6 monthly intervals, i.e. 2 per year.

# 4.16 Improvement requirements of ISO 9001

## Getting better

The output of the performance evaluation process (see Section 4.15) should be a series of opportunities for improvement and ISO 9001 requires that improvement actions be taken. Companies need to improve to prosper and whilst this is the last main clause in ISO 9001 it is still one of the most important.

## 10.1 General

Companies need to find and select opportunities for improvement and to implement these as necessary to meet customer requirements and to enhance customer satisfaction. This covers:

- Improving products, services and processes.
- Improving the QMS.
- Correcting, preventing or reducing undesired effects.

In part, this requirement is also covered in Clause 10.3 (see below) and in both cases the requirements are driven by risk-based thinking, i.e. prevent or reduce undesirable effects (negative risks) and promote or enhance desirable effects (positive risks).

Improvement can be by:

- Corrective action, i.e. a reaction to deficiencies in the product, service or system.
- Continual improvement, i.e. a series of incremental improvements to the product, service or system (see sidebar).
- Step change, i.e. a major breakthrough improvement in the product, service or system.
- Innovation, i.e. a creative improvement in the product, service or system.
- Re-organisation, i.e. transformation of the company to improve the product, service or system.

Companies need to improve products, services and systems to stay competitive and this clause simply formalises this market requirement into the QMS.

- **Tip -** This clause requires companies to think not just about the current customer requirements but to also focus on future customer requirements and how these can be met.
- **Tip -** It is acceptable to prioritise improvement action, i.e. it would be unrealistic to expect simultaneous improvement in all areas. The prioritisation should be carried out using the quality objectives as part of the assessment.
- **Tip -** Auditors will be looking for evidence that the company has set objectives for improvement, has evaluated the risks and related these to the customer requirements and customer satisfaction.

ISO 9001 now considers improvement in a wider context instead of simply 'continual improvement'.

## 10.2 Nonconformity and corrective action

Clause 8.7 (see Section 4.14) deals with 'Control of nonconforming outputs' but this clause deals with nonconformities in the broader context, i.e. it includes nonconformities in the QMS. This clause also deals with the process of corrective action to be taken after a nonconformity has been identified.

Corrective action is required for problems that have already occurred. Companies

**Clause 10 Improvement**

Improvement
- Improve processes
- Improve outputs
- Improve QMS

Nonconformity & corrective action
- React to nonconformities in QMS & outputs
- Assess action to eliminate cause of nonconformity
- Implement any action needed
- Review action taken
- Retain documented information on action

Continual improvement
- Continually improve QMS
- Use performance evaluation data & management review to identify opportunities

### ISO 9001 improvement requirements

Improvement is at the heart of ISO 9001. Companies need to improve to keep market share and to stay in business. This clause brings together the improvement process for both the QMS and for the products (outputs).

must set up a corrective action procedure to both rectify system nonconformities and to prevent reoccurrence.

Companies are required to:

- React to nonconformities, i.e. take action to control and correct the nonconformity and to deal with the consequences.
- Evaluate the need for corrective action to eliminate the cause (Plan), i.e. review the nonconformity, determine the causes and see if there are other similar potential nonconformities.
- Take corrective action as a result of evaluation (Do).
- Review the corrective action for possible effects on the product or QMS (Check).
- Change the QMS if required (Act).

Obviously this follows the PDCA cycle for nonconformities and is consistent with the rest of ISO 9001.

- **Tip** - If required, the risks and opportunities identified during planning will need to be updated after a nonconformity.
- **Tip** - Companies must retain documented information of:
  - Nonconformities found and actions taken as a result of these.
  - The results of any corrective action taken.
- **Tip** - This is a key part of ISO 9001 and auditors will be very interested in the corrective action process. It will not be enough to simply deal with the effects, i.e. react to the nonconformity and deal with the consequences. Auditors will look for evidence of some type of 'root cause analysis' (see Section 7.4 on cause and effect diagrams) to demonstrate the PDCA cycle for corrective action. This should be part of the retained documented information.
- **Tip** - Obviously, this process needs to be followed for any QMS nonconformities identified as part of internal or external audits.
- **Tip** - Preventive action was a specific requirement of previous versions of ISO 9001. The concept of risk-based thinking has now largely taken the place of preventive action.

## 10.3 Improvement

The final clause of ISO 9001 is one of the most important as it deals with continual improvement of the QMS itself. Companies should seek to continually improve the suitability, adequacy and effectiveness of their QMS and the quality information from Clause 9 provides the basis for this improvement. Companies need to show how this information is used to improve the QMS.

Companies need to decide what information is needed to verify that the QMS is working and how the source data for this information are to be collected.

After the data are collected, they should be analysed to provide information on the performance of the QMS and the opportunities for improvement.

**Continuous is not the same as continual!**

Both words describe duration but 'continuous' describes duration without any interruption whereas 'continual' describes duration that continues over a long time but with intervals of interruption, e.g., 'a continuous line of people' implies no breaks in the line but 'he made continual mistakes' does not imply that everything he did was wrong.

'Continuous improvement' implies improvement without any interruption at all. Can I go home now?

# 4.17 Documented information

## The write stuff

Sections 4.7 to 4.16 have outlined the basic requirements of ISO 9001 and creating, maintaining and retaining the required documented information is a key step in the successful implementation of ISO 9001. It is also the area that most companies fail to get right. They write (or have written for them) huge manuals and then suffer the burden of keeping these up-to-date and conforming to the statements in them. The best rule is to 'be general, be brief and be happy'. The less you say is the better in most cases.

- **Tip** - 'Boilerplate' manuals where all you do is substitute your company name are generally too complex and introduce too much documentation. They may seem an easy way to create the information required but in the long run they are not.
- **Tip** - Many consultants will put a 'boilerplate' ISO 9001 manual on top of existing functioning systems. This can lead to excessive work (but not for the consultant) to get the 'new' systems to work. It can also lead to disillusionment with the ISO 9001 process. Beware of low-cost consultants - you will get a 'boilerplate' system and find it difficult to get it to work.
- **Tip** - If you've been in business for more that 3 years, the chances are that you already have a QMS. It may or may not work effectively, it may or may not meet ISO 9001, but it will be there. This will be the processes that have been developed to operate the company and to service the customers.
- **Tip** - Existing systems can be adapted to ISO 9001. The best method is to document the existing processes and systems. These can then be considered against the ISO 9001 requirements and new elements introduced only where the existing processes do not meet the requirements. This is more work for the consultant but less disruptive for the company.

ISO 9001 requires documented information but does not specify how this should be set out, written or stored. This can be on paper (text or flow charts) or on computers (text or flow charts). This is an operational decision for the company.

| Maintained Documents (5) | Retained Records (23) |
|---|---|
| ■Scope of the QMS (4.3).<br>■Operation of processes (4.4.2).<br>■Quality policy (5.2.2).<br>■Quality objectives (6.2.1).<br>■Control of production and service provision (8.5.1). | ■Operation of processes (4.4.2).<br>■Monitoring and measurement Resources (7.1.5.1).<br>■Measurement traceability (7.1.5.2).<br>■Competence (7.2).<br>■Operational planning and control (8.1).<br>■Requirements review (8.2.3).<br>■Design and development:<br>• Planning (8.3.2).<br>• Inputs (8.3.3).<br>• Controls (8.3.4).<br>• Outputs (8.3.5).<br>• Changes (8.3.6).<br>■Externally provided products and services (8.4.1).<br>■Control of production and service provision (8.5.1).<br>■Traceability (8.5.2).<br>■Property belonging to customers/external providers (8.5.3).<br>■Control of changes (8.5.6).<br>■Release of products and services (8.6).<br>■Control of nonconforming outputs (8.7.2).<br>■Monitoring, measurement, analysis and evaluation (9.1.1).<br>■Internal audit (9.2.2).<br>■Management review (9.3.3).<br>■Nonconformity and corrective action (10.2.2). |

**The documented information required by ISO 9001**

The documented information required for ISO 9001 should be either maintained, e.g. process and procedure information, or retained, e.g. records of processes and procedure operation.

## Maintained documents

Several sets of documented information are required to be 'maintained' by ISO 9001 and these are listed on the left.

This documented information can be maintained via a 'company quality brochure' (see Section 4.8). This is a useful method of keeping all the maintained documents in one place and can simply refer to the relevant processes, procedures and documentation but not physically include them. As a summary document, it is relatively stable and does not need updating unless the company's activities and processes change considerably.

- **Tip** - Include an overall process map (see Section 4.5) to help interested parties, e.g. staff, customers, suppliers and auditors, understand how the company operates and to provide an essential overview of the QMS.
- **Tip** - The 'company quality brochure' is NOT the QMS. The QMS exists whether or not the brochure exists. The QMS is part of the operations of the company. The brochure simply documents, in one place, what the QMS is doing.

## Retained documents

There is still a large amount of documented information that is required to be 'retained' as records by ISO 9001 and these are listed on the left.

This might appear daunting at first but most of this information will already be available in a well-run company as part of the normal record keeping. Most companies will already be keeping these sort of records and ISO 9001 will not mean extra effort apart from making sure that the records can be accessed if required.

## Process information

Process maps (see Section 4.6) will provide most of the detail to support processes but procedure flow charts (see Section 4.6) are invaluable to specify the flow of actions, products or information in each process. They can give details of:

- What action is taken or checked.
- How the action is taken or checked.
- When the action is taken or checked.
- Who carries out the action or check.
- What is to be done if the process is not to the standard.
- What records are kept and where they are kept.

Flow charts can form the detailed description of the business processes. They will allow gaps in the system to be clearly seen and also highlight areas of the system which need review and improvement.

Most of the processes that need documentation will already exist as part of the company's operations, even if they are not written down. Simply creating the flow charts will, in most cases, show conformance to ISO 9001 (although it is recommended to cross-reference the processes to the ISO 9001 requirements).

- **Tip** - Flow-charts are good for documenting processes because they reveal the decisions and actions to be taken at each decision point.

## Works instructions

In addition to the process information, there is often a need for 'Works Instructions' for specific products and these are to make sure that everybody knows what they are expected to do at each work station or in each part of the process. These should be very specific and also provide information and instructions on what to do if the product or process shows faults or nonconformities. Works Instructions can be written as simple text, as a flow-chart or in any way that makes it easy for the person carrying out the job to follow the process.

- **Tip** - Works Instructions can include checklists to ensure that all the actions have been taken.
- **Tip** - Works Instructions can be designed to be copied, laminated and displayed near workstations. These are the 'front-end instructions' that the operators see each day as a constant reminder of their duties.

## Training manuals

Works instructions can also be used to generate a 'Works Training Manual' as a training aid and reference document.

- **Tip** - A Works Training Manual will not only add value and consistency to the system but can also show compliance with the competence requirements of ISO 9001 (see Section 4.12).

## Get it right

Keep the documented information simple to make the QMS easy to operate and maintain - documentation should be a help and not a hindrance.

Do not write/try to describe the 'ideal' process. This will lead to excessive work trying to change the actual process to match this ideal.

Instead, describe what is actually done in the real process and then check this against the requirements of the standard. The real process can then be changed to match the requirements.

In many cases, documents are continuously added to the system, but it is also necessary to remove documents from the system. Forms and procedures can become obsolete and redundant and when this happens they should be changed or removed entirely.

Management review should be part of the process of making the system as simple as possible and removing or reducing the costs of operating the system.

- Say what you do.
- Do what you say you do.
- Prove it.
- Improve it.

# 4.18 Quality planning

## Planning prevents problems

ISO 9001 places great emphasis on planning both for the complete system and for operations (see Sections 4.13 and 4.14). This does not have to be an arduous process and quality plans provide a simple method of both carrying out and recording the planning process to show conformance with the standard.

Quality plans can be at the global level, where the complete process is planned, or at the detailed level where the control for a specific product is planned. The uses are slightly different but the time spent planning is rarely wasted.

- **Tip** - As with any planning process, quality plans are 'live' documents that need to be updated to reflect any changes such as:
  - Product changes.
  - Process changes.
  - Inspection or method changes.
  - Additions or subtractions from the plan.
- **Tip** - Many companies carry out either global or local planning but few carry out both types of planning. This is a shame as both types of planning are useful for seeing the gaps in control.

## The global quality plan

At the global level, a quality plan summarises and specifies the procedures and resources applied to a product or process. It refers to the processes and provides the overview that is needed to check that the process is adequately controlled.

A global quality plan should specify:

- The activity.
- The task/operation.
- The interval/type of sample.
- Any reference documents to be used in the checking.
- The checks to be made.
- The person responsible for either carrying out the checks or ensuring that they are carried out.
- The location of the records that result (if any) from the check.

A sample of a global quality plan is shown on the right to show typical planning for an injection moulded company.

- **Tip** - The quality plan does not have to provide the detail of the checking and can freely refer to other documents. They are designed as a summary to spot any gaps in the control of quality.

Quality plans provide an opportunity to tie together all of the processes involved in quality.

## The detailed quality plan

At the detailed level, a quality plan summarises and specifies the methods and frequency of checking the significant characteristics of a product (including the requirements for checking before production release). The significant characteristics may be:

- Materials related and referenced to a materials specification.
- Dimensional and referenced to a drawing.
- Fit and function and referenced to a test method.

A detailed quality plan should specify:

- The product identity.
- Any customer-specified requirements, either dimensional or 'fit and function'.
- Any internally specified requirements.
- Details of what is to be checked.
- The frequency of checking.
- The sample size.
- The method of checking.
- The person responsible for either carrying out the checks or ensuring that they are carried out.
- The location of the records that result (if any) from the check.
- The action to be taken if non-conforming product is found.

**Note:** The detailed quality plan is only one element in ensuring that quality is built in to the product. It links to other processes such as the design, product and process FMEA (see Sections 8.11 and 10.2), gauge planning and other quality methods but again provides an overview of the methods used to ensure quality for the specific product.

- **Tip** - The quality plan must include input from the customer, design and development, purchasing and production.

Quality plans can refer to documents that are not part of the formal quality system but are simply part of the normal operations at the company.

| Activity | Task/operation | Interval/type of sample | Reference documents | Checks to be made | Person responsible | Where recorded |
|---|---|---|---|---|---|---|
| Customer details | Record of customer details. | All customers. | Account Application Form. | All information available and clear. | Accounts Manager. | Computer records. |
| Quotation details | Record of quotation. | All quotations. | Quotation process map. | Quotation details accurate. | Sales Office Manager. | Quotation Record File. |
| Order placement | Enter order onto computer system. | Each order. | Customer Order Form. | All information available and clear. | Sales Office Manager. | Computer records. |
| Supplier selection | Supplier assessment. | Initial and before ordering. | External provider Assessment Questionnaire. | External provider Assessment Questionnaire. | Purchasing Manager. | External provider Assessment Questionnaire. |
| " | " | Each delivery. | Purchase Specification.<br>Purchase Order.<br>Purchasing process map. | Inwards Goods Inspection as required by Purchase Specification and Quality System Procedure. | Warehouse Manager. | Purchase Order Files.<br>Delivery Note. |
| Purchasing | Order placement. | Each order. | Purchase Specification. | | Purchasing Manager. | Purchase Order. |
| Inwards goods | Inwards goods inspection. | Each delivery. | Inwards Goods Inspection Checklist.<br>Purchase Order Specification. | As required, quantity and condition. | Warehouse Manager. | Delivery Note. |
| Handling, Storage | Handling and storage of delivered items. | All products. | None. | All items to be handled and stored in a manner so as not to cause damage. | Warehouse Manager. | None. |
| Handling | Handling of product during processing. | All products. | Works Instructions. | All items to be handled in a manner so as not to cause damage. | All staff. | None. |
| Customer supplied materials | Handling of customer supplied materials. | Each supply. | As required by customer and process. | As required by customer and process. | Production Manager. | As required. |
| Injection moulding | Initial machine setting. | Before tool acceptance and production release. | Product QC drawing.<br>Machine operating instructions.<br>Machine setting instructions.<br>Scientific moulding instructions. | Check and determine machine settings for production running. | Production Manager.<br>Senior Machine Setter.<br>Inspector. | Machine setting records (held by Senior Machine Setter). |
| " | Machine capability study. | Before tool acceptance and production release. | Machine capability study instructions | Machine capability study with selected machine and tool. | Production Manager.<br>Senior Machine Setter.<br>Inspector. | Machine capability study records (held by Senior Machine Setter). |
| Injection moulding | Injection Moulding Machine (IMM) checking. | Weekly. | Works Instructions for selected IMM. | Calibration of all readouts. | Production Manager. | None. |
| " | " | Daily - At shift handover. | Works Instructions for selected IMM. | Visual check to Works Instructions for selected IMM. | Incoming Shift Manager.<br>Incoming Operator. | None. |
| " | " | In-process. | Works Instructions for selected IMM. | Visual check to Works Instructions for selected IMM. | Operator. | None. |
| " | " | Random. | Works Instructions for selected IMM. | Visual check to Works Instructions for selected IMM. | Production Manager. | None. |
| Injection moulding | Moulded product checking. | Production start. | Product QC drawing.<br>Machine setting sheet.<br>Product/machine release instructions. | Initial sample check of product for production run clearance. | Inspector.<br>Operator. | None. |
| " | " | Daily - At shift handover. | Product QC drawing.<br>Machine setting sheet. | Handover of running machines.<br>Visual check to Works Instructions | Incoming Shift Manager.<br>Incoming Operator. | None. |
| " | " | In-process SPC. | Product QC drawing.<br>Machine setting sheet.<br>SPC instructions. | Works Instructions. | Operator. | SPC charts for product. |
| " | " | Random. | Product QC drawing.<br>Machine setting sheet. | Works Instructions. | Inspector. | None. |
| " | Process capability study. | After stable running (and within 3 weeks of production start). | Product QC drawing.<br>Process capability study instructions. | Full process capability study with selected machine and tool. | Production Manager.<br>Senior Machine Setter.<br>Inspector. | Process capability study records (held by Senior Machine Setter). |
| Product packaging | Packaging check | Production start. | Product packaging specification. | Initial check of product packaging. | Production Manager. | None. |
| | | Daily - At shift handover. | Product packaging specification. | Check of product packaging. | Incoming Shift Manager.<br>Incoming Operator. | None. |
| Finished product | Despatch area. | All. | Final Inspection Checklist. | As per Final Inspection Checklist. | Warehouse Manager. | "PASSED" label indicates OK for despatch. |
| Delivery | Delivery of completed product. | All. | None. | All items to be handled in a manner so as not to cause damage. | Warehouse Manager. | Delivery Note. |
| Corrective action reports | Various. | All. | Product standards. | Normal requirements. | Production Manager. | Production Office. |

## A global quality plan for injection moulding of a simple product

A quality plan lists the relevant controls for a product. This provides an overview of the quality processes to check that there are no gaps. Note that many of the ‘quality documents’ are not stored as part of the quality system but as part of the normal operations of the company. The plan shown is the ‘global plan’ but product-specific plans can be generated for each product to list the specific checks (dimensions, etc.) to be made.

# 4.19 Risk-based thinking

## Making the sub-conscious visible

In the past, ISO 9001 viewed 'preventive action' as a method of preventing nonconformities before they happened and this was one of the most important clauses in previous versions. Preventive action is a form of risk-based thinking where the risk is assessed and action taken to reduce the risk of nonconformities. Therefore, risk-based thinking has always been present as an implicit consideration in ISO 9001. The 2015 version of ISO 9001 simply makes risk-based thinking an explicit part of the requirements in most areas (see box on the right).

One of the major functions of the QMS is to prevent nonconformities through management. In fact, the whole QMS can be seen as a preventive tool to manage risk to the company and the customer and it is only appropriate that risk-based thinking now runs through the complete standard.

- **Tip** - In quality management terms, risk is the uncertainty involved in providing customers with products that conform to specification and in enhancing customer satisfaction.

The concept of risk-based thinking may appear strange to many companies but this is because the process is carried out sub-consciously. Most companies already carry out risk-based thinking without thinking about it. Making this a visible process makes it more effective and increases the benefits of assessing risk.

- **Tip** - Risk-based thinking is required by ISO 9001 but there is no requirement for a formal risk management system or risk register. Formal risk analysis can be very useful for many processes (see Sections 8.11 and 10.2).
- **Tip** - An audit of risk-based thinking for ISO 9001 will not generally be possible as a separate step. It will be implicit in the whole audit and auditors will collect objective evidence throughout the audit to show that risk-based thinking has taken place.
- **Tip** - There are software 'solutions' for risk-based thinking but in most cases this can best be done by embedding risk-based thinking in the QMS.

## The risk-based thinking process

Risk-based thinking can be summarised as follows:

- Identify and analyse the risks and opportunities to decide:
  - What risks are acceptable.
  - What risks are unacceptable.
- Plan actions to:
  - Avoid the risk.
  - Accept the risk (for an opportunity).
  - Remove the source of the risk.
  - Reduce the severity of the risk.
  - Improve the risk controls.
  - Share the risk.
  - Decide to retain the risk.
- Implement the plan.
- Check implementation.
- Improve the plan.

This is simply the PDCA cycle applied to risk (uncertainty) and opportunities and is consistent with the ISO 9001 approach.

Unsurprisingly, there is an ISO standard on risk management. This is ISO 31000:2009 'Risk management - - Principles and guidelines'

Risk-based thinking should be built into the system so that risk is considered from the beginning to the end. The risks should also be reviewed after any significant change in circumstances.

Risk-based thinking makes preventive action an integral part of all strategic and operational planning.

### Where risk-based thinking appears in ISO 9001

- Introduction - Risk-based thinking is introduced and explained.
- Clause 4: The company is required to determine the QMS processes and to address risks and opportunities.
- Clause 5: Top management is required to:
  - Promote awareness of risk-based thinking.
  - Determine and address the risks and opportunities that can affect product conformity to requirements.
- Clause 6: The company is required to identify risks and opportunities related to QMS performance and take appropriate actions to address them.
- Clause 7: The company is required to determine and provide the necessary resources.
- Clause 8: The company is required to manage operational processes.
- Clause 9: The company is required to monitor, measure, analyse and evaluate the effectiveness of actions taken to address risks and opportunities.
- Clause 10: The company is required to correct, prevent or reduce undesired effects and improve the QMS and update risks and opportunities.

**Note:** If ISO 9001 uses the words 'suitable' or 'appropriate' then this is a good hint that risk-based thinking is required.

## Showing risk-based thinking

### Using existing information

Risk-based thinking can be shown by:

- SWOT and/or PEST analysis.
- Meeting minutes.
- Customer feedback.
- Competitor or market analysis.
- Company planning processes.
- Management review.
- Staff competence reviews and plans.
- Process maps which include risks and opportunities.

All of these data will already be in existence, it is simply a matter of explicitly considering risk in the processes.

Proving that risk-based thinking is working (the actions) can be shown by:

- Revision or updating of company objectives.
- Action plans (and completion) for:
  - Improved QMS performance.
  - Improved staff competence (training plans).
  - Improved resources.
  - Process risk reduction (improvement plans).
  - Improved monitoring and measurement performance.
  - General improvement plans, e.g. infrastructure, etc.

All of these data will already be in existence and can be used to demonstrate action to reduce risk but it is better if risk reduction is mentioned in the action plans.

### The risk register

This approach is more formal than using existing information but is more explicit in terms of auditing. The risk register is a document or spreadsheet that describes:

- Description of the risk.
- Risk type (company, project, process).
- Probability of the risk event occurring (high, medium or low).
- Severity of the effect if the risk event occurs (high, medium or low).
- Actions to be taken to reduce the risk, this can be either reducing the probability or the severity.
- Who is responsible for action.
- Status and date completed.

A risk register can be company-wide or local but a company-wide document is better to show compliance with all the various clauses of ISO 9001. The risk register needs to be kept up to date to remain relevant.

- **Tip** - Risks can (and should) be removed from the register when reduced to acceptable levels as decided by the company.

### FMEA

A more formal method than the risk register is Failure Modes Effects Analysis (FMEA). This was developed to assess risk both in design and development and in processes (see Section 8.11 and Section 10.2). However, it can be used for general risk management and is proven to reduce risks. Risk assessment through the FMEA procedure enables:

- Concentration on taking action to reduce major risks.
- Effective use of resources.
- Avoiding having to try to deal with all risks, i.e. insignificant risks can be excluded.

FMEA is perhaps 'overkill' for general business risks but it is an excellent for design, development and processes.

- **Tip** - Use a 'risk register' for the general risks and FMEA for the specific risk areas of design, development and processes.

The use of SPC trend analysis can also be classed as risk management, i.e. if the trend indicates that nonconformities will be produced in the future then action on the process would be 'risk-based thinking'.

Never confuse 'risk' with 'uncertainty'.

Risk is quantifiable and the use of probabilities and statistics can help us to assess a risk level.

Uncertainty is not quantifiable and it is not possible to know anything about the probabilities or the statistics or to assess a risk level.

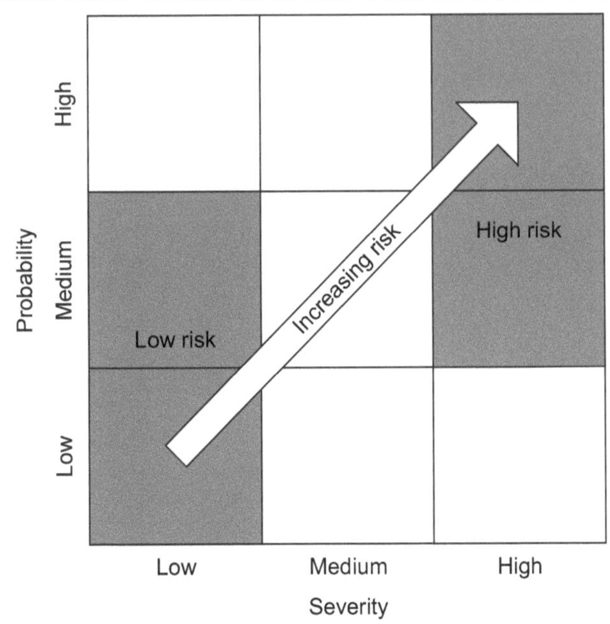


### A quick method of assessing risk

A simple 3 x 3 matrix can rapidly show where risk is and help to set priorities. Identify a risk, consider the potential effect (high, medium or low), consider the probability of the risk (high, medium or low) and you have an immediate assessment of the risk.

# 4.20 Internal auditing

## Say what you do

The audit process is one of the most important processes in maintaining an effective QMS. ISO 9001 (see Section 4.15, Clause 9.2) requires a process to internally audit the efficiency, effectiveness and maintenance of the QMS. The audit process ensures that the process maps and procedures 'say what they do' and that the process owners are 'doing what they say they do'.

Auditing is:

- Not difficult or complicated.
- Common sense.
- About being sceptical until shown the evidence.

## The purpose of auditing

The audit process checks that the QMS is operating as documented and as intended.

- **Tip -** Unless a QMS system is audited and the results fed back to the people then nothing will happen.
- **Tip -** Auditing is the inspection process to ensure that people react as we want them to.

Auditing also has several other purposes and these are:

- Convergence of manual and process: In developing a QMS there will always be differences between what the system says and what people do. This is because people are slow to change, they do things in different ways, new systems take time to become established and even because the QMS procedures were wrong or incomplete. A good audit will find these differences and provide the mechanism to align the written and the actual processes until they are the same.
- Foundations for change: The audit process provides the foundations for change and allows modification of the QMS as the business needs change.
- Expansion: The audit process allows the QMS to expand to include new areas and new requirements.
- Simplification: A less common but an even more valuable use of the audit process is to simplify the system. The audit process can be used to:
  - Remove operations or forms that no longer serve a useful purpose.
  - Combine forms to reduce numbers.
  - Make sure that information is only recorded once (at the point of origin).
- **Tip -** Simplification of the system can be the most valuable aspect of auditing.
- **Tip -** Auditors benefit from knowing the audit process and are better prepared for an audit of their own processes.
- **Tip -** You cannot audit yourself. Do not even try.

Audits should also assess the effectiveness of the QMS and the overall business performance, i.e. does the procedure meet the intent of the system and is it beneficial to the business.

## The process of auditing

### The audit team

The audit team should consist of at least two people. Two people can reinforce one another, provide more questions and different views on treating any concerns.

- **Tip -** For more experienced auditors this can be reduced to a single person.

'People do not react as you expect, they react as you inspect'.

Internal auditing is not a chance to get even or show how much you know. Remember 'Do unto others as they will do unto you' - next time they could be auditing you! Audit is a question session where the auditor asks questions and looks for proof that the documented information is being followed. It is not a chance to trip someone up. It is a chance to understand what happens and to get a better knowledge of the business.

Auditing can not only help make the current system run but it is a vital part of getting better as the competition gets stronger.

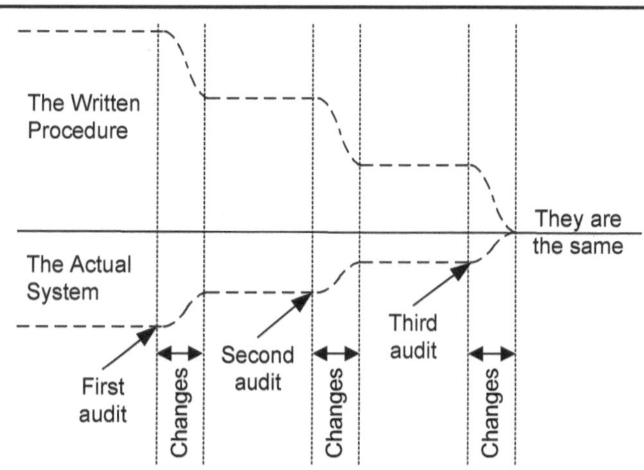


Convergence - the changes through time with auditing

Regular auditing will lead to changes in both the system and the documents until the documented information describes what the actual system does. This is an opportunity to both simplify and improve the system.

## The auditor's questions

The most important auditor phrase is 'Show me' and this is used whenever the process owner makes a statement. It is not a command but a request that the process owner shows the auditor the documents, files, forms or other information to back up what is being said. Only when you have seen the documents can you really say that you have audited the procedure.

### Follow the process/procedure

Follow the process/procedure from start to finish and look for those points where a record is kept or an action performed. Ask to see the proof that these have been done. When it has been proved then move on to the next point. If there are multiple documents involved then look at a random sample to make sure that they are completed correctly.

### Use the audit checklist

If the process/procedure has an audit checklist then use this as guidance on what to look for. This is the minimum to ask for and can be expanded as necessary.

- **Tip -** The reaction of the person being audited is a key indicator. If they are evasive or unsure then they probably know that they are not doing it right. Keep asking questions at this stage but do not let it become emotional.

## The audit report

The audit report should be completed via a form which contains details of:

### Details of nonconformity (auditor)

This should include:

- The evidence of nonconformity.
- The ISO 9001 requirement.
- A clear statement of the nonconformity.

This should be signed and acknowledged by the auditee.

### Action plan (auditee)

**Root cause analysis**

This is the root cause analysis to determine the reason for the nonconformity.

**Correction and completion date**

This is what is going to be done now to fix the nonconformity.

**Corrective action and completion date**

This is what is going to be done in the future to prevent the nonconformity.

Typical actions could be:

- Ensure that the procedure is followed in future by:
  - Instructing others to ensure the procedure is followed.
  - Putting in relevant controls to make sure the procedure is followed.
- Modifying the procedure to fit reality.

### Review of action plan (auditor)

Auditors should review the action plan to check that it adequately addresses the nonconformity.

### Verification of the action plan (auditor)

This should be an assessment of the implementation and effectiveness of the action plan by the auditor. Auditors should also check that the concern is no longer present. Verification can take place immediately or wait until the next scheduled audit for follow-up. This decision is taken based on the importance of the nonconformity.

### Close out

The audit report can then be closed out, dated and retained.

- **Tip -** Where possible change the written procedure rather than the actual system. Unless a vital part of ISO 9001 is being omitted then it is easier to change the words than the system.

Nothing in an audit report should come as a surprise to the process owner when they read it!

Internal auditors should also have an understanding of ISO 9001, rather than simply focusing on the written procedures.

It is also possible to make comments for the process owner in an audit report. Examples might be:

- 'The area is untidy and needs cleaning - quality production in this area is in danger'.
- 'Why don't we get rid of this form it serves no real purpose any more?'

These can be informal but are a good method for improvement suggestions. Do not ignore them; today's comment may be tomorrow's nonconformity.

### Classification of nonconformities?

Some auditors classify nonconformities as major, minor, etc. They are all nonconformities so why use different classifications? The relative importance can be indicated by the completion date:

An important nonconformity will:

- Cost a significant amount of money.
- Directly affect the customer or the company's reputation
- Put ISO 9001 registration is at risk.

These should be completed within 6 weeks (immediately in serious cases) and follow-up and close out should happen directly after corrective action.

A less important nonconformity will:

- Involve human error - e.g. the majority of forms are filled out correctly but one or two have been forgotten.
- Not directly affect the customer or the company's reputation.
- Be where the system is not being followed to the letter but an attempt is being made.

These can have a longer completion date and follow-up and close out actions can take place at the next audit.

# 4.21 External auditing

## A valuable external view

An external audit is essential for external certification and is one of the most important parts of the process of certification.

External audits now take a broader view than the old 'checklist' approach, although some auditors still use them and the Auditing Practices Group (see sidebar on the right) has produced a generic list of auditing questions (see their paper on 'Processes').

- **Tip** - The aim of an auditor is to assess the effectiveness of the QMS and the organisation's overall performance. They are not the enemy and are not there to trip you up.

## The process approach

The focus of the audit is on processes and their effectiveness and efficiency.

The process approach requires the development of a global process map (see Section 4.5) and then associating the individual elements of ISO 9001 with the main business processes. Compliance with the standard is verified by checking the processes and their implementation.

- **Tip** - A spreadsheet (see Section 4.7) can make it easy for the auditor to assess compliance. Do not make auditors guess which processes cover which parts of the standard - they do not know your system as well as you do.
- **Tip** - Some processes may be used to show compliance with more than one part of the standard and some parts of the standard may be met in more than one process. There need not be an exact 1:1 relationship between processes and the standard.

## What are auditors looking for?

Any QMS is large and complex and auditors will not have time to audit the complete system in one visit. They will therefore generally concentrate on specific areas. They are looking for proof that the QMS meets ISO 9001 and is also effective and efficient. The main points that they will focus on are:

### Quality policy and objectives

ISO 9001 requires a quality policy and measurable objectives that are known and understood by all staff. Staff also need to be able to understand how these relate to their job and how they can help achieve quality. Auditors will look not just at the policy and objectives but also check that staff have the required awareness and understanding of the relevant objectives.

Interviews of this type will tell an auditor a lot about what a company is really doing to communicate the message of quality and how the strategic message is communicated through the company.

- **Tip** - This does not mean that staff have to 'parrot' the policy and objectives. They can explain in their own language but they need to understand what the policy and objectives mean in terms of their job.

### Progress on quality objectives

Auditors will look at the quality objectives to verify their achievement (or not). Objectives should be SMART (see sidebar in Section 4.10) and auditors will need to see measurable and definite progress. The objectives should also be regularly updated to reflect changes in the company and may be modified or deleted when achieved.

- **Tip** - Auditors are looking for real progress towards the quality objectives. It is one of the few things that measures the effectiveness of the complete QMS.

### Leadership

Leadership can be shown via the management review process but it is better if the management review is conducted as part of an overall company-wide management review. This integrates the QMS into the management of the company, as ISO 9001 is designed to do.

- **Tip** - Auditors will want to see that reviews are carried out, that reviews consider a full range of inputs and outputs, that reviews are recorded and that the results of the reviews are communicated to staff.

### Customer focus

Auditors will also look carefully at the methods a company uses to focus on the customer. This is mainly in 3 areas:

- Customer requirements: Auditors will look at the methods used to define the

Old-style audits were about ticking the boxes of the standard.

This was fundamentally changed with the process-based approach.

At least one internal audit of each process/procedure and a full management review should have been completed before any external audit is carried out.

If these have not been completed then it is unlikely that there will be sufficient internal records to show compliance.

One of the best resources for understanding what auditors want to see is the series of guidance papers issued by the ISO Auditing Practice Group. These are freely available from a variety of sources including: www.irca.org/en-gb/resources/Auditing-papers.

It is strongly recommended that the full set of guidance papers is downloaded and read before applying for external audit.

customer requirements (stated, unstated, legal and regulatory and internal) and how these are documented. Auditors will also check the accessibility, completeness and change control for the customer's requirements.

- Customer satisfaction: Auditors will look at the methods used to define and collect information. Even more importantly, they will look at what happens to this information, i.e. how is it used and is it used effectively? Monitoring customer satisfaction is worthless unless management acts on the information to meet the company's objectives.
- Customer complaints: Auditors will look for a systematic approach to complaint handling and a process that attempts to not only resolve the specific complaint but also take action to prevent reoccurrence.

### Nonconformities

How nonconformities, both product and system, are treated and the methods used gives an insight into how methodical a company is. Nonconformities need good control and failures here can be very damaging to a company.

- **Tip** - Auditors look for a link between the nonconformities and corrective action.

### Risk-based thinking

Companies should be prepared to provide information and records of their risk-based thinking (see Section 4.19), the output of the thinking and what actions have been taken to reduce risk and maximise opportunities.

- **Tip** - Auditors often look for a link between the risk-based thinking and improvement process (see Section 4.16).
- **Tip** - Disaster recovery planning can be considered as part of risk-based thinking.

### Improvement

Companies need to be able to demonstrate improvement in products, processes and in the QMS. Improvement should be across the company and not simply in production areas but, above all, it should be measurable and measured. Information from customer feedback, from nonconformities (and action plans) and from process monitoring can all be used to show improvement but there must be evidence of improvement.

- **Tip** - There is no requirement to improve everything at once. Improvement should be targeted on the important issues.
- **Tip** - Evidence is needed here - nice words will not be enough.

### Internal audit

Internal audit is part of the evaluation and improvement process and an effective internal audit programme is an essential part of the process.

For some processes there may not be a documented procedure. If a process does not have a related procedure then it is still possible to carry out an effective audit. Simply follow the process (the flow charts referred to in Section 4.6 are invaluable here) and note:

- The objectives of the process.
- The inputs and outputs of the process.
- The consistency of the process.
- If the objectives are being achieved or not (look at the outputs for evidence).
- The information flow.
- Assess if the process performs effectively and efficiently.
- Assess if it meets the requirements of the standard.

These notes will show an effective internal audit process and will help the auditor to audit similar processes.

Never ever lie to an auditor. The consequences will be disastrous.

If your system is faulty then 'take it on the chin' and confess quickly.

I once forgot to include a whole group of products in the purchasing specifications and the auditor found my error. I confessed quickly (and profusely) and got a nice long completion date to fix it.

### What do they want to see?

The first things an external auditor will want to see is the 'maintained' documented information:

- Scope of the QMS (4.3).
- Operation of processes (4.4.2).
- Quality policy (5.2.2).
- Quality objectives (6.2.1).
- Control of production and service provision (8.5.1).

The auditor will probably request sight of these before they even agree to visit.

The company quality brochure (see Section 4.17) will provide most of this information in one place.

It is also useful to have:

- A global process map for the main processes showing their interaction (including the main inputs and outputs).
- Procedures (or references to them) for processes where the company has defined them.

An external auditor will also need to see the retained records (see Section 4.17) showing that the requirements of ISO 9001 have been followed.

If you do not have all of these then do not start an external audit.

# 4.22 Quality management standards - where are you now?

## Is it working for you?

This assessment of quality management standards is not about meeting the requirements of the standard - that is assumed to be part of any external certification. This assessment is about meeting the often unstated objectives of a quality management system:

- The QMS must function as an effective method of assuring quality rather simply as yet another logo for the company letterhead.
- The QMS documentation must be fit for purpose, be easy to understand and modify and be available to all staff.
- The QMS should support and drive improvement.
- External certification of the QMS should be seen as an opportunity rather than a threat.

These are the elements of a QMS that will determine if the system is effective or simply another administrative burden.

## Completing the chart

This chart is completed and assessed as for those presented previously.

Documents make the difference.

Good documentation makes updating easy and makes the system more relevant.

**Quality management standards**

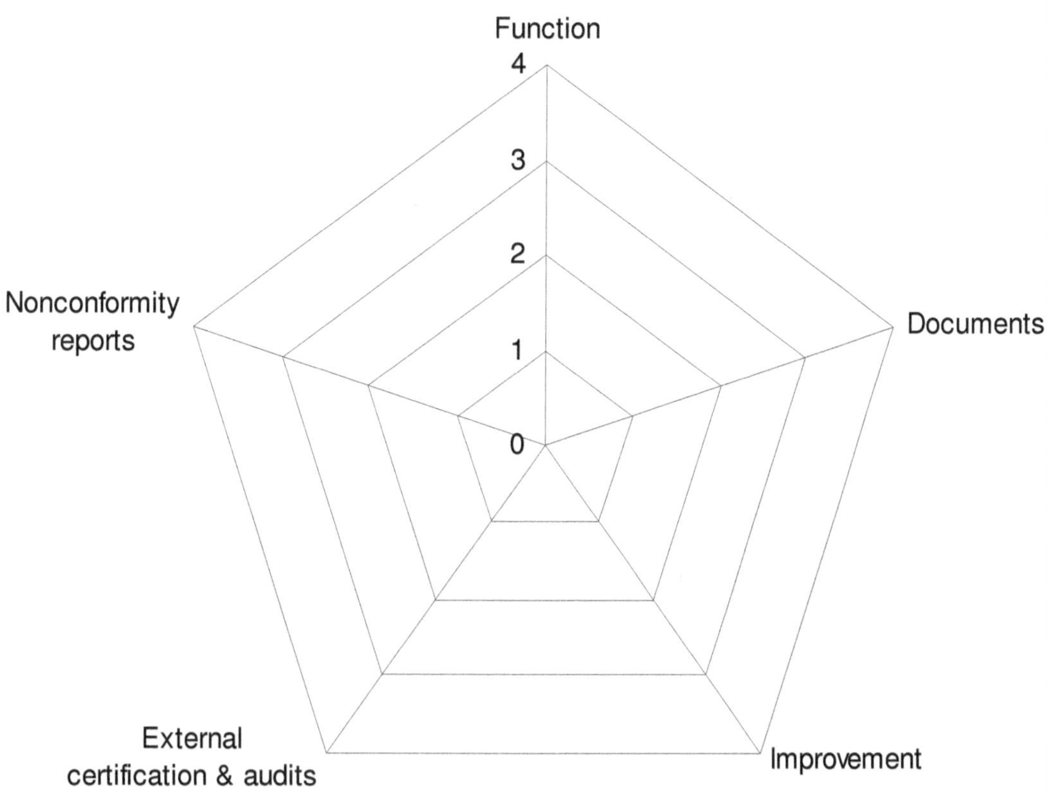


The QMS needs to become a tool for improvement.

If it is simply seen as a 'tick box' exercise then you have wasted your time and money.

### Use the scoring chart to assess where you are in quality standards

The numbers from the scoring chart can be transferred to the radar chart for a quick visual assessment of where you are in terms of quality management standards.

Having an externally certified QMS is not the same thing as having an effective QMS.

| Quality management standards | | | | | |
|---|---|---|---|---|---|
| **Level** | **Function** | **Documents** | **Improvement** | **External certification & audits** | **Nonconformity reports** |
| **4** | QMS is seen by all staff as a method of assuring quality & assisting in quality improvement. QMS is seen as a tool rather than as an imposition. | Documentation is minimised, flexible & designed to be easily modified & updated. All users have access to all documentation & ability to update. | QMS is often used as a tool & driver for improvement. QMS always provides good evidence & system for improvement. | External certification obtained & kept up to date via regular external audits. External audits treated as an opportunity to learn & improve processes. | No nonconformity reports raised in the last complete system audit. |
| **3** | QMS is seen by most staff as a method of assuring quality. QMS is seen as assisting the achievement of quality but not as a tool. | Documentation is relevant, flexible & designed to be easily modified. Users have access only to selected sections of documentation & ability to update. | QMS is sometimes used as a tool for improvement. QMS sometimes provides evidence for improvement. | External certification obtained & kept up to date via regular external audits. External audits treated as a simple check. No effort made to learn from the process. | Less than 3 nonconformity reports raised in the last complete system audit. |
| **2** | QMS is seen by few staff as a method of assuring quality. QMS is not seen as assisting the achievement of quality. | Documentation is moderate in size & inflexible. Users have access only to restricted sections of documentation & no ability to update. | QMS is rarely used as a tool for improvement. QMS cannot provide evidence for improvement. | External certification obtained & kept up to date via regular external audits. External audits treated as an imposition to be endured. | Less than 5 nonconformity reports raised in the last complete system audit. |
| **1** | QMS is seen as a paperwork exercise only & irrelevant to assuring quality. QMS is seen as an imposition that gets in the way of real work. | Documentation is large, clumsy & difficult to modify & update. Access to documentation is restricted. | QMS is never used as a tool for improvement. Any improvement takes place outside the QMS. | External certification obtained but audits are not regular. External auditor treated as an enemy & misled about actual system. Documentation updated only for audit process. | More than 5 nonconformity reports raised in the last complete system audit. |
| **0** | QMS is simply a method of gaining another logo. QMS process is driven by customers who state: 'if you are not certified then we won't buy from you'. | Documentation is large, clumsy & difficult to modify & update. Access to documentation is highly restricted. | QMS is not considered relevant for any improvement efforts. Improvement not part of the site ethos. | No external certification achieved for site. | More than 10 nonconformity reports raised in the last complete system audit OR No system audit undertaken in the last 12 months. |
| **Score** | | | | | |

# Key tips

- A QMS is a system and must be designed as a system.
- Define what you want your QMS to deliver before starting out.
- Set the objectives of the QMS.
- Define the savings and benefits (even if they are only about the logo).
- A QMS to ISO 9001 provides a framework for the QMS but must be specific to the company and associated with product quality standards and control methods.
- Installing a QMS to ISO 9001 does not guarantee that the system will work to improve quality.
- ISO 9001 is based on 7 quality management principles and on the Plan-Do-Check-Act cycle.
- Current ISO 9001 requirements are for a process approach rather than the procedure approach previously used.
- The process approach requires mapping and defining the processes used in the company.
- Map the global processes used at a company before attempting to map the individual processes - it is essential to get an overview first.
- Process maps can be by operational areas or any other convenient division.
- Flow charts are acceptable to document procedures and can be better at showing the actual information flow than 'text-based' documents.
- Make it easy for compliance checking (and auditors) by tracking processes to ISO 9001 requirements.
- Some processes may cover more than one requirement of ISO 9001.
- ISO 9001 has organisation context, leadership, planning, support, operational, performance evaluation and improvement requirements.
    - 'Organisation context' is about understanding what the company is doing and why.
    - 'Leadership' is about providing top-management support for quality.
    - 'Planning' is about addressing risk and opportunities and establishing the objectives of the QMS.
    - 'Support' is about providing the resources and conditions to deliver quality.
    - 'Operation' is about actually designing and producing the product to meet the customer's requirements.
    - 'Performance evaluation' is about assessing if the QMS is delivering what it is designed to do.
    - 'Improvement' is about getting better.
- Documenting a QMS needs several layers of information ranging from the quality policy to the works instructions and training information.
- Documentation should be kept simple and easy to operate - use flow charts for documentation where possible.
- Quality planning needs both a global plan for the overall process and a detailed plan at the product level.
- Every process needs a quality plan to specify what is being checked.
- Corrective action is about stopping errors that have already happened from happening again.
- Risk-based thinking is about assessing risks and reducing the possibility of errors happening before they occur.
- Internal auditing should be regular and should be used to improve both the processes and the system.
- It is easier to change documents than to change the process.
- Internal auditing can help improve the current system run better.
- Nonconformity reports are opportunities for improvement.
- External auditing is a valuable external view, it should be viewed as an opportunity to improve and not as a threat.
- External auditing is no longer simply about ticking boxes, it is about processes and their effectiveness and efficiency (not the same thing).
- External auditors are there to help you improve.

# Chapter 5

# Statistical process control (SPC)

Statistical process control is one of the most important components of effective quality management. It acts as a 'feed forward' control to allow a processor to actually control quality rather than the typical inspection-based approach which is reactive and after the event.

The sad thing is that SPC has been around for many years and yet has never really been taken seriously by the plastics processing industry. Individual companies have used SPC to great effect and have gained control of their process but the industry as a whole is still remarkably reluctant to use the methods and they are not often seen at processors.

This is, in part, due to widespread misunderstanding of how the process works and the very quick benefits that it delivers. Most of the standard texts on SPC are written by statisticians who move very quickly over the basics and get very involved in the mathematics and formulas without explaining how SPC works in practice. Many people have a fear of mathematics and their eyes glaze over at the first sight of an equation. In reality, operating SPC requires nothing more than a basic grasp of arithmetic and the ability to see patterns in the results that are plotted as a result of this. Setting up a basic SPC system does need some higher skills but most of it is very simple and still only needs basic arithmetic.

Anybody who can operate Excel can set up an SPC system and reap the benefits almost immediately.

SPC naturally involves numbers and many off-the-shelf and bespoke systems have been developed (even I developed one in the dim and distant past). Many of these are excellent (sadly, not mine) and when the amount of data becomes large they can be useful to improve data handling. Their fundamental problem is that they take away all of the basics and present the users simply with numbers and little background information. This can divorce the user from the actual process and, by reducing interaction with the actual process, can reduce the value of the SPC system.

Do not be frightened by the concept of SPC, it is simple to apply and gives huge benefits in terms of process control.

For the benefit of those whose eyes do glaze over at the sight of an equation I will try to keep the mathematics to a minimum (although it is really elegant mathematics). I will show some equations but you do not need to understand these at all, simply skip them if in doubt.

I will present some of the background information on the statistics but, again, this is not essential and will be kept to the minimum - the important point is how SPC can be put into operation and how it can quickly reveal what the process is doing.

"Build a man a fire, and he'll be warm for a day. Set a man on fire, and he'll be warm for the rest of his life."

**Terry Pratchett**

# 5.1 What is SPC?

## The development of SPC

SPC is not a new process and was developed many years ago (see Section 1.4) but implementation in plastics processors has been largely driven not by the desire to improve quality but by the demands of customers. One of the major drivers was Ford in the 1980's as they sought to improve their quality to match that of the Japanese. SPC is now being adopted in areas as diverse as the UK National Health Service to monitor projects and outcomes and increasingly in service industries. SPC is a well-developed tool that plastics processors can use to improve processes and quality.

## How does SPC work?

SPC works by measuring the characteristics of a process to control the output of the process. This focus on controlling the process rather than trying to inspect the output of the process is one of the most powerful techniques available to plastics processors.

The process is shown on the upper right and the difference between SPC and conventional quality control is clear.

SPC is about understanding and controlling the process to improve efficiency, improve quality, reduce waste and cut costs. The basic idea of SPC is that information on the performance of the process is gained from the inspection of a small number of products. This inspection of products is NOT designed to give information on the product for a pass/fail decision. It is designed to give information on the process so that the output of the process can be controlled.

This is a critical point and is often misunderstood. The inspection needed for SPC is often thought to be similar to the inspection for conventional quality control but it is fundamentally different, it is process control and not product control and if the process is in control then product control automatically follows.

## Defect detection vs. defect prevention

SPC is about defect prevention by process control whereas conventional quality control is about defect detection. The

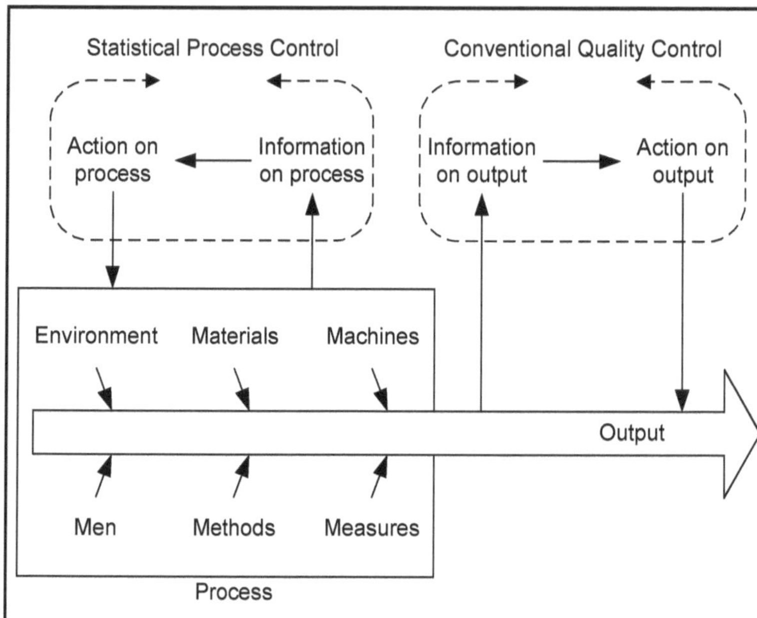


### Control the process and you control the product

SPC acts by controlling the process rather than by trying to control the output of the process (the focus of traditional inspection). Inspection is product-based but SPC only uses products to tell us about the process.

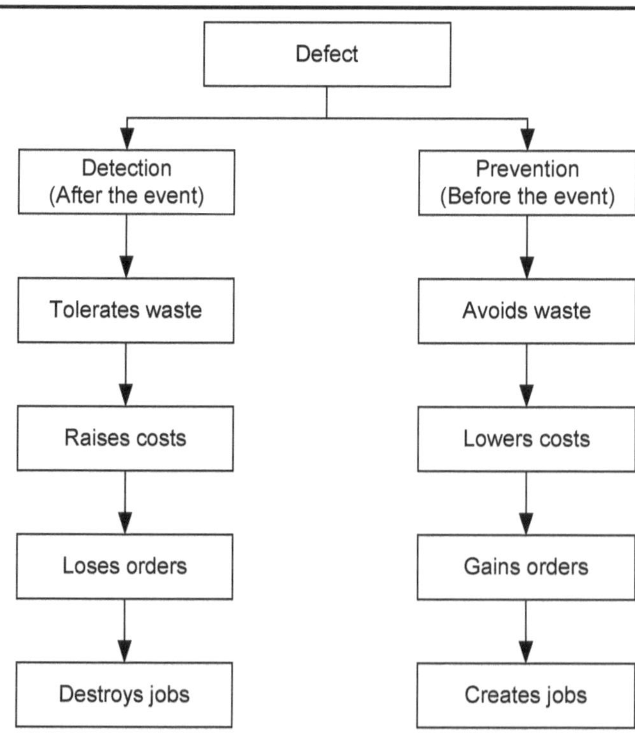


### The difference between detection and prevention

Detection is backward-facing: it attempts to find unacceptable output after it has been produced and to separate this from the good output. Prevention is a forward-facing strategy: it improves quality by seeking to improve or correct the process.

results of defect detection and defect prevention are shown on the lower left.

Defect detection tolerates waste by rejecting bad products that have already been produced, raises costs by wasting materials, time and labour, loses orders from increased costs and potential poor quality and eventually destroys jobs and companies. Defect detection is backward-facing and can be likened to steering a car by looking only in the rear-vision mirror.

Defect prevention avoids waste by controlling the process to control the product, it lowers costs by ensuring that all the products produced are fit for purpose, it gains orders from lower costs and consistent quality and eventually creates jobs and profitable companies. Defect prevention is forward-facing and is where we should be looking if we are steering a car.

## Does it work?

SPC does work and it is effective in a range of industries because processes are fundamentally all the same. The only question is 'What should we control to control the process?' SPC has been proven in all types of industries; it has been used extensively in the automotive industry using a variety of manufacturing processes ranging from casting to plastics processing from basic manufacturing through to assembly processes.

This is a method of controlling plastics processing that I have used extensively in extrusion, injection moulding, blow moulding and assembly to control processes and products.

This is not a theoretical method but an intensely practical method for controlling product quality.

## What about all the mathematics?

SPC relies on mathematics, but it is more accurate to call it statistics and the foundations of SPC are well proven in statistical terms. Knowing the detailed statistical theory helps to understand the operation and basics of an SPC system but even this is not necessary to operate the systems or to gain the benefits. It is entirely possible to run a fully functioning SPC system with absolutely no knowledge of mathematics or statistics.

At the operational level, the only requirement is some basic arithmetic, i.e. the ability to add, subtract, multiply and divide. In many cases, it is possible to use dedicated systems to do most of the arithmetic and the only requirement is then the ability to see any underlying patterns in the results, a willingness to learn and an understanding of the process itself. This is the minimum that we would expect of our employees.

SPC identifies potential faults before any inspection takes place.

Despite this, the benefits of SPC can be improved by providing basic statistical training to all employees so that they gain an insight into what is happening and why they are doing the calculations that they are doing.

## Defect or nonconformity?

From this point onwards we will try to talk in terms of 'nonconformity' rather than 'defect'. The reason for this is that 'defect' can have a legal meaning and get the lawyers excited. 'Nonconformity' simply means not conforming to specification or process limits and does not imply a 'defective' product or any product liability issues. For the same reason, quality people are advised to discuss 'concerns' rather than 'problems'.

SPC is the use of statistical methods to identify, monitor and control the inevitable variations that occur in any manufacturing process.

### So long ago and still so true!

"The long-range contribution of statistics depends not so much upon getting a lot of highly trained statisticians into industry as it does on creating a statistically minded generation of physicists, chemists, engineers and others who will any way have a hand in developing and directing the production processes of tomorrow."

**Shewhart and Deming, 'Statistical Method from the Viewpoint of Quality Control', 1939**

### Some initial definitions (but don't worry about them)

Variables

$X$ : an individual measurement.

$\bar{X}$ : the average or mean of a sample of measurements.

$\bar{\bar{X}}$ : the average or mean of the whole population.

$\tilde{X}$ : the median, the middle value of a sample of measurements.

s: the standard deviation of a sample of measurements.

$\sigma$ : the standard deviation of the whole population.

R : the range of a sample of measurements (largest - smallest).

$\bar{R}$ : the average range of a sample of measurements.

Attributes

N: the batch size.

n : the sample size.

f : the number of nonconforming items allowed in a sample.

c : number of nonconformities allowed in a sample.

# 5.2 Variable and attribute data, special and random variation

## Types of data

The phrase 'SPC' is used as a global term to cover a range of tools and techniques that have some common fundamentals:

- They seek to control the product by controlling the process.
- They use statistical analysis and tools.

However the tools and techniques used in SPC vary depending on the type of data being examined. The two types of data are:

- Variable data: These are data which can be measured and may include variables such as dimensions, weight or any other continuous variable.
- Attribute data: These are data which cannot be expressed as a number or measure but are either acceptable or not acceptable, i.e. go/no go, good/bad, correct/not correct, present/not present.

These two types of data are discussed below and in greater detail later in this chapter.

## Variable data

This type of data is continuously variable (subject to the accuracy of the measuring device) and is probably the most common application for SPC in plastics processing. Variable data will have numerical values of the form 0.1, 0.2, 0.3, etc. and the values are used as the basis for the statistical analysis. As a general rule the number of samples measured to get variable data is quite small.

Measuring the weight of an injection moulding would produce variable data and would use the techniques suitable for variable data.

## Attribute data

This type of data is of the 'Yes' or 'No' type where a nonconformity is either present (Yes) or not present (No). Attribute data are not numerical or measured and do not have a value. The number of samples that do not meet the requirements is used as the basis for the statistical analysis. As a general rule the number of samples assessed to get attribute data is large and much larger than the number of samples used for variable data.

Assessing the colour of an injection moulding as acceptable or not acceptable against a colour sample would produce attribute data and would use the techniques suitable for attribute data.

Control by variables uses measurements to control the process.

Control by attributes uses assessment (accept or reject) to control the process.

## Common and special causes

Despite what most manufacturing people would hope and like to believe, every process (and the output product) varies. Variability is a fact of life and SPC both acknowledges this and provides the tools to deal with it.

The variation in a process can be of two distinct types and these need to be separated to allow the process to be managed and improved. These two types of variation can be termed 'common' and 'special' causes.

## Common causes

Common causes are the variations in the process that are due to natural random causes. These variations are predictable and are characteristic of the limitations of the current process. This type of variation can never be totally removed or corrected but it can be reduced by process improvements and SPC provides a measure of the size of the normal random variation in the process.

Common causes result in predictable variations.

Special causes result in unpredictable variations.

SPC accepts the presence of common causes and a process is considered to be 'in control' if the only variation is from common causes. If a process is 'in-control' then the performance of the process can be predicted.

**Note:** 'In-control' should not be accepted as preventing process improvement. In fact, the measurement of common causes and their reduction provides an excellent measure of progress made in process improvement.

## Special causes

SPC is designed to detect the presence of special or assignable causes and any changes in the process. Special causes are the significant and identifiable causes or events which can lead to the production of products that do not meet the specification. SPC for either variables or attributes looks for patterns that are not random and then provides the tools to identify and rectify the special cause.

If a process is 'in-control' then the performance of the process can be predicted.

If a process is 'out of control' then the performance of the process cannot be predicted.

The best method for finding the special cause is to use the 'cause and effect chart' (see below and Section 7.4). This looks at the most common causes of variation and these are:

- Men: the things that people do will have a large effect on the product quality, e.g. are operators careful or are they slipshod, do they care about the job and the product?
- Methods: the ways that people do things will have a large effect on the product quality, e.g. people may think they all work to the same procedures but do they? Does every operator have a different method of assessing the product?
- Machines: the reliability and consistency of the machines we use will have a large effect on the product quality, e.g. is the machine right for the job? Is it an old machine that breaks down often?
- Materials: the types and consistency of the materials we use will have a large effect on the product quality, e.g. is the material the right one? Did the buyer get a discount for 'off-specification material?
- Environment: the environment will have a large effect on the product quality, e.g. is the factory temperature consistent? Does the water temperature vary with the outside temperature?

A process is considered to be 'out of control' if a special cause is present. The performance of a process that is 'out-of-control' cannot be predicted.

A process that exhibits only random variation is in statistical control.

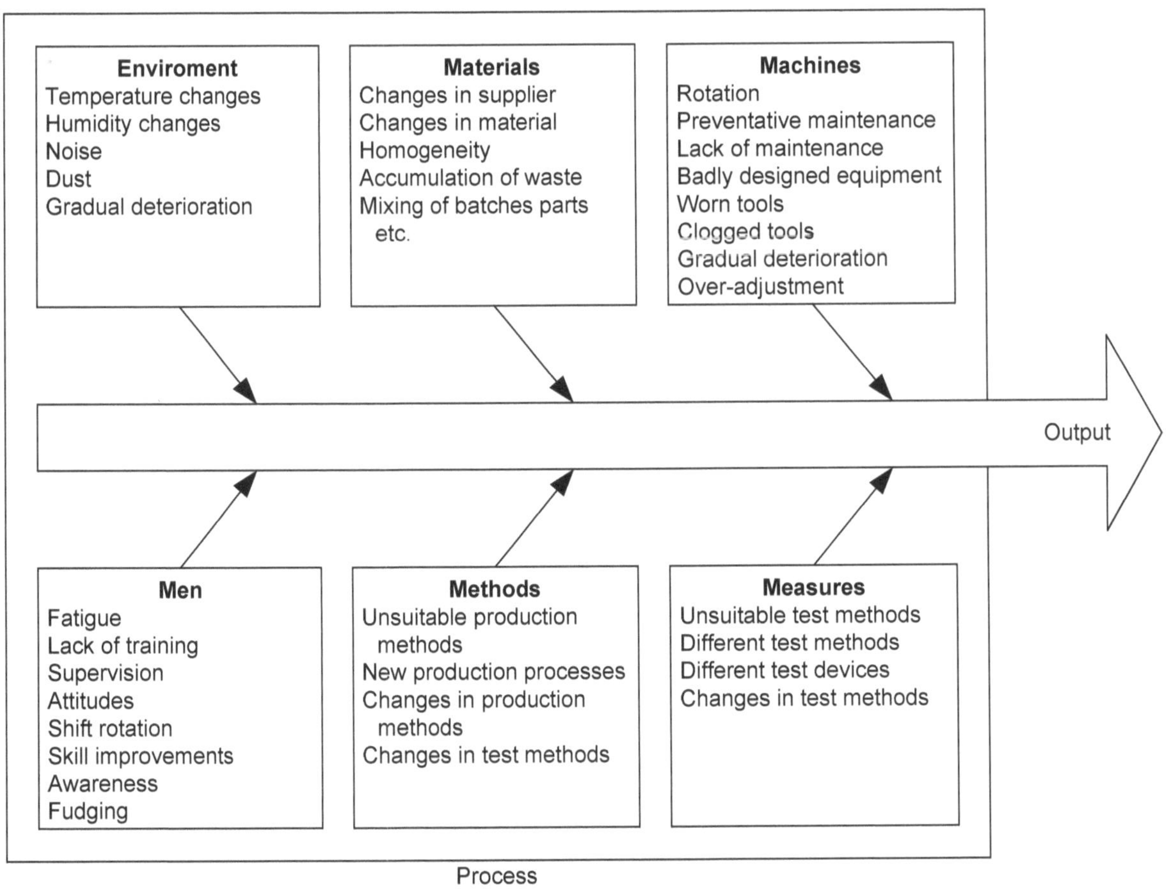


## The cause and effect chart

The cause and effect chart can be used to identify the source of special cause variations. Identifying and removing special causes is one of the main functions of SPC. In the absence of special causes the process will be predictable and will vary only due to common causes. Common causes can be reduced by the process of 'never-ending improvement'.

# 5.3 Sampling and the normal distribution

## Sampling - the basics

SPC relies on sampling to gain an insight into the process and it is necessary to understand a little about sampling to understand SPC.

Taking a sample of a population as a method of measuring the properties of the overall population is often thought of as an easy process but it is often fraught with error.

The average height of a population might initially be thought to be an easy thing to determine but unless you are going to measure the complete population then it is not so easy to do. When the UK Office for National Statistics (ONS) states that the average man in the UK is 175.3 cm tall and the average woman is 161.6 cm then it is certain that they didn't measure the height of everybody in the UK to find the average. This would simply be too difficult and in any case they didn't measure me!

In reality, the ONS has taken a sample of UK men and women and used this sample as being representative of the whole population. Samples can provide good estimates of the properties of the complete population but must be chosen correctly to avoid bias. For example, if the sample was taken in certain areas then the average for the sample might be slightly lower than the average for the population or if the sample was taken outside a basketball match then the average for the sample might be slightly higher than the average for the population.

Samples need to be chosen very carefully to be representative of the population but the alternative is to measure the complete population and this is often an insurmountable task. This is also true in SPC where, to avoid measuring the properties of the whole population, we are measuring the properties of a small sample of products and using the results from this to assess the properties of the complete population of products.

When taking samples it is important that the samples are 'unbiased' and represent the true properties of the population.

- **Tip -** A population is all of the parts in a batch or run. A sample is a small number of parts taken from a population.
- **Tip -** Take samples with care and make sure that they are random and represent the complete population.

## Sampling and histograms

Before starting to use SPC we also need to know a little about 'distributions'.

If we took a sample of parts all produced at the same time from a machine and measured a single property, e.g. the weight, then we would inevitably get a variation in the results. As the measurements were all made at the same time then this variation should be mainly due to common causes.

If these sample results were plotted on a histogram and the centre of each bar was joined up then they would start to show the distribution of the results.

If we measured enough points to get the results for the whole population then, in

The normal distribution is the foundation of SPC.

A population is all of the parts in a lot, batch or run.

A sample is a small number of parts taken from a population.

Each dimension or measured feature will have its own distribution. You cannot assume that every feature has the same distribution.

### The raw data - the histogram for a sample

If we took a sample from a population and measured a single property then we would expect a variation in the measurements. These can be plotted as a histogram (also known as a 'bar chart') to give a 'distribution'.

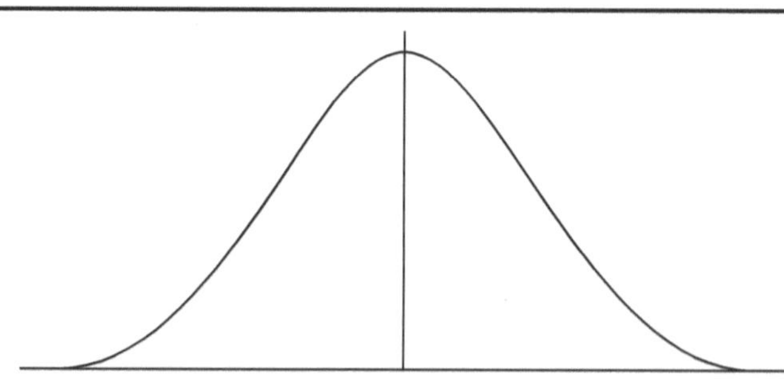

### The population distribution

If we took enough samples and produced a histogram for these then eventually the histogram would start to resemble the familiar normal distribution or 'bell curve' (provided the process remained stable).

most cases, we would end up with a smooth curve and it would probably look like the bell curve or normal distribution. The good thing about the normal distribution is that statisticians have been working with it for many years and know a lot about it.

- **Tip -** There is no a 'single' normal curve; the description is for the general type of curve.

The normal distribution is found in many natural and industrial populations. It can be used to describe (amongst other things):

- People's heights.
- People's intelligence.
- The wealth of nations.
- The size of parts from a process.
- The weight of parts from a process.

The properties of the normal distribution allow us to use information from samples selected from a process to make predictions about the properties of the complete population produced by the process.

## Variations on the normal distribution

The normal distribution isn't always a simple symmetrical 'bell' curve, sometimes it is skewed to one side and sometimes it is truncated (cut-off) at the ends or the middle.

- A right skew might be seen if you measured the height of jockeys - tall ones are less common and they tend to be smaller than the total population.
- A left skew might be seen if you measured the height of basketball players - short ones are less common and they tend to be taller than the total population.
- A truncated distribution might be seen if you sorted items using a 'go-no go' gauge before measuring them.

If the population is severely skewed or truncated then some modification to the SPC techniques may be needed. Fortunately for many industrial uses the normal distribution is applicable and works well.

- **Tip -** The important thing is that you need to be careful that your sample is from the population that you want to control.

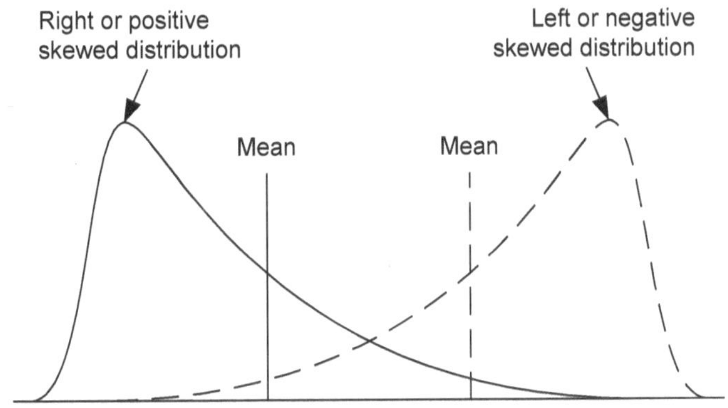


### Skewed distributions

Right or positive skewed distribution

The mean is at the right of the peak. An example of this might be the income of a population where there were many people earning at the lower end but relatively few earning at the top end (bankers and CEOs?).

Left or negative skewed distribution

The mean is on the left of the peak. An example of this could be the days that students hand in assignments. There would be very few handed in early and a rush handing their assignments in on the day it was due. It works the same in business: if your boss wants a report on Friday then he rarely gets it on Monday.

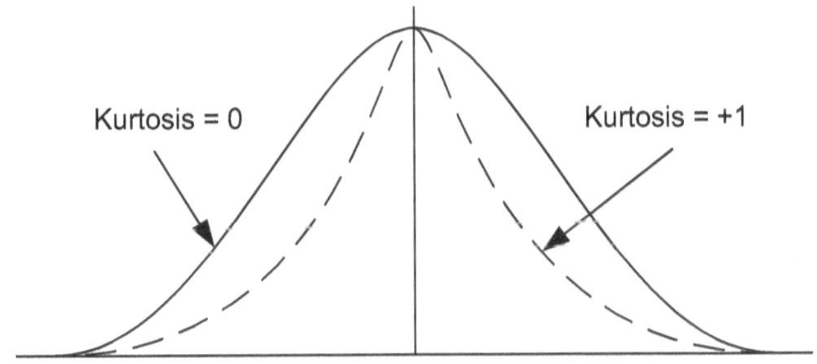


### Variations in kurtosis

Kurtosis is the tendency of the distribution to show a peak. A distribution with a kurtosis of 0 is relatively 'full' and is the standard normal distribution. A distribution with a kurtosis of +1 is much more 'peaked'.

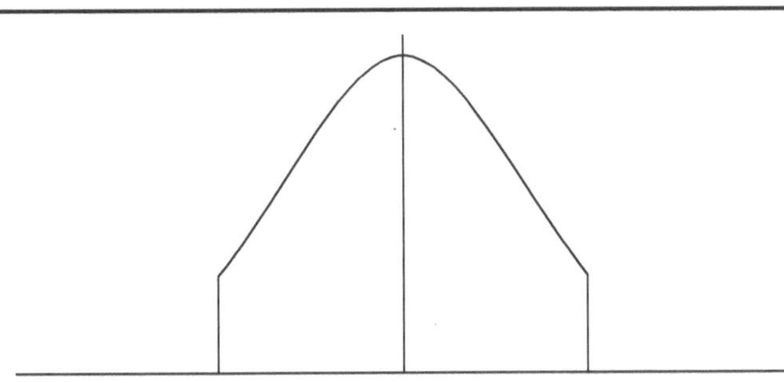

### Truncated normal distribution

Distributions can be truncated either at the left- or the right-hand side. An example of this would be products sorted using a 'go-no-go' gauge. Some products would be removed but the remaining products would still be normally distributed (but truncated).

# 5.4 Can we make it OK? - capability and the normal distribution

## Capable or not capable?

A capability study is simply a method of assessing if it is possible for a given machine or process to produce parts to a given specification. It is not a statement of desire or lack of desire, simply an answer to the question 'Can we make it OK?'.

This is an essential question because if it is not possible to make the part OK, i.e. the process or machine is not capable of producing parts that are within the specified tolerances, then it is futile to try to produce the part. A process that is not capable may produce some parts that are within specification but it will always produce parts that are out of specification. These will need to be 'inspected' out of the batch with all the costs that this incurs.

## Machine and process capability

### Machine capability

This type of study is carried out on new machines or new processes to check if these are capable of producing to the required specifications. These studies are not necessarily carried out under full 'in-control' conditions and should only be regarded as a snap-shot of the full process. They do not necessarily provide a good measure of the long-term performance.

### Process capability

This type of study is carried out over the long term for processes that are in statistical control as shown in the control charts (see Section 5.22). These studies allow long-term prediction of the performance of a process that is subject only to common causes (operators, materials, machine variations etc.).

## Carrying out a machine capability study

Carrying out a machine capability study is relatively easy and will provide vital information on the capability of the process before production starts. Proceed as follows:

- Select 50 samples from a continuous run of product where the process is stable, nominally producing good product and where no changes have been made to the process settings.
- Measure the desired feature and plot the results in a histogram using at least 6 intervals (see Section 7.6).
- If the distribution is approximately that of a normal distribution then proceed. This may require some judgement but in many cases it is reasonably safe to assume that the distribution is normal.

**Note:** If the distribution is not a normal distribution then other techniques are needed.

- Calculate the mean($\overline{X}$ - pronounced as X bar) - this is the centre of the process and is defined by the setting of the process.
- Calculate the standard deviation ($\sigma$ - pronounced sigma) - this is a measure of the variability of the process, i.e. a

A short-term capability study reflects a process at a single time with a single batch of material, a single operator (or team) and the same environment.

Over a longer period there will always be more variation.

'Mean' is simply another word for 'average'.

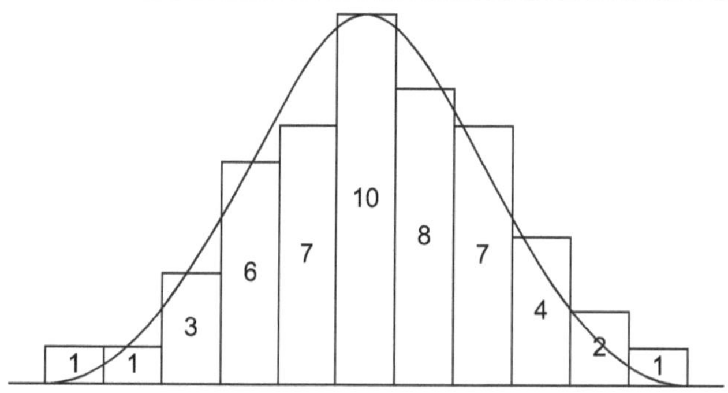


Capability study data

If we took a 50-part sample from a machine and measured a single property then we would expect a variation in the measurements. As for any other sample, these can be plotted as a histogram to give a 'distribution'.

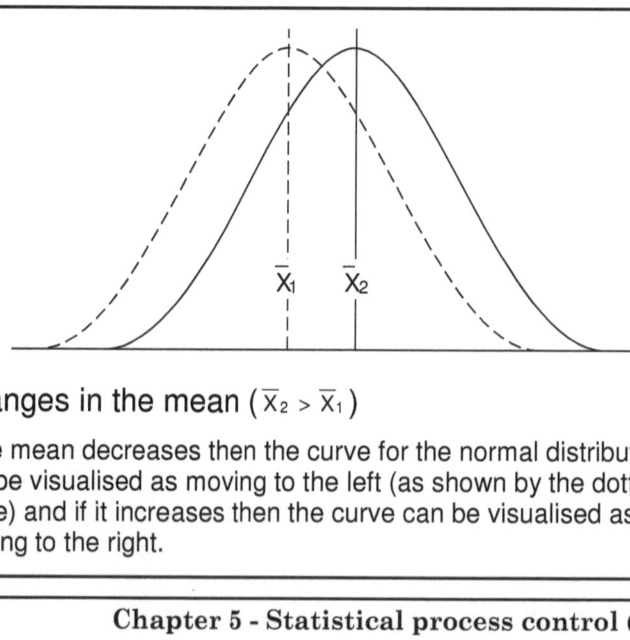


Changes in the mean ($\overline{X}_2 > \overline{X}_1$)

If the mean decreases then the curve for the normal distribution can be visualised as moving to the left (as shown by the dotted curve) and if it increases then the curve can be visualised as moving to the right.

highly variable process will have a high $\sigma$ value.

**Note:** It is easiest to use a spreadsheet to carry out these calculations. In Excel the functions used are AVERAGE for $\overline{X}$ and STDEV for $\sigma$.

## Describing the curve

Fortunately, despite the complex shape of the normal distribution, it is possible to fully describe it by using only 2 numbers:

- The mean or average ($\overline{X}$) - this is the location of the centre line of the normal distribution and describes the location of the curve. If the mean is increased then the curve moves to the right and if it is decreased then the curve moves to the left. Changing the mean changes the location of the normal distribution.
- The standard deviation ($\sigma$) - this is the amount of spread of the curve. A high standard deviation means that the curve is very broad and a low standard deviation means that the curve is very slim. Changing the standard deviation changes the width of the normal distribution.

These are the only numbers needed to describe a fully symmetrical normal distribution and it is variations in the shape and location of the curve that we are looking for.

## The percentages of the normal distribution

After the mean and standard deviation are known, it is possible to use the properties of the normal distribution to find the probability of finding a given value in the process. It is not necessary to know the exact statistical proofs but for a normal distribution:

- $\overline{X} \pm \sigma$ contains 68.26% of the population.
- $\overline{X} \pm 2\sigma$ contains 95.44% of the population.
- $\overline{X} \pm 3\sigma$ contains 99.73% of the population.
- $\overline{X} \pm 4\sigma$ contains 99.99% of the population.

This is illustrated in the diagram on the right.

Therefore, if the mean ($\overline{X}$) of a population is 8 and the standard deviation ($\sigma$) is 1, then 99.73% of the population will have values between $\overline{X} \pm 3\sigma$ or between 5 (= 8 - 3) and 11 (= 8 + 3).

If the tolerance is 8 ± 1, i.e. 7 to 9, this is $\overline{X} \pm \sigma$ and we know that only 68.26% of the population will lie between these limits. This also means that 31.74% (100 - 68.26) of the population will be outside the tolerances. If this were a process then nearly 32% of the production would be outside specification.

- **Tip** - It is possible to calculate the amount of the population lying between 2 limits (see Appendix 4).

50% of the population will be contained within the limits $\overline{X} \pm 0.6745\sigma$.

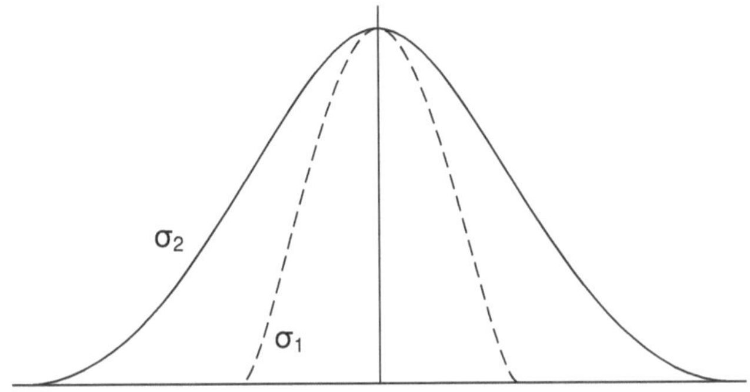


Changes in the standard deviation ($\sigma_2 > \sigma_1$)

If standard deviation decreases then the curve for the normal distribution can be visualised as becoming slimmer (as shown by the dotted curve) and if it increases then the curve can be visualised as becoming broader.

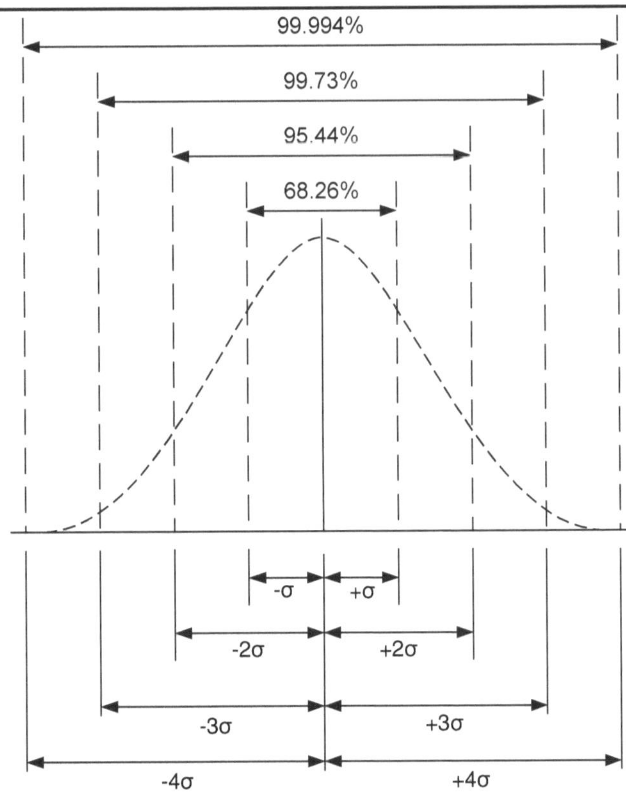


The percentages of the normal distribution

Despite the complex shape of the normal distribution, it is possible to describe it using only the mean and the standard deviation. These two numbers describe how much of the distribution is within certain boundaries.

# 5.5 Capability - spread

## Comparing spread

Section 5.4 showed that it is possible to describe the normal distribution using only the values for the mean ($\overline{X}$) and the standard deviation ($\sigma$) where the mean describes the location and the standard deviation describes the spread.

After the mean and the standard deviation of normal distribution for a process have been established, it is possible to use these to compare the likely output of the process with a specified dimension and a set of tolerances (as required by the customer or as established for good operation of the process).

The ideal result for the spread is a value that fits easily into the tolerance band. It is not necessarily 'as small as possible' as broad statements such as this inevitably raise costs. We want a spread that fits easily into the tolerance band and allows us to produce good products at all times.

## Assessing the spread of a distribution

Using the data from Section 5.4 for the standard deviation ($\sigma$) and the relevant tolerance limits from the process specification, calculate the 'capability' or $C_m$ (relative to the specified tolerances) from the formula:

$$C_m = \frac{USL - LSL}{6\sigma}$$

where:

- USL = Upper Set Limit or Upper Acceptable Tolerance.
- LSL = Lower Set Limit or Lower Acceptable Tolerance.
- $\sigma$ = the standard deviation of the sample.

**Note 1:** We use $C_m$ (the machine capability) here rather than $C_p$ (the process capability) as it is assumed that the results being used are from a machine capability study.

**Note 2:** The value of $6\sigma$ ($\pm 3\sigma$) is set so that we can be confident of producing 99.73% of the parts within the tolerance band (see Section 5.4).

**Note 3:** It does not matter if the tolerances are 'bilateral' (sometimes called 'symmetric'), i.e. A ± x, or 'unilateral' (sometimes called asymmetric), i.e. A + x, − y. The important thing is the difference between the upper tolerance and the lower tolerance. At this stage we are only dealing with the total spread of the distribution and how it fits into the tolerance band.

$C_m$ compares the spread of the distribution with the tolerance band.

## What does $C_m$ mean?

The value of $C_m$ is a measure of the 'process spread' relative to the specified tolerances.

It is easy to see that $C_m$ will increase as the spread ($\sigma$) decreases relative to the tolerance band. There are three main cases for the value of $C_m$:

- If $C_m < 1.00$ then $6\sigma$ is greater than the width of the tolerance band. This means that out-of-tolerance parts will always be produced at either the top or the bottom of the tolerance band (or at both simultaneously). In this case it is not possible to produce all parts within tolerance. Out-of-tolerance parts will undoubtedly be produced. The diagram below shows how the distribution can be located on the tolerance band.
- If $C_m = 1.00$ then $6\sigma$ is equal to the width of the tolerance band. This indicates a process capable to $\pm 3\sigma$. Only a small amount of out-of-tolerance parts (0.27%) will be produced if the process is exactly centred and the average is equal to the

Descriptive statistics describes the elements of a population as a whole or describes data that represents a sample of the population.

Inferential statistics draws conclusions about a population based on information from a sample of the population.

The link between descriptive and inferential statistics is probability.

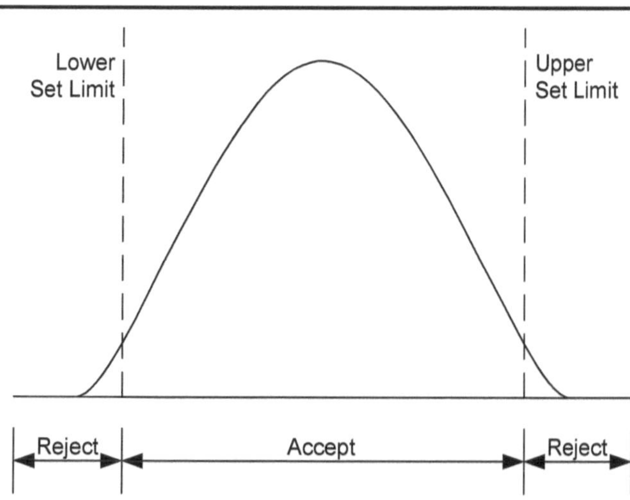


$C_m < 1.00$ - Poor spread

Parts will always be produced outside the tolerance limits. The process may be centred on the tolerances but it will never be possible to produce all the parts inside the tolerance. This process should not be started for production.

specified value. Any small change in the average makes it extremely likely that more out-of-tolerance parts will be produced. The diagram on the upper right shows how the distribution can be located on the tolerance band.

- If $C_m > 1.33$ then $6\sigma$ is less than the width of the tolerance band. This indicates a process capable to $\pm 3\sigma$. Out-of-tolerance parts will not generally be produced and exact centring of the process is not necessary for the production of over 99.73% good parts. When $C_m$ is > 1.33 the process is said to be 'capable'. The diagram on the lower right shows how the distribution can be located on the tolerance band.

To show acceptable process spread, $C_m$ must always be > 1.33. This is to allow for minor special cause variations in materials, men, methods, machines and environment. This is effectively saying that for a process to be regarded as capable then the spread of the normal distribution ($\pm 3\sigma$) should be no more than 75% of the specified tolerance.

$C_m$ only describes the spread, it does not describe the location and, even with a small spread ($C_m$ >1.33), it is possible to produce out-of-tolerance parts if the location of the distribution is poor. Defining the location of the distribution is covered in Section 5.6.

## Increasing the demands

Some companies and customers demand a higher requirement for machine capability to account for the inevitable later effects of common and special causes. In this case, the requirement for $C_m$ is often increased so that $C_m$ is defined by:

$$C_m = \frac{USL - LSL}{8\sigma}$$

This is effectively placing the limits for the process at $\pm 4\sigma$ and requiring that 99.99% of the production will fit within the specification limits. This means that the spread of the process at the machine capability stage has to be much less than at the process capability stage.

## Capability ratio

It is also possible to describe the spread in terms of the Capability Ratio (CR). The CR is simply the inverse of $C_m$ and should be less than 0.75, i.e. 1/1.33, for a process to be capable.

The CR defines how much of the specified tolerance is taken up by the distribution, i.e. a CR of 0.75 is the same as saying that distribution of the population takes up 0.75 of the tolerance band.

## Process capability

Process capability spread is assessed in much the same manner, although the data are gathered differently, and will be dealt with in Section 5.22.

If you use $12\sigma$ ($\pm 6\sigma$) to calculate $C_m$ then we can be confident of producing 99.9999998% of the parts within the tolerance band.

This is the origin of the '6-sigma' philosophy.

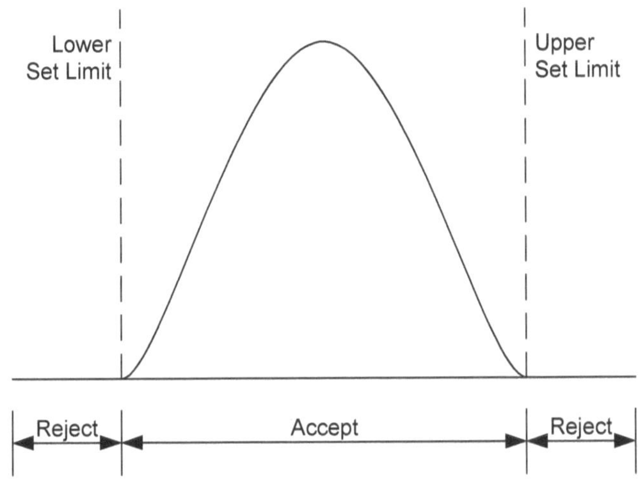


$C_m$ = 1.00 - Acceptable spread

It will be possible to produce 99.73% of the parts inside the tolerance limits but only if the process is exactly centred on the nominal and does not vary for special causes. Parts will only be produced in tolerance if the location is exactly centred.

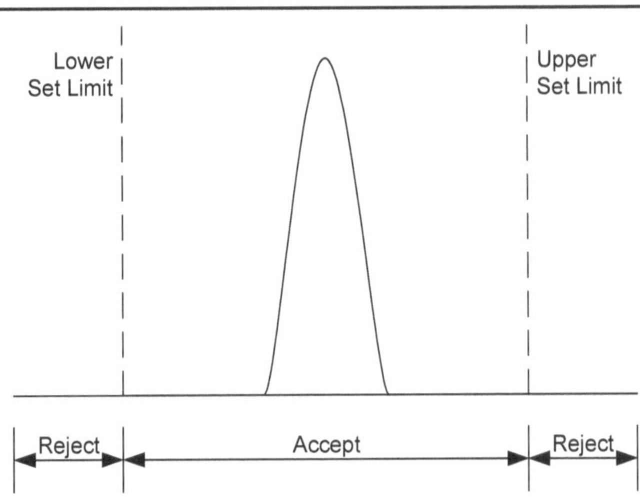


$C_m$ > 1.33 - Good spread

The process spread is less than the tolerance limits. It will be possible to produce all parts inside the tolerances even with some special cause variations. All parts will be produced in tolerance for a wide range of locations.

# 5.6 Capability - location

## Comparing location

Section 5.4 showed that it is possible to describe the normal distribution using only the values for the mean and the standard deviation and Section 5.5 showed how $C_m$ could be used to describe the spread relative to a specified set of tolerances.

However, the spread is not the only important factor in meeting the specification. The location of the distribution is equally important. A small spread is good but only if it is combined with a location that is close to the nominal specification value.

The ideal result for location is a mean which is equal to the nominal value of the dimension or value. This will give a centred distribution and increase the potential for good products.

## Assessing the location of a distribution

Using the data from Section 5.4 for the average ($\overline{X}$) and the relevant tolerance limits from the process specification, calculate $Z_{Upper}$ and $Z_{Lower}$ from the formulae:

$$Z_{Upper} = \frac{USL - \overline{X}}{3\sigma}$$

and

$$Z_{Lower} = \frac{\overline{X} - LSL}{3\sigma}$$

where:

- USL = Upper Set Limit or Upper Acceptable Tolerance.
- LSL = Lower Set Limit or Lower Acceptable Tolerance.
- $\sigma$ = the standard deviation of the sample.
- $\overline{X}$ = the mean of the sample.

Find the lower value of $Z_{Upper}$ and $Z_{Lower}$ (or $Z_{Min}$).

This is termed $C_{mk}$ and is also known as the 'machine capability index'. $C_{mk}$ is a measure of the 'process location' but also gives valuable information on the spread.

**Note 1:** As with $C_m$, we use $C_{mk}$ (the machine capability index) here rather than $C_{pk}$ (the process capability index) because the results being used are from a machine capability study.

**Note 2:** It does not matter if the tolerances are 'bilateral' (sometimes called 'symmetric'), i.e. A ± x, or 'unilateral' (sometimes called asymmetric), i.e. A + x, – y. The important thing is the location of the average of the sample with respect to the upper set limit and the lower set limit. At this stage we are only dealing with the location of the distribution and how close it is to the set limits.

$C_{mk}$ is primarily influenced by process setting and can often be improved by adjusting the process.

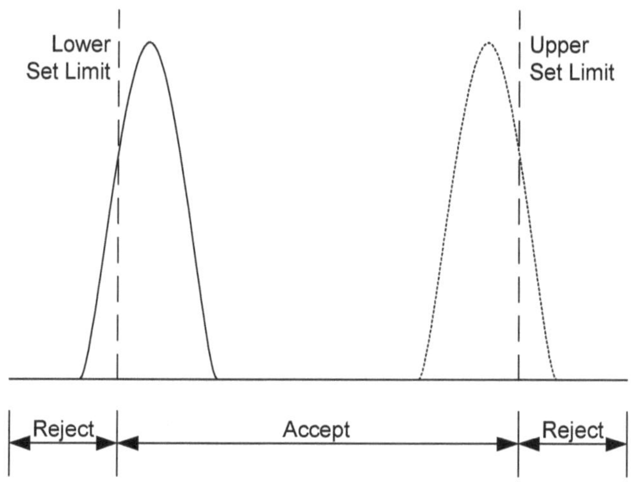


$C_{mk}$ < 1.00 (from either $Z_{Lower}$ or $Z_{Upper}$)

The process is outside the tolerances at either the upper or lower set limits (depending on which of $Z_{Upper}$ and $Z_{Lower}$ is the lowest). It will never be possible to produce all the parts inside the tolerance. This is good spread but poor location.

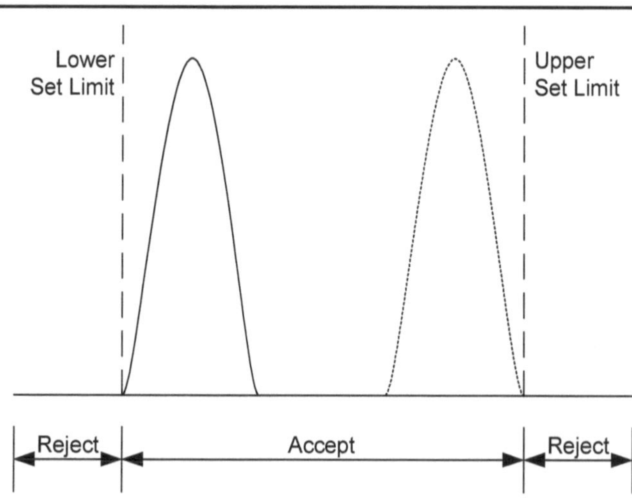


$C_{mk}$ = 1.00 (from either $Z_{Lower}$ or $Z_{Upper}$)

The process is exactly on the tolerance at either the upper or the lower set limit. It will only just be possible to produce in-tolerance parts if the process does not change. Poor location - it is highly likely that parts will be produced outside the tolerance limits.

## What does $C_{mk}$ mean?

The value of $C_{mk}$ is a measure of the 'process location' relative to the specified tolerances.

It is easy to see that $C_{mk}$ will decrease as the location of the process average ($\overline{X}$) approaches set limits (either the USL or LSL). There are three main cases for the value of $C_{mk}$:

- If $C_{mk} < 1.00$ then it will not be possible for the process to produce parts within tolerance. The process is located such that out-of-tolerance parts are produced. These may be either higher or lower than the tolerance depending on whether $Z_{Min}$ is defined by $Z_{Upper}$ (out-of-tolerance parts will be greater than the USL) or $Z_{Lower}$ (out-of-tolerance parts will be lower than the LSL). The diagram on the upper left shows how the distribution can be located on the tolerance band.
- If $C_{mk} = 1.00$ then the process will be located such that any small change in the process will result in out-of-tolerance parts being produced. This indicates a process capable to $\pm 3\sigma$. It is extremely likely that out-of-tolerance parts will be produced where the out-of-tolerance part is greater or less than the tolerance (depending on whether $Z_{Min}$ is defined by $Z_{Upper}$ or $Z_{Lower}$). The diagram on the lower left shows how the distribution can be located on the tolerance band.
- If $C_{mk} > 1.33$ then the process will be located such that it is well within the tolerance band width. Out-of-tolerance parts will not be produced and exact centring of the process is not necessary. When $C_{pk} > 1.33$ the process is said to be 'capable'. The diagram on the upper right shows how the distribution can be located on the tolerance band.

To show acceptable process location, $C_{mk}$ must be > 1.33. This is to allow for variations in materials, men, methods, machines and environment.

$C_{mk}$ primarily describes the location of the spread but can also give valuable information on the location, although this is more easily described by $C_m$.

If both $Z_{Upper}$ and $Z_{Lower}$ are < 1.00 then parts will be produced that will be out of tolerance with respect to both the USL and the LSL. In this case, the values for $Z_{Upper}$ and $Z_{Lower}$ can be used with $P_z$ tables to calculate the proportion of the output that will be beyond the specification limits (see Appendix 4).

- **Tip** - Isn't it better to improve the process to get it better rather than to calculate how many out-of-tolerance parts you are going to be producing?

## Process capability

Process capability location is assessed in much the same manner, although the data is gathered differently, and will be dealt with in Section 5.22.

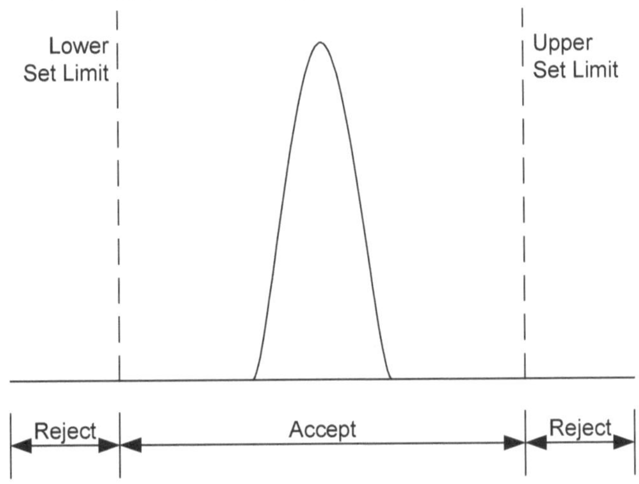


$C_{mk} > 1.33$ (for both $Z_{Lower}$ and $Z_{Upper}$)

The process is well within both the upper and lower tolerance limits and process variations can occur with no out-of-tolerance parts being produced. Good location - all parts will be produced inside the tolerance limits.

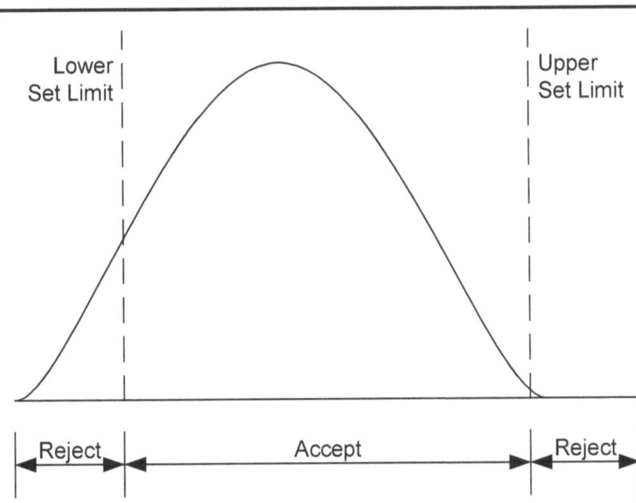


The worst of all possible worlds

For this process, $C_m$ is < 1.00 and $C_{mk}$ < 1.00 on both sides. This process will produce out-of-tolerance products on both the upper and the lower tolerances and the location can never be adjusted to produce good parts.

# 5.7 Are we making it OK? - variables and attributes control charts

## Continuous control

Capability studies of the type discussed in previous sections are extremely important when a process is being initially assessed for performance, i.e. it is a machine capability study and the process is in control. For running production it is necessary to have a method of continuously assessing if the process is in control and the only way to do this is to use control charts on the process.

Control charts are based on the statistics of distributions and are used to regularly assess the process output to assess the process performance. The theory of control charts can be found in many texts[1].

## The benefits of control charts

Control charts have many benefits and can be used to:

- Separate common and special causes to decide between management action (common causes) and local action (special causes).
- Allow operators to control the process.
- Improve process control to give less variation in the output, lower the unit cost and increase the process capacity.
- Provide a common language for all staff to discuss and improve processes.

Control charts are not a substitute for action. They provide the evidence that action needs to be taken, they can help identify where action should be taken but ultimately it is the responsibility of management to ensure that effective action is taken.

## The basics of control charts

In Section 5.2 the two types of data (variable and attribute data) were discussed and the type of control chart used varies with the type of data. Whichever type of control chart is used there are common elements:

- Control charts are completed by measuring or assessing products and recording the results.
- Control charts use process data and information to calculate process control limits. These control limits are a measure of the variability of the process due to common causes.
- Control limits are not the same thing as specification limits. The control limits are set by the process, not by the specification.
- When a process is in control then the plotted points will be distributed according to the normal distribution.
- Patterns in the plotted points will indicate the presence of special causes.

The box (below) shows the typical format of a control chart with the control limits and points marked on it.

## Variables control charts

Variables control charts use measurements taken from the process and simple calculations to set the process control limits.

Variables control charts use:

- The mean or median of a sample to assess the setting accuracy of the process.
- The range or standard deviation of a sample to assess the process consistency.

The aim of using control charts is to measure and reduce process variation from both common and special causes.

'Control' doesn't necessarily mean that the product or service will meet your needs. It only means that the process is consistent.

Do not confuse control limits with specification limits which are related to customer requirements not to process variation.

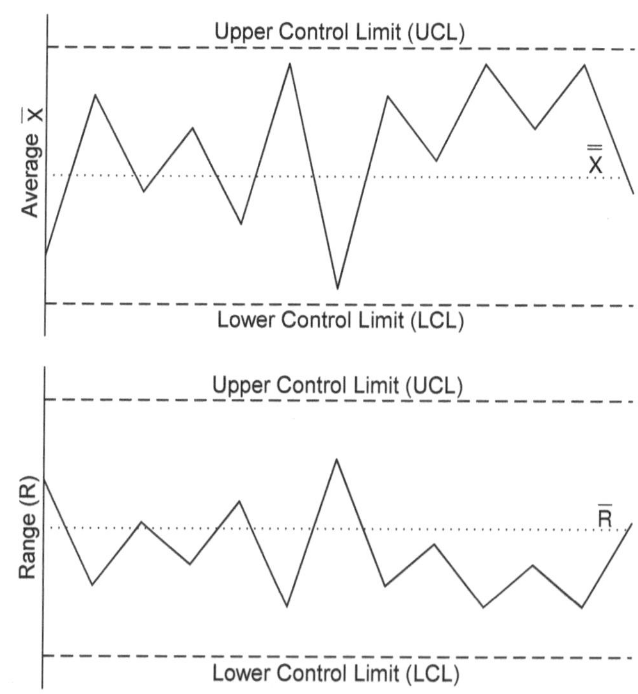


### A typical control chart ($\bar{X}$ and R)

This is a typical variables control chart, in this case using the mean and range. The horizontal lines are the centre line and the Upper and Lower Control Limits. These lines allow assessment of whether the process is in control or not.

## Variables chart types

The variables control charts are:

- Mean and range chart ($\overline{X}$ and R chart) - this chart plots the mean (average) and range (highest measured value - lowest measured value) of a set of samples.
- Mean and standard deviation chart - ($\overline{X}$ and s chart) - this chart plots the mean (average) and standard deviation of a set of samples.
- Median and range chart ($\tilde{X}$ and R chart) - this chart plots the median (middle value) and the range of a set of samples.
- Individuals and moving range chart ($\overline{X}$ and MR chart) - this chart plots the individual values and the moving range of a set of individual measurements.

The most common chart used in industry is the mean and range chart ($\overline{X}$ and R ) and this will be used to describe how variables control charts work.

## Attributes control charts

Attributes control charts work by measuring the number of nonconformities (however defined) in the product. In the terminology of attributes control charts, it is important to distinguish between nonconformities and nonconforming items:

**Nonconformity:** a failure to meet one of the tests or acceptance criteria in a product, e.g. a nonconformity could be a scratch or blemish.

**Nonconforming item:** a complete product that fails to meet one or more of the acceptance criteria. A nonconforming item may have one or more nonconformities, e.g. a nonconforming item may have more than one scratch or blemish nonconformity.

Attribute data are not assessed according to the normal distribution but according to either the Binomial distribution (for nonconforming items) or to the Poisson distribution (for nonconformities) depending on whether the process is being controlled by the number of nonconforming items or the number of nonconformities.

Control or assessment by attributes evaluates and controls the process by how many nonconforming items (or nonconformities) are found. The judgement of whether a product is nonconforming and how many nonconformities it contains may be based on visual assessment, by gauging or by other go/no-go methods.

However, unlike variables control charts, it is not always possible to set a fixed sample size for attribute data and the sample sizes used for attributes are generally much higher than the sample sizes used for variables.

## Attributes chart types

The attributes control charts are:

- Fraction nonconforming chart (p chart) - this chart plots the proportion or fraction nonconforming in a set of samples.
- Number nonconforming chart (np chart) - this chart plots the actual number of nonconforming items in a set of samples.
- Number of nonconformities (c chart) - this chart plots the number of nonconformities in a set of samples.
- Number of nonconformities per unit (u chart) - this chart plots the average number of nonconformities per sample in a set of samples.

- 1. Grant, R.S. and Leavenworth, E.L. 1996. '*Statistical Quality Control*'. McGraw Hill. First published in 1946 with the 7th edition published in 1996, this is still the classic text for statistical quality control.

### The statistics behind control charts

Control charts work on the Central Limit Theorem. This says that where regular samples are taken from a population there is a relationship between the sample means and the original data. For samples of size 'n' taken from a population with a mean of $\overline{X}$ and a standard deviation of σ, the samples will tend towards a normal distribution with a mean of $\overline{\overline{X}}$ and a standard deviation of $\sigma/\sqrt{n}$.

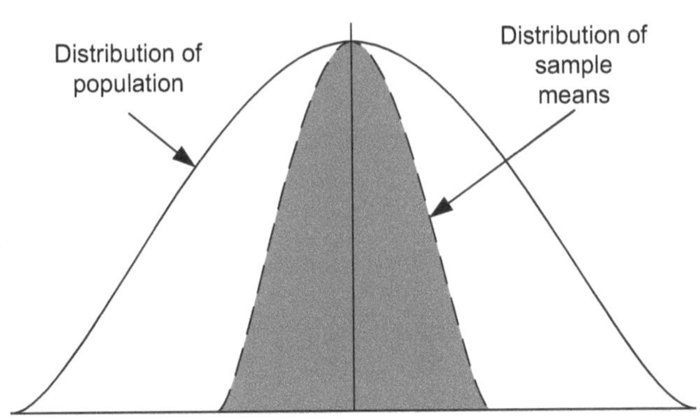


This means that even if the distribution of the population is not an exact normal distribution then the distribution of the samples means will be a much better normal distribution and will behave according to the rules of the normal distribution.

Monitoring the sample means checks if a normal distribution is still operating and whether any special causes are present.

In some countries the control limits used are:

- Upper action limit (UAL).
- Upper warning limit (UWL).
- Lower warning limit (LWL).
- Lower action limit (LAL).

These have largely been superseded by the convention of:

- Upper control limit (UCL).
- Lower control limit (LCL).

The UCL and LCL convention will be used throughout this Workbook.

When control charts are used correctly, it is possible to identify and eliminate special causes.

# 5.8 Variables control charts - range and average

## Planning for control charts

Control charts should only be used for relevant product features to avoid producing excessive documentation and work. Equally, the sampling frequency and size should be relevant to the process timescales, i.e. if you are producing small batches then sampling may need to be more frequent than for long runs.

- **Tip** - Only measure what is necessary to control the process.
- **Tip** - Charting small sample sizes often is better than large samples sizes less frequently. The frequency can always be decreased after the process is in control.
- **Tip** - A sample size of between 4 and 16 is preferred but for this Workbook we will use 5 as the standard sample size.

The decision on the type of chart to be produced (see Section 5.7 for a discussion of the relative merits of each chart type) should be made early in the process but for the purposes of illustration we will use the $\overline{X}$ and R chart.

## Getting the basic data

Using a running and stable process, take 5 samples from the process at 30-minute intervals. Do not adjust the process or make any changes during the run.

- Decide on the control variable (X) that it is desired to control and record the 5 measurements of the control variable.
- Calculate the average ($\overline{X}$) of the 5 measurements.
- Calculate the range (R) of the 5 measurements. This is simply the difference between the highest and lowest measurements) for the 5 samples where:

  $R = X_{highest} - X_{lowest}$.
- Continue until 25 sets of data ($\overline{X}$ and R) are available.

## Setting up the range chart

- Plot the results for each of the calculated ranges on a preliminary control chart in time order.
- Find the average of the 25 values for R. This is $\overline{R}$ and is:

  $$\overline{R} = \frac{\text{Sum of R values}}{25}$$
- Calculate the control limits for R from:
  - Upper Control Limit for R ($UCL_R$):

    $UCL_R = D_4 \times \overline{R}$
  - Lower Control Limit for R ($LCL_R$):

    $LCL_R = D_3 \times \overline{R}$

  where $D_4$ and $D_3$ are constants that vary with the sample size. For a sample size of 5, the values of $D_4$ and $D_3$ are 2.114 and 0 respectively (see Appendix

Control limits are assigned on the statistics and not by the tolerances.

It is essential that the samples are random and all products have an equal chance of being selected.

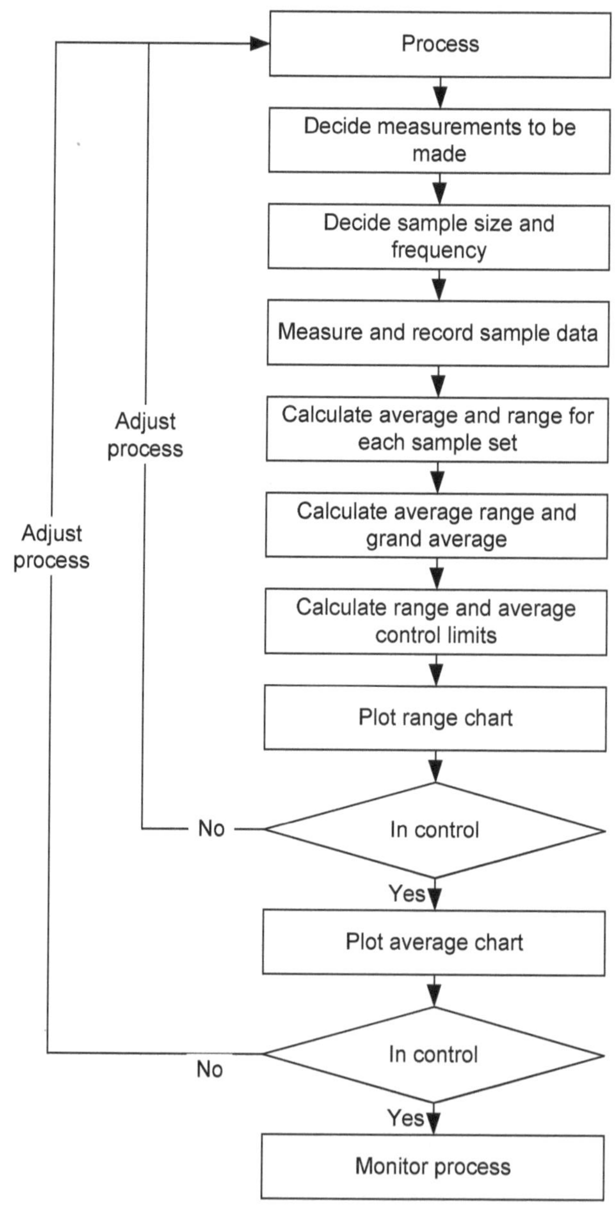


### Setting up a variables control charts - the process

Setting up a variables control chart is easy to do and the flow chart gives the necessary steps. It is important to set up the range chart first and to check that this is in control before setting up the average chart.

2 for values of $D_4$ and $D_3$ for other sample sizes).

- For a sample size of 5, the control limits are:
  - Upper Control Limit for R:

    $UCL_R = 2.114 \times \overline{R}$
  - Lower Control Limit for R:

    $LCL_R = 0 \times \overline{R} = 0$

**Note:** If the sample size is < 7 then the $LCL_R$ will always be 0.

- Mark these trial control limits on the R chart.
- If any of the points on the range chart are outside the trial control limits then discard these points and recalculate the control limits.
- Continue this process until all points are within the trial control limits (provided at there are still at least 20 data points).
- Plot the average for the range ($\overline{R}$) and the Upper and Lower Control Limits ($UCL_R$ and $LCL_R$) on the control chart.

## Setting up the average chart

Plot the results for each of the calculated averages on a preliminary control chart.

- Find the average of the 25 values for $\overline{X}$. This is the grand average ($\overline{\overline{X}}$) and is:

  $$\overline{\overline{X}} = \frac{\text{Sum of } \overline{X} \text{ values}}{25}$$
- Calculate the control limits for $\overline{X}$ from:
  - Upper Control Limit for $\overline{X}$:

    $UCL_{\overline{X}} = \overline{\overline{X}} + A_2 \times \overline{R}$
  - Lower Control Limit for $\overline{X}$:

    $LCL_{\overline{X}} = \overline{\overline{X}} - A_2 \times \overline{R}$

    where $A_2$ is a constant that varies with the sample size. For a sample size of 5, the value of $A_2$ is 0.577 (see Appendix 2 for values of $A_2$ for other sample sizes).
- For a sample size of 5, the control limits are:
  - Upper Control Limit for $\overline{X}$:

    $UCL_{\overline{X}} = \overline{\overline{X}} + 0.577 \times \overline{R}$
  - Lower Control Limit for $\overline{X}$:

    $LCL_{\overline{X}} = \overline{\overline{X}} - 0.577 \times \overline{R}$
- Mark these trial control limits on the $\overline{X}$ chart. If any of the points on the average chart are outside the trial control limits then discard these points and recalculate the control limits.
- Continue this process until all points are within the trial control limits (provided at there are still at least 20 data points.
- Plot the grand average ($\overline{\overline{X}}$) and the Upper and Lower Control Limits ($UCL_{\overline{X}}$ and $LCL_{\overline{X}}$) on the control chart.

## Continue

Continue to measure the values of $\overline{X}$ and R for the control variable. Section 5.9 discusses how to use the charts to control the process.

If the control limits are outside the specification limits then the prime responsibility for improvement is with upper management. (or change the specification).

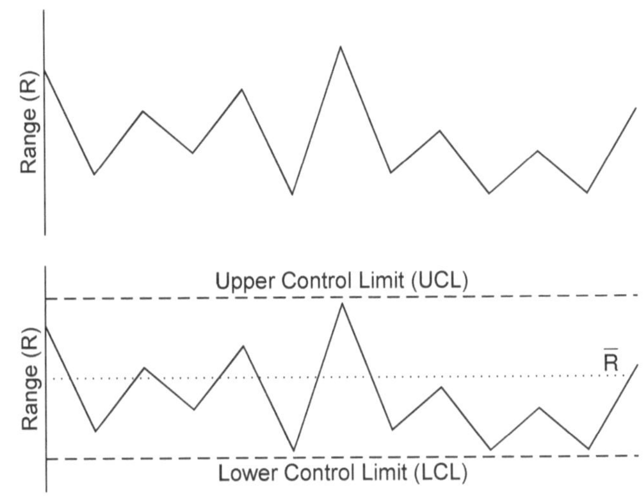


### Setting up a range control chart

Plot the range for a set of samples and then set the trial control limits. If any results are outside the limits then discard these results and calculate the limits again. When there are no results outside the limits then these limits are the starting limits.

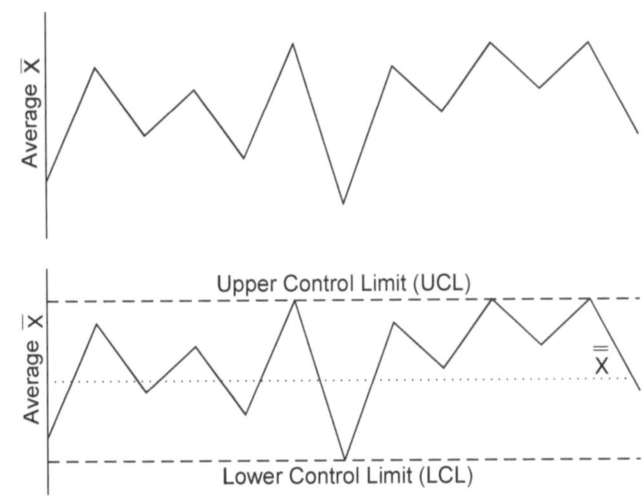


### Setting up an average control chart

After the range chart is complete then calculate the limits for the average chart. If any results are outside the limits then discard these and calculate the limits again. When there are no results outside the limits then these limits are the starting limits.

# 5.9 Using a simple variables control chart

## This is easy!

The simple variables control charts described in Section 5.8 should be continued during normal production and analysed for 'out-of-control' points or patterns. These are patterns that indicate a substantial change in the process.

It is entirely normal for the points marked on the chart to vary over the area of the chart - this is what processes do all the time and is the result of common causes of variation. What we are interested in is when special causes of variation are introduced and start to affect the process. We want to find these quickly and remove them as soon as possible (and definitely before any out-of-tolerance parts are produced).

So what do we record on the charts and what are the patterns that show that special causes have been introduced?

## Recording on the charts

Recording the process performance on the chart is simply a matter of regularly measuring the control variable and recording the results on the chart.

This is done using the previously created control charts with the average and the control limits (UCL and LCL) marked on them.

- Measure 5 samples every 30-minutes (or at the chosen time interval) and calculate:
  - The average of the 5 measurements or $\overline{X}$ (add them up and divide by 5).
  - The range of the 5 measurements or R (subtract the lowest measurement from the highest measurement).
- Plot the average and the range on the relevant control chart during the production run.
- Join the new result to the previous result with a straight hand-drawn line.

## The alarms for variables control charts

It is essential to interpret the range chart before the average chart - if the range chart is not in control then no meaningful information can be obtained from the average chart.

### Interpret the range control chart

Look for the following alarms:

- No points outside the control limits (upper or lower), see the diagram below for an example of an out-of-control point.
- No run of 7 consecutive points above or below the average line.
- No run of 7 consecutive points upward or downward.
- No pattern with more than 2/3 of the points in the middle 1/3 of the control limits.
- No pattern with more than 1/3 of the points in the outer 2/3 of the control limits.

If any of these patterns is present then the process is out-of-control for the range chart. The process should be investigated for special causes and stopped or corrected by adjusting the settings.

If the range control chart is in order then interpret the average control chart

### Interpret the average control chart

Look for the same patterns in the average control chart as in the range control chart.

Look for the following alarms:

- No points outside the control limits (upper or lower).
- No run of 7 consecutive points above or below the average line.
- No run of 7 consecutive points upward or downward.

Even when a process is 'in control' then it is possible to produce a product that is outside the specification.

Statistics are like that.

Any points outside the control limits, once identified with a cause (or causes), should be removed and the calculations and charts redone. Points within the control limits, but showing indications of trends, shifts, or instability, are also special causes.

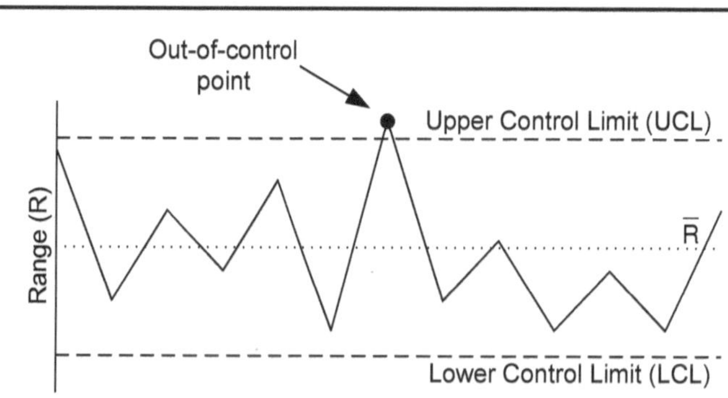


**An out-of-control point**

The easiest out-of-control alarm is where the range (or average) is outside of either the UCL or the LCL. When out-of-control points are seen then action needs to be taken as soon as possible and new measurements taken after the action.

- No pattern with more than 2/3 of the points in the middle 1/3 of the control limits.
- No pattern with more than 1/3 of the points in the outer 2/3 of the control limits.

If any of these patterns is present then the process is out of control for the average chart. The process should be investigated for special causes and stopped or corrected by adjusting the settings.

## If everything is OK on both the range and average charts then do nothing!

## Important points

- Operators should carry out the measurements and record them on the chart themselves - this is not to be done by 'Quality Control', it is an operator job.
- Operators should join the recorded and calculated points together with a straight line to the previous result.
- Operators should always initial and date the control chart when they make the measurements.
- Operators should always record significant events (materials batch changes, operator changes, colour changes, etc.) on the control chart - it makes finding the special causes easier to find and rectify.
- Unless the control chart indicates one of the above patterns then the process should never be adjusted by the operator. The process is in control - there is no need for adjustment.
- **Tip -** To drive this home, make adjusting the machine without a control chart warning a disciplinary offence.
- **Tip -** Do not be in a rush to adjust the machine - always study the process carefully before adjusting the machine. Look for the obvious causes before adjusting the machine.

## What do control charts tell us?

Control charts provide a 'common language' for operators, managers and reporting. They are the glue that holds a factory together.

They can be used to detect special causes from the alarms and patterns and these can be fixed or eliminated by local action.

Control charts quantify the size of the common causes but the detection and elimination of these (the reason for the UCL and LCL) is much more difficult. Eliminating common causes generally requires management action and possibly even investment in the process.

## The important thing at the operator level is action on special causes not on common causes.

## Common causes need management action.

## Recalculating the limits

The initial average and limits should be recalculated whenever $\overline{R}$ or $\overline{\overline{X}}$ changes significantly. When a control chart has been running and all special causes have been removed then the control limits should not be recalculated. The control limits are based on the common causes and not on the special causes. Control limits should only be recalculated after common causes have been reduced by action and only using data from after the common cause was reduced.

## Reading the charts

Reading a control chart is not a difficult exercise and the following sections give examples of the common 'out-of-control' signals and the likely causes for these signals.

## Posting the charts

SPC charts are not 'office' charts, they should always be posted close to the machines and be highly visible.

The control chart must be a complete statement of what is happening in the process. It should include times of checks, operator changes, material changes and any action taken because of alarms seen.

The following sections give details of the alarms and potential causes for the various alarms.

- Look for the simple causes first - do not make it more complex than it needs to be.
- Always look for a pattern similar to (or the reverse of) the chart pattern.
- If no cause is obvious then break the process down in to sub-processes and look for causes in the sub-processes.

The Process Control Chart is a working document - it is not a work of art.

- Mark on it, write on it but keep it safe. It is a record of the process and may be requested by the customer or needed to review the process.
- Unless the Control Chart tells you to take action you DO NOTHING!
- Note on the chart any of the following:
  - materials batch changes.
  - operator changes.
  - machine changes.
  - anything that could have an effect on quality.
  - Any time 'HELP' was called and the action taken by the person helping.

# 5.10 Range chart alarms - normal variation and points outside the control limits

## What is happening?

SPC is 'inductive statistics', where we look at specific examples and propose what is happening with the general population.

We know that the population will include a range of values and that these will be distributed approximately as per the normal distribution. We also know that the properties of a random sample will be related to the properties of the population (see Section 5.7) and will be an even better approximation of the normal distribution.

We do not expect the sample results to be constant but the limits of variation (UCL and LCL) tell us what is a reasonable levels of variation. Variations outside these levels are unlikely to be caused by common causes but by special causes that we need to know about and remove or reduce.

A set of typical $\overline{X}$ and R control charts is shown on the lower right and both charts are 'in control' with respect to the control limits.

What this could typically mean in terms of the presumed distribution of the process and the distribution of the sample results is shown on the upper right.

There are no alarms present in the control charts and this leads us to believe that there is no significant change in the distribution of the population. However, the individual samples will naturally vary and the changes in the sample distribution are also shown on the upper right. These distributions will vary in width (as measured by the sample range) and in location (as measured by the sample average) BUT none of the results are outside the reasonable levels of variation and there is therefore no reason to believe that the underlying distribution of the population has changed in any way.

This process can therefore be considered to be 'in control'.

Many people and operators get worried about this normal variation and assume that they are making errors, they are not. Unless the results show patterns indicating 'out-of-control' conditions then nothing should be done and the process should not be adjusted.

- **Tip** - Operators should be warned to expect variation but also to be aware of the patterns that indicate something has changed.
- **Tip** - 'Over-enthusiasm' for adjustment can often lead to fluctuating processes.

Inductive reasoning is from the specific to the general whereas deductive reasoning is from the general to the specific.

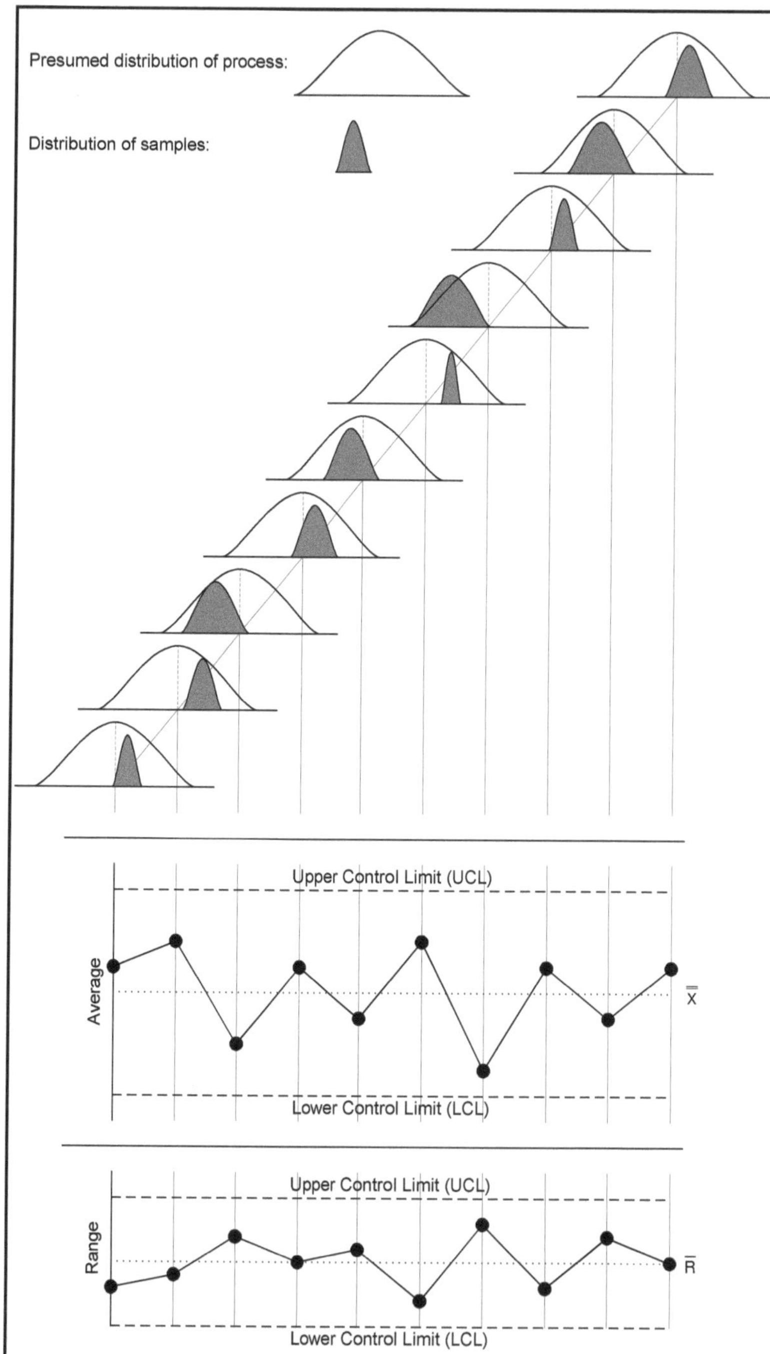


### Sampling will naturally result in variation

Variation in the results for $\overline{X}$ and R is entirely normal (and predictable). Some operators get worried about this but variation is part of the process. We are looking for patterns and results that are outside the normal levels (the UCL and LCL).

## Points outside the control limits (UCL or LCL) in the range chart

A point outside the control chart is the most visible alarm point and indicates that parts are potentially being produced outside the specification limits. It is therefore generally treated with more urgency than some of the other alarms which indicate potential process concerns.

Points outside the control limits on the range are a potential indication of a change in the process as there is only a very small chance that a stable process will produce a point outside the control limits. However, the most common reason for a point outside the control limits on a range chart is generally arithmetic errors and this should be checked first.

## Possible causes and actions

The possible causes for this type of alarm will generally be:

- Arithmetic or calculation errors by the operator.
- Plotting errors by the operator.
- Sample size errors.
- Changes in the measurement system, i.e. different operator or different measurement device.
- Changes in raw materials.
- **Tip** - The first action is to check the value of the point being plotted and the actual point plotted, has the value been correctly transferred to the chart. Plotting errors are unfortunately extremely common.
- **Tip** - Operator calculations for charts should be assisted by calculators or automatic calculation (with error checking).
- **Tip** - If the arithmetic and the sample size are correct then the first area for investigation is the raw material.

**Note:** For sample sizes of < 7, the LCL will always be 0 and it will be impossible for a point to be below the LCL.

The main causes for chart patterns will always be:

- Environment
- Materials
- Machines
- Men
- Methods
- Measures

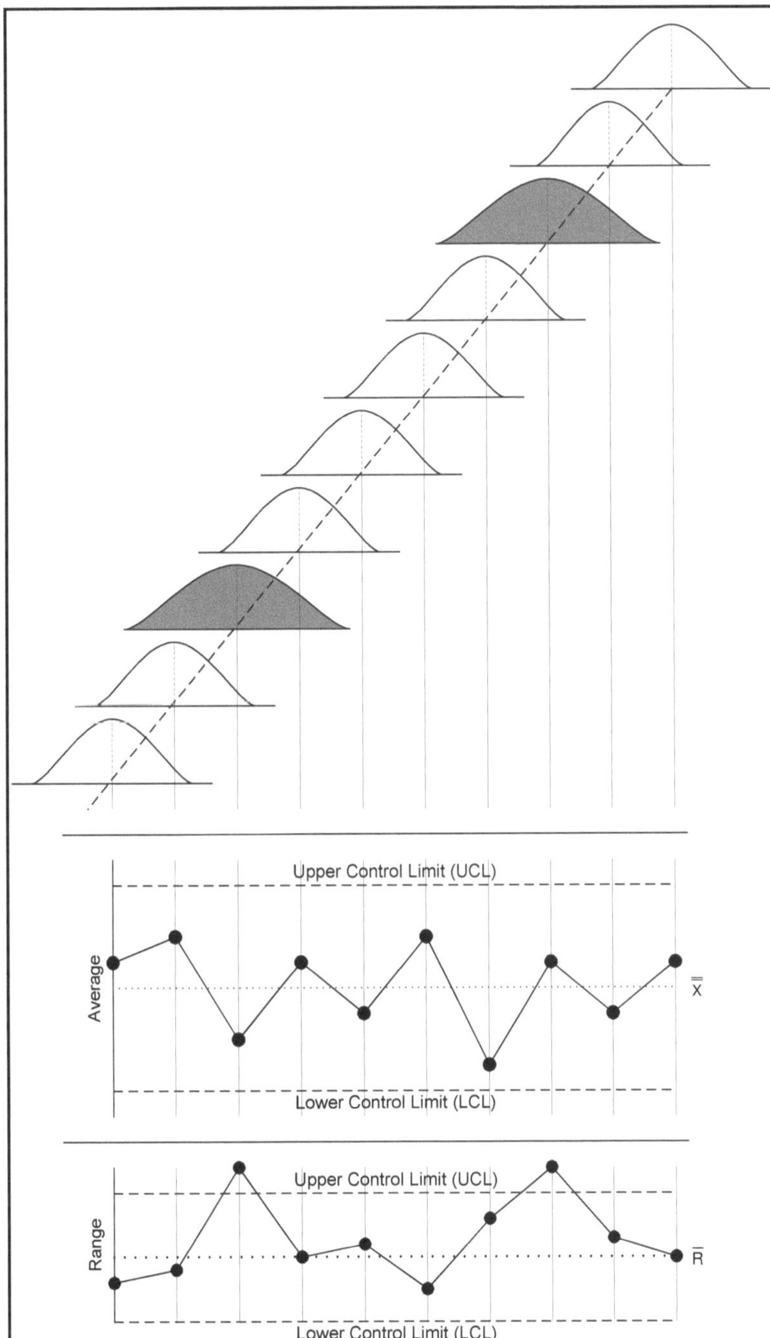


### Points outside the control limits on the range chart

The most common cause for single out-of-control' points on the range chart is simple errors (calculation, plotting, sample sizes). If all these errors are eliminated then the first area for investigation is the raw material.

# 5.11 Range chart alarms - runs (above/below) and runs (up/down)

## No run of 7 consecutive points above or below the average line in the range chart

Recorded results (assuming a normal distribution) should be equally distributed on either side of the average range and a run of 7 consecutive points above or below the average range indicates a potential shift in the underlying process distribution, i.e. the process standard deviation has increased or decreased.

The run can be either above or below the average line depending on what has caused the shift. With this type of alarm, the range will be relatively stable during the period of the shift.

**Note 1:** There may be no out-of-control points on the range chart during the shift.

**Note 2:** The average chart may also start to show large swings but remain centred.

## Possible causes and actions

- Changes in operator - some operators are better than others and operator changes can show as shift in the range.
  - If the shift is up then the operator may be affecting the result as a result of inexperience, poor training or low skill level.
  - If the shift is down then the operator may be affecting the result as a result of experience, good training or better skill level.
- **Tip** - The most common cause of shifts in the range is the operator - look there first.
- Changes in material - some suppliers are more consistent than others and materials changes can show as a shift in the range.
  - If the shift is upwards then the material may be less consistent.
  - If the shift is downwards then the material may be more consistent.
- Changes in machine speed of operation - if a machine is run faster then this can show up as a shift in the range. Higher speeds can often increase variability.
- Changes in method - tool maintenance can reduce the range for a period until the tool becomes worn again.
- **Tip** - Look for short-term effects that match the timescale of the changes in the range chart.
- **Tip** - This pattern in the range chart will also sometimes fail the 'No pattern with 1/3 of the points in the outer 2/3 of the control limits' rule (see Section 5.12).

If the shift is down then it may be possible to improve the process by locking in the improvement.

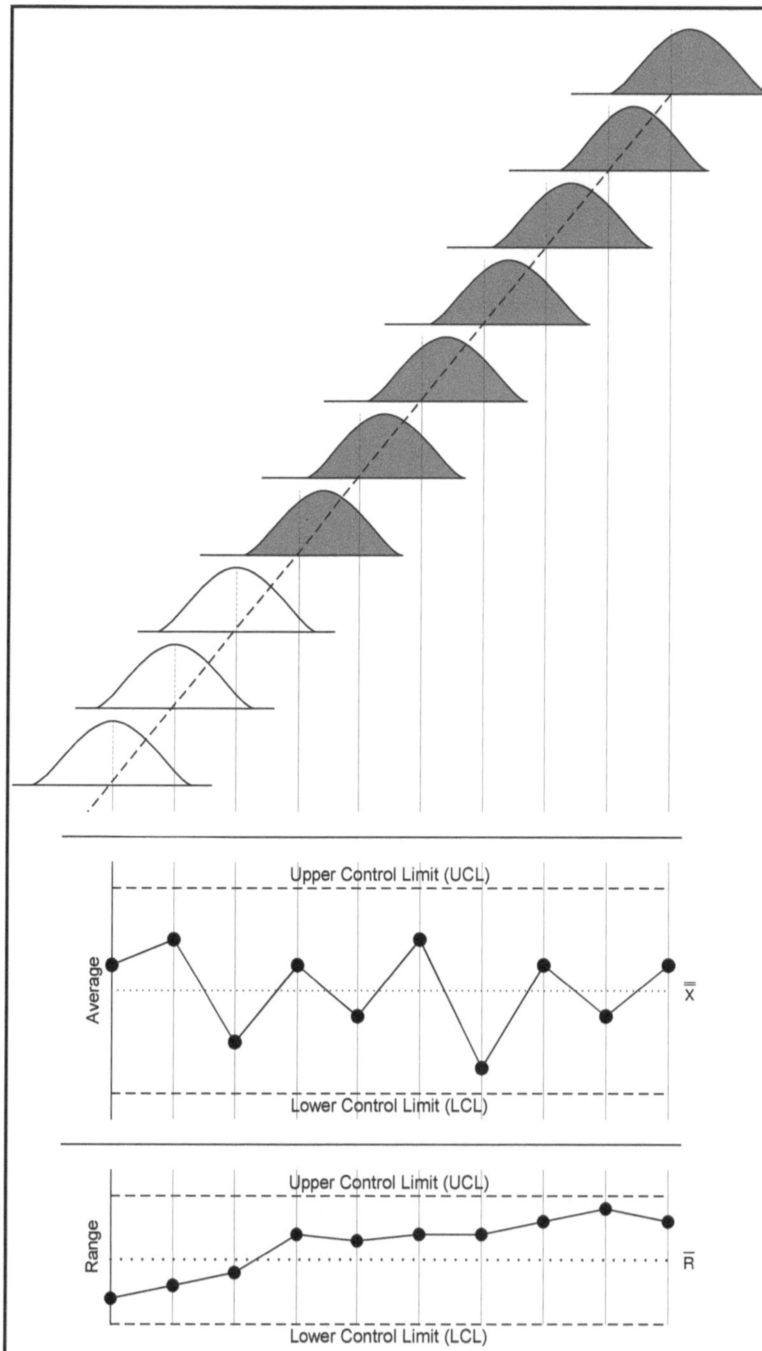


**Run of 7 points above or below the average in the range chart**

A run of points can be either above or below the average range ($\bar{R}$) and indicates a rapid change in the spread of the process. Increasing the spread can lead to out-of-tolerance parts.

## No run of 7 consecutive points upward or downward in the range chart

This alarm will show as an increasing or decreasing value of the range over 7 or more consecutive points. The range chart may go up or down but over the same period there may be no obvious overall trend in the average chart.

**Note 1:** There may be no out-of-control points on the range chart, although these may eventually show on the chart.

**Note 2:** When this type of pattern is present the average chart may also start to show increasingly large swings but it may still remain generally centred.

## Possible causes and actions

- Operator influences - either improvement or deterioration of operator skill.
  - If the range is increasing then the operator may be affecting the result via operator fatigue or poor training.
  - If the range is decreasing then the operator may be gaining experience with the process or good training.
- Machine condition - either improvement or deterioration of the machine.
  - If the range is increasing then the tool may be wearing or becoming clogged and dirty.
  - If the range is decreasing then the machine may be becoming more reliable through regular cleaning or maintenance.
- Control in preceding areas - improvements or degradation in the control in previous areas can give gradual changes in incoming material quality.
  - If the range is increasing then the incoming material may be decreasing in quality (more variable).
  - If the range is decreasing then the incoming material may be increasing in quality (less variable).
- Process controls - changes in the process controls can change the process to give either process improvements or degradation.
- **Tip -** Do not look for the short-term effects, these will often show up as a shift rather than as a drift - a drift is a long-term effect.
- **Tip -** Look for changes that occur over the same time span as the drift in the chart.
- **Tip -** Do not forget that some changes may be delayed in time (shifted along the chart) and the cause may not have been at exactly the same time as the drift in the chart.
- **Tip -** Changes can also be mirrored and not show a direct relationship.

If the run is down then it may be possible to improve the process by locking in the improvement.

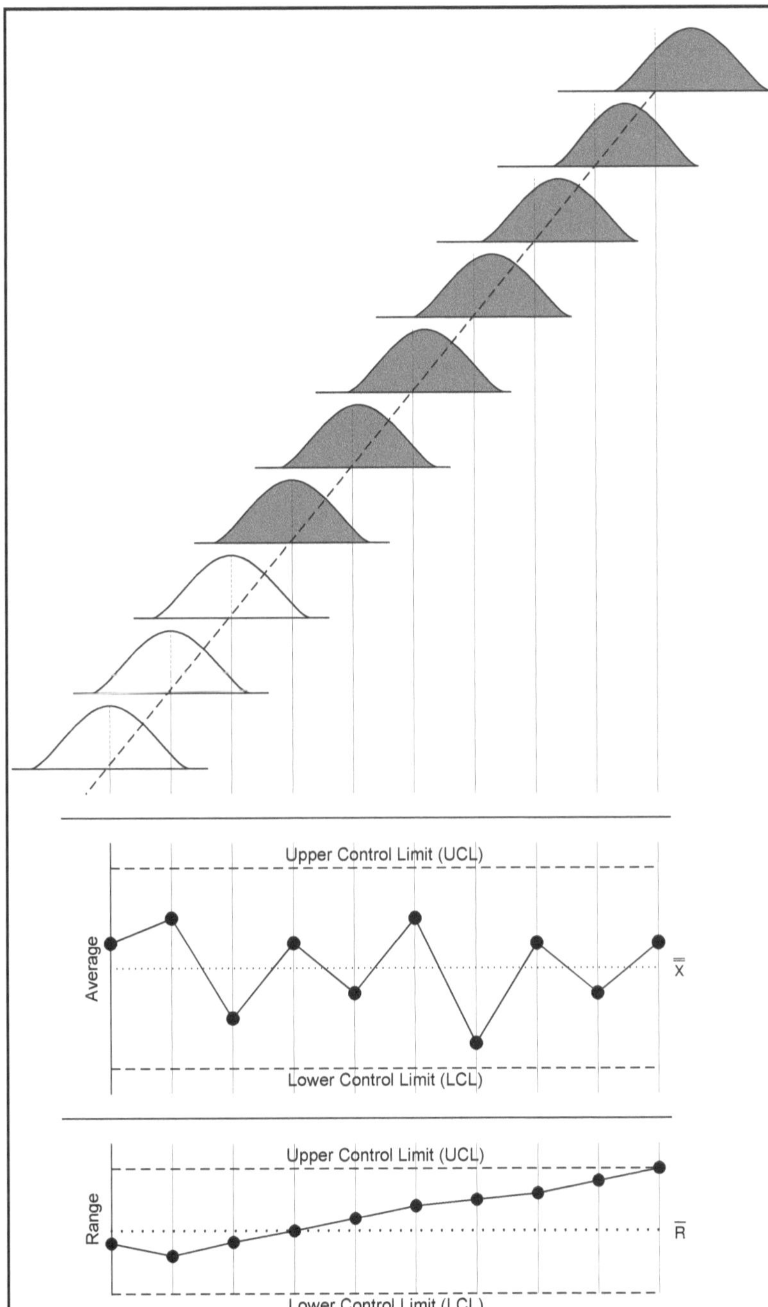


Run of 7 points upward or downward in the range chart

A drift in the range (upward or downward) indicates a slow change in the spread of the process. Increasing the spread can lead to out-of-tolerance parts.

# 5.12 Range chart alarms - 2/3 of points in middle and 1/3 of points in the outer

### No pattern with more than 2/3 of the points in the middle 1/3 of the control limits in the range chart

The normal distribution should show more points in the middle of the chart area than at the outer edges but if the range points are very close to the centre of the chart for an extended period then this indicates a pattern that should be investigated.

**Note 1:** There may be no out-of-control points on the range chart during the period.

**Note 2:** The average chart may also start to show large swings but remain centred.

### Possible causes and actions

- Operators 'fudging' the results to show good numbers - this is more common than realised, operators will sometimes want to keep the process in control and will adjust the numbers to show a low range. This is not always a deliberate policy, simply that when they read the measurement device they 'round' the values so that they are close to the average (see Section 10.15).
- Incorrect calculation of the range limits - if the control limits have been incorrectly calculated then the limits will appear high compared to the measured results and the results will appear to be clustered around the centre line.
- Process improvement - if process control has improved and the process is now more capable then the values on the R chart will decrease and potentially cluster around the centre line and the average will tend to be close to the centre on the $\overline{X}$ chart. In this case the control limits should be recalculated.
- Measuring instrument resolution is not sufficient - if the measurement instrument resolution is too low then it will not be possible to measure correctly. Check that the measurement instrument can measure at least 10 intervals between the UCL and the LCL.
- Sampling is not accurate - inaccurate or preferential sampling can also result in this pattern.
- **Tip -** Look at the numbers and the measurement methods for getting them - unless the process has been improved by management action then this may not be a process problem but a recording problem.

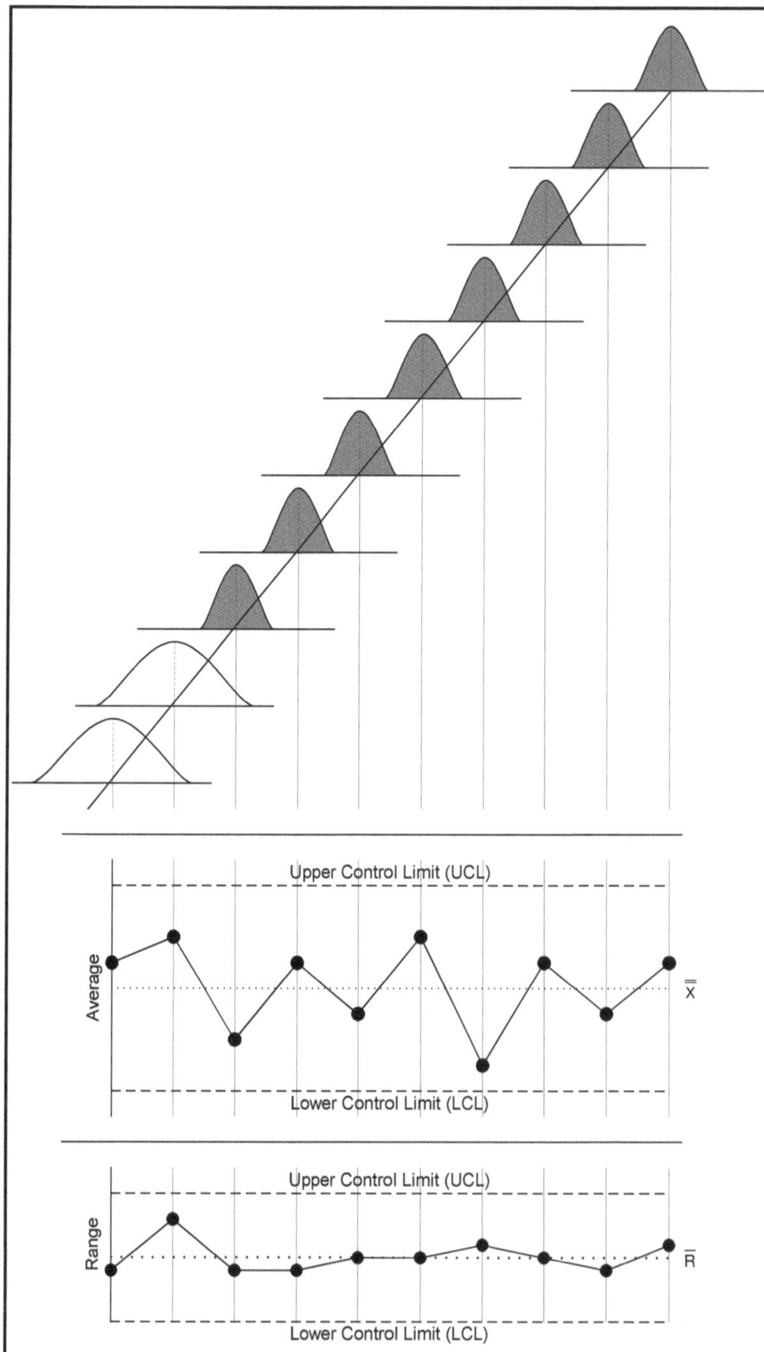


More than 2/3 of points in the middle 1/3 on the range chart

This type of pattern is typically associated with measurement methods (operators and measurement equipment) and with the calculation of the limits. If in doubt, look for recording problems.

## No pattern with more than 1/3 of the points in the outer 2/3 of the control limits in the range chart

The normal distribution should show more points in the middle of the chart area than at the outer edges. If the points are mainly in the outer areas of the chart for an extended period then this indicates a pattern that should be investigated.

During the period of the chart the range may be constantly high (or low) or it may shift up and down over reasonable periods. If the range shifts up and down then the process may be a cyclic one (see Section 5.13). During periods when the range is high (or low) the range is relatively stable and consistent.

**Note 1:** There may be no out-of-control points on the range chart during the shift.

**Note 2:** The average chart may also start to show large swings but remain centred.

## Possible causes and actions

- Changes in operator, e.g. worker rotation - some operators are better than others and operator changes can show as clusters in the outer areas of the range.
  - If the cluster is high then the operator may be inexperienced, have poor training or a low skill level.
  - If the cluster is low then the operator may be experienced, have good training or have a high skill level.
- Changes in material - some suppliers are more consistent than others and material changes can show as a clusters in the range.
  - If the cluster is high then the material may be less consistent.
  - If the cluster is low then the material may be more consistent.
- Changes in method - regular tool maintenance can change the range from a high to a low cluster.
- Changes in speed of machine operation - changing machine speeds (often at shift changes) can change the variability of the process.
- **Tip -** The most common cause of stratification and clusters in the range is the operator - look there first.
- **Tip -** Look for short-term effects that match the timescale of the changes in the range chart.
- **Tip -** This pattern in the range chart could also show up as a cyclic pattern (see Section 5.13) or as a run of points above or below the average line (see Section 5.11).

You already have the best tools for problem solving - looking at and listening to the process will give you invaluable information.

This cannot be done from an office!

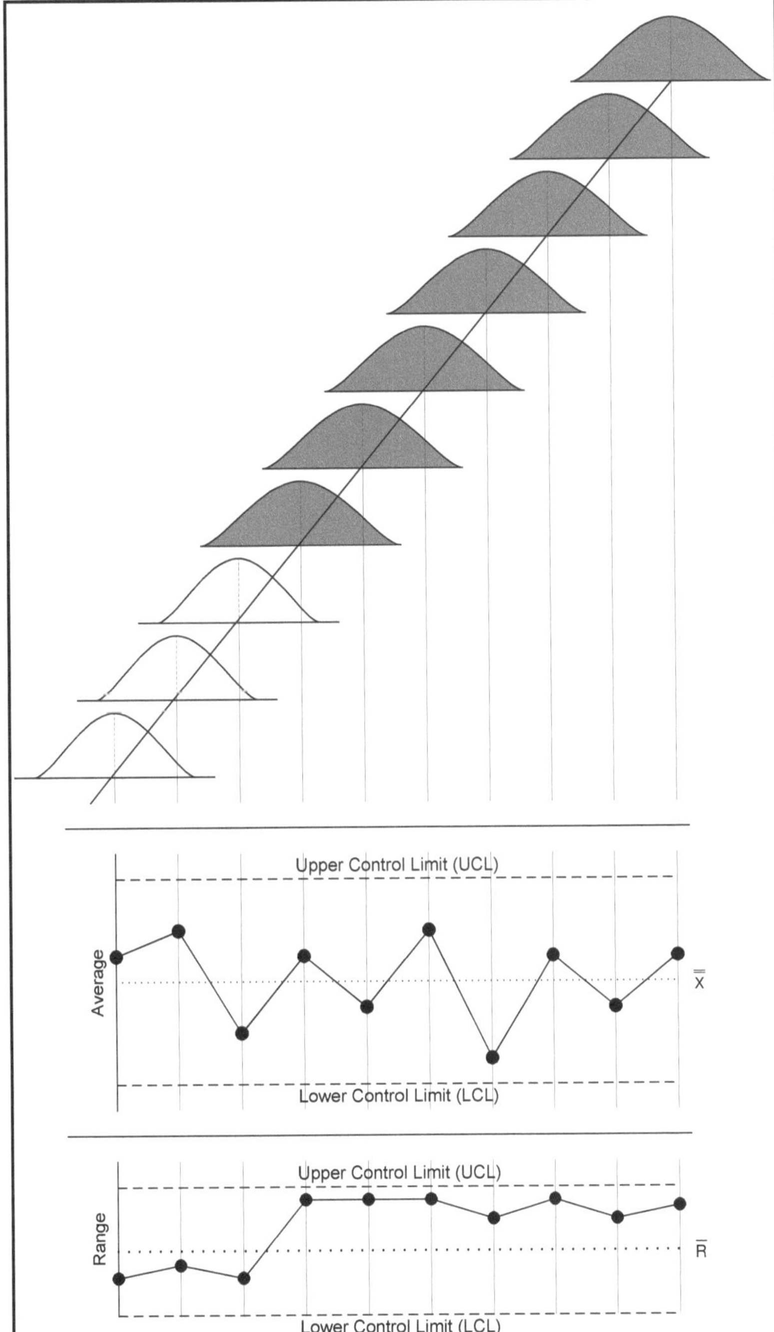


### More than 1/3 of points in the outer 2/3 on range chart

Clustering of results in the outer areas of the chart is typically associated with significant changes in operators (changing skill or experience levels) or material consistency changes. This type of pattern indicates large differences in the process have occurred.

# 5.13 Range chart alarms - other alarms

## Frequent cyclic changes in the range chart

Cyclic patterns will show as changes in the range that are rapid, unpredictable and will generally vary from high to low as shown in the chart on the right.

The major reason for cyclic changes is that the measurements are being made on the output of two machines or processes that are being mixed before the measurements are being made.

**Note 1:** There may be no out-of-control points on the range chart during the shift.

**Note 2:** The average chart may also start to show large swings but remain centred.

## Possible causes and actions

- Different operators using the same measuring instruments and chart - if operators have slightly different assessment methods then this may show up as a mixture. Look for shift changes and get the operators to initial the chart to see if there is an operator pattern to the results.
- Different processes under different conditions being recorded on the same chart - two tools (or multi-cavity tools) may be being used in parallel for the process and being measured as one process. Look further upstream for multiple processes.
- Mixture of materials of different quality or specification - if there are two distinct materials being processed then this can show as a mixture.
- Weather conditions - if the weather conditions are changing, e.g. night/day, hot/cold, etc. then this can affect the process and show as a mixture.
- **Tip** - The most common cause of a cyclic change is the human factor.
- **Tip** - The second most common cause of cyclic variation is suppliers of the process (internal or external).
- **Tip** - Look for variations that are obvious and are part of the process. There is probably a mixture here that you haven't spotted and are recording as a single process.
- **Tip** - Mixtures such as this are often easy to see but are always difficult to resolve (you often need to go upstream to solve them).
- **Tip** - This pattern in the range chart may also fail the 'No pattern with more than 1/3 of the points in the outer 2/3 of the control limits' rule.

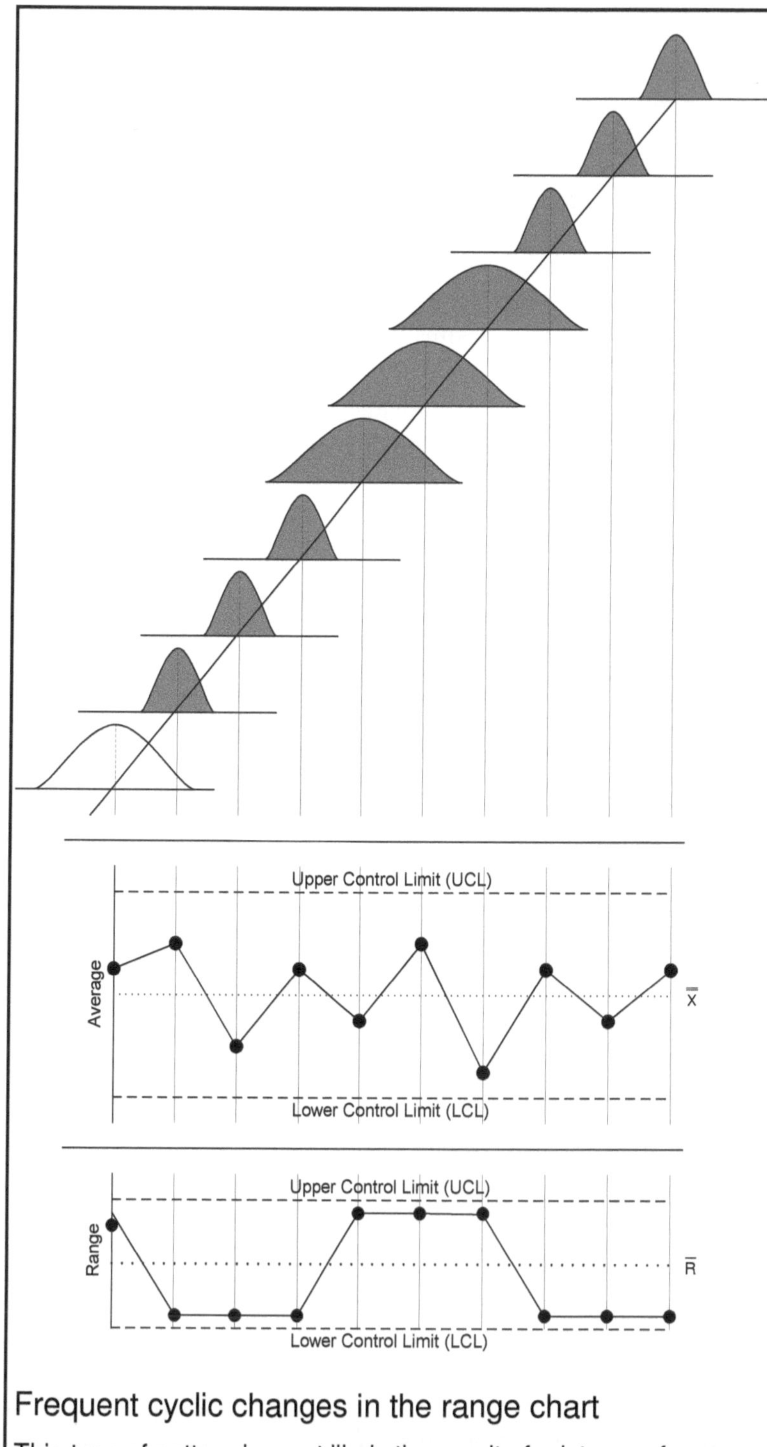


### Frequent cyclic changes in the range chart

This type of pattern is most likely the result of mixtures of some type. It can be mixed different operators, mixed processes, mixed materials or mixed environmental conditions but the important thing is to look further upstream in the process for the solution.

## Patterns and trends

As well as the alarms listed in the previous sections, it is possible for other patterns to be present that will not be revealed by simple runs and out of control points. These can often be revealed as 'bunching' or cyclic results (see opposite page).

In cases such as this it is often useful to divided the points into sub-groups of 2 or 3 points and averaging the results to see if the averages produce a trend.

It is also possible to use moving averages to see other patterns.

## Erratic patterns

Erratic patterns do not necessarily fall into any of the alarms previously mentioned (although they may).

Erratic patterns should generally meet the following requirements:

- At least 10 points are plotted.
- At least 40% of the points are outside the 2/3 limits.
- At least 30% of the points are outside the 1/3 limits.
- The points are not predictable in any way.

This type of pattern can indicate a complex system of causes that are acting together, i.e. more than one cause and/or interacting causes.

- **Tip -** Erratic patterns may mean that you are charting too far from the actual controlling element of the process. Investigate upstream processes to see if these reveal more direct patterns.

Look for the following potential causes of erratic patterns:

- Over-adjustment of the machine by operators - operators may adjust machines when there are no control chart alarms are present. This can happen when shifts change and operators attempt to run machines faster or to their personal preferences.
- Untrained operators - lack of training may mean that some operators do not follow the process guidelines correctly or are variable in their approach to the process.
- Inconsistent raw materials supplies - suppliers may nominally be providing consistent materials but variations can be revealed as erratic patterns.
- Inconsistent or variable measurement devices - if measurement devices allow some 'personal' interpretation of the results then this can show as erratic patterns.
- Errors or malfunctions in the measurement devices - if measurement devices have sporadic errors or malfunctions then this can show as erratic patterns.

## The important thing is action

Detection of patterns indicating changes in the distribution is only the first step in using control charts. The important thing is to take action on the process to remove the special cause that is affecting the process.

"The rules of chemistry and physics are non-negotiable and seldom is a machine following them open to negotiation"

**Kevin Chandler**

**General questions for 'out-of-control' processes**

| | Yes | No |
|---|---|---|
| ■ Have there been any changes to the inputs for the process, e.g. raw material? | ☐ | ☐ |
| ■ Have there been any changes in the operation of the process, e.g. procedure changes, maintenance changes? | ☐ | ☐ |
| ■ Is there potential for the input to contain mixtures, e.g. inputs from different sources or different raw materials? | ☐ | ☐ |
| ■ Have there been any changes in the environment during the period? | ☐ | ☐ |
| ■ Are there environmental factors present, e.g. temperature, humidity? | ☐ | ☐ |
| ■ Do operators use measurement devices of different accuracy? | ☐ | ☐ |
| ■ Are operators using different measurement devices or operational methods? | ☐ | ☐ |
| ■ Are untrained operators working on the process? | ☐ | ☐ |
| ■ Is operator fatigue likely to affect the process? | ☐ | ☐ |
| ■ Do operators frequently adjust the process? | ☐ | ☐ |
| ■ Is there pressure on the operators to report only good news? | ☐ | ☐ |
| ■ Are there predictable changes in the process, e.g. tool wear? | ☐ | ☐ |

Each 'Yes' answer indicates a potential source of a special cause.

# 5.14 Average chart alarms - range control and points outside control limits

## Is the range in control?

It is essential that the range (R) chart is in control before considering the average ($\overline{X}$) chart. This is to ensure that the spread is stable before considering the location. If the spread is not stable then it is useless to try to look at the location.

## The average will vary

Just as with the range, even if the average is in control then it will vary naturally with the variations in the samples taken for the measurements. The same set of typical $\overline{X}$ and R control charts as shown in Section 5.10 is shown on the lower right and both charts are 'in control' with respect to the control limits for both the range and the average.

What this could typically mean in terms of the presumed distribution of the process and the distribution of the sample results is shown on the upper right.

There are no alarms present in the control charts and this leads us to believe that there is no significant change in the distribution of the underlying population. However, the individual samples will naturally vary under the influence of the common causes and the changes in the sample distribution are also shown on the upper right. These distributions will vary in width (as measured by the sample range) and in location (as measured by the sample average) BUT none of the results are outside the reasonable levels of variation for samples and there is therefore no reason to believe that the underlying distribution of the population has changed in any way. This process can therefore be considered to be 'in-control'.

If the average ($\overline{X}$) chart is not in control (as shown by one of the alarm patterns) then it is reasonable to assume that there is a special cause which is responsible for changing the process location from the previous stable location.

As with the range, operators will often not be told this and will get worried about the normal variation that is the consequence of common causes. They may see the centre line ($\overline{\overline{X}}$) as a target that they are expected to achieve and engage in 'fudging' the results to get closer to the 'target'. It should be clearly explained that variation is normal and that unless they see one of the following patterns then they should do nothing.

- **Tip** - Operators should be warned to expect variation and to make no adjustments unless alarms are seen.

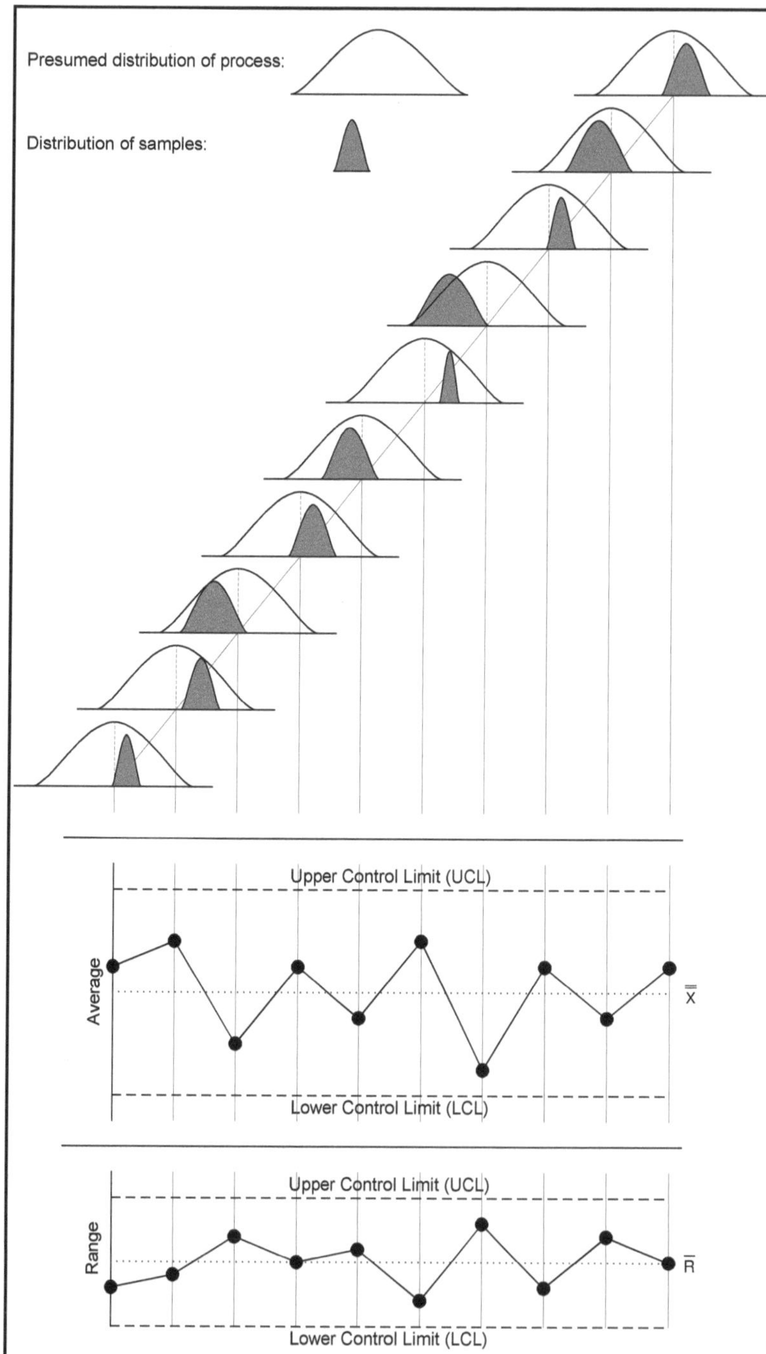


### Sampling will naturally result in variation

Variation in the results for $\overline{X}$ and R is entirely normal (and predictable). Some operators get worried about this but variation is part of the process. We are looking for patterns and results that are outside the normal levels (the UCL and LCL).

## Points outside the control limits (UCL or LCL) in the average chart

A point outside the control limits is a major indication of the presence of special causes and should be investigated as soon as it is seen.

Not all points outside the control limits for the average chart are real movements of the process, one of the most common causes for this pattern is operator error in the calculations and this should always be checked first - there is no sense in carrying out an investigation when the cause is a simple arithmetic error.

**Note:** The range chart should be 'in-control' and go up or down during the period but there may be no obvious overall trend.

## Possible causes and actions

The possible causes for this type of alarm will generally be:

- Arithmetic or calculation errors by the operator.
- Plotting errors by the operator.
- Sample size errors.
- The process has actually shifted - this can be either a single event, i.e. only for the one point or as part of a permanent shift or trend, i.e. for every event thereafter.
- Changes in the measurement system - a different operator or different measurement device can lead to apparent out-of-control points.
- Changes in raw materials - this could be the case for a new or changed material where the new material has different properties to those of the material that it has replaced.
- **Tip -** The first action is to check the value of the point being plotted and the actual point plotted, has the value been correctly transferred to the chart. Plotting errors are unfortunately extremely common.
- **Tip -** Operator calculations for charts should be assisted by calculators or automatic calculation (with error checking).
- **Tip -** If the arithmetic and the sample size are correct then the first area for investigation is the measurement system for the property and the second area for investigation is the source and properties of the raw material.

If the change in the average is a real event, i.e. not an operator error, then the control chart is acting as an early warning device to signal a shift in the average of the process.

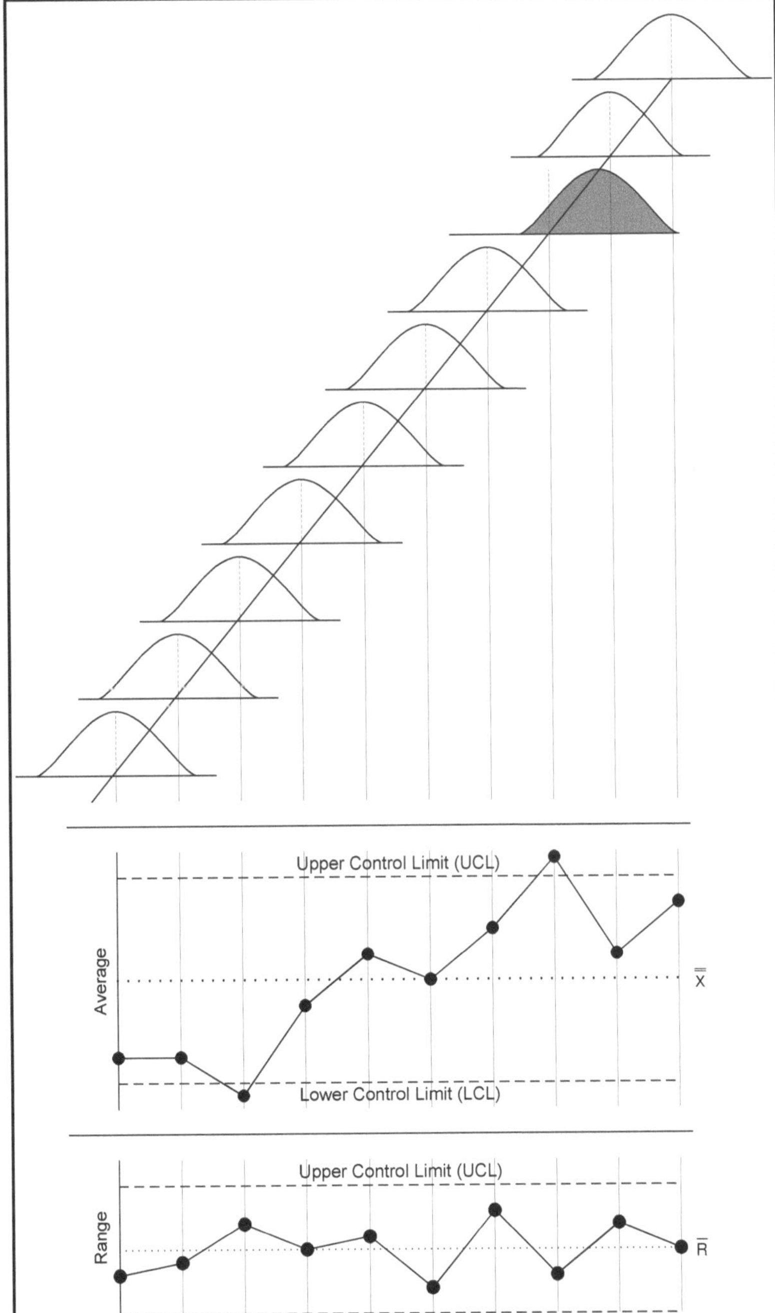


### Points outside the limits on average chart

After the range chart is in control the average chart should be checked. Points outside the control limits in the average chart indicate either a calculation/plotting error, that the process has shifted or that the measurement system has changed.

# 5.15 Average chart alarms - runs (above/below) and runs (up/down)

### No run of 7 consecutive points above or below the average line in the average chart

If there is no change in the process then the value of the sample averages should be consistently spread around the centre line value ($\bar{\bar{X}}$).

If 7 consecutive sample averages are above or below the centre line and no changes have been made to the machine settings then this indicates a high probability that the actual process average has shifted up or down.

**Note:** The range chart should be 'in-control' and go up or down during the period but with no obvious overall trend.

### Possible causes and actions

- Change in proportion of materials or sub-assemblies coming from different sources - if different sources are effectively 'averaged' to give the centre line value ($\bar{\bar{X}}$) then a change in the relative proportions of the source products will shift the process mean.
- Change in inspection method or measurement machine - a different operator or different measurement device can lead to apparent shifts in the process mean.
- New machine or worker starting to produce the products - a new machine or worker may have different characteristics to the previous machine and this may be seen as a shift in the process mean. If a new machine is being used then this should be plotted on a new chart.
- Change in production methods or processes at an earlier process step - when changes are made in earlier processes then this can affect the mean of downstream process, e.g. a change in the mixing process could be reflected as changes in the extrusion process.
- Changes in the process settings (speed, temperature or pressure) could shift the process mean - check the process settings for any changes made at the same time as the shift started.
- **Tip** - Look for any changes in the process (or an earlier process) that matches the timescale for the change in the average chart.
- **Tip** - This pattern may also fail the 'No pattern with more than 1/3 of the points in the outer 2/3 of the control limits' rule (see Section 5.16).

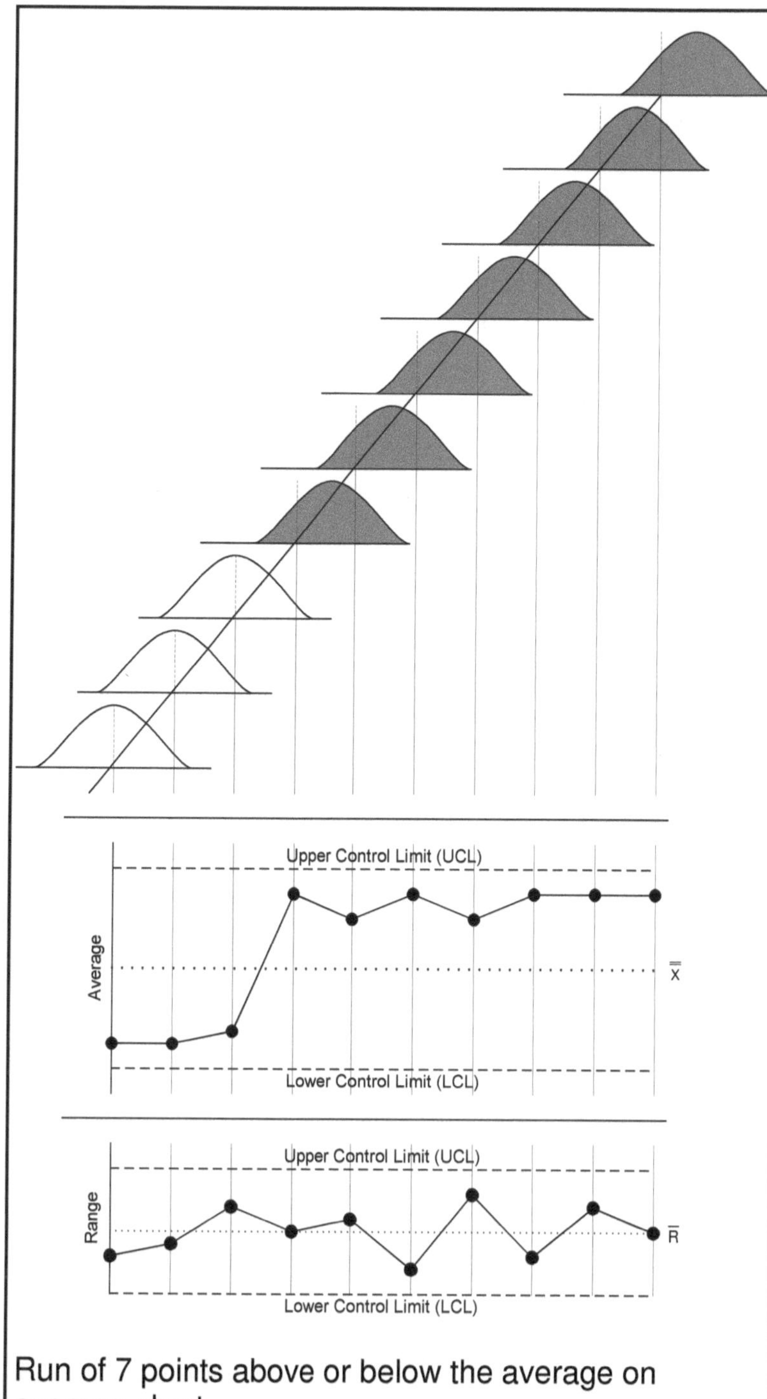


Run of 7 points above or below the average on average chart

A run of 7 or more points above or below the centre line ($\bar{\bar{X}}$) indicates a shift in the process mean.

## No run of 7 consecutive points upward or downward in the average chart

The previous pattern is a shift in the process mean and can generally be linked to a specific event. This type of pattern is a 'drift' which takes place over a long period. As a drift, the source is not normally a specific event but a gradual change in conditions that moves the process mean and during the period of the chart the average gradually increases or decreases.

Action should be taken at the first alarm point ('7 up or down' or 'out-of-control' point) because this type of shift in the process mean is only likely to get worse and eventually lead to out-of-specification products.

**Note:** The range chart should be 'in-control' and go up or down during the period but with no obvious overall trend.

## Possible causes and actions

- Tooling wear or other gradual equipment deterioration - the key here is something that happened gradually to move the process mean. Look for wear and other signs of deterioration in the process.
- Waste product accumulation - clogged or dirty machines can show a gradual shift in the process mean as the waste products build up. Check not only the process but the waste outputs.
- Operator fatigue - this type of pattern can appear as a result of gradual operator fatigue. If this is the cause then the time should be linked to shift or other changes, i.e. the operator will be fresh at the start of the shift.
- Changes in environmental conditions - slow changes in environmental conditions, e.g. temperature and humidity, over the working period can show as slow drifts in the process mean. Look for changes in the environmental conditions over a longer time period, e.g. temperature increases as the site warms up. This can often be related to the time of day the measurements were made.
- **Tip -** Do not look for the short-term effects. A drift is a long-term effect and long-term causes are the most likely.
- **Tip -** To spot trends it can also be worthwhile splitting data into sub-groups of 2 or 3 and using an average of averages or a moving average chart to see any underlying trends.

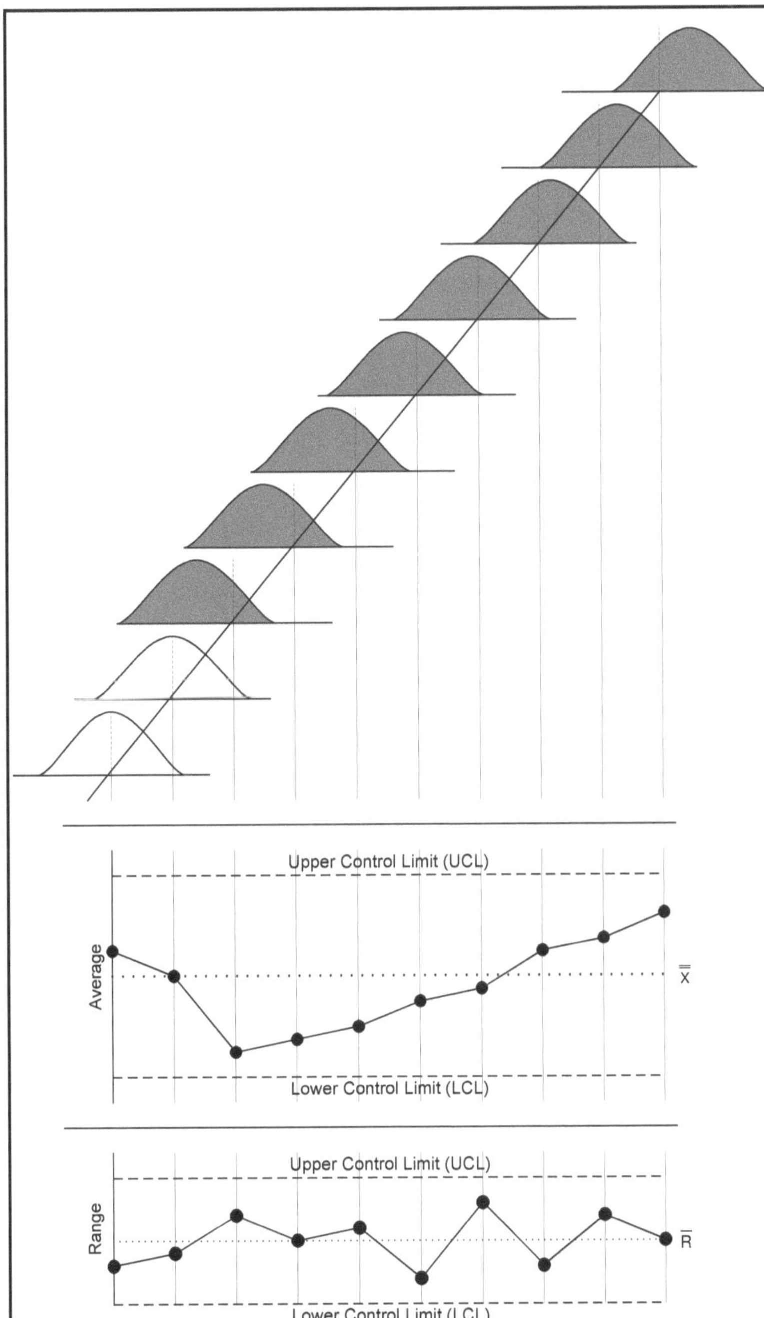


### Run of 7 points upward or downward on average chart

A run of 7 or more points moving upward or downward on the average chart indicates a gradual shift in the process mean and an increase in the potential for producing out-of-specification parts.

# 5.16 Average chart alarms - 2/3 of points in middle and 1/3 of points in the outer

### No pattern with more than 2/3 of the points in the middle 1/3 of the control limits in the average chart

The normal distribution should show more points in the middle of the chart area than at the outer edges but if the average points are very close to the centre of the chart ($\overline{\overline{X}}$) for an extended period then this is a pattern that should be investigated.

When the average is very close to the centre of the chart at all times then there is a pattern that should be investigated.

**Note 1:** There may be no out-of-control points on the average chart.

**Note 2:** The range chart should be 'in-control' and go up or down during the period but with no obvious overall trend.

### Possible causes and actions

- Operators 'fudging' the results to show good numbers - this is more common than recognised, operators will sometimes want to keep the process in control and will adjust the numbers to show a low average. This is not always a deliberate policy, simply that when they read the measurement device they 'round' the values so that they are close to the average (see Section 10.15).
- Incorrect calculation of the control limits - if the control limits have been incorrectly calculated then the limits will appear high compared to the measured results.
- Process improvement - if the process control has improved and the process is now more capable then the values on the range chart will decrease and the average will tend to be close to the centre on the $\overline{X}$ chart. In this case the control limits should be recalculated.
- Measuring instrument resolution is not sufficient - if the measurement instrument resolution is too low then it will not be possible to measure correctly. Check that the measurement instrument can measure at least 10 intervals between the UCL and the LCL.
- Sampling is not accurate - inaccurate or preferential sampling can also result in this pattern.
- **Tip** - Look at the numbers and the methods of getting them. This may simply be a recording problem.

If the average is actually and reliably close to the centre then find out how to lock in the improvement to the process.

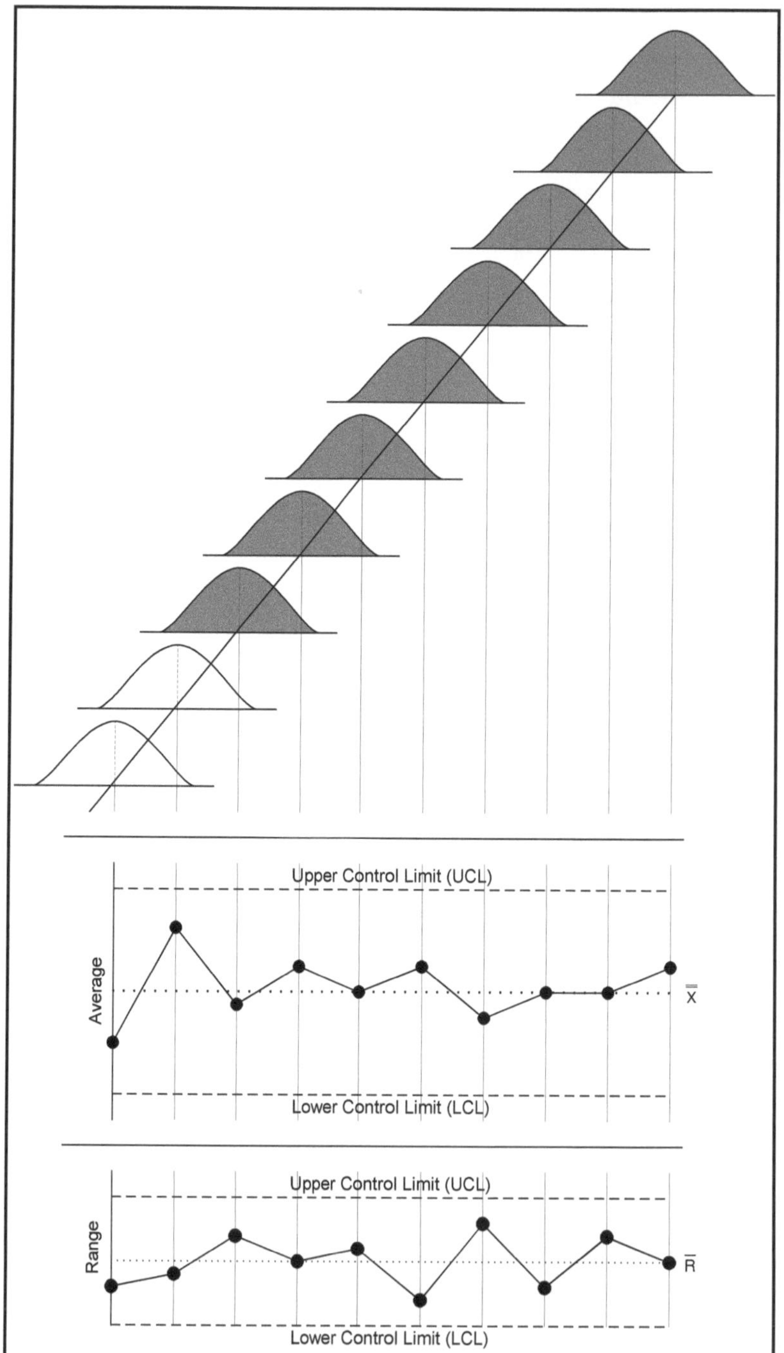


More than 2/3 of points in the middle 1/3 on average chart

When there is a run of points close to the centre of the chart then it is possible that operators are 'fudging' the results to look good or that the control limits have been incorrectly calculated.

## No pattern with more than 1/3 of the points in the outer 2/3 of the control limits in the average chart

The normal distribution should show points distributed between the control limits. If the points are clustered in the outer areas of the chart this should be investigated.

The chart on the right shows stratification and is likely to be the result of measurements produced by similar but slightly different machines or tooling, .e.g. different cavities of a multi-cavity tool. It is likely that each cavity is stable and producing good product but that the measurements are being made on a mixture of the products. Depending on the cavity chosen then the results will be either higher or lower.

**Note 1:** There may be no out-of-control points on the average chart.

**Note 2:** The range chart should be 'in-control' and go up or down during the period but with no obvious overall trend.

### Possible causes and actions

- Multi-process machines and other mixtures - this type of pattern can be caused by measurements made on the output of several processes where the process outputs vary slightly. Look for multi-cavity tooling where measurements are being randomly made on the cavities.
- Mixtures of inputs - this type of pattern can result from variations in the input materials. Check upstream for mixtures from various processes, e.g. materials, components. If there are several upstream processes contributing to the input of the charted process then these should be charted individually.
- Sampling is not accurate - if the sampling is not accurate then it is possible that the samples do not reflect the actual properties of the output.
- Over-adjustment of upstream processes - if upstream processes are constantly being adjusted (without charting) on the basis of random process fluctuations then the charted process could show this pattern of 'hunting' as the upstream changes are made.
- **Tip** - Look upstream first and see if the times of the changes match upstream changes in the process.
- **Tip** - Look at the numbers and the methods of getting them. This may simply be a recording problem.

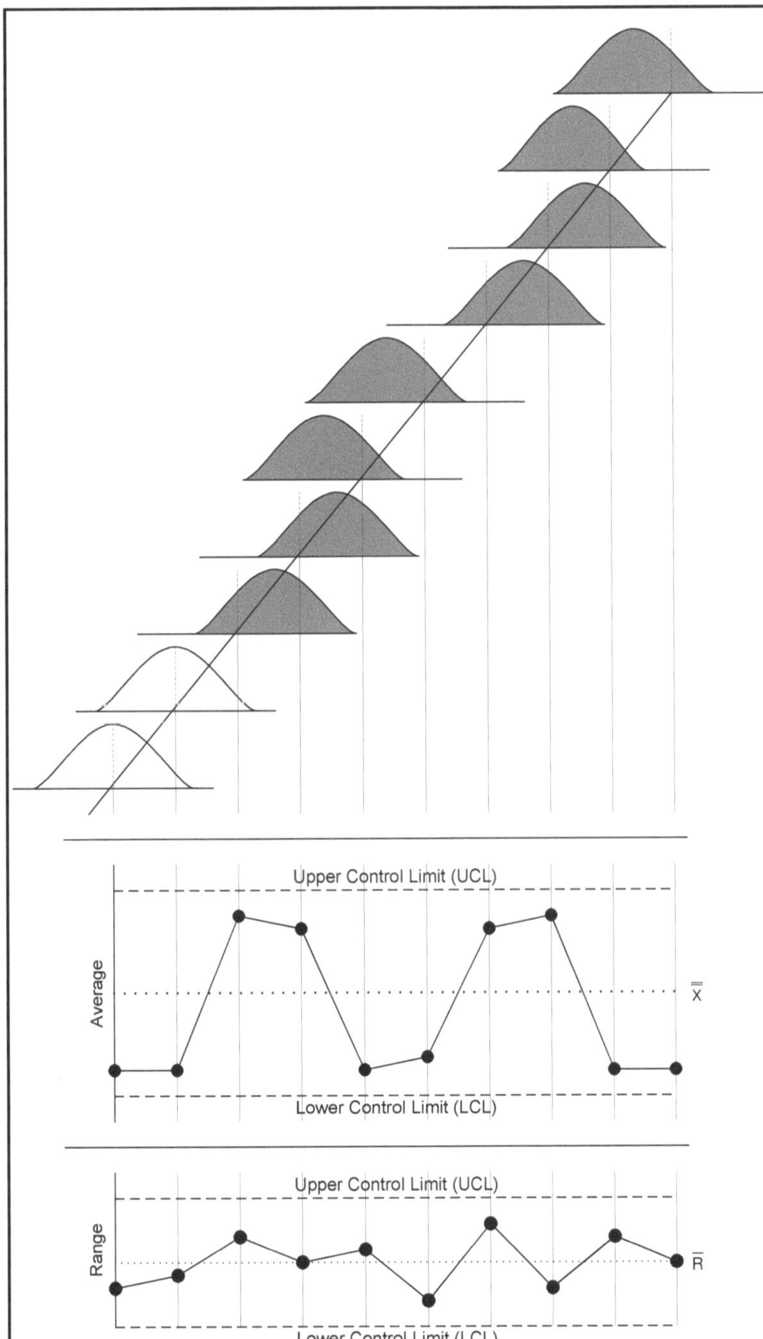


**More than 1/3 of points in the outer 2/3 on average chart**

When there is a run of points close to the outer limits of the chart then it is possible that mixtures are part of the process (either in the process or as inputs). Look first for mixture-related effects.

# 5.17 Other control charts for variables

## Decisions, decisions

All of the examples given so far in this workbook have used the mean and range chart as this is the most common variables control chart.

However, there are times when the other types of variables control charts discussed in Section 5.7 are better for the available data. The selection of which chart to use is determined mainly by the type of data available and the sample size. This is shown in the flow chart on the right and selecting the correct chart to use is relatively simple. If in doubt then use mean and range.

## Alarms for other control charts for variables

The alarms for the other types of variables control charts listed below are exactly the same as for the alarms of the $\overline{X}$ and R chart that have been discussed previously - look for the same patterns and the same special causes.

## The individuals chart

When it is only possible to get one reading at each interval, e.g. melt flow index data, or when the data cannot be sub-grouped by time then a chart for individuals should be used (see Section 5.18).

## The median chart ($\widetilde{X}$ and R)

The median (the middle value of a set of numbers) can also be used as an indicator of the location of the mean. The advantage in using the median is that when using an odd number of samples then no calculation is necessary, i.e. the median is simply the middle value, e.g. if the results were 1, 4, 5, 6 and 8 then the median is the middle value of the ordered numbers, i.e. 5.

If an even number of samples is chosen then the median is the average of the two middle numbers, e.g. if the results were 1, 4, 5, 6, 8 and 9 then the median is the average of 5 and 6 or 5.5. Using an even number of samples removes much of the benefit of using median charts and consequently the sample size is normally an odd number.

When using median charts, all the data are plotted (not simply the median) but the medians are joined up to give the control chart lines. $\widetilde{X}$ and R charts are created and analysed as for $\overline{X}$ and R charts but the following equations are used to calculate and set the centre lines and control limits:

- Centre line for R:

$$\overline{R} = \frac{\textbf{Sum of R values}}{\textbf{Number of samples}}$$

- Upper Control Limit for $\overline{R}$:

$$UCL_R = D_4 \times \overline{R}$$

- Lower Control Limit for $\overline{R}$:

$$LCL_R = D_3 \times \overline{R}$$

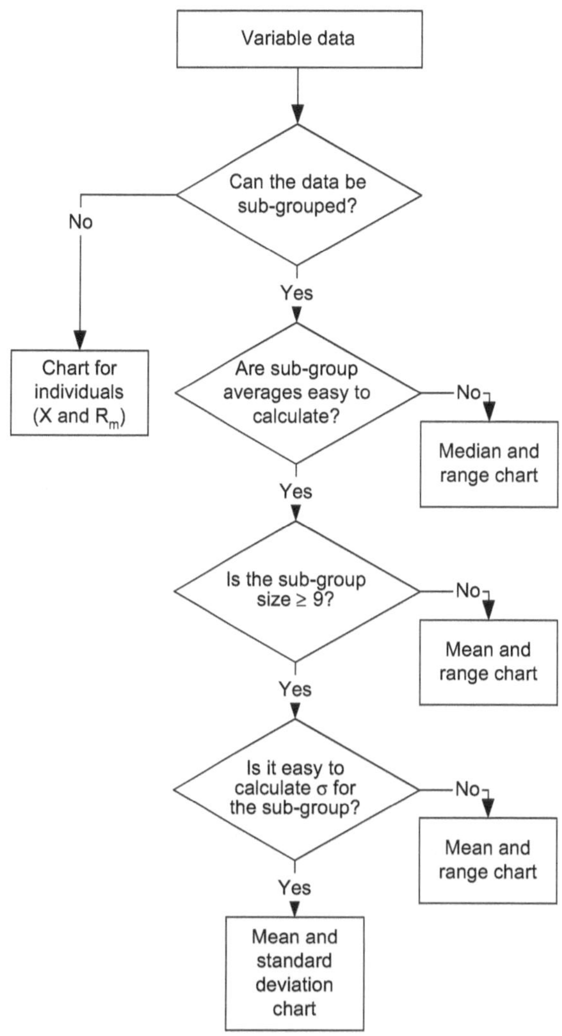


### Which type of variable chart to use?

Although the mean and range is the most commonly used variable control chart, there are times when other charts are better or make it easier to control the process. This flow chart gives a selection process to decide which type of chart to use.

where $D_4$ and $D_3$ are constants that vary with the sample size (see Appendix 2 for values of these constants for a range of sample sizes).

- Centre line for $\tilde{X}$:

$$\bar{\tilde{X}} = \frac{\textbf{Sum of } \tilde{X} \textbf{ values}}{\textbf{Number of samples}}$$

- Upper Control Limit for $\tilde{X}$:

$$UCL_{\tilde{X}} = \bar{\tilde{X}} + \tilde{A}_2 \times \bar{R}$$

- Lower Control Limit for $\tilde{X}$:

$$LCL_{\tilde{X}} = \bar{\tilde{X}} - \tilde{A}_2 \times \bar{R}$$

where $A_2$ is a constant that varies with the sample size (see Appendix 2 for values of this constant for a range of sample sizes).

Despite the advantages of reduced calculations, median charts are not generally preferred because the estimate of the process mean is more variable than that obtained from $\bar{X}$ and R or $\bar{X}$ and s charts and the resulting control limits can be wider than for $\bar{X}$ and R or $\bar{X}$ and s charts. This can result in less efficient process control.

## The mean and standard deviation chart ($\bar{X}$ and s)

The standard mean and range chart ($\bar{X}$ and R) is best for small sample sizes, i.e. n < 8, but for larger sample sizes, i.e. n > 8, the mean and sample standard deviation ($\bar{X}$ and s) provides a better estimate of the process spread. The drawback with using the sample standard deviation (s) is that this is less sensitive in detecting when a single value in the sample is very different from the other values in the sample. The sample standard deviation (s) can be calculated using a spreadsheet or an advanced calculator. $\bar{X}$ and s charts are created and analysed as for $\bar{X}$ and R charts but the following equations are used to calculate and set the centre lines and control limits:

- Centre line for s:

$$\bar{s} = \frac{\textbf{Sum of s values}}{\textbf{Number of samples}}$$

- Upper Control Limit for s:

$$UCL_s = B_4 \times \bar{s}$$

- Lower Control Limit for s:

$$LCL_s = B_3 \times \bar{s}$$

where $B_4$ and $B_3$ are constants that vary with the sample size (see Appendix 2 for values of these constants for a range of sample sizes).

- Centre line for $\bar{X}$:

$$\bar{\bar{X}} = \frac{\textbf{Sum of } \bar{X} \textbf{ values}}{\textbf{Number of samples}}$$

- Upper Control Limit for $\bar{X}$:

$$UCL_{\bar{X}} = \bar{\bar{X}} + A_3 \times \bar{s}$$

- Lower Control Limit for $\bar{X}$:

$$LCL_{\bar{X}} = \bar{\bar{X}} - A_3 \times \bar{s}$$

where $A_3$ is a constant that varies with the sample size (see Appendix 2 for values of this constant for a range of sample sizes).

Control charts either provide evidence for many of the suspicions that production people have about what makes a process work OR they get rid of all the 'folklore' surrounding a process.

### Charts for processes with moving means

In some cases, tool wear, fixture wear or maintenance intervals, e.g. oil changes, can affect the results and change the process over time. When this happens, the mean (or median) is not fixed but increases or decreases with time and the control chart needs to take this into account.

The process is as follows:

- Collect data on the chosen type of control chart as normal over at least one complete process cycle and note any events that may change the process.
- The average movement of the mean =

  Maximum observed $\bar{X}$ - Minimum observed $\bar{X}$ (over cycle)
- Calculate the control limits for $\bar{X}$ from:
  - Upper Control Limit for $\bar{X}$:

$$UCL_{\bar{X}} = \bar{\bar{X}} + 0.5 \times \text{Average movement of the mean} + A_2 \times \bar{R}$$

  - Lower Control Limit for $\bar{X}$:

$$LCL_{\bar{X}} = \bar{\bar{X}} - 0.5 \times \text{Average movement of the mean} + A_2 \times \bar{R}$$

  where $A_2$ is the same constant as for the standard charts. For a sample size of 5, the value of $A_2$ is 0.577 (see Appendix 3 for values of $A_2$ for other sample sizes).
- Calculate the control limits for R from:
  - Upper Control Limit for R (UCLR):

$$UCL_R = D_4 \times \bar{R}$$

  - Lower Control Limit for R ($LCL_R$):

$$LCL_R = D_3 \times \bar{R}$$

  where $D_4$ and $D_3$ are same constants as for the standard charts. For a sample size of 5, the values of $D_4$ and $D_3$ are 2.114 and 0, respectively (see Appendix 3 for values of $D_4$ and $D_3$ for other sample sizes).

In this case, the control limits do not identify out-of-control points but indicate when the tool or fixture needs maintenance, when the oil needs changing, etc.

As a general rule, the tool wear rate in plastics processing is not fast enough to warrant using moving means but this can be a useful technique for wear or maintenance issues.

# 5.18 Variables control charts for individual measurements

## Charts for individual measurement ($\overline{X}$ and MR)

In some cases it is not possible to sub-group the data and it is only possible to take a single measurement at each time. Some examples of this might be:

- Taking the melt flow index of delivered polymer.
- Measuring the impact strength of profiles.
- Measuring the viscosity of paint for a paint line.
- Burst test measurements.

In all these cases, the testing is either destructive and expensive or the values taken at any one time are extremely likely to be very similar, i.e. the range is likely to be very small. This type of result or testing does not generate enough data to use the conventional 'mean and range' techniques which need multiple results and the 'individuals' control chart provides a method of checking for process control.

## Getting the data

- Carry out the testing and record the individual results ($X_i$) for each test. As a starter then try to get at least 20 readings that are arranged in time order.
- Calculate the difference between each successive test result, i.e. the difference between $X_n$ and $X_{n-1}$. This is the moving range (using a sub-group of 2) and is always positive, i.e. use the absolute value. This is the moving range (MR).

**Note 1:** The sub-group size is defined as the number of points used to calculate the moving range. If you use 3 points to find the moving range then the sub-group size will be 3. If in doubt, use successive points and a sub-group size of 2.

**Note 2:** You will always have fewer values for the moving range than for the actual results. If you use a sub-group size of 2 then you will automatically have 1 less value.

- Plot the values for each individual result on the individuals chart (see upper right for an example).
- Plot the values for the moving range on the moving range chart (see lower right for an example).
- Calculate the average from the initial individual results (this is $\overline{X}$ - although it is sometimes called $\overline{\overline{X}}$ for consistency with other control charts) and the average of the moving range ($\overline{R}$).
- Calculate the provisional control limits from:
  - Upper Control Limit for X:

    $UCL_X = \overline{X} + E_2 \times \overline{R}$
  - Lower Control Limit for X:

    $LCL_X = \overline{X} - E_2 \times \overline{R}$
  - Upper Control Limit for MR:

    $UCL_{MR} = D_4 \times \overline{R}$
  - Lower Control Limit for MR:

    $LCL_{MR} = D_3 \times \overline{R}$

  Note that $LCL_{MR}$ is always 0 for a sub-group size of 2.
- When using a sub-group size of 2 the values of the constants are as follows:
  - $E_2 = 2.66$
  - $D_3 = 0$
  - $D_4 = 3.27$

  Values of $E_2$, $D_3$ and $D_4$ for other sub-group sizes are given in Appendix 2.
- These provisional control limits can be used for initial assessment of control but should be recalculated after approximately 100 individual results (assuming the process is in control for this period) to give the final control limits.
- As for other charts, the process standard deviation can be estimated from:

  $$\sigma = \frac{\overline{R}}{d_2}$$

  where $d_2$ is a constant that varies with sub-group size and $d_2 = 1.128$ for a sub-group size of 2. Values of $d_2$ for other sub-group sizes are given in Appendix 2.

## Interpreting for process control

Care needs to be used in interpreting charts for individual measurements for process control because of the relationship between successive points in the charts, i.e. the data are related to one another.

The individuals control chart allows us to use SPC techniques for single measurement data. This is useful for many tests and controls where testing is expensive or difficult.

## Moving range chart

As with standard control charts the moving range control chart is examined first. The only test that should be used for the moving range chart is that of points beyond the control limits. The relationship between successive points means that the run tests, etc. should not be used.

In the example shown (lower right), there is an obvious out-of-control point at Test 16 where the moving range exceeded the control limits.

## Individuals chart

The individuals chart can then be examined using all the standard tests (points outside the limits, spread of points, runs and other standard trends).

In the example shown (upper right), the process is also out of control for Points 9 and 15 and from Point 16 onwards there is a failure of the run test and is out of control for Point 28. It is obvious that there has been a process shift at Point 16 where the Melt Flow Index has generally increased. This could be due to a new batch, faulty material or other special causes. This needs investigation to prevent processing issues.

## Cautionary notes

Control charts for individuals can be very useful but they need to be used with the following cautions:

- Individuals charts are not as sensitive in detecting process changes as other chart types due to the reduced number of samples taken. They can detect shifts in the region of $\pm 3\sigma$ but not less.
- Individuals charts need more care in reading. This is especially true if the distribution is not normal because the central limit theorem will not apply because only a single sample is taken (see Section 5.7). If the process distribution is not a normal then there is a risk of false signals for out-of-control points.
- Individuals charts do not show the piece-to-piece repeatability of the process. In some cases it is better to use the conventional $\overline{X}$ and R chart and to sub-group the samples (by time) to show this.
- The use of only 1 sample per sub-group means that $\overline{X}$ and $\hat{\sigma}$ can be highly variable even if the process is stable (at least until enough data have been gathered).

## Other control charts for the individuals

The individuals control chart described above is not the only method for process control when only single measurements are available. It is also possible to use CUSUM (CUmulative SUM) charts or exponentially weighted moving average (EWMA) charts.

These charts are more sensitive to smaller shifts in the mean because they use the cumulative information of the sequence of values. However, they also assume that a good estimate or known value for the true standard deviation is available.

These charts are outside the scope of this Workbook but for further information consult a more extensive handbook on SPC.

Individuals control charts need care in analysing the results. Do not simply use the standard methods.

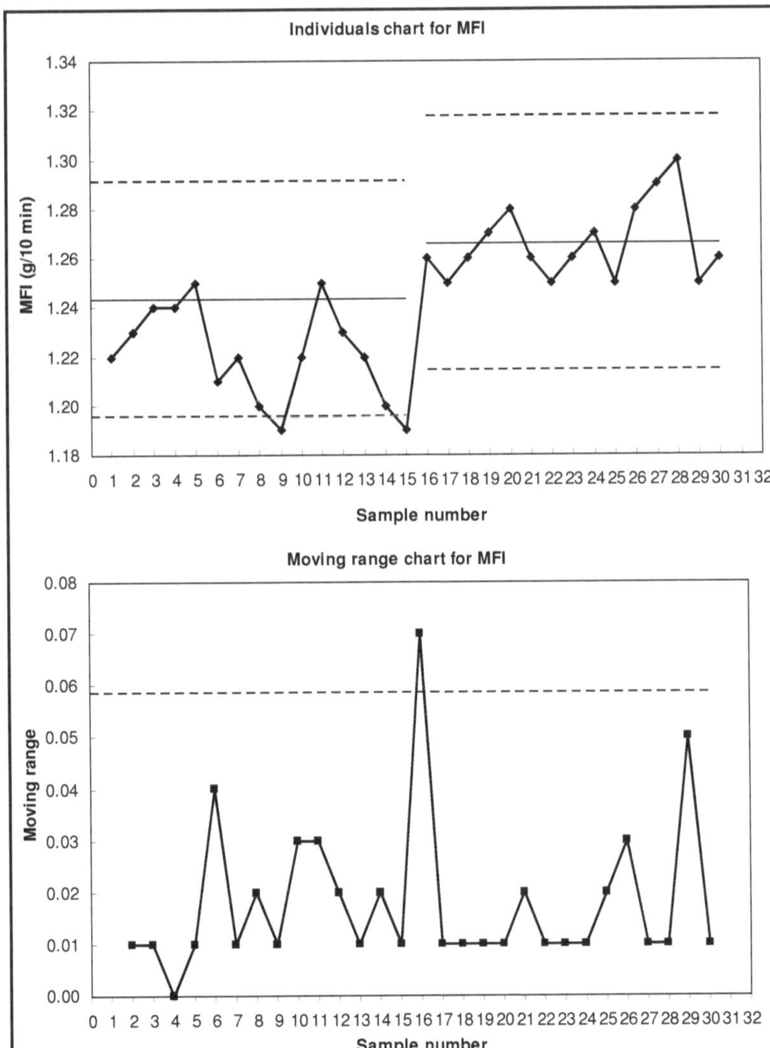


### Individuals chart for Melt Flow Index

The individuals control chart uses the moving range as a measure of the short-term variation. In the example above, the moving range is the absolute difference between each group of 2 samples.

# 5.19 Attributes control charts

## Just like variables control charts

The broad types of attributes control charts have already been discussed (see Section 5.7) and in many ways attributes control charts are very similar to variables control charts. Therefore, the treatment of attributes control charts will be much shorter due to this high degree of similarity.

For those new to SPC, it may appear that selecting the correct type of attribute chart is difficult. This is not the case; the type of data available and the sample sizes effectively define the type of chart to use. This is shown in the flow chart on the right.

As with variables control charts the process of creating a chart is:

- Decide on the process parameter to be controlled.
- Get the data for ≈ 25 samples.
- Plot a trial chart and control limits.
- Check for control.

## Nonconforming items

### Fraction nonconforming (p chart)

The p chart plots the proportion of nonconforming items (p) found in the sample inspected. The use of a fraction nonconforming means that the sample size can vary and this is one of the main reasons for using the p chart, i.e. it can be used when there is a possible variation in the sample size. Despite this, any variation in sample size should be restricted so that sample sizes are within ± 20% of the average sample size ($\bar{n}$) to avoid having to recalculate the control limits for each sample. The sample size for p charts should be > 50 to ensure that at least some nonconforming items are found.

- What is plotted:

p - the proportion of nonconforming items in a sample, i.e. if the sample size = n and the number of nonconforming items in the sample = f, then p = f/n.

- Centre line:

$$\bar{p} = \frac{\textbf{Total number of nonconforming items}}{\textbf{Total number of samples inspected}}$$

= the average proportion of nonconforming items in a set of samples.

- Upper control limit:

$$UCL_p = \bar{p} + 3 \times \sqrt{\frac{\bar{p}(1-\bar{p})}{\bar{n}}}$$

where $\bar{n}$ = average sample size.

- Lower control limit:

$$LCL_p = \bar{p} - 3 \times \sqrt{\frac{\bar{p}(1-\bar{p})}{\bar{n}}}$$

- **Tip** - If the sample size varies greatly then you will have to recalculate the control limits for each sample size. This is a very good reason for keeping the sample size constant if at all possible.

Don't let the formulae confuse you. This is all easy to set up and run.

### Number nonconforming (np chart)

The np chart plots the number of nonconforming items (np) found in the sample inspected. Because the actual number of nonconforming items is plotted, the sample size must be constant. The sample size for np charts should be > 50 to

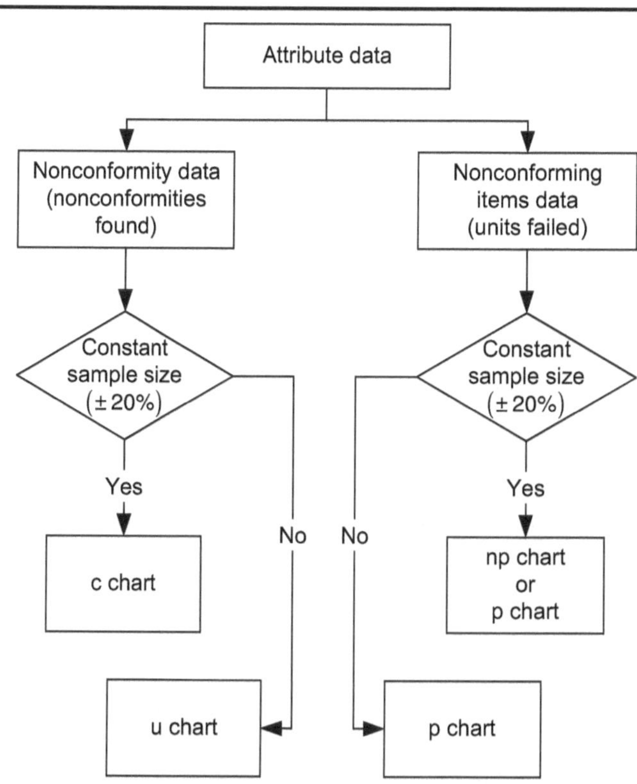


Which type of attribute chart to use?

The number of attributes control charts is often initially confusing but the decision is quite simple. The only decisions are whether the data refers to nonconformities or nonconforming items and whether or not the sample size is approximately constant.

ensure that at least some nonconforming items are found.

- What is plotted:

f - the number of nonconforming items found in the sample.

- Centre line:

$\overline{np} = n \times \overline{p}$ = the average number of nonconforming items in a set of samples.

- Upper control limit:

$$UCL_{np} = \overline{np} + 3 \times \sqrt{\overline{np} \times (1 - \overline{p})}$$

where

$$\overline{p} = \frac{\text{Total number of nonconforming items}}{\text{Total number of samples inspected}}$$

= the average proportion of nonconforming items in a set of samples and n = sample size.

- Lower control limit:

$$LCL_{np} = \overline{np} - \left[3 \times \sqrt{\overline{np} \times (1 - \overline{p})}\right]$$

## Nonconformities

### Number of nonconformities (c chart)

The c chart is similar to the np chart except that it plots the number of nonconformities and not the number of nonconforming items. As for the np chart, the c chart requires a constant sample size, generally > 25 but the sample may be a single unit or a constant sized group of units.

- What is plotted:

c - the number of nonconformities found in the sample.

- Centre line:

$\overline{c}$ = The average number of nonconformities in a set of samples.

- Upper control limit:

$$UCL_c = \overline{c} + \left(3 \times \sqrt{\overline{c}}\right)$$

- Lower control limit:

$$LCL_c = \overline{c} - \left(3 \times \sqrt{\overline{c}}\right)$$

### Number of nonconformities per unit (u chart)

The u chart is similar to the p chart except that it plots the proportion of nonconformities (u) found in the sample inspected. The use of a proportions/unit again means that the sample size can vary but, as for the p chart, any variation in sample size should be restricted so that sample sizes are within ± 20% of the average sample size ($\overline{n}$) to avoid having to recalculate the control limits for each sample. The sample size should be such that at least some nonconformities are found.

- What is plotted:

u - the proportion of nonconformities in a sample.

If the sample size = n and the number of nonconformities in the sample = c, then u = c/n.

- Centre line:

$$\overline{u} = \frac{\text{Total number of nonconformities}}{\text{Total number of samples inspected}}$$

= the average proportion of nonconformities in a set of samples.

- Upper control limit:

$$UCL_u = \overline{u} + \left(3 \times \sqrt{\frac{\overline{u}}{\overline{n}}}\right)$$

where $\overline{n}$ = average sample size.

- Lower control limit:

$$LCL_u = \overline{u} - \left(3 \times \sqrt{\frac{\overline{u}}{\overline{n}}}\right)$$

- **Tip** - If the sample size varies greatly then you will have to recalculate the control limits for each sample size. This is a very good reason for keeping the sample size constant if at all possible.

## Analysing the charts

Unlike variables control charts where there are two charts to analyse, for attributes control charts there is only one chart to analyse. There are less calculations for attributes but variables data provides more information on the process.

One of the good things about attribute charts is that you only have one chart to analyse.

# 5.20 Using attributes control charts

## Judgement is all

Attributes control charts are often based on human judgement of whether a particular feature is acceptable or not. For go/no-go gauges this is often relatively easy but for visual features this is often very difficult, i.e. is this gloss level acceptable or not? Is this a big enough scratch to count as a nonconformity or not?

These judgements will not only vary between individual people but also with time for the same person. Minimising the variations in these judgments is the key to the effective use of attributes charts.

Accurate assessment of attributes must minimise these variations and the basic requirements for accurate assessment are listed in the box on the lower right.

## Setting up an attribute chart

After the type of attribute chart to be used is decided (see Section 5.19) the process of setting up an attribute control chart is very similar to that of setting up a variables control chart and is shown on the lower right on the opposite page.

- Using a running and stable process, assess the selected sample size at the decided frequency.
- **Tip** - The sample size for attributes control charts will be significantly higher than for variables control charts. Do not underestimate how long it will take to assess the samples.
- Do not adjust the process or make any changes during the run.
- Decide on the assessment attribute that it is desired to control and record the assessments (nonconformities or nonconforming items).
- Calculate p, np, c or u depending in the chart type.
- Continue until 25 sets of data are available.
- Plot the results for p, np, c or u on a preliminary control chart in time order.
- Calculate the centre line value from the relevant equation in Section 5.19.
- Calculate the UCL and LCL from the relevant equation in Section 5.19.

**Note:** If the LCL calculation gives value < 0 then set the LCL as 0.

- Mark these trial control limits on the chart. If any of the points on the chart are outside the trial control limits then discard these points and recalculate the control limits.
- Continue this process until all points are within the trial control limits (provided there are still at least 20 data points).
- Plot the average for the centre line and the Upper and Lower Control Limits on the control chart.

## Continue

Continue to assess the selected control attribute and use the attributes chart to control the process.

## The rules for attributes control charts

Interpretation of an attributes control chart is exactly the same as for a variables control chart. The control limits represent the process variation when only common causes are present and any pattern of variation indicates the introduction of a special cause that will affect the process. The rules are therefore:

- No point outside the control limits.
- No run of 7 consecutive points above or below the average line.
- No run of 7 consecutive points upward or downward.
- No pattern with more than 2/3 of the

Standards need to be:

- Realistic.
- Well-defined (especially at the border lines).
- Strongly enforced.

Unless the Control Chart tells you to take action, you DO NOTHING!

**Accurate assessment of attributes needs:**

- ■Well communicated and agreed standards for nonconformities or nonconforming items.
- ■Good reference standards for nonconformities, especially at the border lines between acceptable/not acceptable.
- ■Good reference aids, e.g. samples or photographs of actual nonconformities.
- ■Good conditions for assessment, e.g. good lighting.
- ■Consistent assessment conditions for all staff who assess nonconformities, e.g. do not assess some nonconformities in daylight and other nonconformities in artificial light.
- ■Good training and development of staff in the consistent judgement of nonconformities.

Without these basics, control using attributes will suffer from variations in assessment standards and judgements.

points in the middle 1/3 of the control limits.
- No pattern with more than 1/3 of the points in the outer 2/3 of the control limits.

If any of these patterns is present then the process is out of control for the selected attribute. The process should be investigated for special causes and either stopped or corrected by adjusting the settings.

## If everything is OK then do nothing!

### Important points

The important points noted for variables control charts (see Section 5.9) are all applicable for attributes control charts. The chart is a record of the process that should be completed by the operator and should record all changes to the process.

As with variables, attributes control charts provide the 'common language' for operators, managers and reporting. They are the glue that holds a factory together.

They can be used to detect 'special causes' from the alarms and patterns and these can be fixed or eliminated by local action.

Control charts quantify the size of the common causes but the detection and elimination of these is more difficult. Eliminating common causes generally requires management action and possibly even investment in the process.

### Recalculating the limits

The initial control limits should be recalculated whenever the chart indicates significant process changes. When a control chart has been running and all special causes have been removed then the control limits should not be recalculated. The control limits are based on the common causes and not on the special causes. Control limits should only be recalculated after common causes have been reduced by management action.

### Reading the charts

Reading an attributes control chart is exactly the same as reading a variables control chart although there will only be one chart to consider. The examples given in Sections 5.10 to Section 5.16 for the common 'out-of-control' signals for variables charts can be used to identify the most likely special cause.

## Other applications for attributes control charts

Control charts are not only relevant to the production area - if it is a process and it needs to be controlled then a suitable control chart can be generated.

Attributes control charts are also excellent for many office and similar processes (see Chapter 11):

- An invoice or a cheque will either be correct or not correct (the amount is not relevant).
- A payment will either be made or not made by the due date (the delay may or may not be relevant).
- A delivery will either be made or not made (again the delay may or may not be relevant).

All of these events are go/no-go types of events and can be tracked using attributes control charts.

For most attributes control charts the LCL will be zero - if the chart indicates a change for the better, i.e. a decrease in the value then investigation is just as important. Never loose an opportunity to bolt in a permanent improvement to the process.

The important thing at the operator level is action on special causes not on common causes.

Common causes need management action.

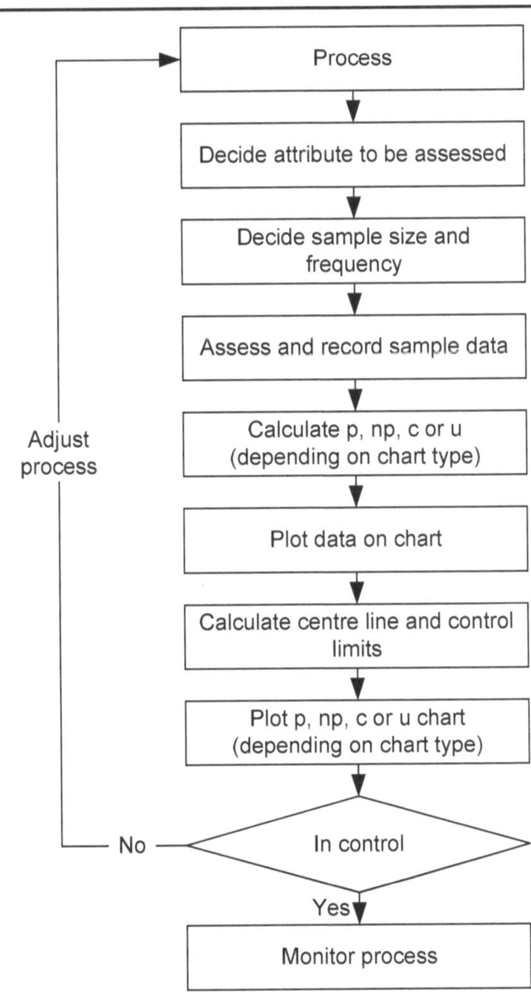


### Setting up an attributes control chart - the process

Setting up an attributes control chart is easy to do and the flow chart gives the necessary steps. For attributes there is only one chart to set up and this makes the chart process easier but does not necessarily make control easier.

# 5.21 Control charts - where are you now?

## Start now!

SPC and control charts are an excellent method to gain control of any plastics processing method. They allow the natural variation of the process (the control limits) to be clearly seen and accounted for, they also give early warning (via the alarms) of any special causes at an early stage. This early warning enables detection and rectification of any special causes before faulty products are produced.

All plastics processors should use control charting of some description to gain control of their process and improve the quality of their product.

SPC can be one of the most useful and vital tools in the drive to improve quality. Start using SPC now to gain essential insights into your plastics processing.

## Completing the chart

This chart is completed and assessed as for those presented previously.

Do not wait, do not think about it.

Start charting the problem processes at your site now. The information gained will help to make processes more consistent and reliable.

**Control charts**

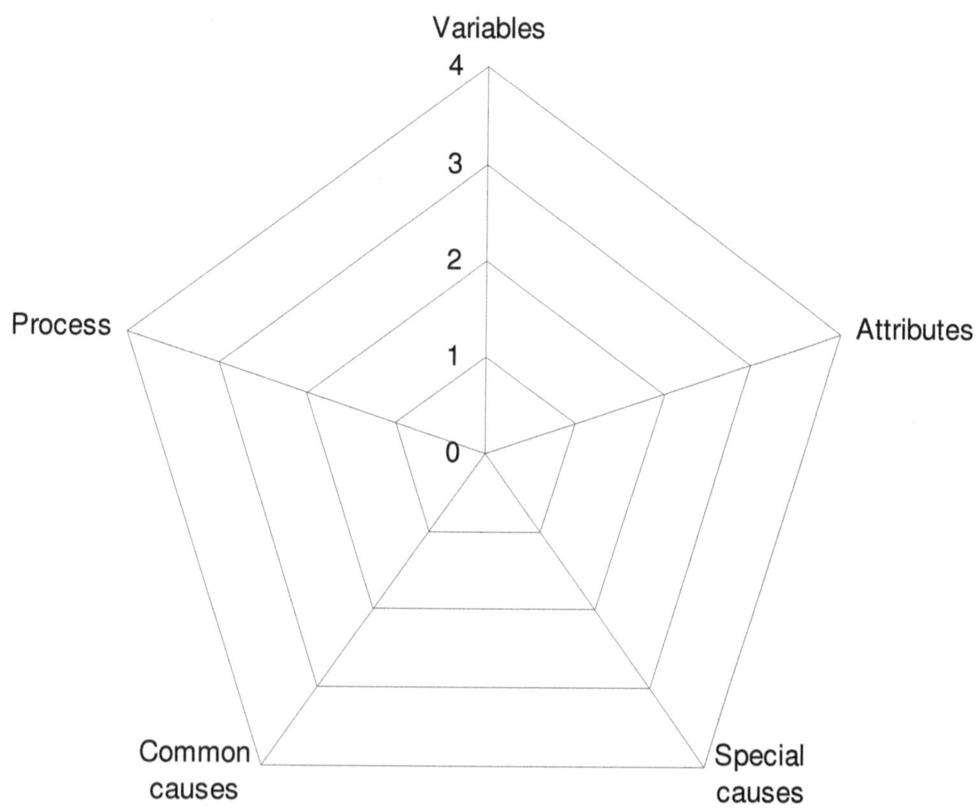


Use the scoring chart to assess where you are in control charts

The numbers from the scoring chart can be transferred to the radar chart for a quick visual assessment of where you are in terms of control charts.

For most processors it is easiest and most valuable to start with variables control charts and to follow on with attributes control charts.

It is not enough to simply chart the process.

The important things are to do nothing when the chart shows no alarms *and* to do something about it when alarms are seen.

| Control charts | | | | | |
|---|---|---|---|---|---|
| **Level** | **Variables** | **Attributes** | **Special causes** | **Common causes** | **Process** |
| **4** | Relevant minimum number of controlling variables identified for process. Control charts for appropriate variables being used. | Relevant minimum number of attributes identified for process. Appropriate attributes being charted. Good sample selection for attribute assessment. | Site has well-defined process for identifying & removing special causes detected from control charts. | Site has well-defined process for identifying & quantifying common causes via control charts. | Process is well controlled via set-up sheets (paper or computer) & these are effectively used to control the process. |
| **3** | Control charts used for variables but excessive number of variables being charted. Higher than necessary costs. | Control charts used for attributes but excessive charting & high costs. Samples for attribute assessment are inadequate. | Site has well-defined process for identifying special causes detected from control charts. Poor process for removing special causes from process. | Site has good process for identifying & quantifying common causes via control charts. | Process is theoretically well controlled via set-up sheets (paper or computer) but these are sometimes ignored in the process set-up. |
| **2** | Control charts used for variables but actual controlling variables are not charted. SPC is ineffective through system design. | Control charts used for some attributes. Samples for attribute assessment are non-existent or very poor. | Site has well-defined process for identifying special causes detected from control charts. No process for removing special causes from process. | Site has poor process for identifying & quantifying common causes via control charts. | Process is theoretically well controlled via set-up sheets (paper or computer) but these are often ignored in the process set-up. |
| **1** | Control charts used for some variables. No clear definition if these are the controlling variables. | Control charts used for few attributes. No clear definition if these are important to the customer. Poor sample selection for attribute assessment. | Site has no process for identifying or removing special causes detected from control charts. | Site has no process for identifying & quantifying common causes via control charts. | Process setting is poorly controlled & actual process set-up can vary widely. |
| **0** | No use of variables control charts to control process. | No use of attributes control charts to control process. | Control charts not used to identify special causes. | Control charts not used to identify or quantify common causes. | Process is set-up by individuals who use their own settings for the process. This is accepted by management. |
| **Score** | | | | | |

# 5.22 The process capability study

## The complete process

Section 5.4 described the use of the machine capability study to initially determine if the machine or process was capable of making the product.

The machine capability study is a 'snapshot' or short-term view of the process where there are no special causes present and where even common causes may be minimised. In reality, any process will vary more than this because of common causes that will appear over time, i.e. changes in operator and raw material supplies, which are not included in the initial machine capability study. The relationship between a machine capability study and the full process capability is shown on the lower right.

For these reasons, the process capability study is carried out to give a long-term view of the process capability.

The machine capability study looks at a 50 samples of product taken at a single time whereas the process capability study looks at long-term results taken directly from the control charts. There is no need for additional testing, all the data comes directly from the existing control charts produced as part of normal operations.

## Variables data

The process for calculating process capability is shown on the lower right on the opposite page. This is straightforward for most types of control chart.

### Mean and range ($\overline{X}$ and R) chart

- Collect the process data from at least 25 sets of sample data. Check that the process is in control for the period before using the data.
- Calculate the grand average ($\overline{\overline{X}}$) for the 25 sets of data - this is the average of all the averages.
- Calculate the average of the range ($\overline{R}$) for the 25 sets of data.
- Estimate the standard deviation for the sets of data ($\sigma$). Provided the process is in control for both $\overline{X}$ and R (see above) then the process standard deviation ($\sigma$) can be estimated from:

$$\hat{\sigma} = \frac{\overline{R}}{d_2}$$

where:

- $\hat{\sigma}$ is the estimate of the standard deviation
- $d_2$ is a constant that varies with the sample size (see Appendix 2 for values $d_2$ for a range of sample sizes).

- Using the estimated standard deviation ($\hat{\sigma}$) and the relevant tolerance limits from the process specification, calculate the 'process capability' or $C_p$ (relative to the specified tolerances) from the formula:

$$C_p = \frac{USL - LSL}{6\hat{\sigma}}$$

where:

- USL = Upper Set Limit or Upper Acceptable Tolerance.
- LSL = Lower Set Limit or Lower Acceptable Tolerance.
- $\hat{\sigma}$ = the estimated standard deviation.

**Note 1:** We use $C_p$ (the process capability) here rather than $C_m$ (the machine capability) as it is assumed that the results being used are from a process capability study.

**Note 2:** The value of $6\sigma$ ($\pm 3\sigma$) is set so that we can be confident of producing 99.73% of the parts within the tolerance band (see Section 5.4).

- Using the value for the grand average ($\overline{\overline{X}}$) and the relevant tolerance limits from the process specification, calculate $Z_{Upper}$ and $Z_{Lower}$ from the formulae:

The process capability study describes what the process is capable of over the long term.

This treatment of a process capability study assumes that the distribution of the results is a normal distribution. For complete accuracy, this should be checked using a probability plot but this is beyond the scope of this Workbook.

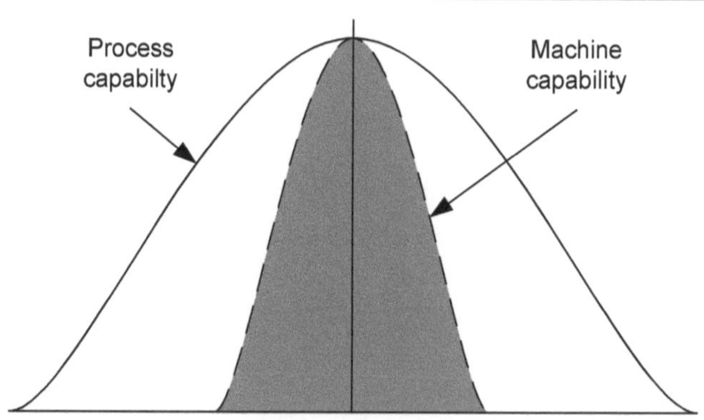


Process capability and machine capability

The full process capability study includes common causes that are not present during the 'snapshot' of the machine capability study. The process capability will therefore have a wider spread than the machine capability.

$$Z_{Upper} = \frac{USL - \overline{\overline{X}}}{3\sigma}$$

and

$$Z_{Lower} = \frac{\overline{\overline{X}} - LSL}{6\sigma}$$

- Find $Z_{Min}$, i.e. the lower value of $Z_{Upper}$ and $Z_{Lower}$. This is termed $C_{pk}$ and is known as the 'process capability index'. $C_{pk}$ is a measure of the 'process location' but also gives valuable information on the spread.

**Note:** It does not matter if the tolerances are 'bilateral' (sometimes called 'symmetric'), i.e. A ± x, or 'unilateral' (sometimes called asymmetric), i.e. A + x, − y. The important thing is the location of the average of the sample with respect to the upper set limit and the lower set limit. At this stage we are only dealing with the location of the distribution and how close it is to the set limits.

Variability itself is not significant; it is natural. What is significant is the size and location of the variability.

## Mean and standard deviation ($\overline{X}$ and s) chart

The process capability data for $\overline{X}$ and s charts is calculated in exactly the same way. For an $\overline{X}$ and s chart the process standard deviation (provided the process is in control for both $\overline{X}$ and s) can be estimated from:

$$\hat{\sigma} = \frac{\bar{s}}{c_4}$$

where:

- $\hat{\sigma}$ is the estimated standard deviation.
- $\bar{s}$ is the average of the standard deviation for the 25 sets of data.
- $c_4$ is a constant that varies with the sample size (see Appendix 2 for values of this constant for a range of sample sizes).

## Median and range ($\tilde{X}$ and R) chart

The median and range chart is not generally used for the calculation of process capability.

## Can we make it OK?

Whether the chart is a mean and range $\overline{X}$ and R chart) or a mean and standard deviation $\overline{X}$ and s ) chart the values of $C_p$ and $C_{pk}$ should both be greater than 1.33 (see Sections 5.5 and 5.6) to reliably provide products that have a good spread with respect to the tolerance band and are well centred with respect to the tolerance limits.

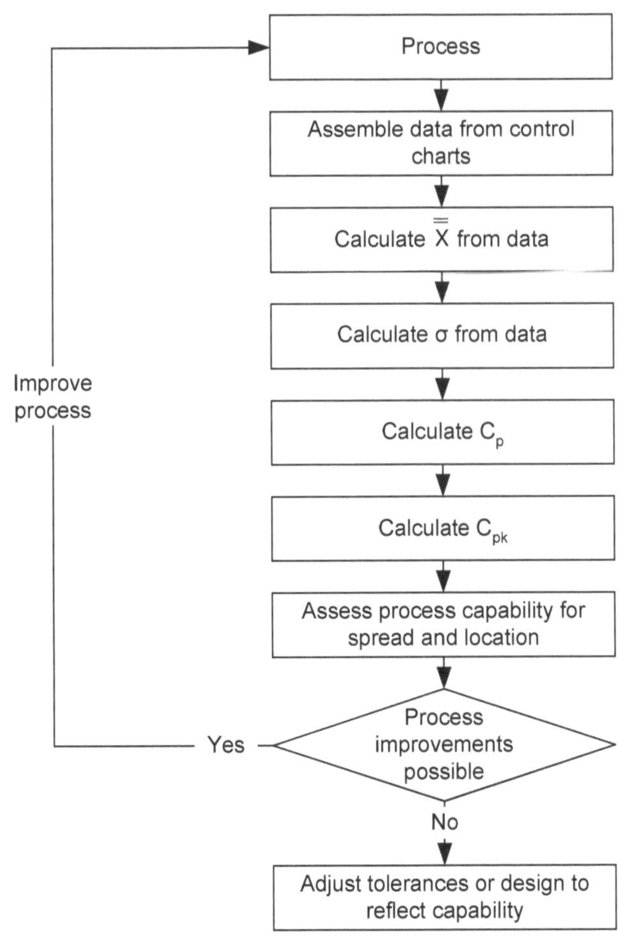


### Calculating process capability

Process capability uses data from the running and 'in-control' process to determine the overall process capability. This uses the familiar values of $C_p$ and $C_{pk}$ as used in the process of finding the machine capability.

# 5.23 Improving process capability

## Minimise unnecessary variation

Control charts effectively divide any changes in a process into either special or common causes:

- Special causes are detected by the control chart as a pattern and the type of pattern gives an indication of the source of the cause. Special causes are not always easy to solve/remove but they are generally easy to identify - they are a change to the system and after identification can be removed to bring the system back to stability.
- Common causes are the basis of the control chart. Each factor affecting the product, i.e. environment, materials, machines, men, methods and measures, will vary slightly over time and the common causes are the sum of these variations.

Improving process capability means reducing the sum of the common causes and reducing the overall variability of the process. This is not about keeping the system in control. This is about reducing variability in all of the inputs to the process to reduce the intrinsic variability of the product and improving the complete system. This process is shown on the right.

## Where do we start?

Special causes are a disturbance to the system and solving special cause issues can almost always be carried out at the local level, i.e. it is more a matter of removing whatever upset the system than improving it.

Common causes are related to all the inputs to the system and reducing the effect of special causes is almost certainly a management issue that needs input and resources from management.

## Define the process and the problem

The first step is to define the process that is to be improved. In many cases the driver for improved process capability is the inability of an existing process to produce to the required tolerances and it is essential to define where the need for improvement actually is. If the real issue is variability of the input materials from a previous process then improving the previous process may well be the easiest and most productive activity.

The diagnosis of the real problem is often rushed in an effort to save time but this is where the success or failure of the project can be decided. Problem definition is a key part of the process.

It is necessary to define the process that is to be improved and the process boundaries. This needs clear definition of what the process is and how it links to other processes.

Inputs can be:

- Controllable inputs - where the site has a high degree of control.
- Non-controllable inputs - where the site has no effective control.

It is essential to recognise the difference and not to waste time on the non-controllable inputs.

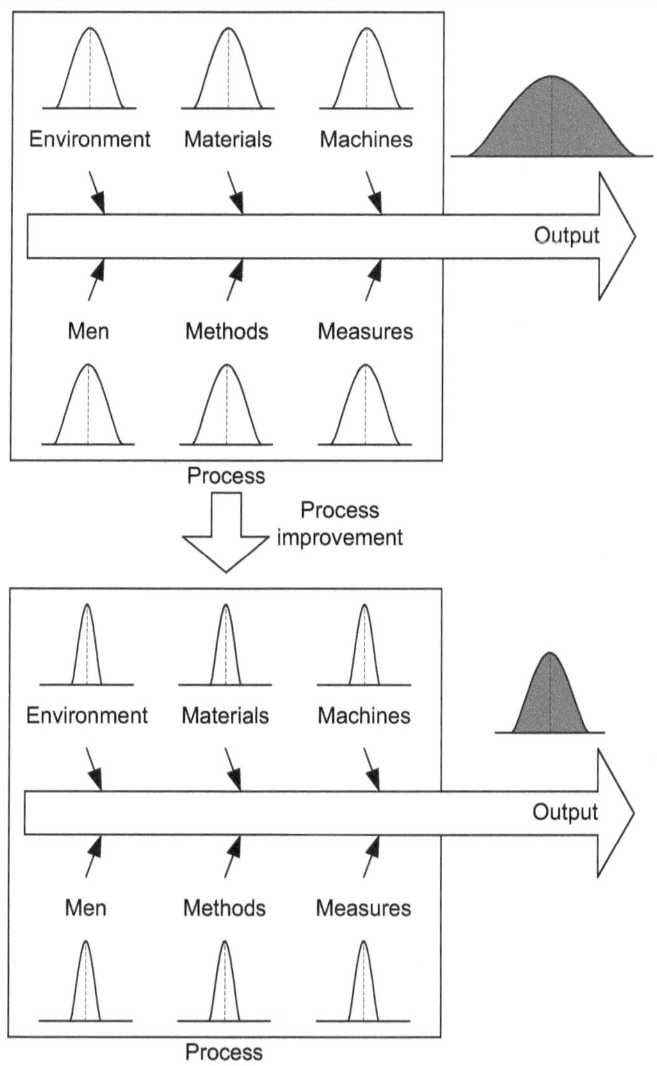


The process of improving process capability

The common causes for the process are the sum of the common causes from the various factors that vary over time. The individual variations add up to the total of the common causes, improving capability means reducing the individual variations.

## Collect data

Data are essential for detailed understanding of the problem. The essential data will include:

- Data on the essential output parameters of the process - it is necessary to understand what you want to improve before you start to try to improve it. Some of these data will come from the problem definition but other data will come from process control charts.
- Data on the essential input parameters of the process - not all the possible input variables will actually affect the essential output of the process and the essential input parameters should be listed.
- Definition of the controllable and non-controllable inputs to the process - not all inputs are controllable and wasting time on non-controllable inputs is not recommended.
- A process FMEA for the selected process (see Section 10.2).
- A current process capability study for the process - this will fix the current situation and will allow benchmarking of any proposed process improvement. A current capability study will also give valuable data on the process.
- Current process control charts - these will highlight the effect of any special causes and give valuable information on the sensitivity of the process to various inputs. Look at the control charts for special causes that have a large effect. Reducing variation in the equivalent common causes can be very effective.

## Propose solution

After the data have been gathered, the important factors should be identified using quality tools (see Chapter 7). Improved control of these factors will reduce variation in the overall process.

## Improve process

The process can then be optimised to cost-effectively reduce variation by:

- Improving control of the input parameters.
- Adjusting the input parameters.
- Reducing the sensitivity of the process to the input parameters.

## Check results

Any change to a process should be monitored by existing control charts to verify that the solution is actually improving the process capability. This should be revealed by a decrease in the process standard deviation as detected by the range on an $\overline{X}$ and R chart.

**Note:** Decreasing the process standard deviation may also require changes to the setting of the process to maintain the location and to optimise $C_{pk}$.

## Develop appropriate process controls

Optimising the process itself is only part of the overall improvement of process capability. The benefits of process optimisation need to be locked in place to ensure sustainable improvement. This may need:

- Staff training.
- Updating of process documents.
- Additional process controls.
- Recalculation of the control chart limits.

## Start again on another process

Do not try to solve all the problems in one big exercise.

It is easier to solve several small problems and get experience with the process.

Improving process capability is project-based and should follow all the standard rules and processes for successful project management (see Section 2.5).

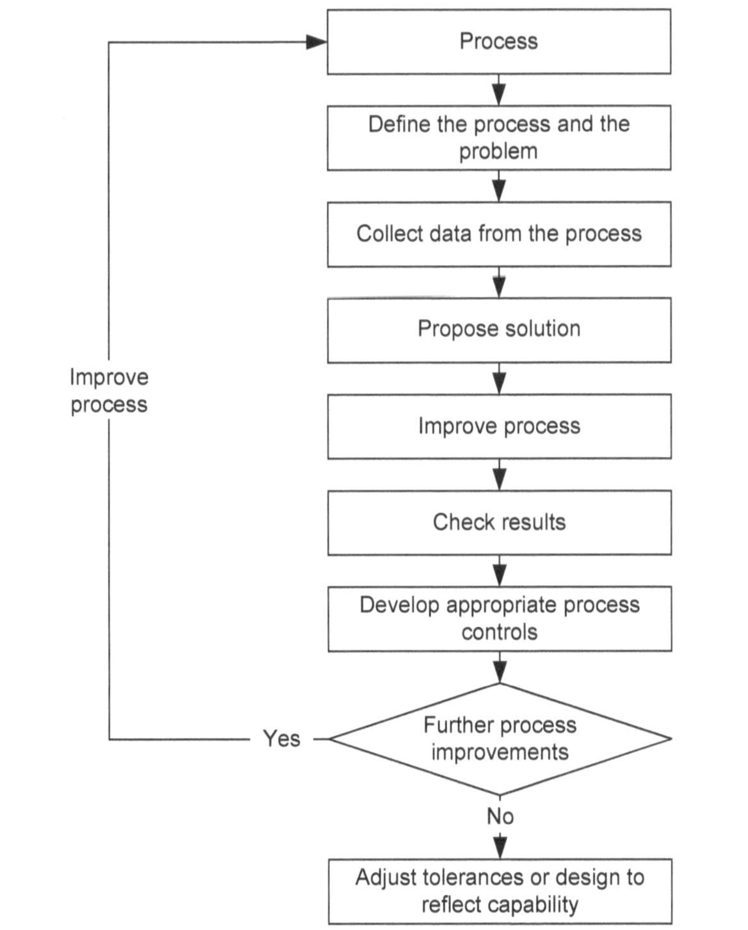


Flow chart for improving process capability

Improving process capability is a process that needs to be managed. Sites should select process capability improvement projects based on the need to meet increasing customer demands and the need for continuous improvement.

# 5.24 Capability studies - where are you now?

## Know your capability

Capability studies tell a company what they can do in terms of delivering quality. Without knowledge of the capability of the main processes and products a company cannot be certain that it is possible to deliver products to specification. The company may be taking on jobs that are impossible to produce correctly and this can be a disaster for any company.

Initial capability can be found from a machine capability study (for tool approval) but long-term capability needs to be assessed from control chart data to include all the potential process variables (men, machines, materials, etc.).

After capability is assessed it is necessary for companies to have action plans to not only reduce the spread ($C_p$) but also to improve the location ($C_{pk}$). These are the essential improvement actions.

Having defined the process capability, it is also necessary to publicise this information to make sure that all staff understand the implications of capability.

## Completing the chart

This chart is completed and assessed as for those presented previously.

We use the concept of capability studies for processes and products because every new product is really a new process that needs to be managed and studied for process capability.

A new tool will have fundamentally different capability to existing tools, i.e. both the spread and location may vary.

**Capability studies**

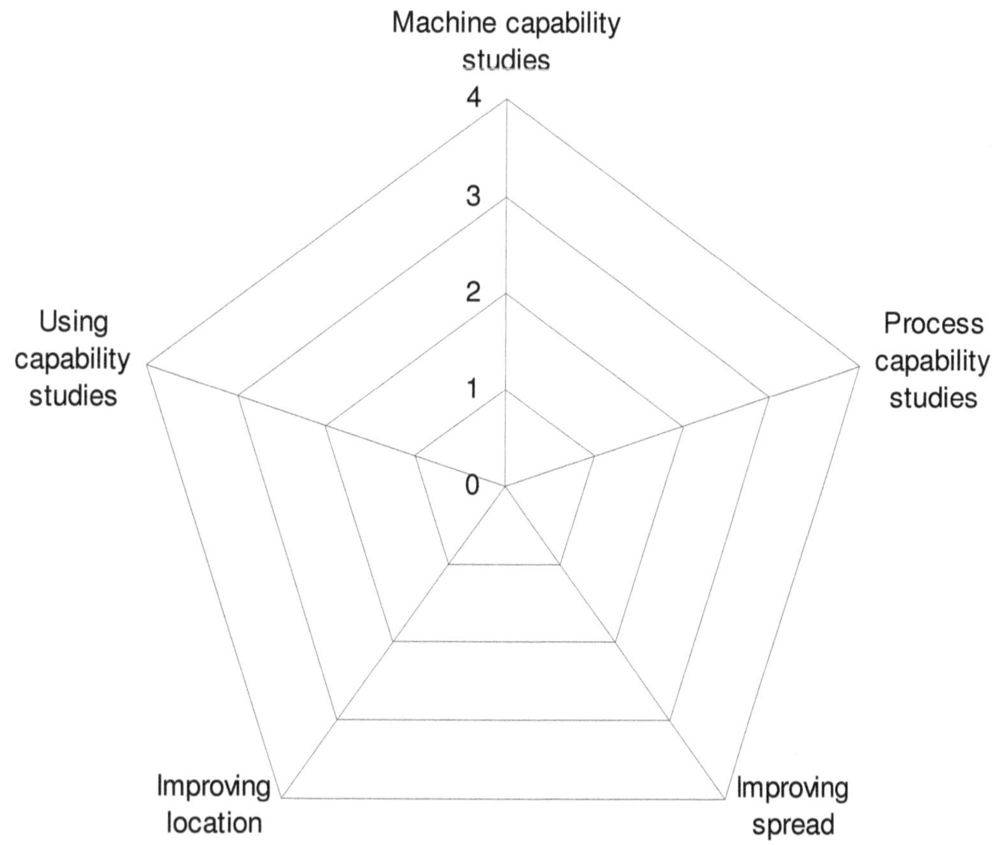


Use the scoring chart to assess where you are in capability studies

The numbers from the scoring chart can be transferred to the radar chart for a quick visual assessment of where you are in terms of capability studies.

It is not enough to simply know the process capability.

It is also necessary to plan to improve it.

Understanding capability studies (at least the broad concept) is essential for sales staff.

They need this to avoid selling products or processes that the site cannot possibly deliver - a prime cause of poor quality that can never be rectified.

## Capability studies

| Level | Machine capability studies | Process capability studies | Improving spread | Improving location | Using capability studies |
|---|---|---|---|---|---|
| 4 | Machine capability studies carried out for all new processes & products. | Process capability studies carried out from control charts after stable running for all new processes & products. | $C_p$ indicates good spread for all processes. Plans in place to decrease spread for all products & processes. | $C_{pk}$ indicates good location for all processes. Plans in place to ensure good location for all products & processes. | Process capability widely known & appreciated throughout site. Capability used by sales staff to advantage. |
| 3 | Machine capability studies carried out for most new processes & products. | Process capability studies carried out from control charts after stable running for most new processes & products. | $C_p$ indicates good spread for main processes. Plans in place to decrease spread for main products & processes. | $C_{pk}$ indicates good location for main processes. Plans in place to ensure good location for main products & processes. | Process capability known only to technical staff (quality & production). Capability not known or understood by majority of staff. |
| 2 | Machine capability studies carried out for some new processes & products. | Process capability studies carried out from control charts after stable running for some new processes & products. | $C_p$ indicates good spread for some processes. Plans in place to decrease spread for some products & processes. | $C_{pk}$ indicates good location for some processes. Plans in place to ensure good location for some products & processes. | Process capability known only to quality staff. Production staff not aware of implications of process capability. Capability not known or understood by majority of staff. |
| 1 | Machine capability studies carried out for few new processes & products. | Process capability studies carried out from control charts after stable running for few new processes & products. | $C_p$ indicates poor spread for most processes. No plans in place to decrease spread for any products or processes. | $C_{pk}$ indicates poor location for most processes. No plans in place to ensure good location for any products or processes. | Process capability known to very few staff. Majority of technical staff (quality & production) unaware of the implications of process capability. Capability not known or understood by majority of staff. |
| 0 | No machine capability studies carried out. Machine capability unknown & assumed to be OK. | No process capability studies carried out. Concept of process capability not known at site. Process capability unknown & assumed to be OK even if rejects are being produced. | $C_p$ unknown for any process. No plans in place to decrease spread. | $C_{pk}$ unknown for any process. No plans in place to ensure good location for all products & processes. | No concept of process capability at site. Tolerances based on industry averages or guessed by sales staff. |
| **Score** | | | | | |

# 5.25 Implementing SPC in plastics processing - the process

## The barriers to implementation

The control chart concept was first developed by Shewhart in 1924 and, despite the proven effectiveness of the approach, it still struggles to find acceptance in Western manufacturing industry. This is particularly true in the plastics processing industry which thinks that its problems are special. They are not, they are simply process control problems of the type the SPC was specifically developed to solve.

The main reasons for a failure to implement SPC are:

- Lack of knowledge in senior managers - senior managers are often confused about what quality is and do not understand that it is both measurable and controllable. This is compounded by the fact that SPC uses obvious statistics (which many managers are afraid of and do not understand) and needs some knowledge of these to be used properly.
- Managers are not aware of the capability of the processes that they are responsible for - many managers have no idea of the inherent capability of their processes and somehow think that quality is what you want it to be and that simply wishing for improved tolerances will make it happen. They assume that their capability is the same as the general industry. This is often without any evidence and often with evidence to the contrary.
- Ill-informed excuses - many people would rather do nothing than face the fact that they have to work to improve quality. Statements such as 'It won't work with our processes' are most often excuses to continue the way we have in the past even when the competition is changing the rules.

These are powerful barriers and need to be either faced head on or otherwise circumvented (see box on opposite page). They should not be used as excuses for a failure to implement SPC but they must be included in the planning process to allow successful implementation.

## The pre-requisites for implementation

Even if the barriers to implementation are overcome, the successful implementation of SPC requires several pre-requisites and these are:

- An adequate and functioning QMS (see Chapter 4) - a QMS provides a framework for successful implementation. It will not guarantee success but it can provide a framework for the implementation of SPC.
- Senior management commitment - the implementation of SPC involves significant changes in the way production works and an acceptance by production of their ultimate responsibility for the production of quality products. Senior management commitment, preferably at the Managing Director level, to the implementation of SPC can provide both the motivation and muscle that will be necessary at critical times.
- **Tip** - Prepare a short presentation on SPC for senior management. Avoid statistics and concentrate on the important messages:
  - that quality is both measurable and controllable.
  - that improving quality will both reduce costs and increase profits.
- An SPC champion/project manager - even with senior management commitment, it is necessary to have a champion in either the Quality or the Production functions. The SPC champion should act as the project manager for the complete implementation of SPC. In most cases, this means the champion should be in either the Quality or Production functions.
- Training - any new initiative requires training and SPC is no different. Quality control and production staff will need training to understand and apply SPC effectively.

## Things to do

Some of the essential things to do in implementing SPC are:

- Start small - big projects often fail because expectations are too high. When implementing SPC is it wise to gain momentum from a series of small successful projects rather than to try to move the whole company. Start with a small 'problem' process or product and

Implementation of SPC is a data-driven process and removes almost all of the current arbitrary judgements that are made - be ready for the change.

Implementation of SPC is a project and a process.

It needs organisation and planning to be successful (see Chapter 2).

use SPC to understand the process and solve the problems. This minimises the effort and allows the team to gain experience and credibility.

- Tackle one problem at a time - choose a small process or process that is the 'problem child' for the site. Look for a project that is within the control of the project team, that is achievable using simple techniques and that can be solved in a reasonable length of time.
- Involve people - every project needs to involve the people who are taking part to be successful. SPC will increase staff satisfaction so get them involved from the start.
- Use data from the process - SPC is data-driven and the more data you have the better. This is particularly true for process capability. An initial snap-shot (using the techniques of Section 5.4) can provide valuable data for the process.
- Train, train, train - a key to staff involvement is to train them so that they can see the reasons for what they are doing. Train anybody who is going to be affected by implementing SPC (upper management, middle management and operators). Training will also reduce the workload for the project leader.
- **Tip -** Consider using an external trainer for this, external trainers can increase the status of the project.
- Continue with existing controls - it is wise to continue to run the existing controls alongside the SPC controls in the early stages. This will validate the process and the staff can then decide when to discontinue the 'old' control methods and inspections.
- Plan for the future - although the start will be a small project, it is essential to have a plan for what happens next even if it is only an outline. Sometimes implementation can accelerate after an initial successful trial and a plan provides the guidance on what to do next.
- Evaluate the results to provide evidence for future work - implementing SPC is a project and a process (see Section 5.26) and every project needs evaluation after it is completed (see Section 5.27).

## Things not to do

Just as important as the list of things to do is a list of 'things not to do'. Doing any of the following will seriously compromise the possibility of successful implementation:

- Thinking that quality is a problem that only involves inspectors - it is a production/process problem.
- Thinking that final inspection works - it doesn't.
- Accepting excuses for failing to start - excuses such as: 'It will never work here' or 'We already have enough work to do sorting these rejects' are not acceptable reasons for failing to start implementing SPC. In fact 'We already have enough work to do sorting these rejects' is the best reason to start implementation as soon as possible.

Implementation must be assessed just as with any project (see Section 5.27).

### Implementing SPC using stealth and envy

We successfully installed SPC at one site not by declaring what we were doing but by 'stealth and envy'.

- The group of 3 (a lunch 'club' of representatives from Technical, Quality and Production) decided that we could no longer continue to produce poor-quality parts and that SPC was the way to improve.
- We selected 1 machine and product that always gave us problems with both inconsistent running and low uptime.
- We collected all the data on the machine and product.
- We took the day shift operator for this machine and trained her (and only her) in SPC.
- We set up control charts for the process and only the selected day shift operator ran them.
- We used the control charts to control the process and she immediately saw the benefits of fewer problems and reduced workload.
- At the end of each day shift we took the charts away and ran the machine as normal.
- After 4 days we found that the afternoon shift operator was envious of our 'trained' operator because she had less problems and her workload was less. The afternoon operator then asked us if he could be trained to run the control charts. We 'reluctantly' agreed to extend the trial to afternoons.
- After another week the night operator asked to be trained and again we 'reluctantly' agreed to extend the trial to nights.
- Within 2 weeks we had all the operators asking why they hadn't been trained and when we would implement SPC in their machines.
- We trained the whole mould shop in SPC and over the next 2 months implemented SPC over every machine and product.

The operators saw the benefits, wanted to be involved and drove the whole process rather than management. Problems decreased and our lunch club enjoyed our lunches a lot more while we plotted our next move.

# 5.26 Implementing SPC in plastics processing - a simple approach

## Control at the point of delivery

One of the most important points about SPC is that it places the responsibility for control of quality where it really belongs, in the hands of the operator and the production area. The Quality Department is responsible for providing training, equipment and guidance on results but the recording of results should be carried out by the actual operators themselves.

## Responsibilities

The Quality Department should be responsible for:

- Initial setting of control limits based on capability studies or tolerance limits.
- Recalculation of limits based on performance.

The Quality Manager should be responsible for:

- Training of operators in both basic SPC theory and practice and in control chart analysis.
- The design, purchase and provision of all necessary materials for measurement or assessment.

The Production Manager for an area should be responsible for:

- Obtaining all relevant information for completion of the information sections of the control chart.
- Distribution of the control charts to the relevant production location.

The Operator should be responsible for:

- Completion of the control chart at the specified intervals. The operator should also note significant events on the chart (e.g. material or environmental changes) to enable pattern detection.
- Signalling any out-of-control conditions detected.

The Setters should be responsible for:

- Machine adjustments as directed by the Production Manager. Setters should note any adjustments made on the control chart and initial the note.

## Initial chart limits

### Capability study method

The capability of the process is calculated (see Section 5.4). The capability study data are used to initially set the centre line, UCL and LCL for the average and the range.

This approach is designed for use with $\overline{X}$ and R charts but can be adapted for any type of control chart.

### Tolerance method

The UCL and LCL for the average are set as the upper and lower tolerances, respectively. The UCL for the range is set as the total tolerance band and the LCL for the range is set as zero. Centre lines for the average and range are the nominal value and half the tolerance band, respectively.

### Ad hoc method

In this method no limits are initially set but measurements are taken during stable production. On completion of at least 10 sub-groups of measurements the limits and average are calculated and inserted on the chart. The results for the completed chart are then used to recalculate the limits.

- **Tip** - The actual method chosen to set the initial limits is not critical because the limits will be revised at the completion of each chart. Using previous capability study data is preferred because this deals with real data. The ad hoc method means operating without limits for some time and this can confuse the operator.

## Chart issue and control

- The Quality Department should inform the Production Manager of the chart

| Control chart responsibilities | | | | |
|---|---|---|---|---|
| | Quality Dept. | Operator | Inspector | Production Manager |
| Chart design | ✓ | | | ✓ |
| Measurement | | ✓ | | |
| Chart plotting | | ✓ | | |
| Chart analysis | | ✓ | ✓ | |
| Action | | ✓ | | ✓ |

### Control chart responsibilities

The responsibilities for creation, completion and action should be clearly defined and communicated so that all staff know their responsibilities in the process of designing, measuring, recording and taking action.

limits before issuing an initial chart and the Production Manager completes all the information sections (Part No. etc.) and marks the control limits on the chart.

- The sampling interval should initially be a maximum of 30 minutes but this can be extended up to 8 hours depending on the stability of the process.
- The Production Manager should ensure that only trained operators are used for SPC work and that the correct measuring equipment is available.

## Chart completion

- The operator takes 5 samples for measurement at the specified sampling interval. The samples should preferably be consecutive but if cycle time does not allow this then they should be produced no more than 5 minutes apart (first to last). In the case of multi-cavity tools, samples should be differentiated by cavity number.
- The operator records the time of the measurement, the shift number and signs the chart.
- The individual measurements are recorded on the chart, the average and range calculated and marked on the chart, and the line from the previous points completed.
- The operator marks any significant events on the chart and initials the chart.
- If no alarms are seen the operator continues work and makes no adjustment to the machine. Alarms and action to be taken are detailed below.
- **Tip -** The important thing about SPC is that it shows you when to do nothing. Unless an alarm is raised the machine settings must not be changed. Alteration of machine settings without an alarm being raised should be a disciplinary offence.

## Alarms

- In the event of an alarm pattern being seen, the operator raises the alarm to contact the relevant setter. The setter and the operator decide on any action to be taken and note any changes made to overcome the alarm.
- **Tip -** All changes made should be noted on the chart and initialled by the setter/operator.
- If the operator and setter cannot agree on the action to be taken the relevant supervisor should decide on any action to be taken. If the alarm is significant the QC should be informed in order to decide if prior work shall be inspected prior to release.

Just as with any project, implementation must be evaluated - (see Section 5.27).

## Chart full

- When the chart is completed the Production Manager should be informed and the completed chart returned to him. The limits should be recalculated and a new chart issued to the operator.
- **Tip -** To avoid an operator working without a chart it is recommended that a new chart be prepared in advance and control limits calculated at a convenient shift change prior to the chart being full.
- The full chart should be retained by the Production Manager for a period defined by the Quality Manager.

One of the best software packages for running SPC at a plastics processing site is NWA Quality Analyst: (see www.nwasoft.com). This is an excellent package but is quite expensive. Still, it will probably save money in terms of development time with Excel™.

It will cope with everything you can throw at it and makes analysis easy but get the basics right first.

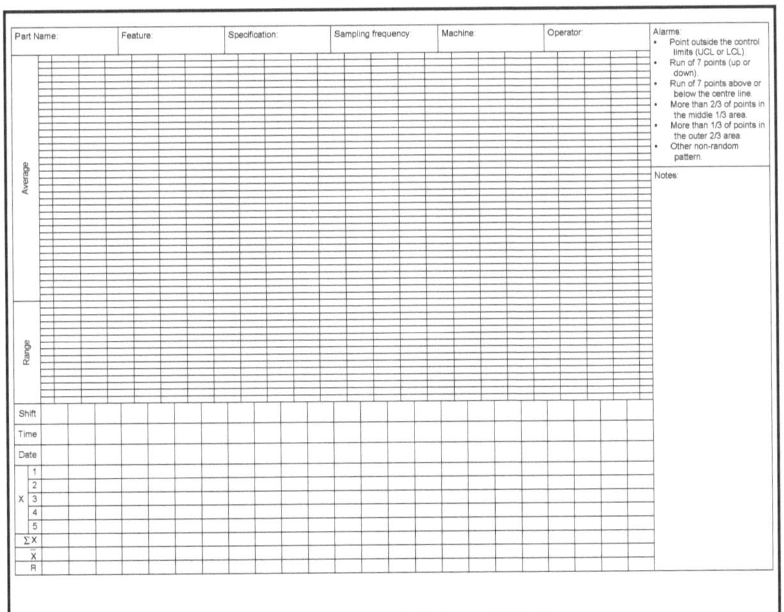


### Sample chart for simple variables SPC

A typical $\overline{X}$ and R variables control chart. This chart has no centre line, UCL or LCL marked on it, these are marked by the Production Manager. A larger-scale version of this chart suitable for copying and use in plastics processors is given in Appendix 3.

# 5.27 Evaluating the effectiveness of SPC

## It is a project

Implementing SPC is a project to change the way a site works, to understand the processes, to improve the processes and to safeguard jobs.

As with any project, this must be evaluated after completion to validate the savings and justify the investment. If implementing SPC via stealth and envy (see Section 5.25), it is worthwhile carrying out a very early evaluation to verify that savings are being made. In this case, evaluation can also be continuous to justify further work and implementation.

## Get the 'before' data

It is important to gather as much data as possible before implementing SPC to verify the savings. Typical data that should be gathered are:

- Uptime (downtime) - machines and products that suffer from quality and process problems will often suffer from more downtime than other machines. They will need more time from the setters as adjustments are made and will not be as productive as equivalent machines producing trouble-free products. Uptime or downtime should be recorded from the production monitoring system depending on the way the system is configured to report.
- Yield (scrap) - machines and products that suffer from quality and process problems will always produce more scrap than other machines. Products will often be out of specification and inspectors will spend more time sorting the good from the bad. The bad products will show up as increased scrap for the machine.
- Problems and time to solve them - machines and products that suffer from quality and process problems will always have more problems than other machines. This will show up as time spent by setters and QC as they attempt to get the machine running correctly (a futile task in many cases) and they will use more specialist technical time than they should.
- Existing process capability - calculating the process capability before implementing SPC will give an insight into the existing process capability.

- **Tip** - An easy way to do this using process data is to use 'open' control charts, i.e. control charts with no centre line, UCL or LCL marked on them. The machine should not be adjusted even if alarms are seen unless other current controls indicate a need for adjustment.
- **Tip** - If you want to know which machine or product to select as the candidate for SPC implementation then ask the machine setters. They will always know which machines are a problem to set.

Post-project evaluation is essential for any project.

The data gathered on the machine and process performance are important for comparison with later data after SPC has been implemented.

## Get the 'after' data

The data gathered after implementation should be exactly the same data as that gathered before implementation. This is to allow comparison on a like-for-like basis.

- **Tip** - Never change the data gathered otherwise you will be accused of manipulating the results.
- **Tip** - Calculate the process capability after the introduction of SPC using the control chart data method. In this case

Be reasonable in any estimates and justify the estimates.

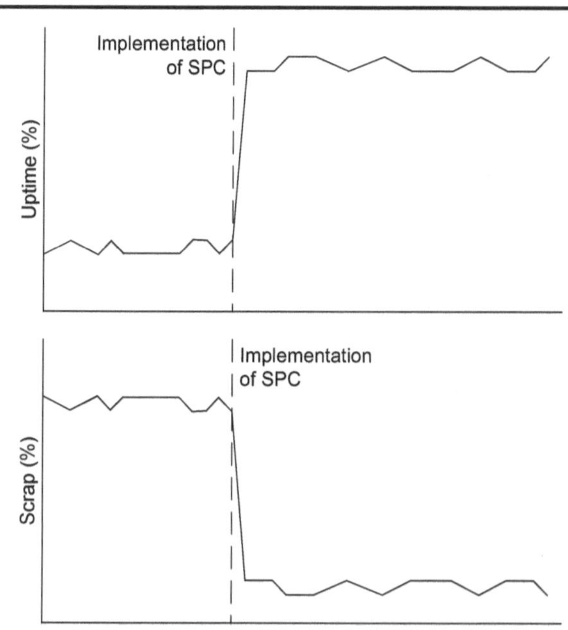


**Track the benefits of implementing SPC**

Using simple charts of uptime and scrap rates can graphically illustrate the financial and operational benefits of implementing SPC in plastics processing. Increased consistency and dependability of operation is one of the greatest benefits.

the chart will be marked with the centre line, UCL or LCL and the machine will be adjusted if alarms are seen. Compare this 'after' capability with the 'before' value to see the improvements achieved.

After implementation present the results clearly and with no bias. Simply state the facts of the case and let the data speak for themselves.

## Record implementation costs

To fully evaluate if implementation of SPC is successful it is necessary to record the costs of implementation. Typical costs that should be recorded are:

- Training costs - training is an essential part of implementation and any internal or external training costs should be recorded as part of the costs.
- Instrument costs - implementing SPC sometimes needs more low-level inspection equipment and the cost for this needs to be included in the total costs. In some cases, more expensive equipment is needed for full control and the cost of this must be justified and included.
- Staff costs - although staff costs are 'sunk' costs, they should always be quantified and included even if they are only estimates. Staff time in attending training, which may potentially involve overtime for afternoon and night shifts should also bc included.

The important thing about recording costs is that an effort is made to assign any reasonable costs to the project to be able to assess the financial value of the project.

In most cases, SPC implementation projects will show an excellent return on investment (ROI) but this does not mean that they cost nothing and the costs must be reported.

## Record savings that can be attributed to the project

Implementation of SPC will normally result in reductions in operating costs and particularly in the cost of quality (see Chapter 3).

The before and after data (see above) can be used to produce an estimate of the savings achieved from implementing SPC. In many cases the financial benefits will need to be estimated but they should still be produced even if they are estimates.

- **Tip** - Many of the benefits of implementing SPC are intangible, i.e. easier setting, less time required for fault finding, etc. These will need estimates.
- **Tip** - One of the greatest benefits of SPC is increased consistency and dependability of operation. Consistent operations give better schedule conformance and allow the introduction of JIT to reduce inventory and produce many other benefits.
- **Tip** - Good cost of quality data can be useful for estimating savings but it is often not sufficiently detailed to go down to the machine level for areas such as failure costs.

## Report savings

After recording the savings, they should be reported to provide a justification for extending the implementation and for further investment in the process.

A Quality Dashboard Report can be an excellent method for doing this.

Track the results for every project and use them to justify further investment.

# 5.28 Implementation - where are you now?

## Implementation is the thing

Despite all the benefits that SPC can bring to a range of plastics forming processes, it is not a theoretical discussion. SPC only helps to deliver better quality at lower stress levels if it is actually implemented.

Trialling SPC is a noble act of faith but actually installing a full SPC system is when the real rewards are achieved. SPC works to reduce costs, reduce complaints and give a better understanding of how to control the process. What is not to like?

Even after a full SPC system has been installed, it is essential to evaluate the effects of the system. Any installation of SPC will cost money and take time and it is therefore appropriate that an evaluation of the effectiveness of the system should be made. Evaluation allows the system to be tuned to deliver the greatest benefits for the lowest cost - surely the aim of any plastics processor.

## Completing the chart

This chart is completed and assessed as for those presented previously.

I have installed SPC at 5 plastics processing sites and 1 rubber processing site.

It was relatively painless at all of them and the evaluation was always extremely positive.

SPC gave the operators control over their process. Instead of feeling that they were simple operators, they began to feel and act as controllers of their process rather than the other way around.

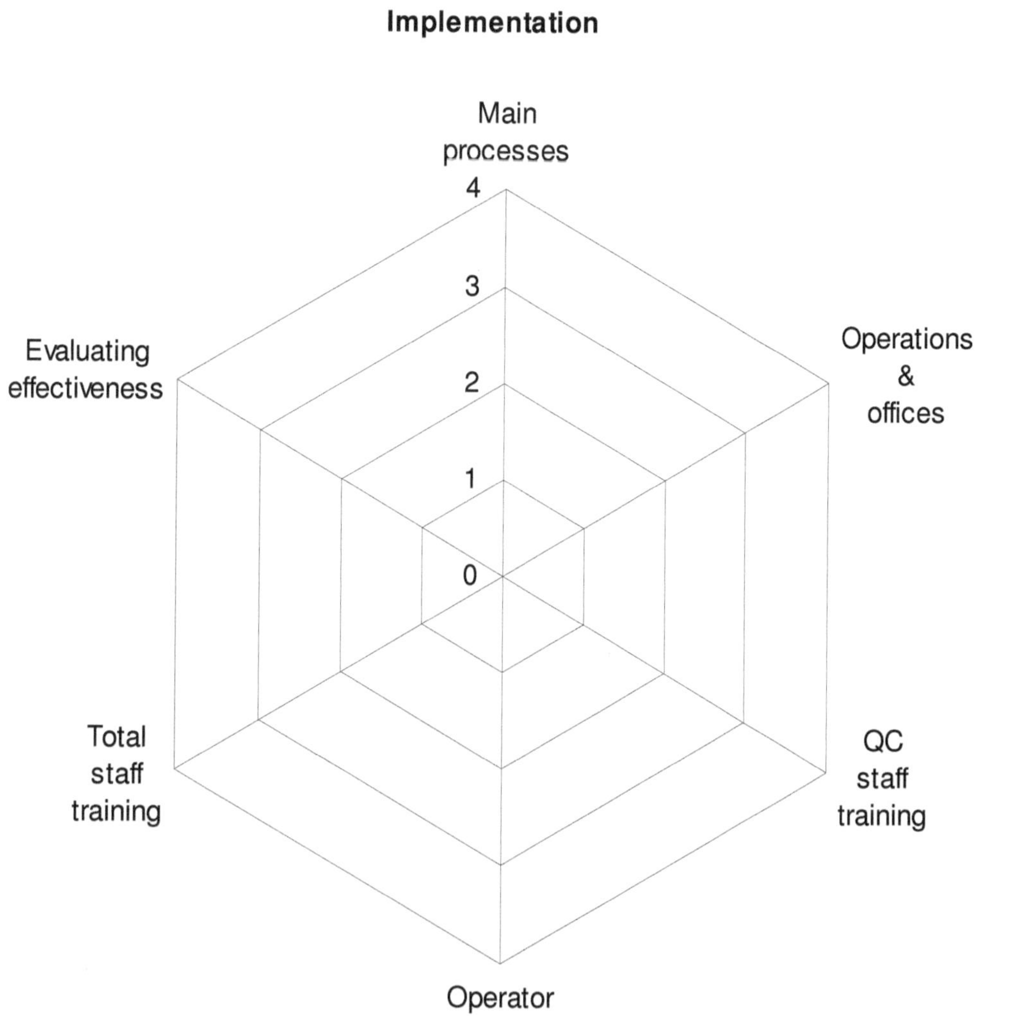


**Use the scoring chart to assess where you are in implementation**

The numbers from the scoring chart can be transferred to the radar chart for a quick visual assessment of where you are in terms of implementation.

Do not forget to implement SPC and control charts in operations and office functions - it is just as relevant.

Training is important as part of the implementation process and motivates the staff to understand and run their processes.

| Implementation | | | | | | |
|---|---|---|---|---|---|---|
| **Level** | **Main processes** | **Operations & offices** | **QC staff training** | **Operator training** | **Total staff training** | **Evaluating effectiveness** |
| 4 | SPC in operation for all processes. | SPC in operation for all processes. | Full QC staff training carried out on control chart setting & alarm analysis. All QC staff competent in carrying out capability studies & setting up control charts. | Full operator training carried out on control chart alarms. All operators competent to complete charts & detect alarms. | Full staff training carried out on basics of SPC. All staff aware of implications of SPC. | SPC in operation & fully evaluated for effectiveness. |
| 3 | SPC in operation for main processes only. | SPC in operation for main processes only. | Some QC staff training carried out on control chart setting & alarm analysis. Majority of QC staff competent in carrying out capability studies & setting up control charts. | Some operator training carried out on control chart alarms. Majority of operators competent to complete charts & detect alarms. | Some staff training carried out on basics of SPC. Majority of staff aware of implications of SPC. | SPC in operation with some evaluation for effectiveness. |
| 2 | SPC in operation for some processes. | SPC in operation for some processes. | Partial QC staff training carried out on control chart setting & alarm analysis. Few QC staff competent in carrying out capability studies & setting up control charts. | Partial operator training carried out on control chart alarms. Few operators competent to complete charts & detect alarms. | Partial staff training carried out on basics of SPC. Few staff aware of implications of SPC. | SPC in operation with little evaluation for effectiveness. |
| 1 | No SPC in operation. Trials carried out but no full implementation. | No SPC in operation. Trials carried out but no full implementation. | Limited QC staff training carried out on control chart setting & alarm analysis. Only selected QC staff competent in carrying out capability studies & setting up control charts. | Limited operator training carried out on control chart alarms. Only selected operators competent to complete charts & detect alarms. | Limited staff training carried out on basics of SPC. Only selected staff aware of implications of SPC. | SPC in operation but not evaluated for effectiveness. |
| 0 | No SPC in operation & no trials planned. | No SPC in operation & no trials planned. | No QC staff training carried out & none planned. | No operator training carried out & none planned. | No staff training carried out & none planned. | No SPC in operation & no evaluation of effectiveness. |
| **Score** | | | | | | |

# Key tips

- SPC is about preventing out-of-tolerance parts and nonconformities rather than about detecting out-of-tolerance parts and nonconformities .
- SPC controls the process to control the product.
- SPC, like most methods of quality control is based in statistics, the only difference is that the statistics are more visible in SPC.
- Every process and product varies with time but variability can be divided into common and special causes. Common causes are the natural variation of the system. Special causes are variations of the system from assignable and controllable events.
- Common causes give predictable variations. Special causes give unpredictable variations.
- Special causes can normally be identified and rectified at the local level.
- Common causes are difficult to identify and need management action to reduce them (they can never be totally removed, only reduced).
- In many cases the normal distribution can be used to represent the distribution of common causes.
- The capability of a process defines the ability of the process to make products to the specification.
- Machine capability defines the ability of a machine or process to make products to the specification, it is a short-term snapshot of the process and does not include all the variables.
- Process capability defines the ability of the complete process to make products to the specification, it includes all the long-term variables.
- A knowledge of capability and the normal distribution allows prediction of the amount of product that will be produced within specification.
- Capability can be defined in terms of both location (where the distribution is in relation to the specification limits) and spread (how big the distribution is in relation to the specification band).
- Control charts can be produced both for variables (numbers) and for attributes (go/no go, present/not present).
- Control charts for variables can be created for any measurable feature. The control limits are assigned on the statistics and not on the basis of the tolerances.
- Control charts for attributes can be created for any decision-based feature. The control limits are assigned on the statistics and not on the basis of the tolerances.
- Patterns or alarms on control charts indicate the presence of special causes. The rules for patterns and alarms are the same for variables and attributes control charts.
- The type of pattern or alarm on the control chart gives a good indication of the source of the special cause.
- Control charts not only tell us when to take action (when there is an alarm pattern) but also when to do nothing (when there is no alarm pattern). If there is no pattern or alarm indicated on the control chart then the process should not be adjusted.
- Improving process capability is part of continual improvement and is based on reducing the effect of the common causes. This requires management action.
- Implementing SPC is a project and a process. It needs planning, control and post-project evaluation.

# Chapter 6

# Acceptance sampling

The AQL approach to quality is needed for quality control and inspection of products where it is impossible to perform standard SPC process control. This is primarily where purchased products or materials are being inspected at Inwards Goods and where AQL is used as an acceptance testing protocol because it is impossible to control the process.

AQL is based on sampling a 'lot' or 'batch' of product and inspecting the samples to see if they meet the specification (or not). The number of nonconforming samples is used to decide if the complete batch will be accepted or not. AQL is not inspection to weed out any bad products - if the number of nonconforming samples exceeds the limits then the complete batch is rejected back to the supplier.

Obviously a sample is simply a snapshot of the complete population (see Section 5.1) and removing any nonconforming products from the samples does not improve the quality of the batch. Removing nonconforming products is not the aim of AQL. It is to form a view on the overall quality level of the batch and to act on this view with some statistical basis.

A possibly apocryphal story is told about a customer who used AQL as an Inwards Goods inspection method. They notified their (very reputable) Japanese supplier that under their AQL rules they would accept 10 nonconformities/10,000 products. On the first delivery of 10,000 products they opened the boxes and found 10 products in a small bag with a note which read 'We are not sure why you want 10 nonconforming product but they are in this bag'. The supplier had good control and was sure that all their product met the specification but had produced 10 nonconforming products specifically for the customer.

AQL can be used as method of setting inspection levels for in-process inspection or as the final inspection of a product immediately prior to despatch to the customer. This is ultimately a futile approach because failing to control the process will always lead to failing products.

AQL should not be seen as a substitute for effective process control. Finding out that you have produced a nonconforming product after the event is a waste of time and money (but you should know that by now).

"You can't inspect quality into a product."

**Anon**

# 6.1 Sampling systems approaches to quality

## How many bad ones can you accept?

Imagine you have just had a delivery of components from a supplier and want to check if these are acceptable for use in production. It is obvious that you will have to inspect these in some way but:

- How many components should be inspected (the sample)?
- What number of nonconforming items in the sample is enough to reject the entire lot?
- After the number of nonconforming items in the sample has been determined how can the number of nonconforming items in the lot be estimated?
- What is the possible or probable percentage of error in the estimate of the total number of nonconforming items?
- How can the number of nonconforming items in various samples be charted over time in order that unfavourable trends may be detected and the causes corrected?

These are very complex questions to answer and involve some fairly high-level application of probability and statistics theories. Fortunately, most of the hard work has already been done by statisticians and all we have to do is set up the system and read the values from tables.

**Note 1:** Sampling a constant percentage of the lot might seem intuitively correct but actually suffers from some severe problems and is very unreliable. A constant percentage is too small for small lots and over-inspects on large lots. It may seem surprising but a 10% sample on a lot of 10,000 is statistically far more stringent than a 10% sample on a lot of 50.

## A short history of acceptance sampling

Acceptance sampling was developed when Dodge and Romig were working at Bell Laboratories in the early 1930's. They developed sampling systems to provide inspection plans with a statistical approach to sampling and to allow an assessment of the risks to both the producer and the consumer.

The resulting Dodge-Romig tables enabled a small proportion of randomly chosen components from a lot to be tested. The results from this sample could then be used to accurately predict the behaviour of the rest of the lot. These tables were further developed by Dodge and Romig and were used extensively by the US military during World War II as MIL-STD-105. Although the standard was initially used as a control method for ammunition suppliers, where full inspection was not possible, it was later extended and used for many other suppliers and products.

The success of the Dodge-Romig sampling systems for incoming inspection led to very similar systems being used all around the world and they still form the basis for most sampling systems where they are referred to as AQL systems. AQL meant 'Acceptable Quality Level' when I was young but this has since been redefined as 'Acceptance Quality Limits'. This more correctly defines the method as it is largely used for acceptance of product rather than as a sign that the product is of acceptable quality.

- **Tip** - It is possible to use acceptance sampling as a production control tool for continuous production but control charts are far preferable.
- **Tip** - Sampling systems are designed to assess the quality of the lot from the quality of the sample. They are not designed to find the bad products.

## The basics

Inspection of products by sampling is normally inspection by attributes. A unit of the product is classified as conforming or nonconforming or the number of nonconformities in the unit of product is counted. Inspection by attributes allows go/no-go inspection and the idea of a sampling plan is to provide a solid statistical basis for acceptance or rejection of a lot based on the results obtained from limited sampling.

- **Tip** - Sampling systems can also be used for variables but this is less common (see ISO 3951 for details) and we will only consider sampling by attributes.
- **Tip** - Inspection is defensive quality control and filters products already

The last version of MIL-STD-105 was MIL-STD-105E (1989) but many other similar tables and concepts were developed for BS, ASTM and ANSI/ASQ. The current international standard is ISO 2859-1:1999.

All of these standards have the same statistical basis although they differ in other areas such as terminology.

For the purposes of this chapter we will refer to ISO 2859-1:1999.

Sampling by attributes classifies the samples as either conforming or nonconforming with nothing in between.

Sampling by variables uses actual results, e.g. the length of a product.

Sampling can also be termed 'Statistical Product Control' as it uses statistical techniques to determine sample sizes and acceptance levels.

Despite this, 'Statistical Product Control' is still an 'after the event' process of quality control. It is not the statistics that make it good or bad, it is when the control takes place.

produced, it does not magically inject quality into the product!

## The operating curve

Every sampling plan has an operating curve (OC) that describes the probability of accepting a lot of products based on the results of inspecting a random sample taken from the lot.

The ideal sampling plan would involve agreeing a specified % of nonconforming items between the producer and the consumer (see Section 6.2 for all definitions), inspecting a small sample of the delivered product and knowing that this sample was totally representative of the complete lot.

If the % nonconforming items in the sample was higher than the agreed % then the complete lot would be rejected and if the % nonconforming was less then the complete lot would be accepted. The operating curve for this ideal situation is shown on the upper right.

Unfortunately the selection of samples from the lot is governed by the laws of probability and this type of operating curve will only be seen for 100% inspection (which doesn't work anyway due to inspector fatigue and error).

In real sampling systems, where the laws of probability operate on the selection of samples, the operating curve is similar to that shown on the lower right. In this case if the % nonconforming was 2% then there would be a 52% chance of the lot being accepted but if the % nonconforming was 3% then there would only be a 24% chance of the lot being accepted.

Operating curves vary depending on the exact details of the sampling system used but, because of the probabilistic nature of selecting samples from a lot, they are at the heart of sampling systems. There is always a probability of rejecting a lot, based on the samples inspected, when the overall quality of the lot is actually better than that required - this is called 'the producer's risk'. Equally there is a probability of accepting a lot, based on the samples inspected, when the overall quality of the lot is actually worse than that required - this is called 'the consumer's risk'.

The benefit of the Dodge-Romig tables and similar approaches is that both 'the producer's risk' and 'the consumer's risk' can be quantified from probability theory. These risks can then be negotiated between the producer and the consumer along with the resulting costs of conformance and inspection.

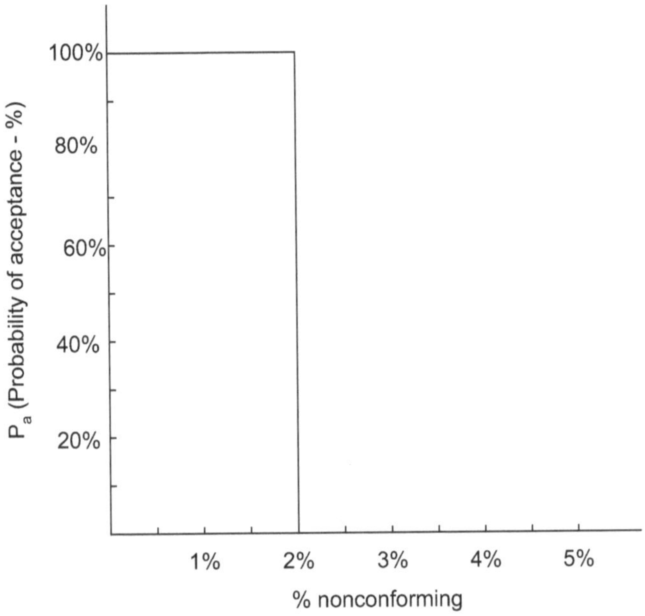


### The ideal operating curve

For the ideal operating curve if the % nonconforming was higher that 2% then there would be a 0% chance of accepting the lot of product and if the % nonconforming was lower than 2% then there would be a 100% chance of accepting the lot

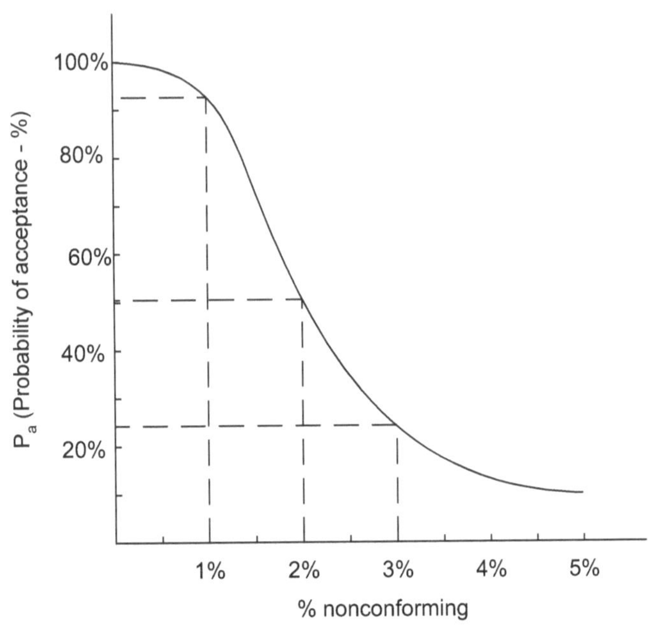


### A typical operating curve

This shows that if the % nonconforming is 3% there is a 24% chance that the lot will be accepted, if the % nonconforming is 2% there is a 52% chance of acceptance and if the % nonconforming is 1% there is a 93% chance of acceptance.

# 6.2 Sampling systems - general

## This is all about probability

Sampling systems are all based on probability theory and do not provide absolute certainty that a lot is either good or bad. This is the debit side of inspecting only a few samples from the complete lot.

The main features of an operating curve are shown in the diagram on the upper right and these are:

- The Acceptance Quality Limit (AQL) is the maximum % nonconforming that is acceptable to the consumer. If the % nonconforming is less than this then the lot is regarded as acceptable quality. If the % nonconforming is more than this then the lot is regarded as doubtful or poor quality.
- The Consumer's Risk Quality (CRQ) is the maximum % nonconforming that the consumer will accept. If the % nonconforming is more than this then the lot is regarded as poor quality.

The statistical nature of sampling plans means that it is possible for a lot that conforms to fail the inspection (or vice versa). These possibilities are shown in the diagram on the lower right and the possibilities are:

- Actually conforming and declared conforming - In this case the plan is correctly assessing the product quality.
- Actually nonconforming and declared nonconforming - In this case the plan is correctly assessing the product quality.
- Actually conforming and declared nonconforming - In this case, the lot is falsely rejected, i.e. the lot really conforms to the specification but fails the inspection. This is the producer's risk (α-risk) and is shown on the operating curve as the area at the top left.
- Actually nonconforming and declared conforming - In this case, the lot is falsely accepted, i.e. the lot does not conform to the specification but passes the inspection. This is the consumer's risk (β-risk) and is shown on the operating curve as the area at the bottom right.

These risks (α-risk and β-risk) can be quantified and clearly defined through the choice of the sampling plan from the sampling system. This knowledge of the risks and the sharing of the risks between the supplier and consumer (in return for reduced inspection levels and costs) is the great strength of sampling systems.

An AQL level defines the worst a manufacturer can allow production to become before action is required.

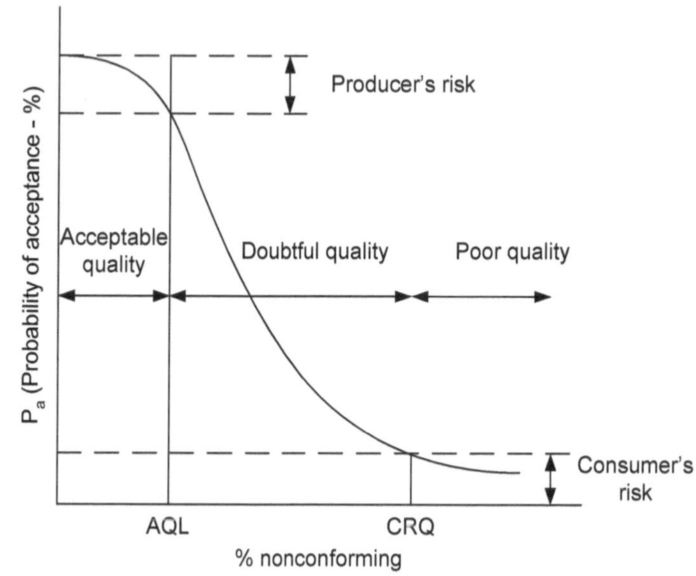


### The main features of the operating curve

Every sampling plan has a different operating curve but the main features remain constant. The important points are the AQL and the CRQ because these define the producer's and consumer's risk levels.

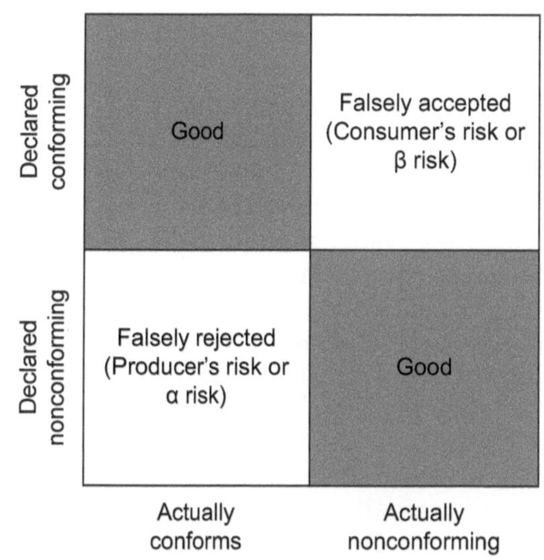


### The possibilities with sampling plans

Sampling plans use the results from a small sample to assess the status of the complete lot. It is still possible that the sampling plan will either falsely accept or falsely reject a lot but the plans allow an accurate assessment of these risks.

## The ISO 2859 family

The international standard for sampling systems is ISO 2859 and if sampling systems are going to be used to set inspection levels then it is strongly recommended that a copy of this standard is obtained. However, ISO 2859 is not a single standard but a series of standards covering sampling systems for a variety of situations. The standards are:

- **Part 1: Sampling schemes indexed by acceptance quality limit (AQL) for lot-by-lot inspection.** This standard is primarily designed for use with a series of lots coming from regular suppliers using the same production process. This is the most widely used standard of the series.
- **Part 2: Sampling plans indexed by limiting quality (LQ) for isolated lot inspection.** This standard is primarily designed for use with single lots of supplies or with new suppliers where the switching rules between normal, tightened and reduced inspection levels (see below) cannot be used for lack of previous information.
- **Part 3: Skip-lot sampling procedures.** This is used for very good suppliers where confidence in the supplier is very high.
- **Part 4: Procedures for assessment of declared quality levels.** This is used to verify a declared quality level for reviews or audits, it is not suitable for assessing lots of delivered products.
- **Part 5: System of sequential sampling plans indexed by acceptance quality limit (AQL) for lot-by-lot inspection.** This is designed to supplement Part 1.
- **Part 10: Introduction to the ISO 2859 series of standards for sampling for inspection by attributes.** This provides a general introduction to sampling plans and guidance on the selection of which type of sampling plan to uses, i.e. which part of the standard to use.

Most companies will use sampling systems for inspecting regular deliveries from established suppliers and will therefore use ISO 2859-1 as the main standard.

- **Tip** - Inspection by attributes can also include go/no-go gauges so you can also assess dimensions.

Definitions

■**Producer** - The organisation or person who manufactures the batches of goods.

■**Consumer** - The organisation or person who purchases or receives batches of goods from the producer.

■**Producer's risk** - The probability that an acceptable lot (at the AQL or better) will be rejected. This is sometimes termed the α-risk.

■**Consumer's risk** - The probability that a rejectable lot (at the CRQ or worse) will be accepted. This is sometimes termed the β-risk.

■**Nonconformity** - Non-fulfilment of a specified requirement. Nonconformities can be graded as follows:

- **Class A nonconformity** - A nonconformity which judgement and experience indicate is likely to result in safety critical failure, regulatory or standards infringement or complete functional failure. For general purposes, an AQL of 0.15 should be set for all Class A nonconformities.
- **Class B nonconformity** - A nonconformity that is likely to result in non-critical failure or will reduce materially the function of the product. For general purposes, an AQL of 1.0 should be set for all Class B nonconformities.
- **Class C nonconformity** - A nonconformity that is not likely to reduce the function of the product but will affect the visual appearance and consumer acceptance of the product. For general purposes an AQL of 4.0 should be set for all Class C nonconformities.

■**Nonconforming item** - Item with one or more nonconformities. These can be Class A, B or C nonconformities.

■**% nonconforming** - % of nonconforming items in a sample or lot.

■**Process Average (p)** - Average % nonconforming number or nonconformities per hundred items over the long term, i.e. the average quality.

■**Acceptance Quality Limit (AQL)** - This is the maximum process average which is acceptable to the consumer. Different AQL's may be designated for groups of nonconformities considered collectively or for individual nonconformities. An AQL for a group of nonconformities may be designated in addition to AQL's for individual nonconformities or subgroups within that group.

■**Consumer's Risk Quality (CRQ)** - The maximum % nonconforming that the consumer will accept. This effectively defines the consumer's risk level and is normally set at 10%. In the original MIL-STD-105 (and in many US standards), this is termed the Limiting Quality (LQ).

■**Lot (N)** - A collection of units of product from which a sample is to be drawn and inspected. This may differ from a collection designated as a batch for other purposes, e.g. production or shipment.

■**Sample (n)** - A randomly selected subset of a lot (see above) that is inspected.

**After ISO 2859-1:1999**

# 6.3 Sampling systems - the choices

## A multitude of decisions

A sampling system contains many sampling plans and it is necessary to decide which of these plans to use for a specific component. These decisions define the operating curve and the Accept/Reject numbers for the sample inspection.

## Inspection levels

The inspection level sets the relative amount of inspection to be carried out. ISO 2859 provides 7 inspection levels:

- The special inspection levels (S-1 to S-4). These should be used when the sample size must be small and larger sampling risks can be accepted.
- The general inspection levels (I, II and III). Level I is the least demanding and Level III is the most demanding.

The inspection level chosen defines the code letter for the relevant lot size and these are shown in the table on the right. The code letter defines the number of samples inspected and the Accept/Reject numbers for the lot.

- **Tip -** Unless otherwise specified, general inspection level II should be used.
- **Tip -** The inspection level can be set by the producer or the consumer and can be varied depending on the application.

## Sampling plan types

Sampling plans can use single, double or multiple sampling. This decision changes how samples are drawn from the lot and inspected.

### Single sampling plans

In single sampling, the accept/reject decision is based on one random sample from the lot. The sampling plan specifies the number of samples to be inspected (n) and these are randomly drawn and inspected. The value of 'n' depends mainly on the size of the batch (and also on the inspection level - see above). If the number of nonconformities is equal to or lower than the Accept number then the complete lot is accepted. If the number of nonconformities is equal to or higher than the Reject number then the complete lot is rejected.

Single sampling is the preferred option for most sampling needs.

### Double sampling plans

This is a two-stage process and the accept/reject decision for the lot may not be made until two samples (not necessarily of the same size) have been inspected.

In double sampling, the inspector starts by inspecting a sample ($n_1$) from the batch - this is normally smaller than the number of samples chosen for single sampling.

- If the number of nonconformities in $n_1$ is equal to or less than the first Accept value then the complete lot is accepted.
- If the number of nonconformities in $n_1$ is equal to or greater than the first Reject number then the complete lot is rejected.
- If the number of nonconformities in $n_1$ is between the first Accept and the first Reject numbers then a second sample of size $n_2$ is inspected.
- If the cumulative number of nonconformities (the sum of those in $n_1$ and $n_2$) is equal to or less than the second Accept number then the complete lot is accepted.

We have used the terms 'sampling system' and 'sampling plan' very specifically.

ISO 2859 etc. are sampling systems that contain many different sampling plans.

When inspection is carried out, it is carried out to a specified sampling plan from the relevant sampling system.

Any inspection system has costs associated with it. Taking these decisions affects the costs.

| Lot size | Special inspection levels | | | | General inspection levels | | |
|---|---|---|---|---|---|---|---|
| | S-1 | S-2 | S-3 | S-4 | I | II | III |
| 2 to 8 | A | A | A | A | A | A | B |
| 9 to 15 | A | A | A | A | A | B | C |
| 16 to 25 | A | A | B | B | B | C | D |
| 26 to 50 | A | B | B | C | C | D | E |
| 51 to 90 | B | B | C | C | C | E | F |
| 91 to 150 | B | B | C | D | D | F | G |
| 151 to 280 | B | C | D | E | E | G | H |
| 281 to 500 | B | C | D | E | F | H | J |
| 501 to 1,200 | C | C | E | F | G | J | K |
| 1,201 to 3,200 | C | D | E | G | H | K | L |
| 3,201 to 10,000 | C | D | F | G | J | L | M |
| 10,001 to 35,000 | C | D | F | H | K | M | N |
| 35,001 to 150,000 | D | E | G | J | L | N | P |
| 150,001 to 500,000 | D | E | G | J | M | P | Q |
| > 500,001 | D | E | H | K | N | Q | R |

**Lot size code letters for sampling plans (ISO 2859-1)**

Each inspection level has different code letters to specify the number of samples to be inspected and the Accept and Reject numbers. The code letter also changes with the lot size. Unless otherwise specified, use general inspection level II.

- If the cumulative number of nonconformities is equal to or greater than the second Reject number then the complete lot is rejected.

Double sampling plans can reduce costs by reducing the number of samples to be inspected but they can be more difficult to administer for inspectors.

### Multiple sampling plans

These are similar to double sampling plans but the inspection may involve up to 5 stages before an accept/reject decision is made. These are again cheaper to operate but can involve a lot of use of the tables and are even more complex to administer.

- **Tip** - Unless otherwise specified, a single sampling plan should be used as these are easier to control and understand.

## Inspection types

In addition to the types of sampling plan, there are also different types of inspection depending on experience with the supplier and previous inspection results. Typical operating curves for normal, tightened and reduced inspection are shown on the right.

### Normal inspection

This is the starting inspection type and is used when there is no reason to suspect that the supplier will have difficulty in achieving the specified AQL.

### Tightened inspection

This is used when evidence from normal sampling shows that the supplier is having difficulty meeting the specified AQL. Tightened inspection generally requires more samples to be inspected and the Accept/Reject values may also change.

### Reduced inspection

This is used when evidence from normal sampling shows that the supplier is easily meeting the specified AQL. Reduced inspection generally requires fewer samples to be inspected and the Accept/Reject values may also change.

### Switching between inspection types

It is normal to start inspection at the normal type and to then switch between normal, tightened and reduced types depending on the inspection results. The changes are made on the basis of 'switching rules' (see diagram on the lower right).

These rules are very similar for most sampling systems except for the switch from normal to reduced. ISO 2859-1 uses a switching score (see ISO 2859-1 for full details), whereas MIL-STD-105E used the simpler concept of '10 lots accepted and passes a Limit Number'.

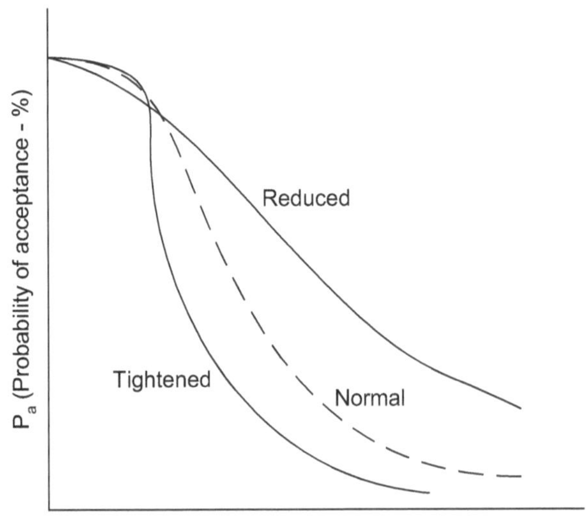


**Changing the inspection type changes the OC**

When the inspection type is changed from normal, tightened or reduced to another inspection type then the OC will change. The diagram shows some typical changes in the OC with changes in inspection type.

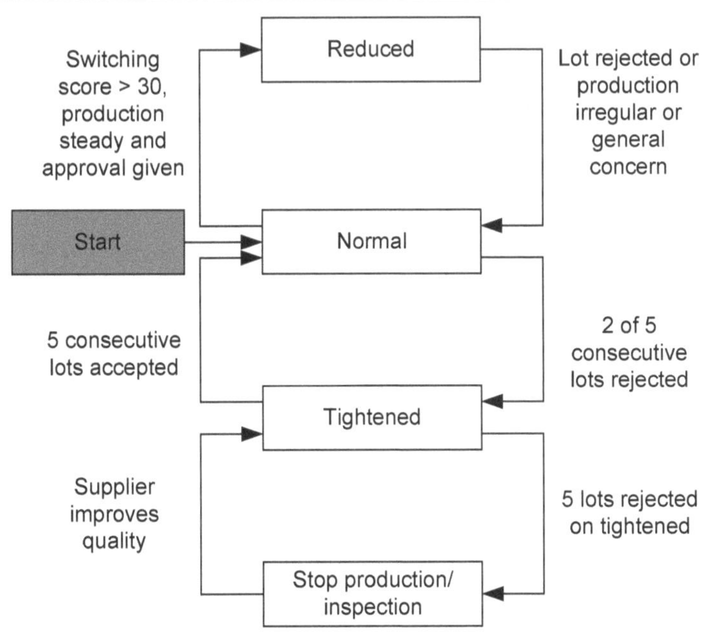


**Switching rules for ISO 2859-1**

The rules are based on experience with previous inspections. Switching from normal to reduced is based on a 'switching score': as lots are accepted the switching score increases, when it is > 30 inspection is switched from normal to reduced.

# 6.4 Using sampling systems

## Setting up the sampling scheme

AQL sampling systems are relatively easy to implement but there are a number of decisions to be made and there is a process to be followed. This is:

- The first task in setting up a sampling plan is to select the relevant AQL. It is possible to use any AQL level but it is far preferable to use the pre-calculated AQL tables from ISO 2859-1:1999.
- The chosen AQL level is a matter for discussion between the producer and the consumer. A lower AQL level means that there is less risk of the consumer being delivered unsatisfactory product but also increases the level of inspection required at either the producer or the consumer (and raises costs). There always needs to be a balance between costs and risks.
- If in doubt, use the following AQL levels depending on the product being inspected:
  - AQL = 0.15 for safety and functionally critical components. This gives a high level of confidence that the complete batch will be acceptable. This means that it is only acceptable for 0.15% of the components to be nonconforming or to contain nonconformities.
  - AQL = 1.0 for components with some critical aspects (functional or visual) but not safety-critical. This means that it is acceptable for 1.0% of the components to be nonconforming or to contain nonconformities.
  - AQL = 4.0 for non-critical components (functional or visual). This means that it is acceptable for 4.0% of the components to be nonconforming or to contain nonconformities.

**Note 1:** As discussed in Sections 5.7 and 6.2, there is a difference between nonconformities and nonconforming items. A nonconformity is non-fulfilment of part of the specification and a nonconforming item is an item with one or more nonconformities. Different AQL's can be set for individual nonconformities or for groups of nonconformities considered together. These AQL's may depend on whether the nonconformity (or group of nonconformities) is critical or non-critical.

- Select the inspection level to be used.
- **Tip** - Unless there is a very good reason to doubt or trust the supplier then general inspection level II should be used.
- Select the type of sampling plan to be used.
- **Tip** - Unless there is a good cause to select another sampling plan then single inspection should be used.
- Select the switching level to be used. For new suppliers the initial sampling should be based on 'normal' sampling. This may be changed by following the switching rules as inspections are carried out (see Section 6.3).
- Determine the size of the lot. This can be based on a delivery lot, a production lot (there may be more than one in each delivery) or other consistent definition of a lot. In ISO 2589-1 this will give a 'code letter' (A-R) for the lot size (see box in Section 6.3).
- Use the relevant sampling table (see Section 6.5 for an example table) to find out the number of components to be inspected and the Accept (Ac) and Reject (Re) numbers.

  The 'Accept' number means that if the number of nonconforming items or nonconformities is equal to or less than this number then the complete batch is accepted. The 'Reject' number means that if the number of nonconforming items or nonconformities is equal to or greater than this number then the complete batch is rejected.

**Note 2:** An arrow down (▼) means use the first sampling plan below the arrow. An arrow up (▲) means use the first sampling plan above the arrow. The arrows force movement to a lower or higher sample size and the new sample size must be used. This may result in 100% inspection in some cases.

**Note 3:** Always read the notes at the bottom of the table for special instructions. These may be vital for acceptance or rejection.

- Select the specified number of components.
- **Tip** - It is essential that the sample selected is random.

The same samples can be inspected for all the specified nonconformities.

In this case, the Accept/Reject decision should be based on each criterion separately, i.e. the sample should meet the AQL for each potential nonconformity or group of nonconformities and a lot should be rejected if it fails to meet the AQL for one or more criteria.

Use the samples from AQL assessment to apply forward control:

- Examine nonconforming items to see how and where the nonconformity could have happened.

**Note:** Look at good products as well.

- Examine the plant and the handling systems to look for swarf, abrasion, etc.
- Look on the floor, under the machines and inside the machines.
- Look anywhere that the machine could tell you what was happening.
- Take steps to remove any potential cause.

- It should not be concentrated near the start or end of production but should be spread throughout production. This is essential to provide a statistically valid assessment of the lot.
- If inspecting for visual attributes then do not look at the components when they are being selected to avoid biasing the sample selection.
- Never mix two small lots into one large one, they may not be from the same production run.

- Inspect the specified number of components using the relevant check list and record the number of nonconformities/nonconforming items.
- If the number of nonconformities/ nonconforming items is equal to or higher than Re then reject the complete lot.
- If the number of nonconformities/ nonconforming items is equal to or lower than Ac then accept the complete lot.
- Decide on what is going to be done with the complete lot. This may be:
  - Acceptance of the complete lot into the process.
  - Rejection of the complete lot back to the supplier.
  - Scrapping of the complete lot (if the supplier is internal).
  - Reworking of the complete lot (if the supplier is internal).
- Record the results of the inspection in the relevant checking form for later use in deciding how to apply switching rules.
- If appropriate, take action to improve quality (see Chapter 9.6).

## Product specifications

Sampling systems and sampling plans need to be firmly based on a well defined Product Specification.

- **Tip** - If a full Product Specification is not available then it should be created as soon as possible.

Only by using a complete Product or Purchase Specification can the appropriate functional and visual aspects be determined and realistic inspection and sampling plans assigned.

- **Tip** - Inspection by attributes is generally visual inspection. It is good practice to provide inspectors with samples of 'good' and 'nonconforming' products to assist in their judgement. These can be provided as photographs

but samples are better.

## Inwards goods specification

Sampling plans are primarily used for the inspection of inwards goods and inspection to a full Product Specification may be beyond the capabilities of inwards goods inspection, i.e. it may require specialised gauges or equipment.

In such cases, it is clearer and more convenient to create an abstract of the full Product Specification listing only those checks required on inwards goods inspection. This is the 'Inwards Goods Specification' and should contain not only features to be checked but also the AQL applicable to that feature and notes on whether any nonconformities found are to be considered singly or as a group.

## Check lists

To help inspection personnel, check lists (see Section 7.3) should be used for each product being inspected. This will enable simple marking to be made and totals to be calculated for ease of accept/reject decisions.

## Records

Check lists and other inspection results should be retained for at least 3 years to:

- Enable notification of acceptance.
- Provide conformance to product liability legislation.
- Serve as a record for ISO 9001.

As the quality of the product improves the value of inspection drops. If there are very few nonconforming items or nonconformities then, after some time, the inspectors assume that there are not many faults and virtually stop looking for them.

The higher the quality of the product the less effective inspection of any type is.

Acceptance sampling is most commonly based on attributes as these are relatively easy to use. It is, however, possible to use variables data for acceptance sampling. This is beyond the scope of this Workbook.

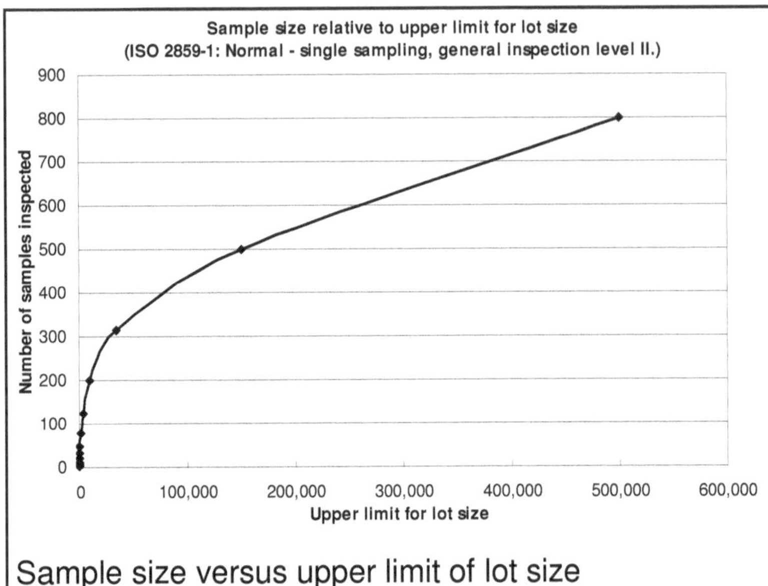


### Sample size versus upper limit of lot size

The number of samples inspected does not increase linearly with the upper limit for the lot size. This makes acceptance sampling very efficient for use with large lot sizes because the sample size is much smaller than sampling a constant percentage of the lot.

# 6.5 Sampling plans - the details

## Tables, tables and more tables

Sampling systems are all based on sampling plans and the decisions outlined in Section 6.4 result in a multitude of different tables. ISO 2859-1 contains tables for:

- Sample size code letters.
- Single sampling plans
  - Normal
  - Tightened
  - Reduced
- Double sampling plans
  - Normal
  - Tightened
  - Reduced
- Multiple sampling plans
  - Normal
  - Tightened
  - Reduced

In addition it contains tables for:

- Producer's risk for normal, tightened and reduced inspection: This set of tables gives the probability of rejection for lots of AQL at various inspection levels.
- Consumer's risk for normal, tightened and reduced inspection: This set of tables can be used when the number of lots does not allow the switching rules to be used and it is desired to set the consumer's risk quality (CRQ) and the consumer's risk.
- Average outgoing quality limits for normal inspection.
- Average outgoing quality for tightened inspection.
- A variety of other tables for using and assessing sampling plans.

The tables on the right are typical of those used in MIL-STD-105E and later in ISO 2859-1 for single sampling at a general inspection level II. These are not exactly as shown in ISO 2859-1 as we have used the relevant lot size directly instead of referring to the code letter for the lot size. These are presented simply to allow readers to see what typical sampling plan tables look like and to get an idea of what the actual standard contains (see Section 6.4 for guidance on how to read and use these tables).

- **Tip** - If you are going to use sampling plans for the control of purchased goods then purchasing a copy of ISO 2859-1 is strongly recommended.
- **Tip** - In the case of a high-quality supplier, i.e. the last 10 consecutive batches have been accepted on reduced sampling then it is possible to introduce 'Skip Lot Sampling' (see ISO 2859-3) where only a proportion of the batches are examined. This should not be introduced without formal approval at a management level.
- **Tip** - If you are on tightened inspection and the acceptance number has been exceeded but the rejection number has not been reached then accept the batch but do not reinstate normal inspection until 5 consecutive batches have been accepted.
- **Tip** - If you are on reduced inspection and the acceptance number has been exceeded but the rejection number has not been reached then accept the batch but reinstate normal inspection.

Sampling plans attempt to find the bad products in the lot but SPC attempts to control the process to avoid the production of bad products. It is self evident which is better.

The Process Average (p) should be significantly better than the AQL for most producers.

The setting of an AQL does not imply that the producer has the right to knowingly supply nonconforming items or nonconformities product.

**Normal inspection - single sampling, general inspection level II.**

| Lot Size | Sample Size | AQL (% nonconforming items or nonconformities per 100 items) | | | | | |
|---|---|---|---|---|---|---|---|
| | | 0.15 | | 1.0 | | 4.0 | |
| | | Accept | Reject | Accept | Reject | Accept | Reject |
| 2 to 8 | 2 | | | | | ↓ | |
| 9 to 15 | 3 | | | | | 0 | 1 |
| 16 to 25 | 5 | | | | | ↑ | |
| 26 to 50 | 8 | | | ↓ | | ↓ | |
| 51 to 90 | 13 | | | 0 | 1 | 1 | 2 |
| 91 to 150 | 20 | | | ↑ | | 2 | 3 |
| 151 to 280 | 32 | | | ↓ | | 3 | 4 |
| 281 to 500 | 50 | ↓ | | 1 | 2 | 5 | 6 |
| 501 to 1,200 | 80 | 0 | 1 | 2 | 3 | 7 | 8 |
| 1,201 to 3,200 | 125 | ↑ | | 3 | 4 | 10 | 11 |
| 3,201 to 10,000 | 200 | ↓ | | 5 | 6 | 14 | 15 |
| 10,001 to 35,000 | 315 | 1 | 2 | 7 | 8 | 21 | 22 |
| 35,001 to 150,000 | 500 | 2 | 3 | 10 | 11 | ↑ | |
| 150,001 to 500,000 | 800 | 3 | 4 | 14 | 15 | | |
| > 500,001 | 1,250 | 5 | 6 | 21 | 22 | | |

**Tightened inspection - single sampling, general inspection level II.**

| Lot Size | Sample Size | AQL (% nonconforming items or nonconformities per 100 items) | | | | | |
|---|---|---|---|---|---|---|---|
| | | 0.15 | | 1.0 | | 4.0 | |
| | | Accept | Reject | Accept | Reject | Accept | Reject |
| 2 to 8 | 2 | | | | | | |
| 9 to 15 | 3 | | | | | ↓ | |
| 16 to 25 | 5 | | | | | 0 | 1 |
| 26 to 50 | 8 | | | | | | |
| 51 to 90 | 13 | | | ↓ | | ↓ | |
| 91 to 150 | 20 | | | 0 | 1 | 1 | 2 |
| 151 to 280 | 32 | | | | | 2 | 3 |
| 281 to 500 | 50 | | | ↓ | | 3 | 4 |
| 501 to 1,200 | 80 | ↓ | | 1 | 2 | 5 | 6 |
| 1,201 to 3,200 | 125 | 0 | 1 | 2 | 3 | 8 | 9 |
| 3,201 to 10,000 | 200 | | | 3 | 4 | 12 | 13 |
| 10,001 to 35,000 | 315 | ↓ | | 5 | 6 | 18 | 19 |
| 35,001 to 150,000 | 500 | 1 | 2 | 8 | 9 | ↑ | |
| 150,001 to 500,000 | 800 | 2 | 3 | 12 | 13 | | |
| > 500,001 | 1250 | 3 | 4 | 18 | 19 | | |

**Reduced inspection - single sampling, general inspection level II.**

| Lot Size | Sample Size | AQL (% nonconforming items or nonconformities per 100 items) | | | | | |
|---|---|---|---|---|---|---|---|
| | | 0.15 | | 1.0 | | 4.0 | |
| | | Accept | Reject | Accept | Reject | Accept | Reject |
| 2 to 8 | 2 | | | | | ↓ | |
| 9 to 16 | 2 | | | | | 0 | 1 |
| 16 to 25 | 2 | | | | | ↑ | |
| 26 to 50 | 3 | | | ↓ | | | |
| 51 to 90 | 5 | | | 0 | 1 | ↓ | |
| 91 to 150 | 8 | | | ↑ | | 1 | 2 |
| 151 to 280 | 13 | | | | | 2 | 3 |
| 281 to 500 | 20 | ↓ | | ↓ | | 3 | 4 |
| 501 to 1,200 | 32 | 0 | 1 | 1 | 2 | 5 | 6 |
| 1,201 to 3,200 | 50 | ↑ | | 2 | 3 | 6 | 7 |
| 3,201 to 10,000 | 80 | | | 3 | 4 | 8 | 9 |
| 10,001 to 35,000 | 125 | ↓ | | 5 | 6 | 10 | 11 |
| 35,001 to 150,000 | 200 | 1 | 2 | 6 | 7 | ↑ | |
| 150,001 to 500,000 | 315 | 2 | 3 | 8 | 9 | | |
| > 500,001 | 500 | 3 | 4 | 10 | 11 | | |

# 6.6 Sampling systems - where are you now?

## Effective sampling

Sampling systems and sampling plans can both improve the efficiency of product inspection and reduce the inspector's workload whilst at the same time providing a rigorous statistical basis for the sampling and inspection process.

Sampling systems and sampling plans allow an assessment of the risk to both the producer and the consumer in accepting or rejecting a lot of product and these risks need to be quantified, discussed and agreed. The rational and statistically based approach to risk provided by sampling plans is only possible if sampling plans are firmly based on a relevant product specification and use the appropriate Acceptance Quality Level and sampling plan design.

## Completing the chart

This chart is completed and assessed as for those presented previously.

**Sampling systems**

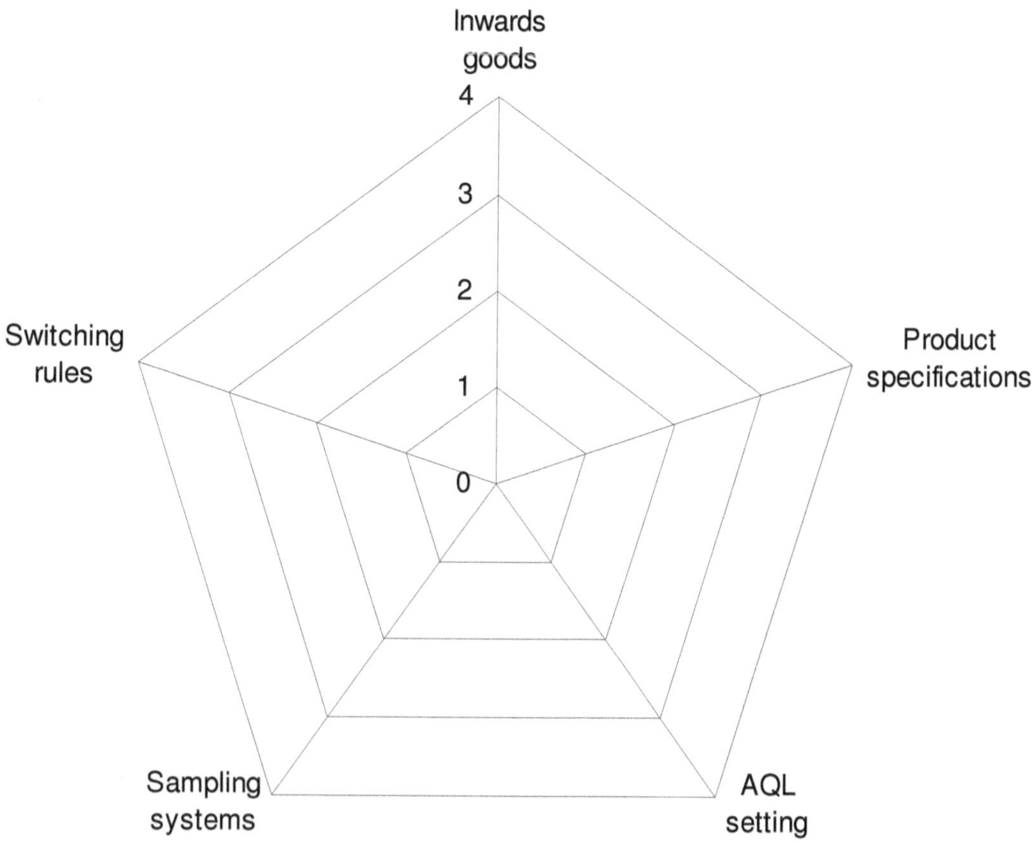


Use the scoring chart to assess where you are in sampling systems

The numbers from the scoring chart can be transferred to the radar chart for a quick visual assessment of where you are in terms of sampling systems.

"An individual sampling plan has much the effect of a lone sniper, while the sampling plan scheme can provide a fusillade in the battle for quality improvement."

**E.G. Schilling**

## Sampling systems

| Level | Inwards goods | Product specifications | AQL setting | Sampling systems | Switching rules |
|---|---|---|---|---|---|
| 4 | Sampling systems used for all incoming product. | Full product specifications available for all products. All product specifications adapated for ease of use by inspectors. | Appropriate AQLs set for all potential nonconformities and groups of nonconformities. AQLs related to product specification. | Excellent setting of inspection levels, sampling plan types and inspection types. All sampling plan decisions recorded and justified. | Switching rules known, results recorded, decisions justified and followed. |
| 3 | Sampling systems used for some incoming product. Other products use constant % of product or other measure. | Full product specifications available for all products. Most product specifications adapated for ease of use by inspectors. | Appropriate AQLs set for most potential nonconformities and groups of nonconformities. AQLs broadly related to product specification. | Good setting of inspection levels, sampling plan types and inspection types. Some sampling plan decisions recorded and justified. | Switching rules known and followed. Most results recorded and justified. |
| 2 | Inspection based on constant % of product or other measure. All incoming product inspected. | Full product specifications available for all products. Some product specifications adapated for ease of use by inspectors. | Appropriate AQLs set for few potential nonconformities and groups of nonconformities. AQLs not related to product specification. | Poor setting of inspection levels, sampling plan types or inspection types. No sampling plan decisions recorded and justified. | Switching rules known but not always followed. Few results recorded and justified. |
| 1 | Inspection based on constant % of product or other measure. Not all incoming product inspected. | Full product specifications available for some products. No product specifications adapted for ease of use by inspectors. | Poor or inappropriate setting of AQLs for potential nonconformities and groups of nonconformities. AQLs not related to product specification. | Inappropriate setting of inspection levels, sampling plan types or inspection types. No sampling plan decisions recorded or justified. | Switching rules known but rarely followed. No results recorded and justified. |
| 0 | No sampling systems used for inwards goods. No effective inspection of incoming product. | No product specifications available. | No AQLs set for any product. | No setting of inspection levels, sampling plan types and inspection types. No sampling systems used. | Switching rules not known or followed. |
| **Score** | | | | | |

# Key tips

- Sampling systems developed from the original Dodge-Romig tables in the 1930s to MIL-STD-105 in the 1940s to ISO 2859-1 in the 1990s.
- Sampling systems are needed for economic inspection of products that have already been produced. The inspection is after production has been completed.
- Most sampling systems are based on inspection by attributes but it is possible to use sampling systems for inspection by variables.
- Sampling systems allow a rational approach to deciding the number of samples to be inspected, the allowable number of rejects and the associated risks to the producer and consumer.
- Sampling plans have an operating curve that describes the properties of the sampling plan. The operating curve can be used to define the Acceptance Quality Level (AQL) and the Consumer's Risk Quality (CRQ).
- It is possible for a sampling plan to either falsely accept or falsely reject a lot of product. These are the consumer's risk and the producer's risk, respectively.
- Individual sampling plans are chosen from a sampling system.
- The producer and the consumer must decide on the parameters of the chosen plan to provide the appropriate AQL for selected nonconformities, the inspection level, the type of sampling plan (single, double or multiple) and the inspection type (tightened, normal or reduced).
- Switching between normal, tightened and reduced is controlled by 'switching rules' that vary with the sampling system used.
- The operating curve will vary with each choice made by the producer and consumer and this will also change the costs associated with inspection.
- Sampling plans assess the quality of the complete lot based on the number of nonconformities or nonconforming items found in a sample drawn at random from the lot.
- After assessing the quality of the lot based on the samples then the complete lot is either accepted or rejected - sampling plans do not simply form a judgement on the quality of the sample, they form a judgement on the quality of the complete lot.
- Sampling plans and the quality assessment of the sample need to be based on clear and unambiguous specifications of what is acceptable and what constitutes a reject.

# Chapter 7

# Tools for quality management and improvement

The preceding chapters have covered quality systems and statistical methods such as SPC and AQL. This has included many references to quality improvement and how this should be a driving force in quality management.

Quality improvement is largely driven by projects (see Chapter 2) but:

- How do you choose which projects to carry out?
- How do you know which data to collect?
- How do you treat the data you get from the process to actually get information?
- How do you present the information to get management involved in the quality improvement process?

Quality improvement involves making use of the tools for quality management and improvement. The main tools can be divided into two groups: these are the '7 Basic Tools' and the '7 New Tools'.

The 7 Basic Tools were first grouped by Kaoru Ishikawa (after whom the Ishikawa diagram was named and who also developed Quality Circles (see Section 2.6). Ishikawa did not invent all these tools but was the first to group them as quality improvement tools and develop their use in quality management. This was an achievement in itself.

The 7 New Tools were collected by the Union of Japanese Scientists and Engineers (JUSE) in the mid-1970s. Again, JUSE did not develop these tools but was the first to group and develop these tools as a group. Despite their usefulness, the 7 New Tools are not as commonly used as the 7 old tools and deserve more attention and use.

The 7 Basic Tools are mainly based on numbers, they may be slightly statistical in nature but need no high-level statistical knowledge to use effectively. The 7 New Tools are more oriented towards planning and management but are nonetheless extremely useful for quality management and improvement.

This chapter reviews all these tools and shows how they can be used to improve quality at all levels of a company.

"You must learn to talk clearly. The jargon of scientific terminology which rolls off your tongues is mental garbage."

"A conclusion is the place where you got tired thinking."

"Knowledge is a process of piling up facts; wisdom lies in their simplification."

**Martin Henry Fischer**

# 7.1 Quality management and improvement tools

## Quality improvement needs information

The statistical tools and methods described in the preceding chapters are only part of the toolkit for quality management and improvement. Quality management and improvement needs information. Many companies are awash with data but have little information.

Data are simply a set of numbers but information is something that can be used to manage and improve the business. Never fall into the trap of just looking at data, they will tell you nothing.

The '7 Basic Tools' can be used to transform data into information. These tools are:

- Flow charts.
- Check sheets.
- Cause and effect charts.
- Pareto analysis.
- Histograms.
- Scatter diagrams.
- Control charts (SPC charts).

**Note 1:** Control charts are covered at length in Chapter 5 and although they are one of the most useful of the quality tools they will not be covered in this chapter.

**Note 2:** Some lists of the 7 basic tools include 'stratification' as one of the tools but flow charts is often accepted as the alternate seventh tool. Stratification simply means breaking down the data into groups or sub-groups until they make sense. This will be shown in Section 7.3 for check sheets where defect data will be stratified to show patterns not clearly seen in the un-stratified data.

The '7 New Tools' can be used to organise information so that it is easily accessible and clearly understood for analysis and planning. These tools are:

- Relations diagram.
- Tree diagram.
- Affinity diagram.
- Matrix diagram.
- Matrix data analysis.
- Arrow diagram.
- Process decision program chart.

All of these tools are useful but they must be used appropriately. The diagram on the right gives some broad areas where the tools are applicable but using the correct tool is essential. No one would attempt to cut wood with a screwdriver, so you would not be wise to use Pareto analysis when cause and effect charting is the best tool. Quality professionals learn to choose their tools wisely and, like any professional, learn to keep their tools sharp, renewed and maintained.

**Note 3:** Matrix data analysis is mainly used as a prioritisation tool. The tool is very complex and rarely used. We will not cover this subject in this Workbook.

- **Tip -** These tools can also help in general cost reduction, provided you choose the right one for the job.
- **Tip -** Most of these tools have also been incorporated into the 6-Sigma methodology and 6-Sigma practitioners will be familiar with these and many other statistical tools.

Before studying the actual techniques, some points should be made about the general use of the tools.

"If the only tool you have is a hammer, everything starts to look like a nail."

**Abraham Maslow**

Using these tools can be used as evidence of quality improvement for ISO 9001.

| The 7 basic quality tools | |
|---|---|
| Identification tools | Analysis tools |
| • Flow chart<br>• Check sheet | • Histogram<br>• Statistical process control<br>• Scatter plot |
| Identification and analysis tools | |
| • Cause and effect chart<br>• Pareto analysis | |

| The 7 new quality tools | | |
|---|---|---|
| Organising tools | Relationship tools | Planning tools |
| • Relations diagram<br>• Tree diagram | • Affinity diagram<br>• Matrix diagram<br>• Matrix data analysis | • Arrow diagram<br>• Process decision program chart |

### The applications of the 7 basic and 7 new quality tools

The 7 basic tools are mainly number-based, i.e. statistical tools but the 7 new tools are mainly thinking-based, i.e. management and planning tools. Using the right tools can make quality planning and improvement a natural part of the business.

## Accuracy and precision

There is a tendency to assume that accuracy of any information is improved by adding as many numbers as possible after the decimal point. This is 'precision' and it is not the same thing as 'accuracy'. In discussing figures, round numbers are more easily remembered than spurious 'precision'. When using the quality tools, high accuracy (even with low precision) is far preferable to high precision with low accuracy (for a full discussion of accuracy and precision see Appendix 5).

You should not hesitate to use the word 'about' when trying to convert data into information, e.g.:

No of products last week = 243

No of products this week = 289

Increase = 18.93004%

or

About 20 % increase.

Which number best focuses your attention on the real issue?

## Sample sizes

Beware of making any statement where judgements are made without knowing the sample size. As shown in Section 5.3, a small sample is not necessarily representative of the whole population. It would be unrealistic to try to assess the average age of the UK population by asking five people in the street on a weekday their age. Most school children are at school, workers at work; and so on. The sample is not representative of the population and the average will be well adrift of the correct number. The result may well be precise to many decimal places but it would not be accurate.

For any sample to be significant then there must be a minimum number of articles, people, or tests. Be warned - making a judgement based on a sample size of 1 is not only foolish but downright dangerous, yet it happens every day in industry.

- **Tip** - How often has your boss jumped to conclusions based on a single event and how often have you been sent in the wrong direction by these judgements and decisions? It has happened to me, do not do this to your people

## The improvement action cycle

A key point in the management of information is the need to carry out the 'action cycle' of measurement, recording, analysis and action (see diagram below). This is similar to the PDCA cycle but I prefer this one. Far too often the first two stages (measurement and recording) are carried out but there is a failure to carry out the final two stages of analysis and action. The full action cycle should be followed to get the best results from the tools.

Try not to get your only exercise by jumping to conclusions.

## Go to it!

Start using the tools today. Select a quality problem, gather data and measurements, record the data, analyse the data using the tools and act to improve the quality. Then measure again .....

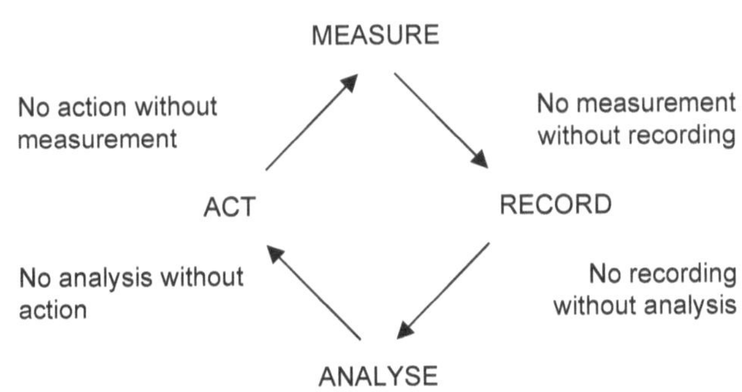


### The improvement action cycle

One of the first things to be done is to set the scene for using improvement tools by instituting the formal action cycle shown above.

The cycle can be used via the following statements:

- We will take no measurements without recording the results.
- We will not record results without analysing them.
- We will not analyse results without acting on them.
- We will not act without measuring the results of our actions.

This cycle provides the formal method for improving every aspect of our business based on evidence and analysis. It can be used in all areas of the company.

# 7.2 Flow charts

## Go with the flow

By visually interpreting a procedure into a flow chart, complex processes can be made easy to understand at a glance. A flow chart is simply a diagram or a picture of a procedure that is to be carried out. It is nothing exotic or special and everybody should use them at some stage in their work. The old saying that 'a picture paints a thousand words' is never truer than when trying to write out a procedure to be followed. This procedure may be for anything from how to process an order to how to set up a machine or even the more formal procedures such as those used for ISO 9001. Any process that needs to be carried out in a definite order should be capable of being described and simplified by the use of flow charts.

Flow charts enable everybody to see how the process works. Temporary or new staff can be trained to do things the right way rather than how they feel like it at the time. Training time is reduced, errors are reduced and everybody knows what is happening.

A flow chart is used to break a process down into the simple component steps and decision points. It should give an exact picture of all steps of the process that can be followed by anyone.

Creating a flow chart may seem simple but it is often difficult because all the process steps have to be evaluated and every response given in terms of definitive answers, generally of the form 'yes/no' or 'pass/fail'.

- **Tip** - If it is difficult to draw the flow chart when you understand the system then consider how difficult it is for employees to understand how it all works.
- **Tip** - If it is difficult drawing a flow chart at any point then it is at this very point that the employees will have difficulty in knowing exactly what to do, they will make up their own minds, be inconsistent and make costly mistakes.

The black and white decisions are easy to describe on a flow chart but the grey areas that are difficult to chart are the ones that need to be considered for system improvements, staff training or even new systems and controls.

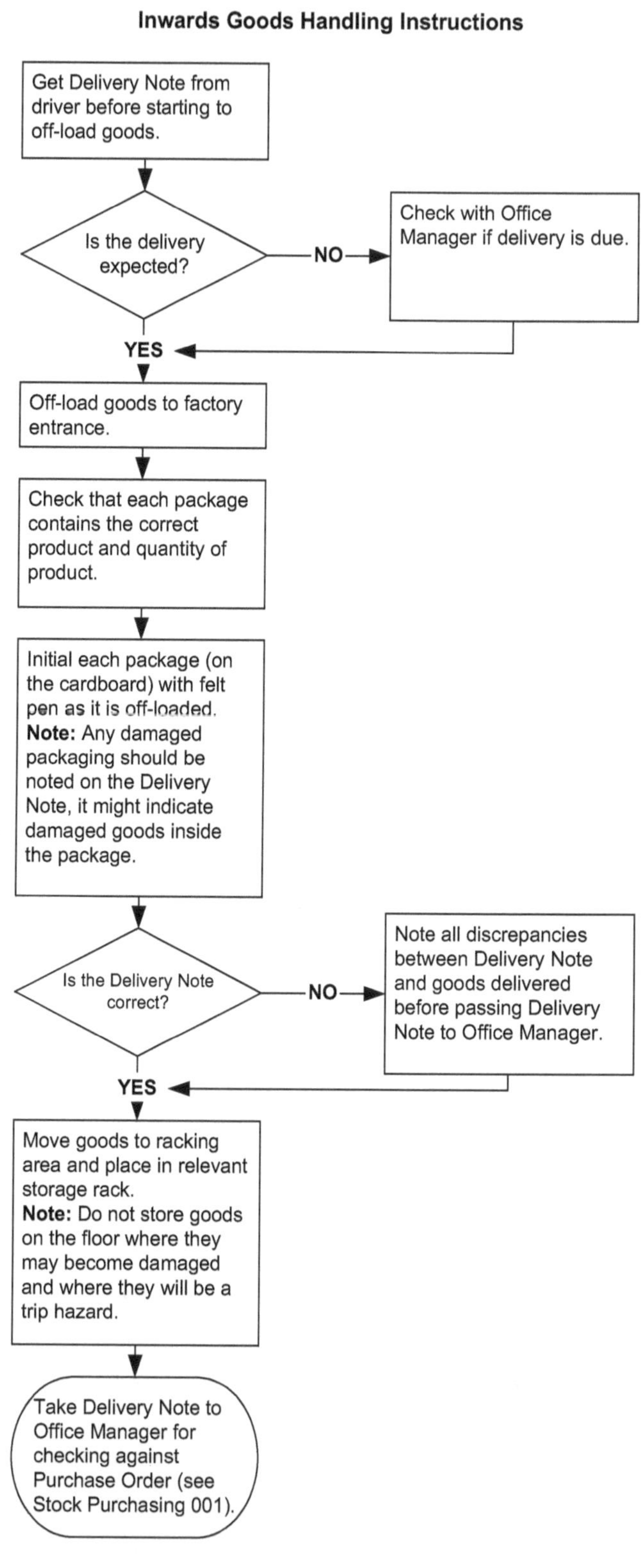


### Example flow chart for Inwards Goods handling

This is an example of a flow chart used to define the Inwards Goods handling process at a plastics processor. The flow chart makes the process simple to understand and easy to follow consistently. Similar charts work for all processes.

## Creating the chart

The steps to creating a flow chart are:

- Set the boundaries for the process to define where the chart starts and stops.
- List the major steps in the process using brainstorming or experienced process operators.
- Get the steps in the right order and check the decisions that need to be made at each step.
- Test the chart on the process to make sure that it really works.

For existing processes this should be for the actual process (with the operators or users). The actual can then be compared with the idealised flow chart (as it was thought to be by the person who developed the process). The two charts can be compared to find out where they are different. This is usually the source of a problem.

- **Tip -** The Post-it note is a vital tool in flow charting. Use them to get the basics in place before committing it all to paper.
- **Tip -** Avoid using computer programs to generate flow charts until they are finalised. Operating the software takes effort away from getting clarity. It can also only be operated by 1 person and generating a flow chart should be a group activity.

## Benefits

A flow chart will highlight the need for clear, unambiguous procedures that fully define everybody's responsibilities. Difficulties may be encountered because the process is not understood, there have been changes to the process which have not been recorded or because there really is no common process, i.e. everybody does it differently. The discipline of drawing a flow chart can solve all of these problems.

Flow charts often make very complex written procedures easy to understand at a glance. The logic and clarity of flow charts make them appropriate for many operations. Rather than write a long detailed procedure first, why not try making a flow chart and put it on the wall. This will improve the clarity of decision-making.

### How does the process really work?

Flow charts are an excellent method for showing (or discovering) how a process really works. They are invaluable for:

- Finding the inevitable 'gaps' in processes and procedures.
- Finding potential areas for improvement.
- Communicating what the process really does to everybody involved in the process.

If a procedure cannot be described in terms of a flow chart then the procedure will not work.

### Cross-functional flow charts

Conventional flow charts trace the process but make no attempt to look at the interactions of the departments involved. A cross-functional flow chart looks at the process and how it cuts across the company's structure. Managing the interfaces between the functions is critical to effective process operation.

### Quality procedures

Produce flow charts for ISO 9001, they are far easier to understand than conventional 'text' (see Section 4.6 for an example). Errors and inconsistencies can be seen quickly and rectified faster. We always use flow charts to map processes and procedures for ISO 9001 and they invariably come as a surprise to everyone (except the actual process operators). Changes then happen rapidly to smooth the procedure and to relate it to what was really needed in the first place.

Map what actually happens and not the ideal.

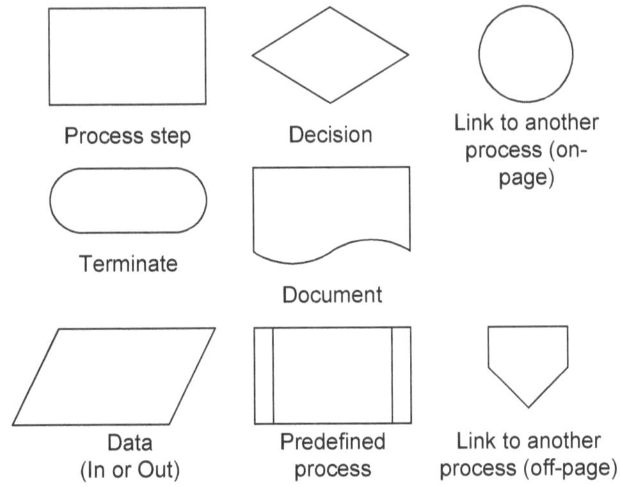


### The conventional flow chart shapes

These are the standard symbols to be used for flow charts. Get to know them and use them. The symbols will then allow all flow charts to be consistent and they also help to get extra information on to the chart.

# 7.3 Check sheets

## A deceptively simple tool

A check sheet is a method for rapidly gathering information on a problem in real time and at the point of generation of the information. There is a range of types of check sheets and they can also be called tally sheets or concentration diagrams depending on the exact type.

Check sheets use simple forms that are completed at the machine or process to both collect and analyse data. The simplicity of check sheets sometimes hides their usefulness and broad applicability and check sheets can be used in almost any situation where data needs to be gathered.

Check sheets provide a simple and effective method of displaying data gathered at source because they rely only on simple marks on the check sheet and not on recording actual measurements.

## Types of check sheets

There are several different types of check sheets (see upper right) and they all have different uses. The main types are:

### Classification

This type of check sheet classifies nonconformities, actions or events on the basis of time, cause or other classification. The classification checklist is one of the most common and useful check sheets. A first example is shown on the lower right and a second example is shown on the opposite page on the upper right. The first is based on the nonconformity type and the second is based on the time of the nonconformity in terms of the shift when it occurred (it could just as easily be based on smaller time periods).

Classification check sheets can be used to progressively 'drill down' into a problem to gain more information on importance and timing.

### Frequency

A frequency check sheet may not, initially, appear to give as much information as the classification type but they are useful as a first sift of nonconformities to identify the issues. An example is shown on the middle right on the opposite page. Frequency check sheets can be quickly used to create Pareto charts that identify the most profitable areas for action (see Section 7.5).

### Location (concentration)

A location check sheet marks the issues on a map, drawing or photograph to give a highly visual display of the nonconformity location. The location check sheet does not just focus on the number of nonconformities but on their location. This can be used to look for nonconformity concentrations and to show where work is needed, e.g. crime reporting logged on a map of an area may tell you where not to go! In plastics processing, if surface finish is an issue then a location check sheet can show where tooling needs to be modified.

Check sheets provide the means to gather valuable data in real time directly at the machine or process by the operator.

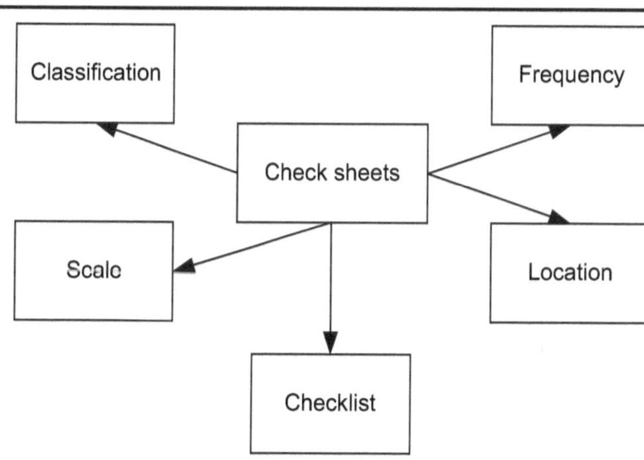


**Types of check sheet**

There are 5 different types of check sheet and they are all useful. The checklist is a special type of check sheet and some people do not consider it to be a true check sheet because it is used not for recording nonconformities but for preventing mistakes.

| Issue / Day | Mon | Tues | Wed | Thurs | Fri | Sat | Sun | Totals |
|---|---|---|---|---|---|---|---|---|
| Start-up | 𝍸 𝍸 𝍸 | | | | | | | 15 |
| Short shot | \|\| | | | | | | | 2 |
| Flash | \|\| | \| | \|\| | \|\| | | | | 7 |
| Silver Streak | \| | | \| | \|\| | \|\| | \| | | 7 |
| Sinkage | \| | \|\| | \| | \| | \|\|\| | \|\|\|\| | \| | 13 |
| Colour | 𝍸 | 𝍸 𝍸 | 𝍸 𝍸 𝍸 | 𝍸 𝍸 | 𝍸 𝍸 𝍸 \|\| | 𝍸 𝍸 𝍸 \|\|\| | 𝍸 \|\|\|\| | 84 |
| Totals | 26 | 13 | 19 | 15 | 22 | 23 | 10 | 128 |

**Typical classification check sheet 1**

A typical check sheet for nonconformities in injection moulding. Colour is an obvious problem in terms of nonconformities but it is not clear what the problem actually is.

## Scale (frequency distribution)

Scale check sheets use a measurement scale of 'boxes' or ranges (similar to that of a histogram) to group measurements quickly to assess the shape of a frequency distribution.

The measurement range is divided into interval boxes and each time a measurement is made it is recorded in the relevant interval box. This quickly builds up into a frequency distribution as the process is being run and can reveal if the process is normal or non-normal and if the process is skewed.

- **Tip** - Mark the upper and lower specification limits on the check sheet to quickly see the location of the distribution.
- **Tip** - It is useful to start with a few actual readings (in number format) to get an idea of the size of the intervals to be used (use the same rules as for a histogram).

## Checklist

A checklist is a special type of check sheet and some people do not consider these to be true check sheets, i.e. they are not one of the 7 basic tools. The other types of check sheets are designed to gather or display information for later processing but check lists are designed to make sure that an operator has taken all the important steps in a process. They are the 'memory joggers' or 'mistake-proofing' for a process and can be very important in complex processes, e.g. they are used extensively in medicine and aviation applications.

- **Tip** - If you have ever written a 'To-Do' list or a shopping list then you have used a checklist. There is nothing special about them!
- **Tip** - Use check lists to ensure that machines are turned on correctly, set correctly and turned off correctly.

## Using check sheets

Check sheets have many uses in plastics processing and the limitation is not so much in the method as in the imagination of the user. They can be used for:

- Nonconformity quantification by machine, product, worker or other potential cause in real time.
- Nonconformity tracking and location.
- Rapid process assessment.

| Issue / Day | Shift 1 | Shift 2 | Shift 3 | Totals |
|---|---|---|---|---|
| Start-up | 𝍸 𝍸 𝍸 | | | 15 |
| Short shot | II | | | 2 |
| Flash | II | II | III | 7 |
| Silver Streak | II | III | II | 7 |
| Sinkage | III | 𝍸 | 𝍸 | 13 |
| Colour | 𝍸 𝍸 II | 𝍸 𝍸 II | 𝍸 𝍸 𝍸 𝍸 𝍸 𝍸 𝍸 𝍸 𝍸 𝍸 𝍸 𝍸 | 84 |
| Totals | 36 | 22 | 70 | 128 |

### Typical classification check sheet 2

These are the same data as shown in check sheet 1 but divided by shift. Shift 3 obviously has an issue with colour. Does it have more colour changes? Does it use the same standards?

| Issue | Frequency | Notes | Totals |
|---|---|---|---|
| Start-up | 𝍸 𝍸 𝍸 | | 15 |
| Short shot | II | | 2 |
| Flash | 𝍸 II | | 7 |
| Silver Streak | 𝍸 II | | 7 |
| Sinkage | 𝍸 𝍸 III | | 13 |
| Colour | 𝍸 𝍸 𝍸 𝍸 𝍸 𝍸 𝍸 𝍸 𝍸 𝍸 𝍸 𝍸 𝍸 𝍸 𝍸 𝍸 IIII | | 84 |

### Typical frequency check sheet

The frequency check sheet does not generally give as much information as the classification check sheet but it can be a useful 'first sift' of information to identify the biggest issues.

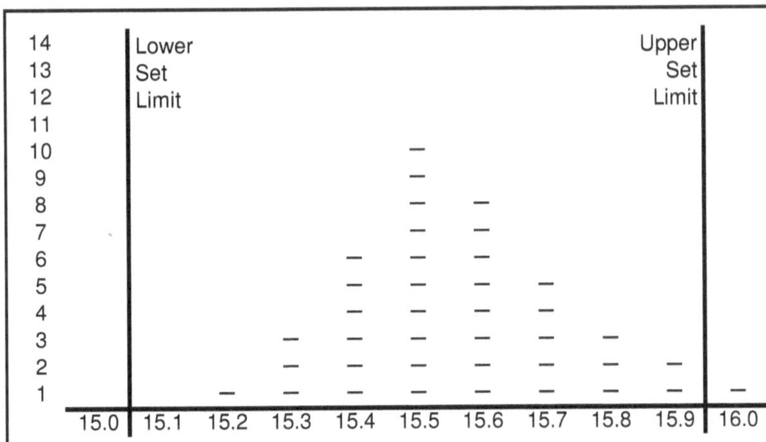


### Typical scale check sheet

The scale check list gives a quick idea of the frequency distribution (without any calculations). It is easy to see that the distribution here is slightly higher than nominal.

# 7.4 Cause and effect charts (root cause analysis)

## Finding the root cause

Cause and effect diagrams (also known as Fish-bone Diagrams, Ishikawa Diagrams or root cause analysis) are used to generate and note ideas or theories for problem solving in production or other processes. The 'effect' (or 'fish head') is the factor that needs to be increased (or decreased), the large fish bones indicate major categories or potential causes and the small fish bones are the minor categories or potential causes. In many ways the cause and effect diagram resembles a Mind Map but it is more specific in application.

The cause and effect diagram allows a team to focus on the content of the problem and not on the history, it focuses attention on the causes rather than simply on the symptoms and also allows a team to start to build a consensus for what are the likely causes and possible solutions to the problem.

The cause and effect diagram has been expanded by CEDAC (Cause and Effect Diagram Adding Cards) to increase the interactivity of the process. In this case the diagram is posted in a visible place to allow other staff to add their comments to the diagram and get the widest possible input.

- **Tip** - CEDAC can be used to get a wider range of inputs to a problem.

## The frameworks

To start and gain confidence, the basic framework can be initially based on 'the Ms' or 'the Ps'. These basic frameworks are shown on the right and provide the initial structure for the diagram. It is general to use the Ms for a production process and the Ps for a service but these can be altered as the team gains experience with the method.

- Ms - Materials, Machines, Manpower, Methods, Measures and Environment.
- Ps - Process, Product, Plant, People, Procedures and Environment.

## Generating cause and effect diagrams

The best way to gain value from the cause and effect diagram is to:

- Select a concern identified as being significant. This could be from Pareto analysis, check sheets or any other method.
- Using a flip chart, the concern or effect is drawn to the right of the major fish bone as the 'fish head'. It is important that the team is in agreement that this is the real concern and that it is not simply a symptom of the real effect.
- Using the basic frameworks as a guide, use the employees involved with the process or service to identify the various causes that are likely to create the concern. Brainstorming techniques or previous lists generated by the team can be used to quickly generate a list of possible causes.
- **Tip** - It is essential that the people making the diagram know what really happens in the process and are not

**5 Whys**

This is asking 'why does this happen' at least 5 times. This method tries to avoid looking at symptoms and looking for the root causes.

If presented with an oil leak on a machine it is not enough to clean it up. Using '5 Whys' can reveal that the real cause was that the machine was not subject to regular maintenance.

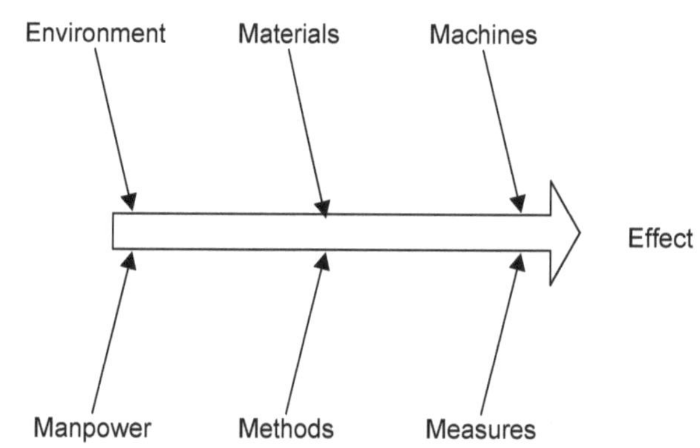


The Ms framework - Materials, Machines, Manpower, Methods, Measures and Environment

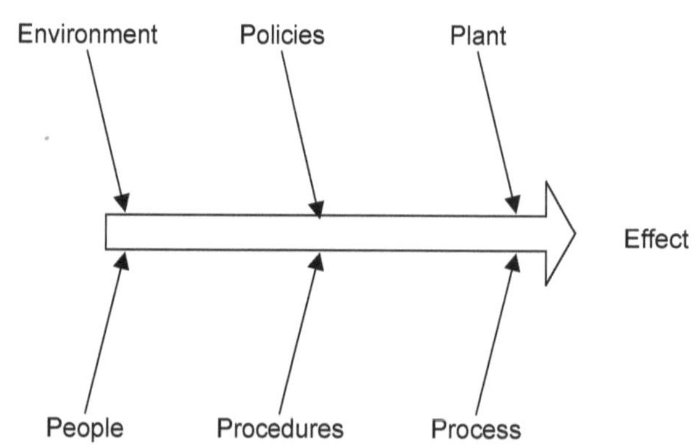


The Ps framework - Process, Product, Plant, People, Procedures and Environment

completing the diagram as if it is an ideal world.

- **Tip -** Try not to censor ideas at the brainstorming stage.
- Group the ideas into logical branches of related causes.
- **Tip -** The major cause headings are not sacred and you can change these to reflect the particular effect that you are seeking to minimise or maximise.
- Probe to seek the 'root cause' rather than simply accepting that everything is a cause. Often some of the initial causes identified are actually symptoms of another cause. Use '5 Whys' to get to the root causes.
- Look for root causes that appear in more that one major cause category.
- **Tip -** Use Post-it notes as part of the process (CEDAC) because they can be moved around easily as the diagram is developed.
- Use the employee's knowledge to identify the most significant or likely factors and to rank these on a 1 to 10 scale. The people doing the ranking should know what is really happening in the process.

## Taking action

After the diagram of possible causes is developed and refined, proposals can be developed to test and verify the importance of the suggested causes.

Ways of doing this are:

- Analysis of past data using Pareto analysis (see Section 7.5).
- Studies of current production using check sheets (see Section 7.3), scatter diagrams (see Section 7.7) or capability studies (see Section 5.4).

After the important causes are identified the team can start to develop the methods and actions to minimise or maximise the effect.

## Examples

Two examples of cause and effect charts are shown on the right using the M's and the P's. These are for a production environment (general plastic parts) and for a service environment (installation of PVC-U windows). Real cause and effect charts will be more complex and have more details on the main branches as well as scores for the relative importance of the cause in the creation of the effect or methods for checking the importance of the cause.

## Results

I have used this method, working with four line operators to identify and make a simple process modification that cost £450 to implement and saved £450 per week every week and is still saving £450 per week to this day.

The operators knew their process better than I ever could and they also knew ways to make it better - but no-one had thought to ask them or listen to their suggestions.

Use this with operators not managers. Operators know the real process not what was written in the capital expenditure request.

Materials, machinery and methods will be the most likely causes.

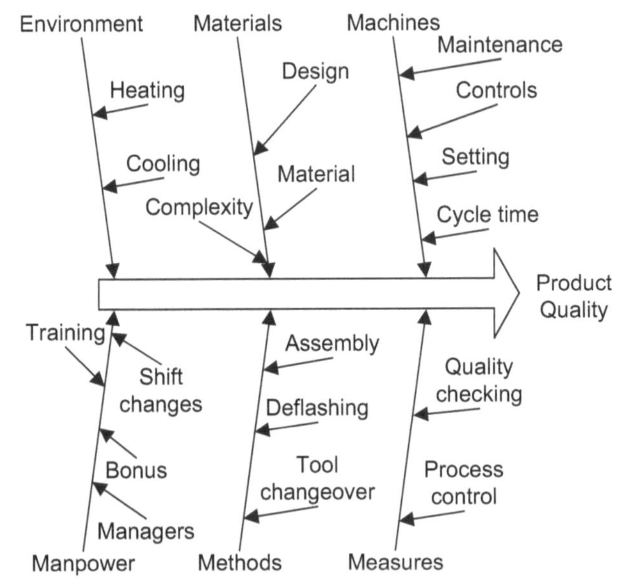


Example cause and effect chart for production - product quality of plastics products

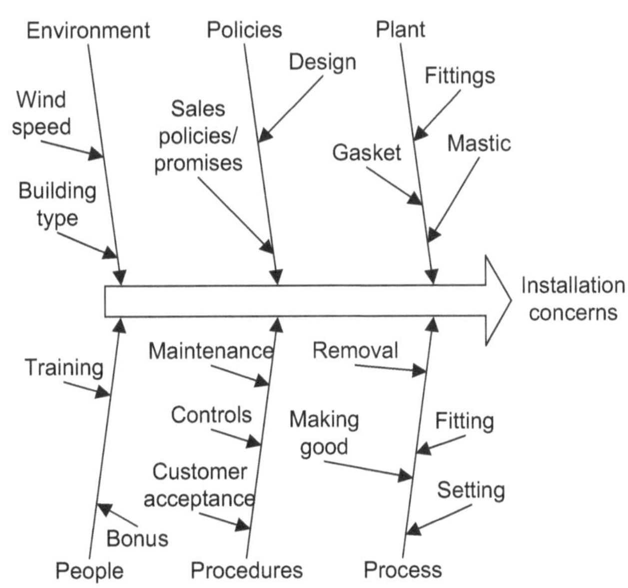


Example cause and effect chart for service - installation concerns with PVC-U windows

# 7.5 Pareto analysis

## Sifting the vital few from the trivial many

In any business it is essential to separate the 'vital few' from the 'trivial many' and Pareto analysis provides us with a way to find and identify those 'vital few' that can really make the difference to the business. It really is looking after your vitals.

The Pareto principle is named after the Italian economist Vilfredo Pareto (1848-1923) who studied the distribution of wealth. He found that wealth was not evenly distributed and that a few people controlled most of the wealth. Things have not changed a lot since then and his findings apply just as much to the world today as they did to Italy in the 19th century.

The rule was generalised by Juran (the quality guru) who also coined the phrase 'the vital few and the trivial many' to describe the Pareto principle. The Pareto principle or 80:20 rule can be applied to all areas of business and is used to focus on the real problems or issues. In essence the Pareto principle is a means of separating the vital few from the trivial many. To give some examples:

- 80% of your sales volume will come from 20% of your customers.
- 80% of your profit will come from 20% of your customers. These will not necessarily be the same customers as in the first example!
- 80% of your complaints will come from 20% of your customers.
- 80% of the nonconformities will be in 20% of jobs.
- 80% of the time taken off for sickness will come from 20% of the employees.
- 80% of lateness for work will come from 20% of the employees.
- 80% of the mass in the solar system will be in 20% of the planets. Actually it is 92% of the mass in 20% of the planets (if you include Pluto).

## Instinctive rule

In many ways the application of the 80:20 rule is instinctive and we know that it is right when it feels right. The important point is that when you can separate the vital few from the trivial many then you can begin to concentrate your efforts where the rewards are greatest. You can be more effective in your work when you concentrate on the vital few. It is logical that if you are going to attempt to improve something then you should start with the biggest contributors.

Make sure that you have all the information before making your decision.

Take, for example, stocks in your warehouse. If you have £8,000 of stock of item A and £2,000 of stock of item B and you need to reduce inventory by 20%, i.e. £2,000, then where do you start? The obvious answer is that it is easier and quicker to reduce the stock value of A by £2,000 than to reduce the stock value of B by £2,000 (this would mean stocking no B at all). Pareto is all about finding the 'vital few' and concentrating on these to get the best results.

The idea of having a 'key account manager' to be personally responsible for your top accounts is a reflection of Pareto, only 20% of your customers probably represent 80% of your turnover/profit/complaints and it pays to look after them and to make them feel special (see Section 11.1).

Pareto analysis is useful for taking a 'snapshot' of the information and targeting areas for improvement or concentration.

## Logical groupings

The first step in using Pareto is to gather information on products, complaints, sickness, profit per customer or whatever it is you wish to improve. This information can then be grouped into logical categories. For complaints analysis, an obvious initial grouping would be by product, and for customer analysis the initial grouping would be by customer. Try to get the information into as few categories as possible (about six is best) and if you have small numbers, i.e. lots of once-off customers, then you can use a special category called 'other' for the trivial many.

This information is then plotted as a bar chart by category (see Examples), starting with the highest and going to the lowest. If you have used a category for 'other' then this should be plotted last. The charts are then used to decide where you should take action and will also give information as to the action required.

- **Tip -** Sometimes the cumulative frequency is also plotted.

Pareto charts for production processes can be quickly generated using check sheets (see Section 7.3).

- **Tip** - Do not hesitate to try to break the data down into more Pareto charts if more clarity is needed.

## Examples

### Example 1: Complaints by product line.

This example shows the use of the 80:20 method for dealing with products and complaints. The information was gathered over a 12-month period for the six products and plotted to show the number of concerns per product.

| Product | Complaint | % |
|---|---|---|
| A | 25 | 9 |
| B | 36 | 13 |
| C | 75 | 28 |
| D | 98 | 36 |
| E | 15 | 6 |
| F | 20 | 7 |
| **Total** | **269** | **100** |

**Response:**

- Products C and D need investigation, tooling refurbishment, redesign or perhaps even more radical responses.
- Another chart on the reason for the concerns may give more detailed information for solutions, i.e. is this one problem that can be easily fixed.

### Example 2: Sales by customer

This example shows the use of the 80:20 method for dealing with customer analysis for sales.

| Customer | Sales | % |
|---|---|---|
| Adam | 15,000 | 9 |
| Baker | 5,000 | 3 |
| Cold | 80,000 | 49 |
| Down | 25,000 | 15 |
| Eric | 8,000 | 5 |
| Fame | 16,000 | 10 |
| Other | 15,000 | 9 |
| **Total Sales** | **164,000** | **100** |

**Response:**

- Spend more time with Cold. They are 50% of your turnover but do you give them 50% of your time? Ensure good service and quick response to Cold. As an alternative to this try to plot profit per customer, the margins on Cold's work may be so small that the actual profit on Down's work may be greater. This may be an incentive to try to get more work from Down.
- At nearly 50%, the volume of work from Cold is a large percentage of your turnover; you may need to consider trying to reduce your dependence on this business.
- **Tip** - Always go for the big issues first. They are generally the easiest to improve and you can look very good, very easily.

Pareto analysis does not stop you making instant improvements. If one of the 'trivial many' causes is easily fixed now then it should be done now.

You should never use Pareto as an excuse for not doing something. It is also important that you think about the information that you are gathering.

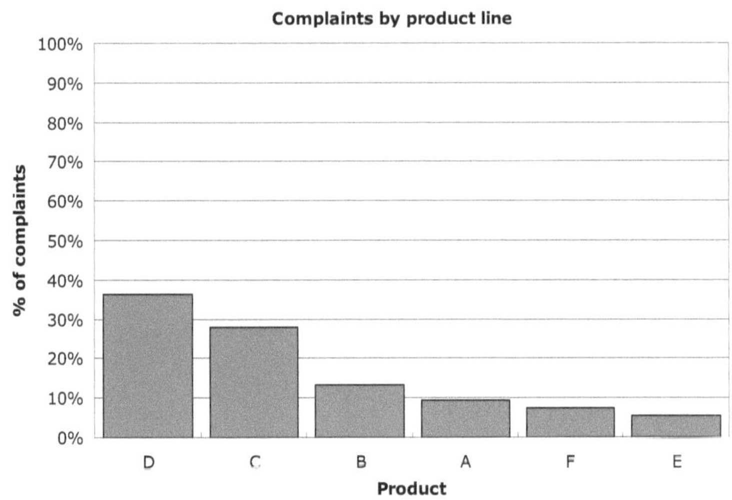


Pareto analysis of complaints by product line

Products D and C need attention for quality improvement action.

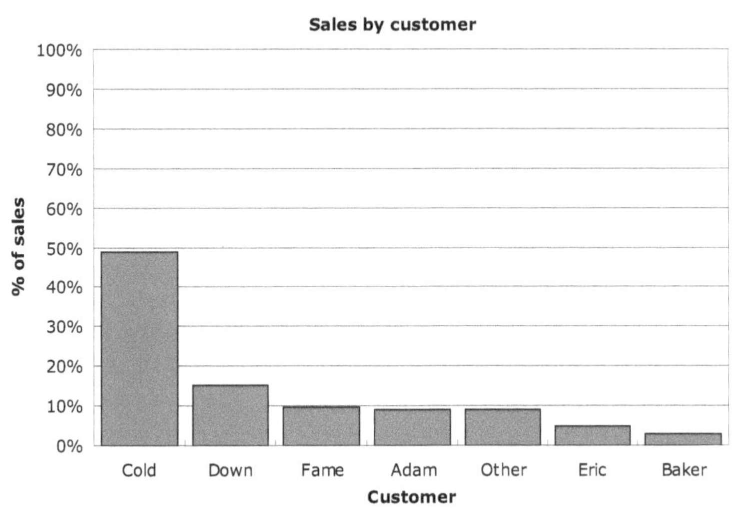


Pareto analysis of sales by customer

Cold and Down should get better treatment than Eric or Baker.

# 7.6 Histograms

## Plotting for success

Histograms can structure data to make them easier to understand and act upon. Whenever we produce something and measure it, be it in manufacturing or in any other process, the value that is measured will vary. One problem we all have is that we make many such measurements and then have a problem with the amount of data that we have. There is an obvious need to structure the data and there is a need for techniques to turn the mass of data into manageable and useful information. These techniques are known as 'data reduction methods' and histograms are one method for structuring data to make them easier to understand and act upon.

Consider the following measurements:

| | | | | |
|---|---|---|---|---|
| 20.1 | 20.0 | 19.7 | 19.4 | 19.5 |
| 19.9 | 19.3 | 19.8 | 19.3 | 19.4 |
| 19.5 | 19.5 | 19.6 | 19.7 | 19.1 |
| 20.2 | 19.6 | 19.7 | 19.0 | 19.6 |
| 19.4 | 19.2 | 20.0 | 19.9 | 19.8 |
| 19.9 | 19.4 | 20.1 | 19.7 | 19.5 |
| 19.6 | 19.6 | 20.0 | 19.8 | 19.7 |
| 20.0 | 19.9 | 19.5 | 19.4 | 19.9 |
| 19.8 | 19.5 | | | |

These data do not tell us a lot and we tend to concentrate on individual numbers and fail to see an overall pattern. Instead of data we really want information.

## Tallying information

Histograms are very simple to create and give a quick picture of the information. For the data given above the easiest way to reduce the data would be to use a 'tally sheet' or check sheet (see Section 7.3) where individual measurements are marked as they are recorded. If the data were known to be reasonably ordered then a scale or frequency distribution check sheet could be used to generate a histogram automatically.

**Note:** When using a check sheet it is best to use 'five bar gates' to make recording and later counting easier. The previous data have been presented in the 'ordered' form below:

| Measurement | Frequency | % |
|---|---|---|
| 19.1 | 1 | 2.4 |
| 19.2 | 1 | 2.4 |
| 19.3 | 2 | 4.8 |
| 19.4 | 5 | 11.9 |
| 19.5 | 6 | 14.3 |
| 19.6 | 7 | 16.7 |
| 19.7 | 6 | 14.3 |
| 19.8 | 4 | 9.5 |
| 19.9 | 4 | 9.5 |
| 20.0 | 3 | 7.1 |
| 20.1 | 2 | 4.8 |
| 20.2 | 1 | 2.4 |

Histograms can be used to convert raw data (which are useless) into real information (which is useful and what we need).

## Plotting a histogram

The real benefit comes when you plot the histogram as frequency versus value. The seemingly 'random' data on the left can be converted via the tally sheet into a histogram. The complete histogram is show on the opposite page. Which of the three presentations gives the most information?

- **Tip -** The minimum number of data points should be 50. We have only used 42 for brevity.
- **Tip -** Try to make the data points collected consecutive so that no bias is introduced.

Histograms present large amounts of data clearly and concisely.

## Adding value

When the tally sheet is complete, this information can be plotted as a histogram to help show the 'big picture' clearly and simply. By using a histogram it is easy to see the highest and lowest values, the centre of the distribution, and to get a 'quick' picture of the underlying information that was previously hidden by the mass of data.

The interval of the measurements is referred to as the 'cell size' and for the example above the size of the 'cell' was effectively defined by the measurements.

"The plural of *anecdote* is not *data*."

**Jen Smeed**

In some cases, this is more difficult to see and the rules for cell size are as follows:

- The cell size can be the same as the unit of measurement but must never be less than this.
- The unit of measurement is often the best guide to the number of cells, provided the scatter of the results is not too great.
- When the unit of measurement is not used then a rough guide for the best number of cells is given by the following rules:

| Number of Measurements | Number of Cells |
|---|---|
| Under 50 | 5 to 7 |
| 50 to 100 | 6 to 10 |
| 100 to 250 | 7 to 12 |
| Over 250 | 10 to 20 |

- After the number of cells has been decided an initial estimate of the cell size can be given by the formula:

  Cell Size =

  $$\frac{\text{Highest Value - Lowest Value}}{\text{Number of cells}}$$

  This cell size is then rounded up or down to give a convenient number to plot and use.
- **Tip** - Cells may also be called 'classes' and naturally, these have a 'class size'.

Histograms give much more information than either simple measurements or averages. Averages tell you nothing about the best and worst cases or the spread of the results, histograms tell you all about these.

Situations where histograms can be useful:

- To record the results of manufacturing trials.
- To check if the distribution of results from production processes is normal or bi-modal.
- To check the shape of the distribution of results from production processes.
- To record the throughput time for products through the factory.
- To examine inventory levels at various times.
- To record elapsed time from receipt of order to delivery.
- To determine number of days credit actually given.
- To determine order value variations across the business.

Histograms help to provide the user with a great deal of information, are easily communicated and inexpensively displayed.

- **Tip** - Do not expect the histogram plot to always be a 'normal' distribution. It can also be skewed to the left or right, contain more than one 'peak' (multi-modal) and even stop suddenly on one side if there is a lower or upper limit to the data. An example of this might be the number of days credit given - if the terms of payment are 30 days then it is extremely unlikely that there will be any points less than 30 days! The data will probably be skewed with the majority of payments at around 45 days and the really slow payers will be out at more than 90 days.
- **Tip** - Histograms can communicate masses of data quickly and accurately. They are much more useful than simple averages.

## Advanced uses

Histograms are the starting point for statistical quality control and produce the information necessary for process capability studies.

Histograms make numbers easy to understand. Decisions about actions become clearer.

Plotting the histogram is not enough - You have to take action on the results otherwise it is just a waste of paper.

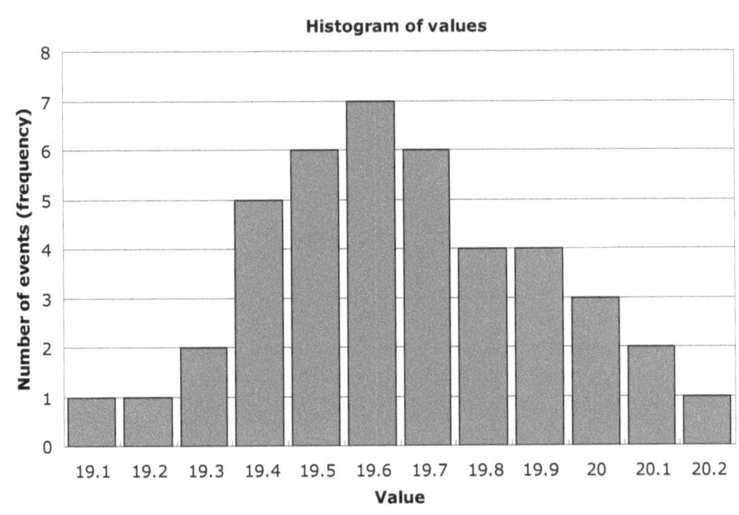


### Histogram plot of values versus frequency

The shape of the seemingly random data becomes obvious and can now be put to use to analyse the data and assess the distribution shape. Data to information is what it is all about.

# 7.7 Scatter diagrams

## Seeking relationships

The cause and effect diagram (Section 7.4) is a method of attempting to list the causes that could influence an effect you see. The diagram is really a list of possibilities and it is natural to want to check if the cause in the diagram really does influence the effect. In other words: are the cause and effect related?

This is the step that many people leave out, they list the causes but then they assume that the first cause they identified is the main and only one.

Just because Event A comes before Event B it is not necessarily true that Event A must cause Event B.

In the Western world we have become very attached to the idea of causality. If I change the oil in my car and one week later the engine seizes up then it must he the oil that caused it! The fact that the car is 25 years old and was dying anyway has nothing to do with it. This is also true where everything has got to be somebody's fault, so you can first blame them, and then sue them for it. The idea of accidents and coincidences seems to have lost favour in the West. Event A does not necessarily cause Event B and they may be totally unrelated. Correlation is not causality!

The scatter diagram is a quick and dirty test for possible cause and effect relationships. It does not prove that one variable causes the other but it does make it clear whether a relationship exists and the degree of scatter gives a good idea of the strength of the relationship.

## Examples

Five examples of scatter diagrams are shown on the right.

In the first diagram the two variables are probably independent and in the last four the two variables are probably related in some way. Scatter diagrams do not prove, or even attempt to prove, causality or a one-to-one relationship but they do show if there is some type of relationship that we can influence.

Scatter diagrams are most easily generated using a spreadsheet to handle the data. Most spreadsheets will allow you to generate a scatter chart as part of the charting package.

## Essential tips

To get the best out of scatter diagrams you should note the following points:

- The easiest way to assess the amount of correlation or scatter is by simply looking at the data points. If you want to put a number on the degree of

'Post hoc ergo propter hoc' (After this therefore because of this) is not always true.

Correlation is not causality.

| | |
|---|---|
| **No relationship**<br>Scatter chart showing no relationship between two variables. The two variables are probably unrelated. Changing one of the variables probably has no effect on the other. | 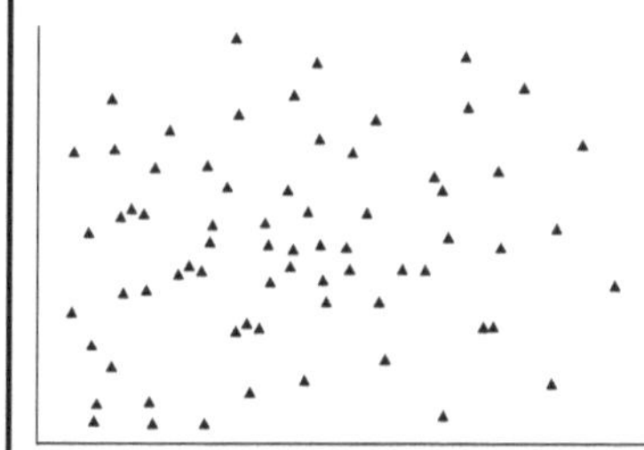 |
| **Good positive relationship**<br>Scatter chart showing a good positive relationship between two variables. Increasing the cause will almost certainly increase the effect. | 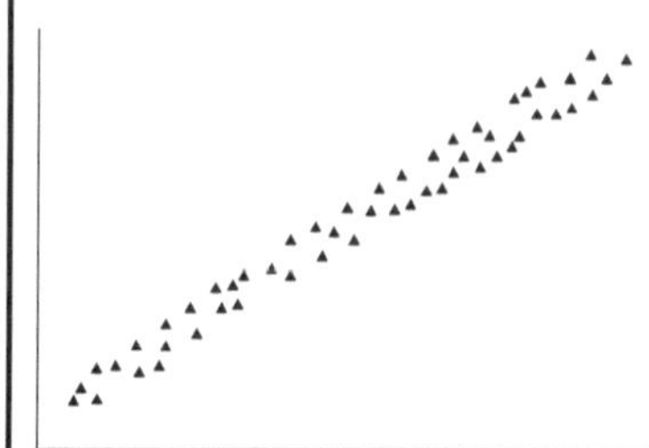 |
| **Poor positive relationship**<br>Scatter chart showing a poor positive relationship between two variables. Increasing the cause will possibly increase the effect. There may be other variables involved that are not being controlled. |  |
| **Good inverse relationship**<br>Scatter chart showing a good inverse relationship between two variables. Increasing the cause will almost certainly decrease the effect. |  |
| **Poor inverse relationship**<br>Scatter chart showing a poor inverse relationship between two variables. Increasing the cause will possibly decrease the effect. There may be other variables involved that are not being controlled. |  |

correlation then you can use the spreadsheet to calculate the correlation co-efficient ($R^2$). $R^2 = 1$ for perfect correlation, i.e. a perfect relationship, and $R^2 = 0$ for no correlation, i.e. no relationship at all. As a general rule, when $R^2 > 0.5$ then there is some degree of correlation. There is an alternative visual method of assessing correlation using Q and N values but this has largely been superseded by spreadsheets.

- If the variables are related but the scatter is very high then the relationship may not be direct. There may be another factor that has not been considered that is varying and affecting the result but it is not being measured or controlled. Finding this other factor (usually by cause and effect diagrams) is generally important in reducing the variation and scatter.
- The line does not need to be straight; it can be curved as well. The examples shown are all linear - where the cause and the effect are probably related by an equation of the form $y = Ax + B$. Non-linear relationships such as $y = Ax^B + C$ are also possible. This doesn't matter because using scatter diagrams is not about precision science. After all we are normally only interested in increasing or decreasing one of the variables.
- Get as many points on the diagram as possible. The more the merrier is the rule for scatter diagrams.
- It is possible to do sophisticated multiple regression analysis to calculate correlation coefficients and statistical significance using a computer but in practice it is simply necessary to establish that there is a relationship.
- After a relationship has been established it may be possible to control the cause (using SPC or other methods) and therefore control or minimise the effect.

Scatter diagrams are useful to find out if a relationship between two variables exists and to solve a problem in the quickest and easiest way. They are a natural extension of cause and effect diagrams and should be used to get the best result.

## Caution: Storks deliver babies

A lovely paper (www.econ.queensu.ca/files/other/storks.pdf) shows the problem of assuming causation between two variables. The relationship between stork breeding pairs and the birth rate is revealed by a scatter diagram and shows a positive correlation between the number of stork breeding pairs and the birth rate in 17 different countries. As the author (Robert Matthews) says, it is "a non-trivial example of a correlation which is highly statistically significant, not immediately explicable and yet causally nonsensical".

The idea is not to plot a graph but simply to get some points on a piece of paper and see if the cause and effect are related.

Stratification can prevent good correlation. Look for stratification in the data.

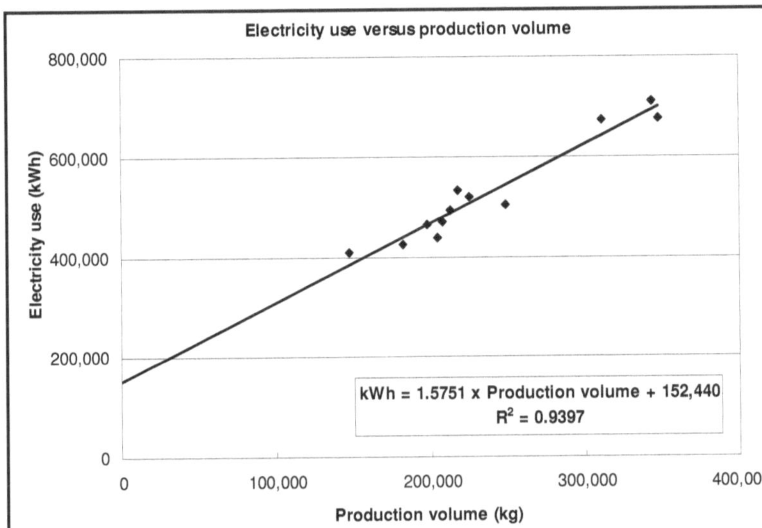


### Electricity use versus production volume for injection moulding

The amount of electricity used in injection moulding is directly related to the amount of material processed. This relationship can be used for monitoring and targeting energy use.

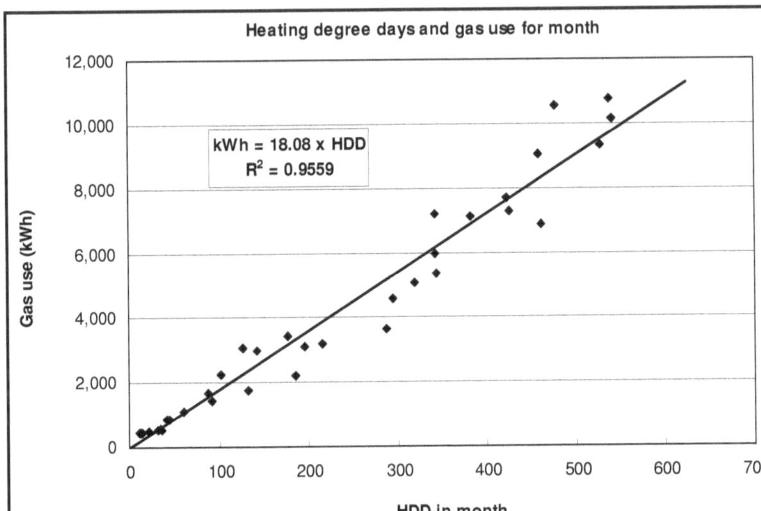


### Gas use for heating systems

The gas used in heating a plastics processing site is directly related to the Heating Degree Days (HDD). This is a measure of how cold it is at a place over time. As for electricity use (above), this can be used for monitoring and targeting energy use.

# 7.8 Relations diagrams

## What are the causes?

Relations diagrams (sometimes called inter-relationship diagrams) are one of the 7 new tools. They are mainly used to organise information and ideas and to identify and rank the causes of a particular issue. They are similar, in many ways, to cause and effect diagrams (see Section 7.4) but have a more flexible format and are broader in scope.

Whereas cause and effect diagrams use a very formal linear layout (M's and P's) a relations diagram uses a much more free format model to try to identify the logical connections between a range of ideas and to then group these according to their relative importance.

This allows links to be made between multiple causes in complex issues or problems and the relative importance of these causes to be assessed, e.g. causes can be assessed as primary, secondary, tertiary or even lower-order.

## When do you use relations diagrams?

Relations diagrams can clarify planning issues (not the timing but simply the importance) for complex issues or implementation effects. They can be used to:

- Give an overall perspective of the situation and the causes.
- To develop a consensus for the priorities and actions needed to solve a problem and implement a solution.
- To clarify the relationships between causes and break large problems down into smaller issues that can be more effectively resolved (especially when one secondary cause can affect more than one primary cause).
- To develop and change thinking in a team.

A relations diagram can be used in conjunction with other tools such as cause and effect diagrams (see Section 7.4), tree diagrams (see Section 7.9) or affinity diagrams (see Section 7.10) to more completely explore areas or issues that have been identified. The relations diagram is particularly useful when the issues are complex and multi-dimensional.

## Making a relations diagram

The process for making a relations diagram is as follows:

- The problem or issue is stated in form:
  - Why does this happen?
  - Why does this not happen?
  - How are we going to implement this project with minimum disruption?
  - How do we improve this product or service?

  Getting the problem statement right is a key issue. You must get agreement on this before you start the process.
- The team (and it should always be a team exercise) independently completes 5 cards or Post-it notes listing 5 or more causes that affect the situation, i.e. why it happens, why it doesn't happen, how it can be improved etc.
- If the team is small then increase the

We have used relations diagrams to assess the best methods to:

- Reduce nonconformities in injection moulding.
- Reduce nonconformities in extrusion blow moulding.
- Improve customer services.

These are only some of the potential uses.

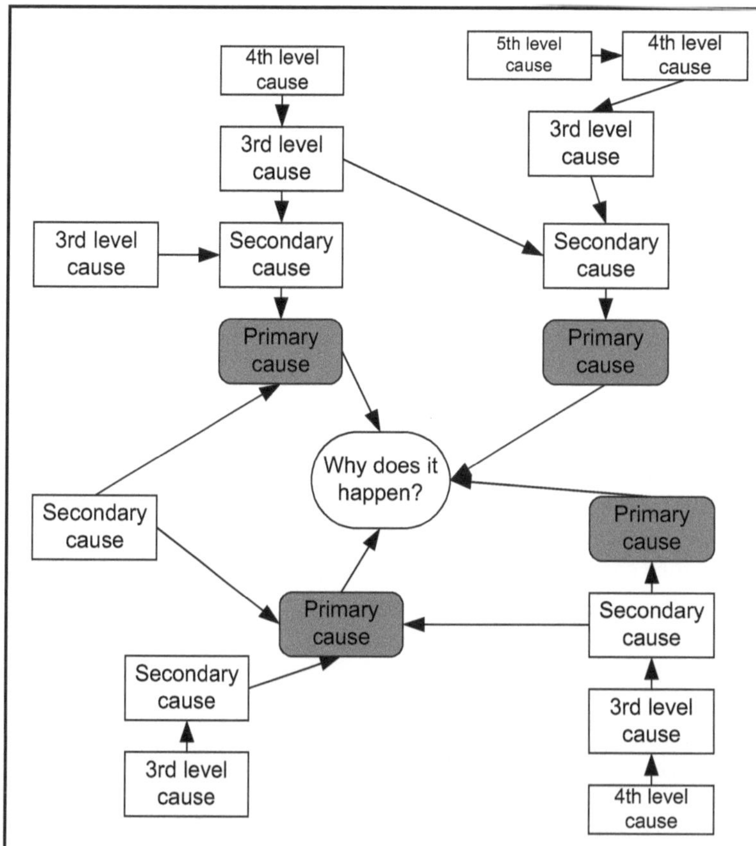


### Typical relations diagram

A typical relations diagram will have a central 'problem statement' and a range of primary, secondary and lower-order causes that affect the main problem. The arrows reflect the relationship (and importance) of the causes on the main problem.

number of causes per team member.

- These causes are placed on a table in no particular order, discussed amongst the team and grouped into a set of natural groups that reflect their relationship. It is quite normal for team members to duplicate causes but do not throw them out at this stage, examine them for potentially differing viewpoints on the issue, i.e. they may say the same thing but examine what they actually mean.
- Each cause is discussed in terms of its effect on the other causes and in terms of the relative importance (primary, secondary, or lower order). At this stage the relationships between the causes (and their relative importance) should start to become clear to the team.

**Note 1:** It is possible that a lower-order cause could affect more than one higher-order cause, i.e. poor training could affect not only product quality but also schedule conformance. This is not a problem, map the relationships even if there is more than one link per cause.

- Related cards continue to be moved around and linked until the team is happy that all the potential causes and relationships have been mapped and assessed for importance.
- The cards are connected with arrows (as shown on the left) to highlight the connections and reveal the driving factors and the potential solution to the problem or issue.

**Note 2:** It is possible to gain an indication of the importance of a factor by counting the number of 'out' arrows (those going out of the factor) and the number of 'in' arrows (those going into the factor).

  - Factors with the highest number of arrows (in and out) are likely to be key causes/factors.
  - Factors with a high number of 'out' arrows are likely to be basic causes/factors.
  - Factors with a high number of 'in' arrows are likely to be the effects that should be solved.

This is only an indication!

**Note 3:** Do not try to make it neat. The important thing is to get the relations right. You can always tidy it up on a computer later.

## Highlighting the drivers

A relations diagram highlights the drivers of a problem or issue. In fact, often it reveals that the real driver was something that nobody had ever considered before. This ability to draw out the 'hidden' drivers makes it one of the most useful of the 7 new tools.

Relations diagrams offer a quick and easy method of analysing many potential causes.

They also allow these to be ranked in importance.

# 7.9 Tree diagrams

## Breaking it down

Tree diagrams are named because they look like a tree! They have main branches that divide into secondary branches and secondary branches that divide into tertiary branches and so - just like a real tree. They are used to break down broad categories into smaller chunks that can be more effectively dealt with, i.e. an already known broad objective or central issue can be broken down into smaller and smaller issues that are actually manageable.

- **Tip** - Q: How do you eat an elephant?

  A: One mouthful at a time.

This ability to logically move from a nebulous poorly defined central issue to the specific manageable concerns is the strength of the tree diagram. It systematically identifies the detailed tactics or operational tasks necessary to achieve a specified general strategy or objective and this systematic process means it is less likely that you will forget any issues or items that will affect the result. The process also:

- Helps teams to agree on the tactics to achieve an objective.
- Identifies potential and actual constraints on the tactics.

## When do you use tree diagrams?

There are 3 very different types of tree diagrams (see box on the right) and these are:

- For planning - where the tree diagram is used to produce a detailed analysis of the processes and steps necessary to achieve an objective. In this case, the tree diagram identifies the logical steps necessary to achieve the desired result. This planning type of tree diagram is used when the objective is known in broad terms, e.g. install new cooling system or install SPC system, but the actual steps necessary to get there are vague and need definition. In this case, the tree diagram identifies the logical steps needed to achieve the objective and communicates the implementation process.

**Note 1:** The tree diagram does not assess times or the critical path, this is done by an arrow diagram (see Section 7.12) after a tree diagram has captured all the essential steps.

- For root cause analysis - where the tree diagram is used for an initial broad analysis of why a problem is happening and to capture all possible causes. This type of tree diagram is used to capture all the possible factors affecting the root cause. When used for root cause analysis the tree diagram is very similar to cause and effect analysis (see Section 7.4) but the tree diagram does not start out with a defined structure and can sometimes capture more information.
- For component analysis - where the tree diagram identifies all the components of a product or a process, i.e. what are the parts necessary to assemble this product or what are the inputs necessary for this process to take place? This type of tree diagram identifies all the components or inputs and can be used to check that all the factors are considered in the product or process planning. A component analysis type of tree diagram is sometimes called a 'gozinto' chart because it identifies what 'goes into' a product or process. It can be very useful for generating a complete Bill of Materials (BOM) for a product or the equivalent of a BOM for a process.

Whichever use the tree diagram is used for they are a very useful thinking tool for assessing causes.

A tree diagram can also be called a hierarchy diagram for obvious reasons.

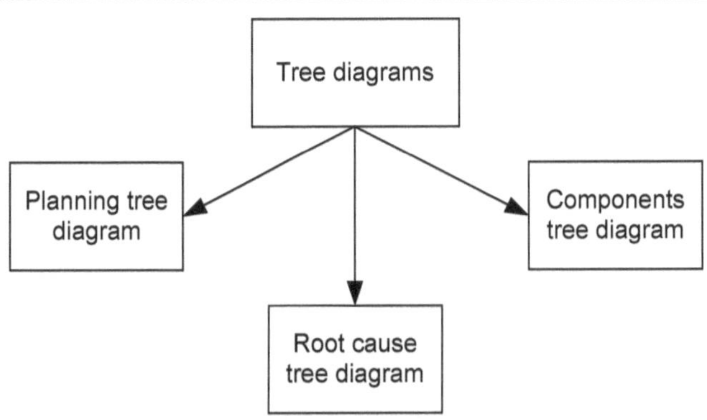


### Tree diagram types

Tree diagrams can be used for detailed planning (the goal is the tree trunk), for root cause analysis (the problem is the tree trunk) or for producing a components list (the final object is the tree trunk). They all use the same basic format.

## Making a tree diagram

The process for making a tree diagram is as follows:

- If a relationship diagram (see Section 7.8) or affinity diagram (see Section 7.10) has been previously produced then the outputs of this can be used to identify the trunk of the tree, i.e. the main issue. If this has not been done, or if the tree diagram is being used for planning then the objective is written as the main trunk of the tree at the top or left of the page.

**Note 2:** Tree diagrams can be either vertical or horizontal, the example shown on the right is horizontal but feel free to use a vertical format (it looks more like a tree).

- Identify the main constraints for the planning process (time, etc.), root cause analysis (machines, etc.) or component analysis (the boundaries of the part or process being considered).
- List the tactics or means of getting to the main objective:
  - For planning, ask 'What do we need to do to achieve this?'
  - For root cause, ask 'What causes this?'
  - For component analysis, ask 'What are the components of this?'
- These are the secondary braches of the tree diagram.
- Continue to break the secondary branches down into tertiary or further branches until:
  - For planning, the task is very small and independent, i.e. no subsidiary tasks.
  - For root cause, the cause has no further sub-causes, i.e. you are at the root causes.
  - For component analysis, the component has no sub-components.
- Check that all of the listed branches are 'necessary and sufficient' for the higher branches.
- Ask the questions:
  - Are all the steps, causes and components necessary for the higher branch?
  - If all the steps, causes and components were present then would they be sufficient to define the higher branch?
- Complete the tree diagram and review for accuracy and completeness.

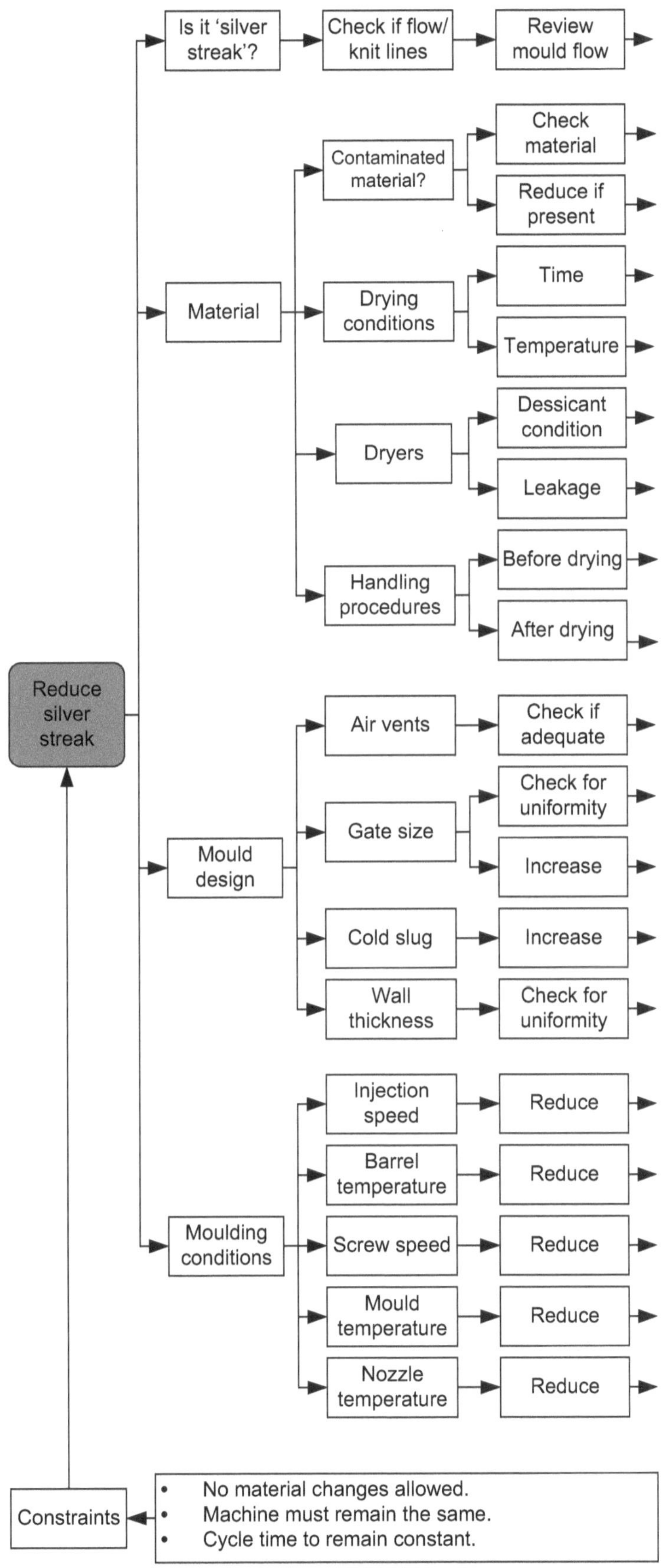


### Typical tree diagram (root cause type) for reducing silver streak

Silver streak is a set of bright lines on the surface of a part (especially visible on dark products). Many sites simply increase the drying but silver streak can be caused by a variety of factors.

# 7.10 Affinity diagrams

## Seeking lots of relationships

Most of the 7 basic tools are used to organise numbers and data but the affinity diagram is useful to organise unformatted verbal or idea data. In a situation where there is lots of verbal data then it can often be difficult to see the main issues and the affinity diagram is the tool for organising these data into natural relationships.

They are particularly useful for handling issues where verbal data are wide and varied, e.g. the output of the customer surveys where the comments can cover a range of inter-related issues. The affinity diagram allows these issues to be grouped to see the higher-level relationships rather than focusing on the lower-level symptoms. This grouping allows further analysis and the actions needed to solve the high-level issue (and hence the low-level symptoms).

## When do you use affinity diagrams?

Affinity diagrams allow all the chaotic verbal data, opinions and symptoms to be considered to reveal the basic actions that need to be taken. Affinity diagrams are similar in many ways to cause and effect diagrams (see Section 7.4) but are more suitable for larger issues where even the main issue is not well defined. They can be an essential first step in defining what the issue really is, whereas a cause and effect diagram assumes that the issue is already well defined. They can be used to:

- Organise disparate verbal data.
- Include the opinions of a large group of people and raise team spirit.
- Give an overall view of what the issues really are and to raise awareness of the issues.
- Allow vague symptoms to be fully described in terms of the main issues.
- Provide a route map for solving large or complex issues.
- Help groups to break out of their traditional thinking about issues or effects.

Affinity diagrams are widely used in 'road-mapping' exercises for technology development, departmental or company

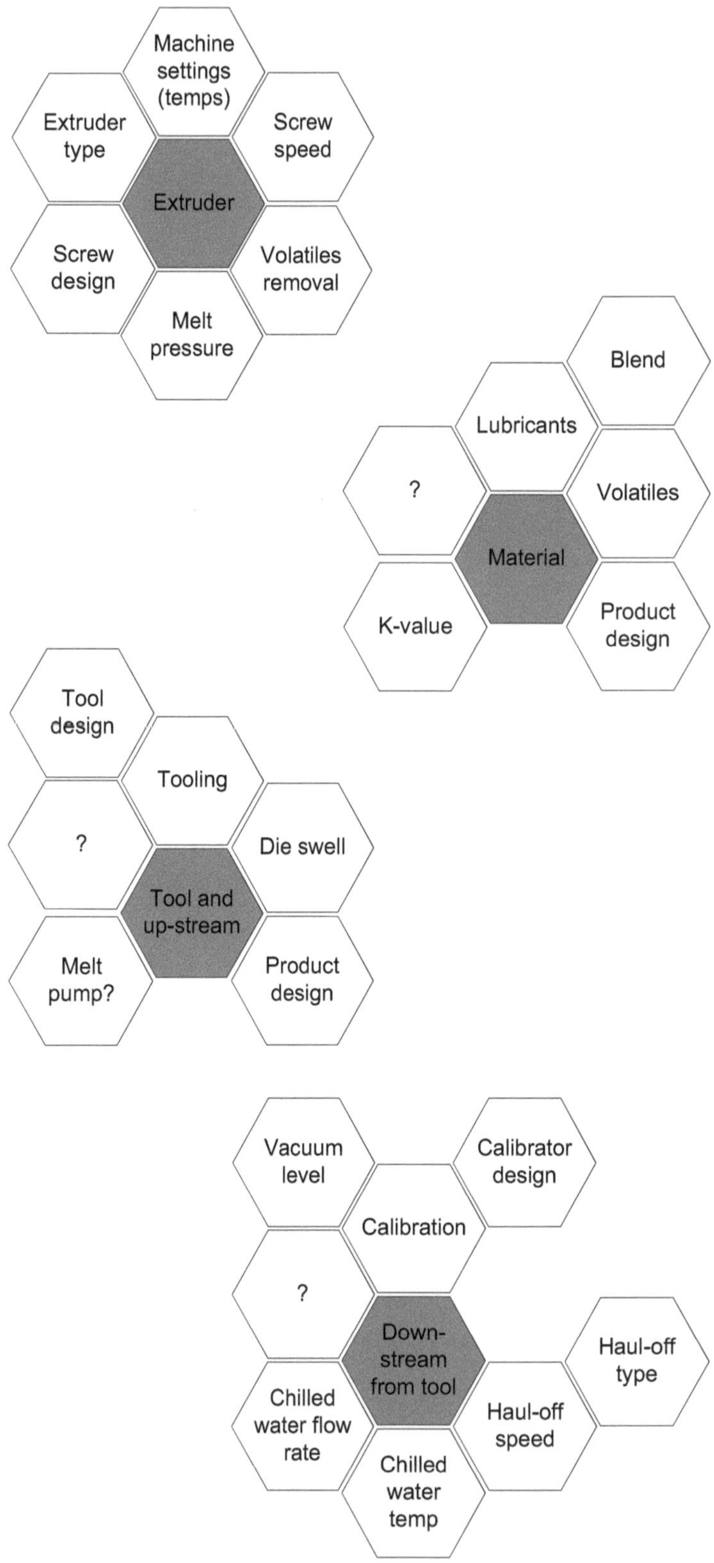


Affinity diagrams for the important factors in dimensional control of extrusions

Affinity diagrams do not put a priority or importance on the cause. They simply group issues/effects into logical groups that can then be investigated for further data or decisions.

development and market development. They can be used to get input from a wide range of people and to reach a consensus on possible actions for the future to resolve the issues or effect.

Affinity diagrams can be used before other tools such as cause and effect diagrams (see Section 7.4), relations diagrams (see Section 7.8) or tree diagrams (see Section 7.9) to initially define the major areas or issues and then define the need for further data or decisions.

## Making an affinity diagram

The process for making an affinity diagram is as follows:

- Decide on the broad topic or issue. It does not have to be very well defined at this stage.
- Collect verbal data from brainstorming, survey results or other input. This will normally be a set of statements, possible causes or drivers. This should be put on Post-it® notes that summarise the data.

**Note 1:** Special hexagonal notes are available but they are expensive and normal ones are just as good.

- Randomly put the notes on a table, white board or door so that the complete team can see them.
- The team moves the notes around freely into groups of similar themes (affinity) with, again, very limited discussion. At this stage the notes can be moved by any team member for any reason and moving the notes is encouraged. At this stage there should be very limited, if any, discussion and no criticism for any of the grouping decisions is allowed.

**Note 2:** If a note 'belongs' in 2 groups then make a separate note for each group.

**Note 3:** It is possible that new notes can be generated at this stage.

- Groups should be allowed to form naturally as the notes are moved around. There should be no 'headings' or statements about the reason for the grouping at this stage.
- Discuss the groups with the team and find the logic behind their groupings.
- Continue grouping the notes until there are about 5 major groups of notes.

**Note 4:** This can take some time!

**Note 5:** If a note does not belong in any group then keep it in by itself and do not count it as one of the 5 major groups.

- Keep the groups together to view the final affinity diagram for the topic or issue.
- Combine and summarise the statements on the notes in each group to make the 'affinity' statement that relates to each group of notes. This can be the text of one of the notes or can be a new statement that effectively groups the notes.
- Complete the diagram.

## Clarifying the issues

Affinity diagrams allow groups to get away from their traditional thinking patterns (which can be driven by their departmental or group association) and to force them to consider the complete issue rather than simply their own perspective on the issue. They provide a method of collecting all of the main ideas quickly and then finding the key driver of an issue or effect.

The input data for an affinity diagram are mainly subjective, i.e. text data, and this allows them to be used across a wide range of business issues where number-based data are not yet available. The 'free-format' grouping of the factors provides a basis for further data gathering to validate the groupings and their effect.

We have used affinity diagrams for:

- Deciding on the priorities for service delivery improvements to make sure that all the customers saw the improvements.
- Road-mapping technology progress and the drivers for new developments. This gave us new insights into how to improve the quality of products by using the best available technology.
- Deciding on the work programme for a large committee so that time and energy were spent on the most effective projects.

Producing an affinity diagram is a creative process.

# 7.11 Matrix diagrams

## Multiple relationships

Matrix diagrams are an extremely powerful visualisation tool to show relationships between two, three or four groups of information. They allow the clear visual display of complex data in multiple dimensions and this makes them very suitable for presenting information to decision makers.

The type of matrix diagram used depends on the number of data sets being used and the main types of matrix diagrams are shown on the right, these are:

- Roof matrix: This is the simplest matrix of all and can be used to compare the effects of elements of a single set of data on the other elements of the same set, i.e. in the example on the right, A1 has a positive effect on A2 and A5 but A3 has a negative effect on A4 and A5. The roof matrix is commonly used in the 'house of quality' (see Section 8.10) and is often used with L and T matrix diagrams.
- L-matrix: This is used to compare two sets of results (A and B) to each other or to compare 1 set of results to itself (A to A). The data can be anything from nonconformities to performance-related. In the example A1 has a weak effect on B1 but A3 has a strong effect on B2.
- T-matrix: This is used to compare two sets of results to a third set, i.e. in the example, B and C can be compared to A but B and C cannot be compared. This example shows the use of symbols instead of numbers and shows that A1 has a strong relationship to B2.
- Y-matrix: This is used to compare three sets of data to each other (but not simultaneously), i.e. in the example A can be compared to B or C.
- C-matrix: This is for a full simultaneous comparison of 3 sets of data, i.e. it is possible to see the variation in A simultaneously with variation in B and C. This is a 3D matrix and can be complex to use and see. It is used less often than the other types of matrix.
- X-matrix: This is used to compare 4 sets of data, i.e. in the example A can be compared to D and C and B can also be compared to D and C but A cannot be compared to B and D cannot be compared to C.

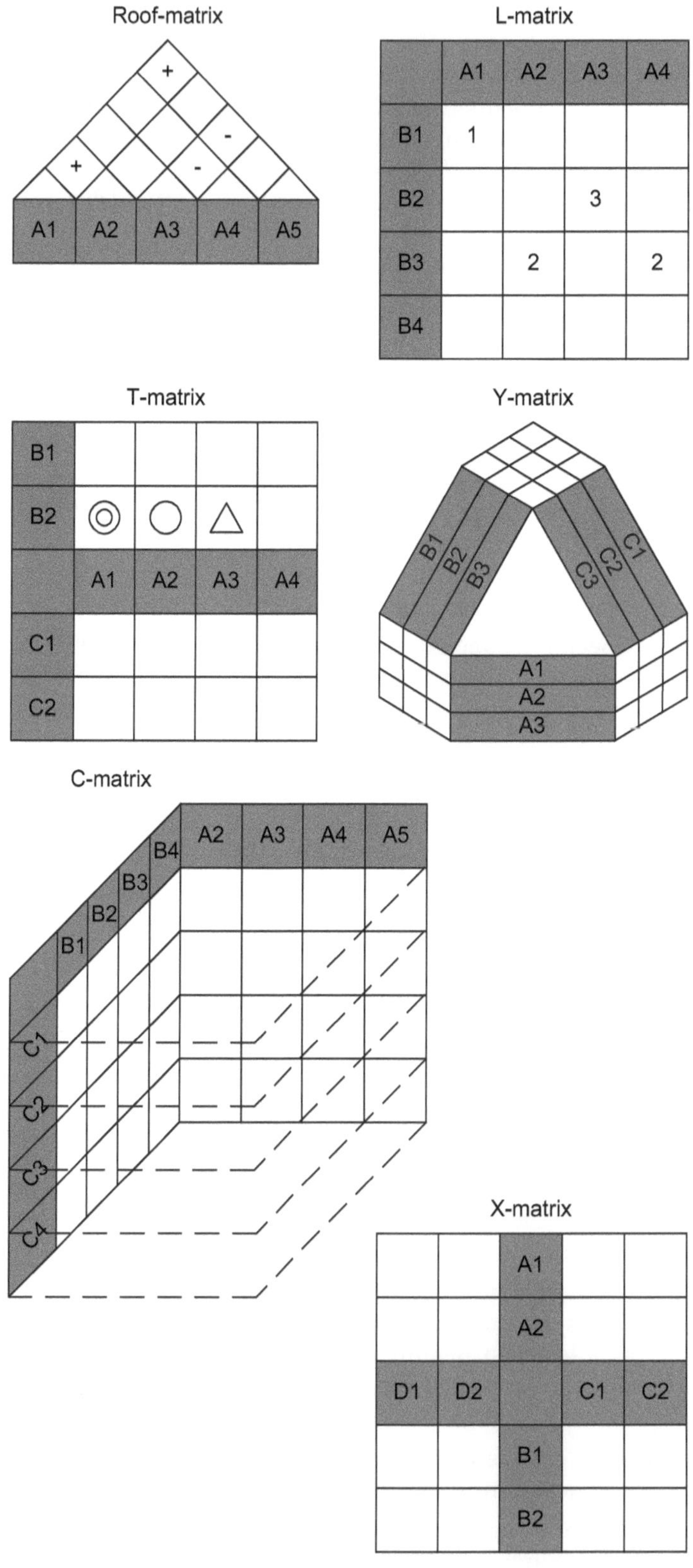


### The 6 basic matrix diagrams

The 6 types of matrix diagram all serve the same purpose, i.e. to compare items on the basis of specific criteria. The only difference is in the number of groups being compared and the type of relationship you are looking at.

## When do you use matrix diagrams?

Matrix diagrams can be used to clearly reveal the structure of an issue from a mass of information. They compare options, provide a clearer understanding of the underlying issues or data and are an excellent method of presenting complex information.

They can be used to compare:

- Service issues with potential causes.
- Maintenance loads and causes for multiple machines.
- Tooling issues and causes for multiple tools.
- Quality issues and potential causes.
- Brainstorming potential effects on product or service quality.

In fact, matrix diagrams provide a simple method for comparing multiple data sets for almost any purpose.

## Making a matrix diagram

The process for making a matrix diagram is as follows:

- The first step is to think about the underlying data to find out what you are comparing. This is needed to select the correct type of matrix diagram.
- Decide on the type of matrix to use (Roof, L, T, Y, C or X)
- The data sets being compared are filled in on the matrix axes. As an example, consider the maintenance load on your injection moulding machines using an L-matrix, the machine numbers would be filled in for data set A and the maintenance issues, e.g. electrical, faults, mechanical faults, hydraulic faults, etc. (to as detailed a level as you want) would be filled in for data set B. For companies with multiple sites you can easily use sites instead of machines to get an idea of the maintenance issues at each site.
- Consider the data for each interaction/ square of the matrix and mark the strength of the interaction in the square using a standard notation. For example, if you have high maintenance loads then mark with a 3, medium with a 2, low with a 1 and none is left blank.
- The completed matrix diagram provides an overview of the maintenance loads on all the machines and will show the machines with the highest maintenance loads (candidates for replacement) and the highest maintenance areas (training or more staff).
- This example could be extended to include the running hours as the C data set in a T matrix. This would provide better data for machine replacement issues, i.e. you may not consider replacing a machine with low running hours.
- The 'coding' of the squares can be flexible and can:
  - Use numbers (0, 1, 2, 3) to indicate the strength of the relationship.
  - Use +/- to indicate the direction of the relationship. In the roof matrix on the left, A1 is positively related to A2 and A5 (increasing A1 will increase A2 and A5) and A3 is negatively related to A4 and A5 (increasing A3 will decrease A4 and A5).
  - Use symbols (◎, ○, △) to show strong, moderate and weak relationships respectively.
- **Tip** - As a general rule, I prefer to use numbers so that I can add them up.

Matrix diagram are poorly understood and little used but they offer a great tool for investigating the relationship between machines, tools and people in producing quality products.

Start using them today.

"Highly skilled engineers have confidence in their ability to figure out the logic of anything they choose. They continually look for order, system and interrelationships."

**Paul, Niewoehner and Elder**

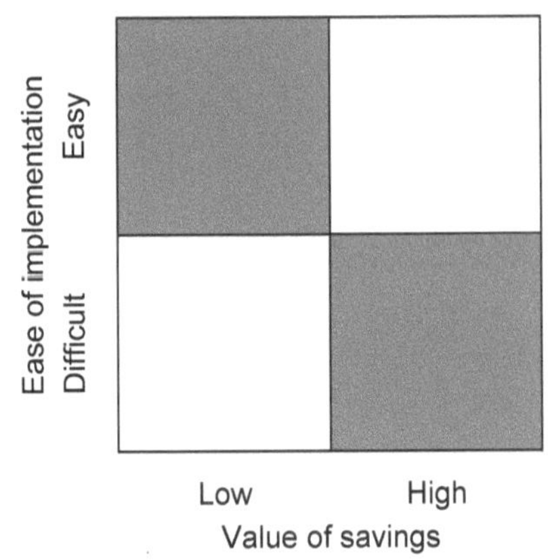


Focus on the projects that have high savings and are easy to implement (top right) and ignore projects that have low savings and are difficult to implement (bottom left).

It is easy to see that we should focus on tasks of high importance and high urgency but when an important task is also urgent then it is generally too late! Try to focus on important tasks while the urgency is still low.

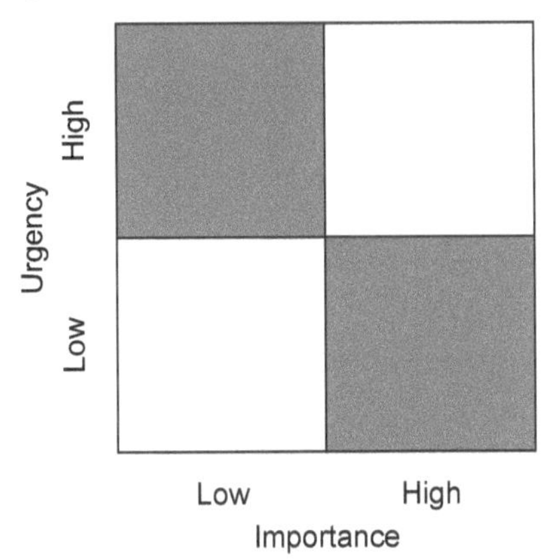


### 2 x 2 matrix diagrams help to visualise the choices

The 2 x 2 grid used by management consultants is a very simple use of a matrix diagram (see Section 2.10). These help to clarify the choices when faced with a group of tasks or projects competing for finance or attention. 'Cash cows' anybody?

# 7.12 Arrow diagrams

## Failing to plan or planning to fail?

Arrow diagrams are used to evaluate and review projects and their planning. They are a development of the familiar Gantt chart (see lower right) but are more flexible. Arrow diagrams and their extensions can show a greater amount of detail in terms of the order that tasks should be completed, the best schedule for the complete project and reveal any potential schedule or resource issues.

Extensions to the basic concept of the arrow diagram have led to development of techniques such as PERT (Program Evaluation Review Technique) and CPM (Critical Path Method), both of which add additional information (and also complexity) but give more information for project managers.

The basic arrow diagram shows:

- How the individual tasks in a program are related, i.e. their dependencies. This promotes an understanding amongst a team of the importance of their particular task.
- The possibility of potential problems with timings and resources.
- The possibility of potential program improvements through resource additions or program changes.
- The best schedule for the project, i.e. the critical path.
- The effect of task delays on the overall project schedule.

Arrow diagrams are an essential tool for project planning and if you fail to plan then you really are planning to fail.

## When do you use arrow diagrams?

Basic arrow diagrams should be used even for very simple projects to visually set out the schedule and gain agreement that it is feasible. However, for complex projects which have many dependent tasks and possible resource conflicts they are essential. If adherence to a project schedule is critical because of penalties for late delivery (or even benefits for early delivery) then arrow diagrams and their extensions (PERT and CPM) are indispensable for schedule adherence.

## Making an arrow diagram

The tree diagram (see Section 7.9) when used in the planning format is a very useful precursor for making an arrow diagram. It provides an overview of the stages of a project and the tasks (and their dependencies) that make up the project. If it strongly advised that a tree diagram is completed to provide the essential input for the arrow diagram.

It is possible to use a software-based approach, e.g. Microsoft Project or similar software, to complete an arrow diagram with all the resources and dates. The disadvantage of the software approach is that the project plan is controlled by the operator of the software and the process is not always completely transparent to the team. In any case, I always have problems getting the software to print the complete diagram correctly (but maybe that is just me).

- **Tip -** If you want to use a software approach then at least capture the basics using a manual method and involve the whole team at the start.

## The basic arrow diagram

The process for making a basic arrow diagram is as follows:

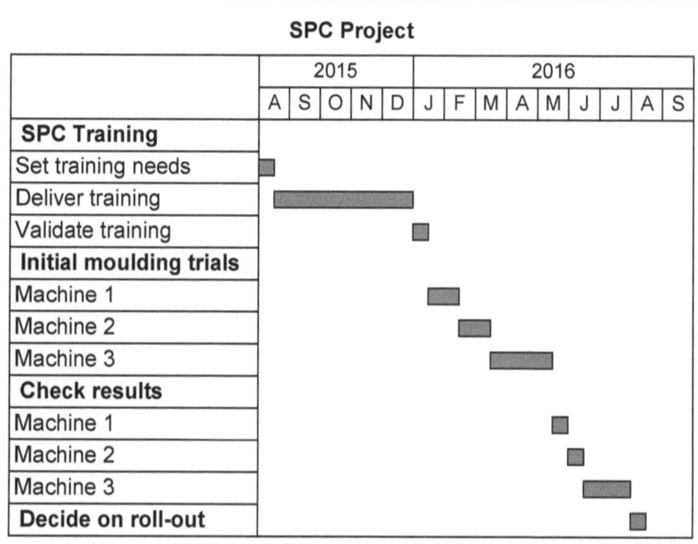


### Simple Gantt chart for trial implementation of SPC

The simple Gantt chart is useful for overall planning but does not show task dependencies and provide the best schedule for a project. The arrow diagram and the extensions (PERT and CPM) allow extra information and better scheduling of resources.

"A project is only ever late because nobody knew it was going to be late until too late."

**Henrique Neto**

- Using the output from a tree diagram, (including the constraints), note each of the main activities/tasks on the top of a separate Post-it note.
- Note all of the tasks necessary for the main activities on additional Post-it notes.
- Arrange the notes in time order (left to right) based on when the task should take place to allow subsequent tasks to be completed. Tasks that can take place in parallel, i.e. at the same time, should be arranged vertically.
- Remove any duplicate tasks, i.e. those necessary for two or more subsequent tasks.
- Connect the tasks with arrows and nodes. The convention for drawing arrow diagrams is:
  - Events are noted as circles. An event is either the finish or start of a task (although you can use 'dummy events' to get the diagram to work).
  - Tasks are drawn on the arrows.
- Move the notes around to reduce the overall time for completion.

**Note 1:** 'Dummy' events can be useful for separating simultaneous tasks or controlling the sequence of events.

**Note 2:** It may be possible to split some tasks up to reduce the total time needed, i.e. it may be possible to start a subsequent task before the preceding task is fully complete.

- Connect the events to complete the diagram.

An example of a simple arrow diagram for a trial implementation of SPC is shown at upper right. These are the same data as used for the Gantt chart on the left.

## Putting times in the diagram

After the standard arrow diagram is completed it is possible to extend this by inserting the time required for each task and using this time to find the critical path for the project. This is shown on the lower right for the trial implementation of SPC with the 'critical path' marked in bold, this is the shortest time for completion of the project and activities making up the critical path must be protected to ensure that the project is finished on time.

The convention for abbreviations used in the critical path calculations are:

- DR - the duration of the task (in days, weeks or months but the units must remain consistent).
- ES - the earliest start time of the task, i.e. sum of the times for the tasks needed before this task can start.
- LS - the latest start time of the task to maintain schedule adherence.
- EF - the earliest finish time of the task, i.e. ES + DR.
- LF - the latest finish time of the task, i.e. the latest finish time of the task to maintain schedule adherence.
- SL - the 'slack' time of the task, i.e. LS - ES or LF - EF.

The arrow diagram can be simple or complex but either way it provides an invaluable planning tool for delivering complex or simple projects on-time - on budget is another topic altogether!

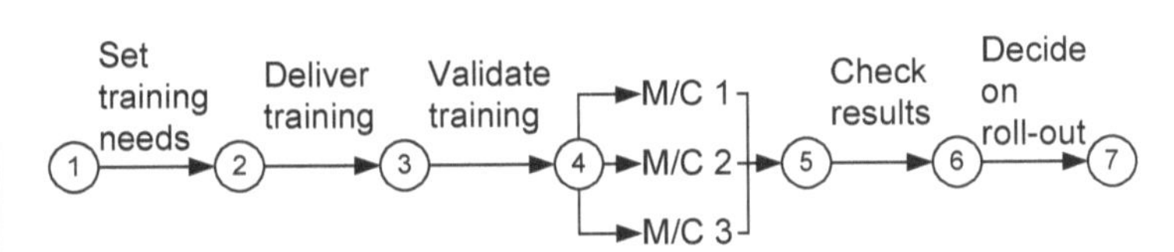


### Simple arrow diagram with no times

The simple arrow chart does not have to have times on it. It can be as simple as you want as long as it shows the tasks in the order in which they must be completed. Note that in this example we have run the machine trials simultaneously with each other.

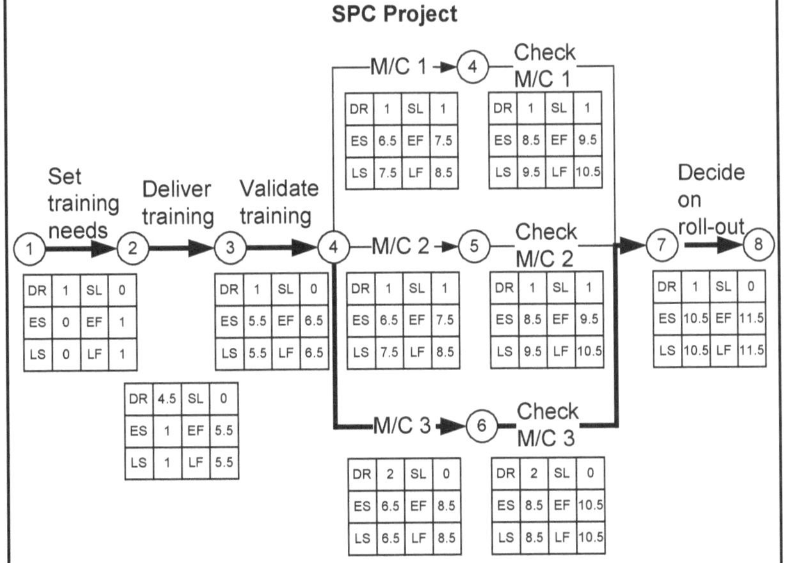


### Arrow chart with added times to locate the critical path

By putting task durations (DR) and other task time data (in months) into the diagram it is possible to find the 'critical path'. This is the set of tasks that determine the project length. These tasks must be 'protected' to allow the project to finish on time.

# 7.13 Process decision program charts

## What are the risks and what do we do about them?

Every quality improvement project and strategy plan has risks that can affect successful implementation of the project or plan. The Process Decision Program Chart (PDPC) helps to systematically identify and assess what might go wrong (the risks) and then provides a method for planning how to overcome the risks and assess the effectiveness of the planning.

In many ways a PDPC is the project and strategy equivalent of the FMEA (see Sections 8.11 and 10.2). An FMEA is a structured method of identifying and assessing the risks for a product or a manufacturing process and then planning the actions to reduce the risk. The PDPC does much the same thing for a project or strategy plan - both methods involve thinking about what can go wrong, how the risk can be reduced and what the action should be if the risk event occurs.

A PDPC can:

- Allow potential problems to be seen before they occur.
- Allow preventive action to be taken so that problems do not occur or are reduced in severity.
- Provide potential solutions to problems before they occur and when you can think about them clearly.
- Reduce the reaction time needed to get the project back on track if problems do occur.

Overall, a PDPC reduces the risk associated with any project or strategy plan and increases the chance of successfully delivering the project or strategy on time.

## When do you use PDPC charts?

A formal PDPC is not always necessary although it is always useful as a method to document the risk reduction process and provide a template for the actions to be taken in the event of any problems.

A formal PDPC is not the important thing, it is the thought process that is important, not the actual output document (although the written PDPC can be very useful if and/or when things go wrong).

A formal PDPC should be generated whenever:

- The risks in implementing a project or plan are high.
- The cost of project or plan delay or failure or plan failure is high even if the risk is low.

It is important to note that a PDPC helps to forecast potential problems based on experience with past problems. It may help with totally unexpected problems, not by providing a defined response (because the new problem won't be in the PDPC) but by providing a structure via the already anticipated problems and their solutions.

PDPCs can also be used to systematically identify those projects and strategies that are likely to be successful and to rank competing projects and strategies in terms of their likelihood of success.

- **Tip -** If a PDPC indicates a lot of problems or risks that cannot be managed economically or easily then it is likely that the project and strategy will not be successfully completed or implemented. The PDPC process can act as an early warning signal for those projects and strategies which initially sound good but which do not stand up to the PDPC analysis.

## Making a PDPC chart

The process for making a PDPC chart is as follows:

- The PDPC is the logical outcome of a tree diagram when used as a planning aid (see Section 7.9) and creating a tree diagram is the first step. This allows the overall goal of the project or plan, the constraints and the sub-tasks to be defined.
- The tree diagram is completed and the main tasks have been broken down into the secondary tasks and the tertiary tasks etc. until there are no subsidiary tasks.
- Each of the lowest-level tasks is considered for any potential problems that may delay or affect completion of the task.

**Note 1:** There may be more than one potential problem associated with an individual task and all potential problems

A PDPC helps you to think about what can go wrong in a project or strategy and how to recover when things inevitably go wrong.

I wish I had used these more in my younger life, it would have saved a huge amount of heartache!

Whilst this section talks about PDPCs in a graphical format, I use a spreadsheet (the same method applies).

This gives me a bit more flexibility in how I structure the output and makes changes easier.

Simply make a tree diagram as part of a spreadsheet and then add the potential problems and counter-measures as another column.

Same idea, different layout.

should be linked to the individual task.

- After all the tasks have been considered for potential problems then each problem is considered in terms of the counter-measures that can be taken to either reduce the likelihood of the problem occurring or the reaction necessary should the problem occur. This is the contingency plan for this task and should be considered in some detail at this stage.
- Contingency plans and counter-measures should be marked clearly on the PDPC, i.e. in the example on the right they are shaded to show that they are the contingency plans and counter-measures.
- Each contingency plan is then considered in terms of how feasible it is to carry out, i.e. is it likely to be effective in reducing the potential problem and is it cost-effective?
- Feasible contingency plans are marked with a 'O' and non-feasible contingency plans are marked with an 'X'.
- The overall PDPC is checked for consistency and that the contingency plans are adequate for the associated risks.

**Note 2:** As for an FMEA, a PDPC is a 'live' document that can and should be updated as the project or plan progresses.

## What are the type of problems we should think about?

A PDPC should attempt to identify:

- The constraints for a project or strategy.
- The expected and potential unexpected outputs from tasks.
- Controllable and uncontrollable events.
- The assumptions made during the planning process and their validity (and what to do if they are invalid).
- The margins for error in the project and what to do if the results are outside the error margins.

Overall, the PDPC process is all about risk management and contingency planning. Anybody who has run a failed project knows that this is important stuff.

"The goal of forecasting is not to predict the future but to tell you what you need to know to take meaningful action in the present."

**Paul Saffo**

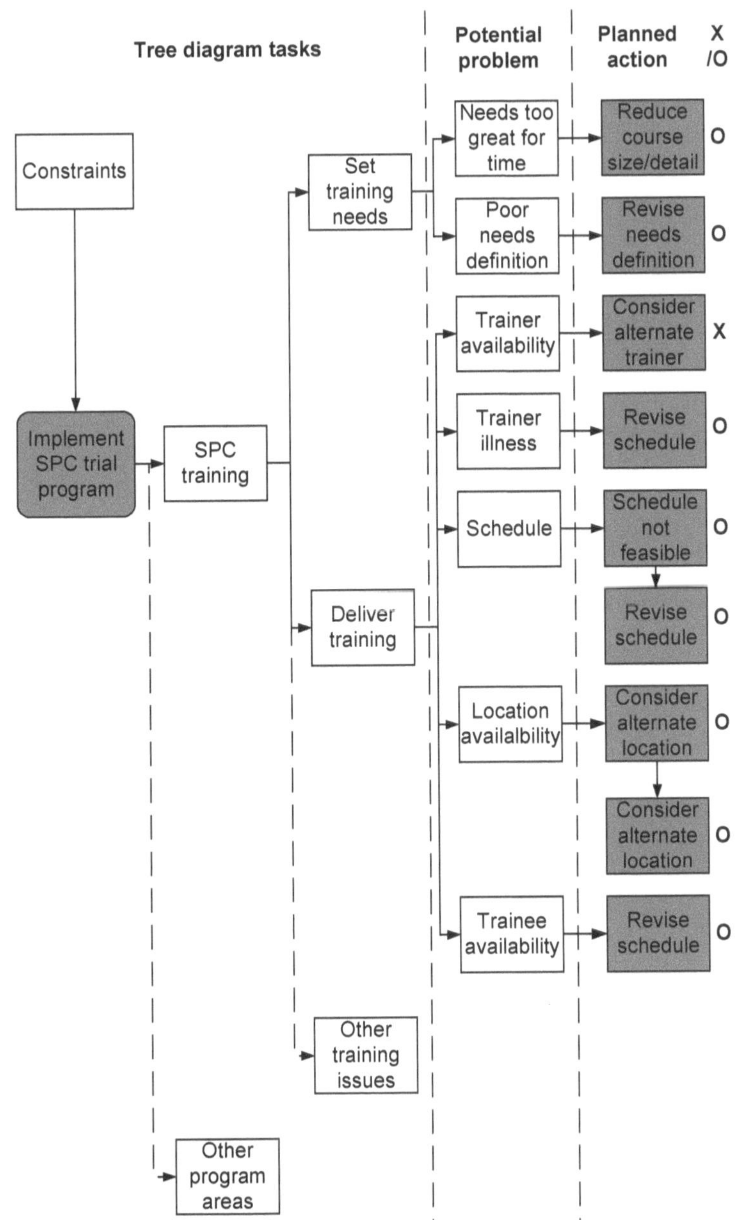


### A partial PDPC for SPC implementation

This is the graphical layout for a partial PDPC based on the arrow diagram shown in Section 7.12 but only 2 of the elements of the program have been shown. Don't you wish they did this sort of thing when they were planning a strategy such as Brexit?

# 7.14 Quality tools - where are you now?

## Using tools to improve quality

The basic and new sets of quality tools are essential for quality improvement. Quality improvement does not happen by itself, it is driven by people who make decisions and take actions. These decisions and actions must always be based on data and information and not on opinion or anecdotes.

Only by using the tools are we able to convert the data and opinions into actionable information and that is really what the tools are designed to do. Data are simply loads of numbers or opinions and does not tell us what to do. Information is what we get when we take the data and structure them. The tools (particularly the new tools) also allow us to set targets and plan improvement projects so that they not only achieve their objectives but also do so within time and budget.

## Completing the chart

This chart is completed and assessed as for those presented previously.

"A bad workman quarrels with his tools."

**Anon**

**Tools for quality improvement**

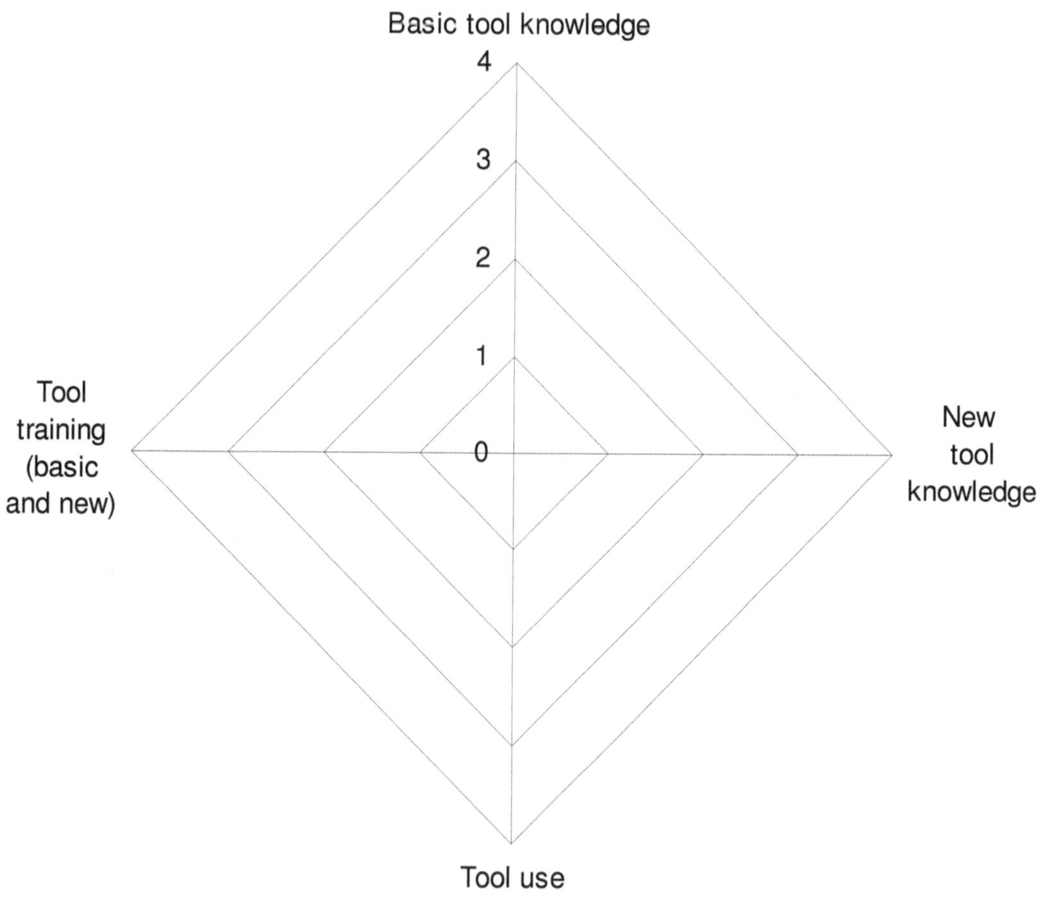


Use the scoring chart to assess where you are in using the quality tools

The numbers from the scoring chart can be transferred to the radar chart for a quick visual assessment of where you are in terms of using the tools for quality improvement.

"Men are more important than tools. If you don't believe so, put a good tool in the hands of a poor workman."

**Anon**

| Tools for quality improvement | | | | |
|---|---|---|---|---|
| **Level** | **Basic tool knowledge** | **New tool knowledge** | **Tool use (basic and new)** | **Tool training (basic** |
| **4** | Excellent knowledge of the basic quality tools in all areas of the company. | Excellent knowledge of the new quality tools in all areas of the company. | All departments are aware of the existence of the basic & new tools & evidence seen of their use to improve quality. | All staff are trained in the use of the basic & the new tools to improve quality. |
| **3** | Good knowledge of the basic quality tools in most areas of the company. | Good knowledge of the new quality tools in most areas of the company. | The majority of departments are aware of the existence of the basic & new tools but no evidence seen of their use to improve quality. | All quality department staff & some other staff are trained in the use of the basic & new tools to improve quality. |
| **2** | Average knowledge of the basic quality tools in most areas of the company. Some departments (other than the quality department) are aware of the existence of the standard tools. | Average knowledge of the new quality tools in most areas of the company. Some departments (other than the quality department) are aware of the existence of the new tools. | Only the quality department uses the basic & new tools to improve quality. | All quality department staff are trained in the use of the basic & new tools to improve quality. |
| **1** | Poor knowledge of the basic quality tools in most areas of the company. | Poor knowledge of the new quality tools in most areas of the company. | The quality department is aware of the existence of the basic & new tools but does not use them to improve quality. | Only limited quality department staff are trained in the use of the either the basic or the new tools to improve quality. |
| **0** | No knowledge of the basic quality tools in any area of the company. The quality department is unaware of the existence of the standard tools. | No knowledge of the new quality tools in any area of the company. The quality department is unaware of the existence of the new tools. | No use of either the basic or the new tools to improve quality. | No training given to any staff in the use of either the basic or the new tools to improve quality. |
| **Score** | | | | |

# Key tips

- The 7 basic quality tools form the basis for transforming data (a set of numbers or concepts) into information that can be used to take action and improve quality.
- The 7 new tools are more suitable for planning and managing the process of taking action. These tools can make the action taken quicker and more effective.
- Flow charts show how processes really work, where they are likely to fail (the gaps) and what is needed to make them more effective.
- If a flow chart is difficult to draw for a process then the process itself is too complicated and likely to fail. Streamlining the flow chart will lead to changes in the process to make it more effective.
- Check sheets have multiple uses. They can be used to classify nonconformities, check the frequency of nonconformities, identify the location of nonconformities, scale nonconformities numbers for a quick histogram for process assessment and act as a checklist.
- Check sheets gather valuable data in real time from the process.
- Cause and effect charts are useful for gaining information from operators by providing a structured method of getting information.
- Pareto analysis is about separating the vital few from the trivial many to target the most productive areas for work and effort.
- Histograms allow the distribution of large data sets to be seen quickly and clearly. They are one of the first steps in SPC.
- Scatter charts reveal the relationship between two or more variables and show if they are related in some way. They do not prove causality, simply correlation.
- Relations diagrams look for the drivers of a particular effect, they look at the causes of an effect and break these down into secondary and lower-order causes to see the relationship between all of the possible causes.
- Relations diagrams can be used to assess the importance of various causes.
- Tree diagrams are used for planning, root cause analysis and component analysis. They break plans, problems of products down to their component parts to allow potential solutions to be found.
- Affinity diagrams are used to organise large amounts of verbal or unstructured data into logical groups to find the key driver of an issue or effect. The key drivers need not be known at the start but can come out of creating the affinity diagram.
- Matrix diagrams are used to show the relationships between groups of variables or information and to reveal the structure of an issue from a mass of information.
- The type of matrix diagram used depends on the number of variables being compared.
- Arrow diagrams use the information from a tree diagram to plan projects. They can range from simple arrow networks with no times or resource calculations to complex PERT and CPM networks. The only difference is in the amount of information contained in the arrow diagram.
- Arrow diagrams provide a project management tool to ensure that projects are completed on time and using the right resources at the right time.
- Process Decision Program Charts (PDPCs) are used for project risk management to assess the likelihood of problems and the actions to be taken in the event of problems.

The tools of quality are as essential in delivering quality products as are the tools and machines used to produce the actual products.

# Chapter 8

# Design quality management

Product quality can be broadly divided into 3 separate areas, the quality of the design, the quality of the raw material input and the quality of the processing into the final product. Note that we have excluded the 'quality' of the user where product misuse causes failure although this is a significant cause of failure in plastics parts.

This chapter covers the methods for ensuring the quality of the design and the design process whereas controlling the quality of the raw materials and the processing are covered in Chapters 9 and 10, respectively. We will not cover the 'quality' of the user as that is beyond anybody's scope of control!

A flawed design can never be processed to produce a quality product. Designs that place high operating stresses on small areas of the product, designs that use the incorrect material, designs that specify (and need) excessively tight tolerances or designs that inadequately signpost the product function are all examples of failures of the design process. They will all affect the quality of the product both physically and as perceived by the user.

It is for these reasons that the quality of the design process is at the heart of quality management. It is the start of the process and produces the template for the rest of the process. Failures at the start, i.e. design, can rarely be easily recovered no matter how good the delivered material or how good the actual plastics processing. Despite this, the process of designing for quality and particularly the process of quality design with plastics is often poorly treated in both plastics processing and design courses.

In my career in plastics product design, I often felt that I was more a 'technical translator' - translating what the product designer wanted into technical terms for the processing people and translating the process limitations into terms that the product designer could understand. Needless to say, I didn't always succeed without upsetting somebody or, on some occasions, everybody.

Good-quality designs are easily manufactured, fault-tolerant and meet the user expectations - this is where the process of delivering quality begins.

**Note:** Some of the sections in this chapter refer to sections of a previous publication 'Cost Management in Plastics Processing'. Where especially relevant these have been reproduced with modifications to reflect the emphasis of this book on quality (although cost and quality are eventually the same).

"The engineer knows that it is easier to analyze a machine than it is to design one. Engineering analysis is simpler than engineering synthesis or design."

**W L Starkey**

# 8.1 What makes a quality design?

## Never mind the quality, feel the width

Product design is the start of producing and delivering quality products. Products have to meet the 3 basic requirements of functionality, aesthetics and manufacturability to be considered a quality product. However, the relative importance of these varies dramatically for every product.

'Industrial' products generally have a higher functional requirement and a lower aesthetic requirement. Typical products are pipes, clips, cable trunking, most non-visible car components, medical disposables, etc. This type of product is driven by fit, function and standards - I can design these.

'Consumer' products generally have higher aesthetic requirement and a lower functional requirement. Typical products are cosmetics packaging, some food packaging, household products, etc. This type of product is driven by creativity, imagination and style - I am hopeless at these.

This is not to say that industrial products can be ugly or that consumer products can be non-functional, it is simply that the balance of the requirements changes with the type of design.

Whichever type of product is considered (although they all must still function and not be too ugly), the manufacturability of the design is key and remains a deciding factor in the 'quality' of the design.

- **Tip -** I am a regular judge for the Design Innovation In Plastics awards and I am always astounded at the creativity of the entrants. Equally I am astounded as to how little they know about design and manufacture with plastics. Some of the products are very beautiful but impossible to produce.

This differentiation between industrial and consumer design makes it difficult to consistently quantify quality at the design level. For industrial products, the functional requirements dominate but for consumer products, the aesthetics and user functionality dominate. Whichever definition of quality is chosen, the ultimate arbiter is the customer.

## Failure in plastics products

Product failure is the ultimate defining line in poor quality. Smithers Rapra (www.smithersrapra.com) has investigated thousands of product failures and the results of this are shown below. The overwhelming majority of product failures (65%) were caused by poor specification/material selection and design. Of the remaining 35%, 20% were

"You can't make a good landing from a poor approach."

**Old pilots saying**

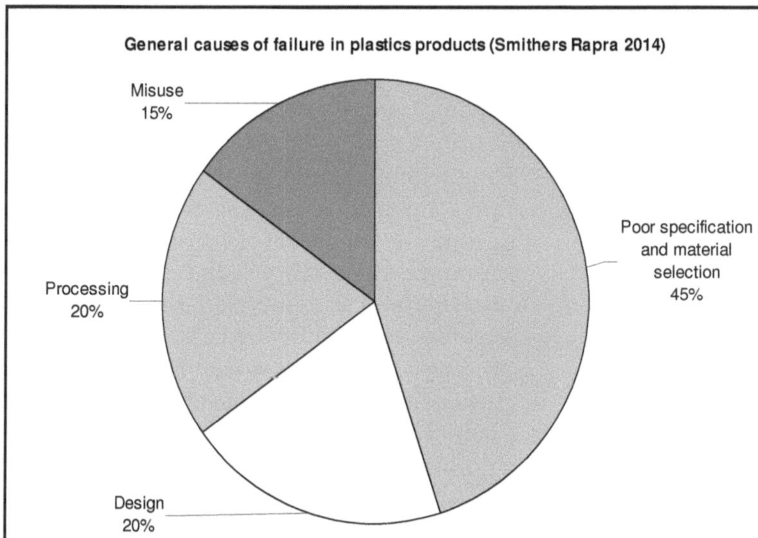


### General causes of failure in plastics products

Failure is the extreme of poor product quality, i.e. when the product no longer fulfils the design function. Smithers Rapra analyses thousands of product failures and the bulk of the failures (65% at a minimum) can be attributed to poor design.

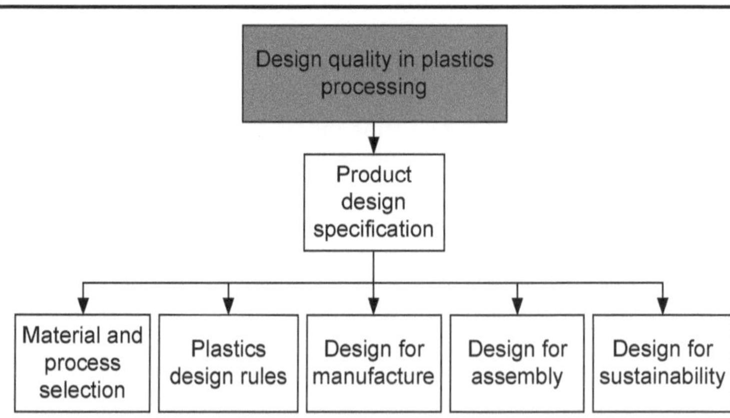


### The fundamentals of quality in design

Aesthetic design is beyond the scope of this book but there are certain tasks that must be completed to achieve design quality. The Product Design Specification (PDS) is the key to getting all the rest of the process correct. Get this wrong and the rest fail.

caused by processing (although in many cases this could also potentially be due to poor design) and 15% by misuse (although again good design can minimise the potential for misuse by either clear signposting of the limitations or the design intent).

**Note:** These are the products that made it into the field before failing. They may, or may not, be truly representative of the full extent of the quality issue, i.e. failure or poor product at manufacturing that did not make it into the field.

Whilst poor quality is broader than failure, the Smithers Rapra results show that design and materials selection are key elements in quality management for plastics processors, get these right and producing quality products becomes much easier.

## The fundamentals of quality in design

Aesthetic design is beyond the scope of this book (and in any case I am not good at it) and instead we will concentrate on the fundamentals of quality in the design process and on the processes and tools needed to deliver a design that is capable of being reliably produced to meet the customer's expectations.

The fundamentals of this process are shown on the lower left and the key document is the Product Design Specification (PDS). This is not the design, it is the definition of what the product is expected to provide. The PDS then drives materials selection, process selection and design for manufacture, assembly and sustainability.

- **Tip -** In today's world, any design which does not consider sustainability at the outset is potentially fatally flawed.

These design fundamentals will be covered in the later sections of this chapter.

## The tools and processes for quality design

Most of the 7 basic and 7 new quality tools can be used at the design stage but there are additional tools and processes that are needed. The essential tools and processes for quality design are shown on the right and using these will provide the foundations for quality design. These are:

- Advanced Product Quality Planning (APQP) - this is a process that was developed in the automotive industry but which can be applied to any design. It is a rigorous planning method designed to deliver the design intent (the PDS) at the production level.
- Quality Function Deployment (QFD) - this is a specific application of a matrix diagram ('The House of Quality') to provide the 'voice of the customer' in the design intent.
- Failure Modes and Effect Analysis (FMEA) - the standard FMEA process can be adapted for use in design, process and machinery development and provides a method of refining the design, process or machine to improve quality.
- Drawings - obviously any product design must be supported by drawings but these are needed at several different levels to allow effective production and control.
- Design of experiments - although more commonly used in the production area, design of experiments is a useful tool for designers seeking robust designs at any level.

These essential tools and processes will all be covered in later sections of this chapter.

The design function has a much greater influence than the manufacturing function on price, quality and cycle time.

80% of the product cost and quality attributes are fixed in the first 20% of the design to production timeline.

Getting it right at the start is absolutely critical.

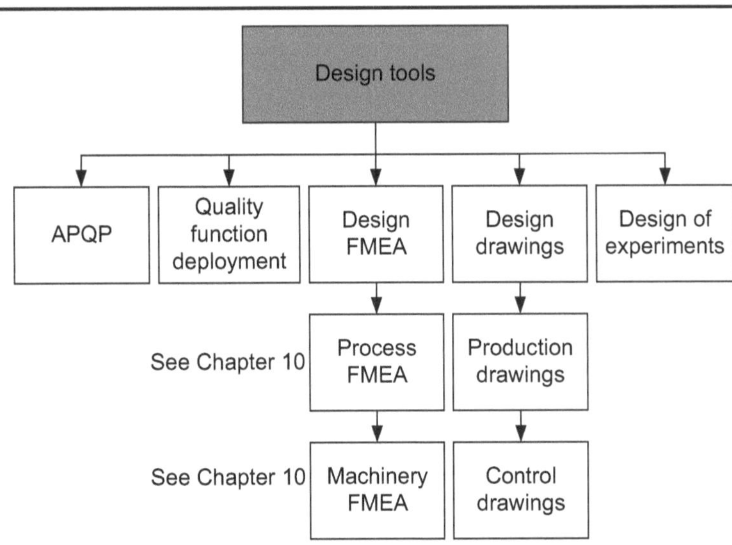


### The tools and processes for quality design

Most of the quality tools discussed in Chapter 7 are used (in one form or another) at the design stage but additional processes and tools are used to provide the foundations for robust high-quality production.

# 8.2 The product design specification

## The essential definition

The product design specification (PDS) is the essential definition of what the product is required to provide.

The PDS is a statement of what the customer wants the product to achieve. In some cases the customer is external and in some cases (e.g. product range extensions for own-brand products) the customer is internal. In every case, a PDS must be prepared to act as a reference for the objectives of the design.

This is initially a statement by the customer of the design intent but it should be expanded by the design and development function to give all the technical detail and limits that are realistically achievable.

## Preparing the PDS

The PDS is normally prepared by the design and development function for agreement/sign-off by the external or internal customer and by the project team. The PDS sets out the technical detail of the requirements to be met to achieve a successful product.

**Note:** The PDS is *not* the design but the specification for the product and must be completed before any design work is started.

The PDS should be circulated to ensure that comment and criticism is given by all members of the project team and other interested parties.

- **Tip** - A PDS must not be confused with all the other specifications which refer to the product, i.e. the type of polymer to be used or the standards that the eventual product must conform to.
- **Tip** - A PDS is not sacred and may be changed but all changes must be noted and recorded. This is a 'living' document.

## Creating the PDS - the checklist

The creation of the PDS involves asking the right questions - if the answers are easy it is probably because the difficult questions have not been asked. These questions should be answered by the internal or external customer in broad functional terms to allow the designer freedom to innovate. A broad checklist of general points is shown on the right. These are designed simply to stimulate thinking and not all of the factors noted will apply to a given product but the subject headings should be scanned to give areas for profitable consideration. It is a series of questions to ask to enable the capture of the essential requirements for the product and not a rigorous design methodology.

## Verifying the PDS

A PDS can be created by one person but should always be checked and agreed by the project team (especially by the customer). The full commitment of the project team is necessary before the PDS is issued.

## Using the PDS

The PDS forms the basis for the design. The PDS is subject to change but should become firm at the pre-production / production stage. The PDS then forms the basic raw material for handbooks, manuals, sales literature and, in conjunction with the final drawings, can become the final specification for the product.

- **Tip** - Create other checklists and standard procedures to standardise later projects and processes. It is quicker and helps to remember to include all the factors.

•This section is a revised and expanded version of Section 2.8 of Kent, R.J. 2012. '*Cost management in* plastics *processing*', Plastics Information Direct.

The PDS for every project will be different. Creation of a standard format or generic PDS for the specific company requirements will make completion of the PDS rapid and will ensure that most of the points are considered.

All of my PDS checklists now include 'Thermal properties'.

Get bitten by thermal expansion early in your career and you never let it happen again.

The increasing importance of environmental issues such as 'carbon footprint' and 'sustainable design' (see Section 8.7) means that these must be considered at the PDS stage.

These are not 'afterthoughts' but integral parts of the design.

## Product design specification checklist

### General

- ☐ Aesthetics
- ☐ Assembly
- ☐ Bought-in parts
- ☐ Complexity
- ☐ Constraints
- ☐ Competition
- ☐ Cost
- ☐ Customer
- ☐ Design life
- ☐ Delivery requirements
- ☐ Disposal
- ☐ Energy
- ☐ Ergonomics
- ☐ Existing designs
- ☐ Experts
- ☐ Export
- ☐ Fatigue
- ☐ Features
- ☐ Finish
- ☐ Friction
- ☐ Function
  - ☐ Main
  - ☐ Secondary
  - ☐ Tertiary
- ☐ Installation
- ☐ Lead time
- ☐ Legal
- ☐ Life in service
- ☐ Manufacturing
- ☐ Manuals/information
- ☐ Maintenance
- ☐ Marketing
- ☐ Modelling
- ☐ Noise
- ☐ Operator
- ☐ Outdoor exposure
- ☐ Packaging - delivery, transport
- ☐ Packaging - point of sale
- ☐ Patents
- ☐ Performance
- ☐ Portability
- ☐ Process
- ☐ Project team
- ☐ Project schedule
- ☐ Prototype
- ☐ Quality
- ☐ Quantity
- ☐ Reliability
- ☐ Safety
- ☐ Shelf life and storage conditions
- ☐ Size
- ☐ Standards
- ☐ Sub-contractors
- ☐ Targets
- ☐ Testing
- ☐ Timescales
- ☐ Tools
- ☐ Toxicity
- ☐ Training
- ☐ Transport
- ☐ Use factors
- ☐ Vibration
- ☐ Waste
- ☐ Weight and wall thickness

### Materials selection

- ☐ Production
  - ☐ Production volume
  - ☐ Weight
  - ☐ Shape and size
  - ☐ Tolerances
  - ☐ Surface finish
  - ☐ Inserts
  - ☐ Undercuts
  - ☐ Holes
- ☐ Acoustic property requirements
- ☐ Assembly requirements
- ☐ Carbon intensity
- ☐ Chemical resistance requirements
- ☐ Density
- ☐ Electrical property requirements
- ☐ Environmental factors: use environment
- ☐ Environmental factors: transport and storage environment
- ☐ Environmental factors: effect of the product on the environment
- ☐ Insulation property requirements
  - ☐ Electrical
  - ☐ Thermal
- ☐ Fire resistance property requirements
- ☐ Magnetic property requirements
- ☐ Mechanical property requirements
- ☐ Optical property requirements
- ☐ Surface finish requirements
- ☐ Thermal requirements
  - ☐ Maximum operating temperature
  - ☐ Normal operating temperature
  - ☐ Minimum operating temperature

  These temperatures should include production, storage and transportation conditions.
- ☐ UV requirements
- ☐ Wear and friction requirements
- ☐ Miscellaneous requirements

### Sustainability

- ☐ Raw materials
- ☐ Manufacture
- ☐ Use
- ☐ End-of-life
- ☐ Disposal
- ☐ Transport
- ☐ Carbon footprint

**Note:** This list is not comprehensive. It should be expanded and modified to clarify the design intent.

# 8.3 Material and process selection

## The basics

Smithers Rapra (www.smithersrapra.com) have also used their experience in the failure of plastics products to divide the recorded failures by failure mode and this is shown on the lower right. They state *'Smithers Rapra's experience has proven that the vast majority of plastic product failures are due to the cumulative effects of synergies between creep, fatigue, temperature, chemical species, UV and other environmental factors.'*

Product failure is not always instantaneous or predictable at the production stage - the majority of failures occur some time after production as a result of combinations of various factors and this highlights the importance of understanding the PDS requirements for the use phase.

Material and process selection is not a simple topic and the two areas are intimately related. Choosing the material limits the available processes and choosing the processes limits the available materials. Material and process selection is an iterative process.

This Workbook is not designed to cover materials and process selection despite their huge contribution to the real and perceived quality of finished products. There are already more than enough textbooks and databases covering this subject.

## Processes and materials

Every process is limited in the materials that it can process and the table on the upper right gives some basic guidance on the materials that can be used for various forming processes.

## Processes and features

Every process has limitations on what can be achieved from the process and the table on the lower right gives some guidance on the features of various forming processes.

## Material data for selection

There are estimated to be over 50,000 distinct types and brands of plastics and this is growing all the time. These have all been developed for a range of different processes and applications.

Some of the best databases for materials data are CAMPUS, Prospector (formerly IDES), MatWeb and the Granta products (my favourite) which use Prospector and other data in a great interface (www.grantadesign.com). All of these give excellent materials selection information.

- **Tip -** Datasheets are mainly designed for comparison and quality control. They are not suitable for full engineering design or final materials selection. They provide guidance but are laboratory documents and do not necessarily represent real service conditions.
- **Tip -** Most datasheets use 'single-point' short-term data. This can be very misleading if taken to represent the plastic's properties over a wide range of conditions. Ask for multi-point data!
- **Tip -** Most datasheets do not consider real-life interactions or long-term effects (which the Smithers Rapra data show are the real causes of failure).
- **Tip -** Environmental stress cracking and chemical attack make up 35% of the failures. Always identify the environment carefully.
- **Tip -** Use databases as an initial screen only, especially if it is mission-critical.

Reading datasheets is an art in itself. Do not be too flippant when looking at them and be aware of their limitations, i.e. what they were designed to do and who wrote them.

"Selecting a material for a part involves more than selecting a material that has the properties to provide the necessary service performance; it is also intimately connected with the processing of the material into a finished part."

**George Dieter**

There are other free resources on polymer properties, polymer processing, the Periodic Table of Thermoplastics and more at tangram.co.uk.

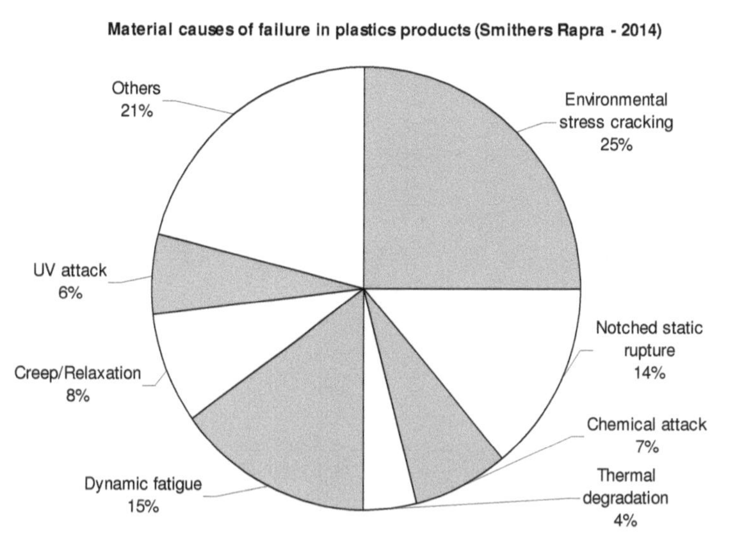


### Material causes of failure in plastics products

Smithers Rapra (see Section 8.1) have also looked at the causes of material failure and state that the majority of failures are due to the cumulative effects of synergies between creep, fatigue and other environmental factors.

Processes and materials - an overview (from www.tangram.co.uk)

This table is for guidance only and is not definitive.

| Material \ Process | Extrusion | Injection Moulding | Blow Moulding | Rotational Moulding | Thermo-forming |
|---|---|---|---|---|---|
| ABS | Yes | Yes | No | Yes | Yes |
| EVA | No | No | No | Yes | No |
| PA | Yes | Yes | No | Yes | No |
| PC | Yes | Yes | Yes | No | Yes |
| PE-LD | Yes | Yes | Yes | Yes | No |
| PE-HD | Yes | Yes | Yes | Yes | No |
| PMMA | Yes | Yes | No | No | Yes |
| POM | Yes | Yes | Yes | No | No |
| PP | Yes | Yes | Yes | No | Yes |
| PS / PS-HI | Yes | Yes | Yes | Yes | Yes |
| PVC-U | Yes | Yes | Yes | Yes | Yes |
| PVC-P | Yes | Yes | Yes | Yes | Yes |
| SAN | Yes | Yes | Yes | No | No |

Processes and features - an overview (from www.tangram.co.uk)

This table is for guidance only and is not definitive.

| Feature \ Process | Extrusion | Injection Moulding | Blow Moulding | Rotational Moulding | Thermo-forming |
|---|---|---|---|---|---|
| Equipment Cost | High | High | High | Moderate | Moderate |
| Tooling Cost | Moderate | High | Moderate | Low | Low |
| Cycle Time | Continuous | <1 min | <1 min | >3 min | < 1 min |
| Economic Quantity | > 5,000 m | >10 K | >10 K | 100 - 10 K | 100 - 10 K |
| Precision | Good | Good | Moderate | Moderate | Low |
| Wall Thickness Control | Yes | Yes | Partial | No | Partial |
| Open-ended Hollows | No | Yes | Yes | Yes | Yes |
| Enclosed Hollows | No | No | Yes | Yes | No |
| Very Small Items | No | Yes | No | No | No |
| Complicated/Intricate Shapes | Yes | Yes | Yes | No | No |
| Large Enclosed Volumes | No | No | Yes | Yes | No |
| Inserts | No | Yes | No | Yes | No |
| Moulded-in Holes | No | Yes | No | No | No |
| Threads | No | Yes | Yes | No | No |

# 8.4 Basic design rules for plastics products

## The basics

Plastics products need to follow much the same design rules whatever the process and these are relatively easy to set out and learn.

However, before this we will go even further back to give some tips about plastics that will seem very obvious but need to be clearly understood by anybody designing or using plastics products. These important tips are:

- Plastics are visco-elastic materials and are very sensitive to temperature and the rate of loading. In fact, there is a principle in polymer physics called the 'time-temperature superposition'. This was developed by Williams, Landel and Ferry (WLF) to explain the changes in plastics properties with short-term and long-term loads and with low and high temperatures. Basically, leaving out all the science, a plastic product will broadly react the same under a long-term load at a low temperature as it will under a short-term load at higher temperature, i.e. the time and the temperature are roughly equivalent.
- Product designs for metal or other materials cannot simply be transposed to plastics. The behaviour of the plastics product will be very different at high temperatures or long times (whatever the data sheet says) and failure may result.
- The chemical resistance of plastics, particularly for environmental stress cracking is highly affected by increased temperature. Raising the temperature increases the potential for environmental stress cracking dramatically.

Understanding these differences between plastics and the more traditional materials is basic to the production of high-quality products that are capable of good long-term service.

## PDS conformity

Whatever the design rules applied, conformance to the PDS (see Section 8.2) is a basic prerequisite. If the design does not meet the PDS then it is not a good one even if it conforms to all the rules! The PDS is the guide for the overall design intent and any design must meet it.

**Note:** We will not consider tolerances at this stage even thought they are basics of the design rules (see Section 8.14).

## The design rules

The design rules for the various processes are relatively material-independent, i.e.

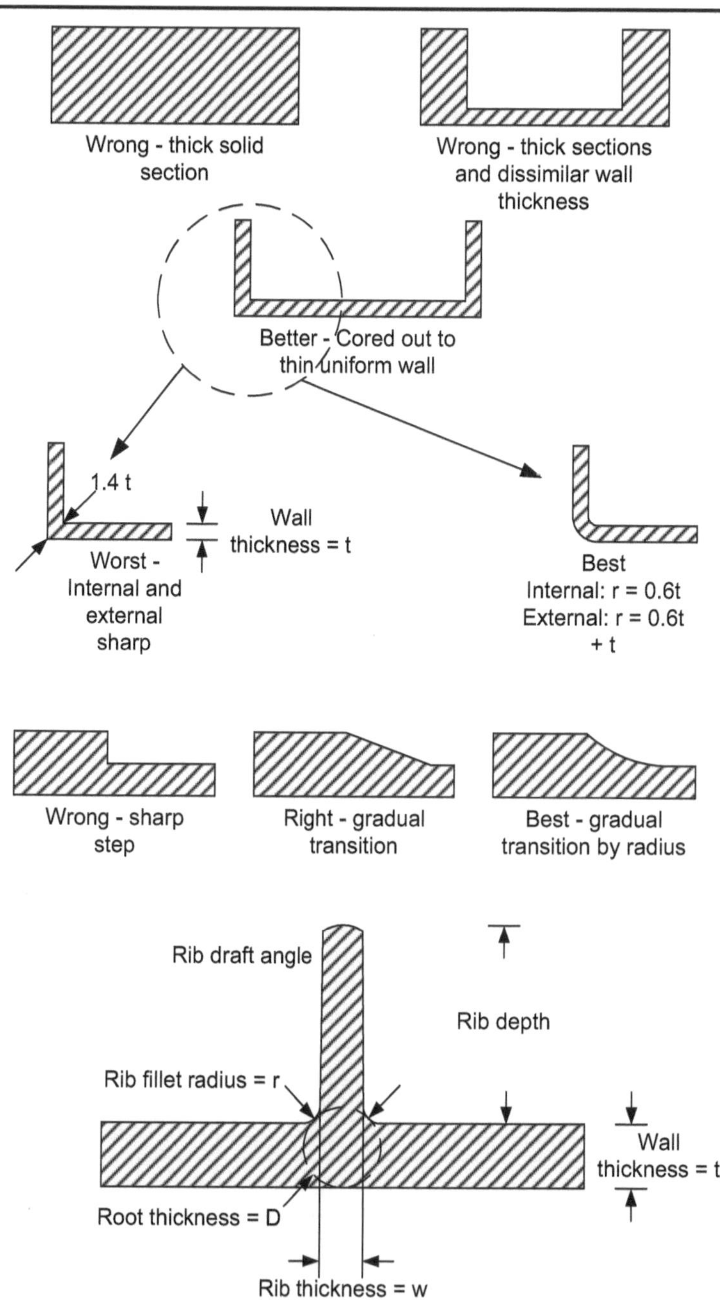


### The basic design rules

Most of the plastics forming processes have similar design rules. Keep the wall thickness thin and uniform, avoid sudden changes in thickness, use radii instead of sharp corners to minimise stress concentrations and keep ribs smaller than the parent material.

they apply to all the thermoplastics. The thermoplastics will vary in properties, e.g. shrinkage rates, and some rules are more important for some materials than for others but they are all important and should only be ignored at your own peril.

- **Tip** - It is strongly recommended that designers use a checklist approach to the basic design to ensure conformance with the basic design rules for plastics.

## Injection moulding design rules

The main design rules for injection moulding are:

- Keep wall thickness as uniform as possible.
- Use gradual transitions between thick and thin sections.
- Wall thickness must suit both function and process.
- Where possible keep wall thickness constant throughout the part and avoid sharp changes in wall thickness.
- Avoid sharp internal corners.
- Internal radii should be at least 0.5 and preferably 0.6 to 0.75 times wall thickness.
- Keep corner wall thickness as close as possible to nominal wall thickness. Ideally, external radii should be equal to internal radii plus wall thickness.
- Rib thickness should be 50-75% of wall thickness.
- Fillet radius should be 40-60% of rib thickness.
- Rib root thickness should not be more than 25% greater than wall thickness.
- Rib depth should not be more than 5 times rib thickness.
- Taper ribs for mould release.
- Always think about creep (deformation under low loads and long times).
- Always look at the environment.
- **Tip** - Further design tips for injection moulding are available at www.tangram.co.uk.
- **Tip** - Perhaps the best texts on injection mould design are those by Pye and by Menges and Mohren. They are classics. Anything by Glenn Beall is also invaluable.

## Extrusion design rules

The main design rules for extrusion are very similar to those of injection moulding with regard to wall thickness, gradual transitions and ribs. This should not be a surprise given that in many cases we are working with the same material. Extrusion has some unique rules with regard to internal walls where the wall position cannot be guaranteed due to the lack of calibration. These are:

- Make uncalibrated internal spaces larger than necessary and adjust after tool tuning with small location pips.
- When using external open spaces then 'bridge' these with reinforcing areas to support the location.
- Avoid small 'closed' internal or external chambers as these can be difficult to calibrate and control.
- Be wary of small protrusions on the surface of the extrusion, these can be difficult to calibrate accurately. Use clearance on the calibrators and allow the protrusion to float.
- Be wary about excessive haul-off forces as they can crush small extrusions.
- Always think about creep (deformation under low loads and long times).
- Always look at the environment.
- **Tip** - Further design tips for extrusion are available at www.tangram.co.uk.
- **Tip** - The classic text for extrusion design is 'Polymer Extrusion' by Rauwendaal.

## Thermoforming

Thermoforming has some specific design rules depending on whether it is thin-wall (packaging applications) or thick-wall (more properly vacuum forming).

- **Tip** - The classic text for thermoforming is 'Thermoforming Technology' by J. L. Throne.

## Rotational moulding

Rotational moulding also has some specific design rules but in many ways is more flexible than other processes.

- **Tip** - The classic texts for rotational moulding are 'Rotational Moulding Technology' by Crawford and Throne and 'Rotational Moulding: Design, Materials and Processing' by Glenn Beall.

Not following the basic design rules or getting them wrong will almost certainly give a poor-quality product.

They are called 'rules' because they were meant to be followed.

There is a lovely Bachelor's Degree Project on 'A checklist for plastic product design' by Johanna Kaartinen from the University of Skövde (Sweden).

She produced a complete checklist for injection moulding but ....

There are some very good free resources available on the Internet. Highly recommended are:

- Bayer 'Engineering Polymers, Part and Mold Design - Thermoplastics'.
- GE Engineering Thermoplastics 'Guidelines for Injection Molded Design'.
- LANXESS 'Part and Mold Design - A Design Guide'.
- BASF 'Design Solutions Guide'.

# 8.5 Design for manufacture

## Design for manufacture

Design for manufacture (DFM) is a design concept to improve the manufacturability of products and is often applied with simultaneous engineering (when the product and manufacturing processes are developed at the same time). The combination of DFM and simultaneous engineering includes a range of proven methods and tools to:

- Produce designs that can be reliably manufactured.
- Reduce development costs.
- Reduce time-to-market.
- Reduce manufacturing costs.
- Improve product quality, reliability, safety and customer satisfaction.

For the purposes of this Workbook, the focus is on the improvements in product quality that can be delivered by DFM. This is also the basis for the APQP process (see Section 8.8). DFM allows quality to be 'designed in' to the product by considering part quality and manufacturing at the beginning of the design rather than designing a product and then hoping that manufacturing can produce it to the required quality.

DFM can be divided into four elements:

### People and teams

DFM includes the concepts of project and design teams similar to those discussed in an earlier book[1]. It uses team-based project management where team members have a variety of technical and managerial skills (see also Section 2.7).

### Simultaneous engineering

Simultaneous engineering uses the team concept to develop the whole product system in parallel to greatly reduce time-to-market. The parallel strands are:

- Product design.
- Manufacturing method design.
- Manufacturing systems and support items.

This concept has been termed 'concurrent engineering' by some software vendors but the major issues are organisation and planning and no computer is needed. Simultaneous engineering is possible with a fertile mind, a pen and paper, a good company structure and willing colleagues.

### Tools

DFM uses the 'alphabet soup' of design and manufacturing tools. The main tools are:

- Failure modes and effects analysis (FMEA) at the design, process and machine levels.
- Taguchi analysis or Design of Experiments (DOE).
- Value engineering (VE).
- Design for assembly (DFA).
- Quality function deployment (QFD).
- Statistical process control (SPC).

Most of these tools are covered in this Workbook at some stage. DFM simply uses these tools in an integrated manner during design, development and production commissioning. The tools are not used in isolation but form part of the overall project management.

### Management

DFM uses a matrix management approach where the management is based more on processes and projects than on the traditional functional divisions.

As a global approach, DFM can be adapted to the needs of individual companies. A fully implemented DFM system will reduce design and development costs but more importantly it will deliver robust high-quality products to market quickly and reliably.

- **Tip -** Train designers in DFM and in the principles of economic product design.

## The basic operation of DFM

DFM has some basic operational principles and these are:

### Understand the process

Every process has design guidelines (see Section 8.4) and ignoring these is a recipe for disaster. Process specialists must be part of the DFM team to highlight potential process difficulties. Avoiding and designing-out process concerns at the start is far better than trying to solve them after the product is in production. This is also relevant for the assembly process (see Section 8.6).

DFM and DFA (see Section 8.6) are sometimes treated as a single 'method' and referred to as DFMA.

We have treated them separately for clarity but many of the concepts are similar.

Sometimes the phrase DFx is use where 'x' is the desired attribute, e.g. DFAA = Design for Automated Assembly.

One design tool that is not as well-known as it should be is 'TRIZ'.

This is an excellent and well-developed tool for finding innovative solutions using a defined invention process.

Fascinating to see in action with many successful case studies.

Aim for wide tolerances that are rigidly enforced rather than tight tolerances that are loosely enforced.

It makes life a whole lot easier.

- **Tip -** Process specialists are essential!

### Understand process tolerances

Process specialists can advise on the achievable tolerances and limits of the process. More importantly, they can stop designers specifying unachievable tolerances (see Section 8.14). DOE (see Section 8.15) can be used to understand how tolerance variations will affect product quality and how to optimise these to produce a robust design for high-quality and low-cost production.

- **Tip -** Tolerances are one of the keys to good DFM.

### Design tooling at the same time

Designing the tooling at the same time as the product will highlight complexity, time and quality issues with any proposed tooling (production or assembly).

- **Tip -** Tooling and process engineers also need to be part of the DFM team.

### Learn from the past

Most products are similar in some way to a previous product and it is important that hard-won experience with previous products is incorporated into new designs.

- **Tip -** If you had a quality/time issue with a fastener in the last design then why are you using the same fastener in the new design?

## DFM guidelines

DFM also has some basic guidelines for good designs and these are:

### Reduce the number of parts

This is a key issue for DFM and DFA (see Section 8.6). Reducing the number of parts generally reduces manufacturing costs and the potential for quality concerns as well as reducing inventory and development costs.

- **Tip -** If a part does not need to move relative to any other part and does not need to be made from a different material then consider it for elimination.

### Use modular designs

If the product design can be broken down into discrete modules then this helps with production, inspection and servicing.

- **Tip -** Modules with good 'connection' design can also be reused in other products and updated easily. This is how good software using DLL's (dynamic link libraries) works.

### Design multi-functional parts

If parts can be designed to be multi-functional this will reduce the number of parts as well as improving their quality. The concept of multi-functionality can be used to check functionality, guide assembly, align and fix subsequent parts, aid inspection or provide user instructions.

- **Tip -** If you have to include a part then make it as 'useful' as possible.

### Design multi-use parts

Reducing the number of parts should not only consider a single product. If similar parts are being used in multiple products then reducing part variation can dramatically reduce development costs, time and improve quality.

- **Tip -** Why try to get two parts right when you can concentrate on one?
- **Tip -** Identify potential multi-use parts and consider what modifications would need to be made to use them in multiple products.

### Design to minimise finishing

Finishing operations such as painting, chrome plating and manual trimming are always very costly and are potential areas for quality concerns. Always attempt to minimise finishing operations at the design stage.

- **Tip -** A two-shot moulding can be more consistent and reliable than a painting operation.

### Design for assembly from the start

Design for assembly (see Section 8.6) is an essential for DFM. Thinking about it at the start pays dividends.

## Want to know more?

One of the best books on the subject of DFM is by David Anderson[2] (see quote above). You won't regret buying it.

•1. Kent, R. 2012. *Cost management in plastics processing.* PID.

•2. Anderson, D. 2014, *Design for Manufacturability: How to Use Concurrent Engineering to Rapidly Develop Low-Cost, High-Quality Products for Lean Production.* CIM Press.

"Never design a part you can buy out of a catalogue."

**David Anderson**

and select quality suppliers!

**'Coffee Jar Design'**

Some years ago whilst working with PP injection mouldings we experienced severe quality problems with screw threads and jamming threads. Examining the product design it was obvious that the thread form was virtually that of a machine thread when functionally it only needed to pull down and lock against a shoulder.

We changed the design to an open thread - similar to those on coffee jar lids - with huge tolerances. The problem went away and moulding was much easier.

# 8.6 Design for assembly

## Design for assembly

The 'design for assembly' (DFA) methodology was pioneered by Boothroyd and Dewhurst at the University of Rhode Island. The method is a design evaluation tool that enables designers to:

- Reduce the part count of an assembly.
- Design products for easier manual or automatic handling in assembly.
- Reduce the labour and time involved in assembly.

DFA is primarily a cost-saving tool that is concerned with reducing the product assembly cost by minimising the part count, the number of assembly operations needed to produce the part and by making these assembly operations as easy and fail-proof as possible.

DFA reduces the number of 'opportunities to fail' in any assembly operation. This automatically leads to improved design and manufacturing quality as well as significant cost reductions.

- **Tip -** Designers need to be familiar with DFA methods to improve quality and to reduce costs.

## The basics of DFA

DFA has some basic principles and these are:

- Minimise the part count in the assembly by combining part functions.
- Parts should have self-aligning and self-locating features so that they cannot be installed incorrectly.
- Parts should have self-fastening features.
- Parts should not be 'left-' or 'right-handed', even if this means adding unused features to the parts.
- Parts should either be symmetrical (for easy orientation during assembly) OR very asymmetric for clear identification (for easy detection and orientation).
- Parts should automatically check that all the previous parts are present and located correctly.
- Parts should be designed to reduce the need for re-orientation during assembly.
- Parts should be designed for handling and insertion (both manual and automatic).
- Parts should be designed for easy gripping and transfer in fixtures using registration locations.
- An assembly should have a 'base' part into which all other parts are located.
- Parts should be designed to be assembled from the top down onto the 'base' part to use gravity rather than fight against it.
- Fasteners should be avoided where possible.

## The DFA method

### Functional analysis

DFA first assesses the part functionality to determine if it is a primary part or a secondary part. The primary part is the 'base' part of the assembly. A secondary part is potentially a non-essential part, i.e. a candidate for elimination.

Functional analysis should also:

- List the parts in the assembly and their order of assembly.
- List the number of interfaces between the individual parts.

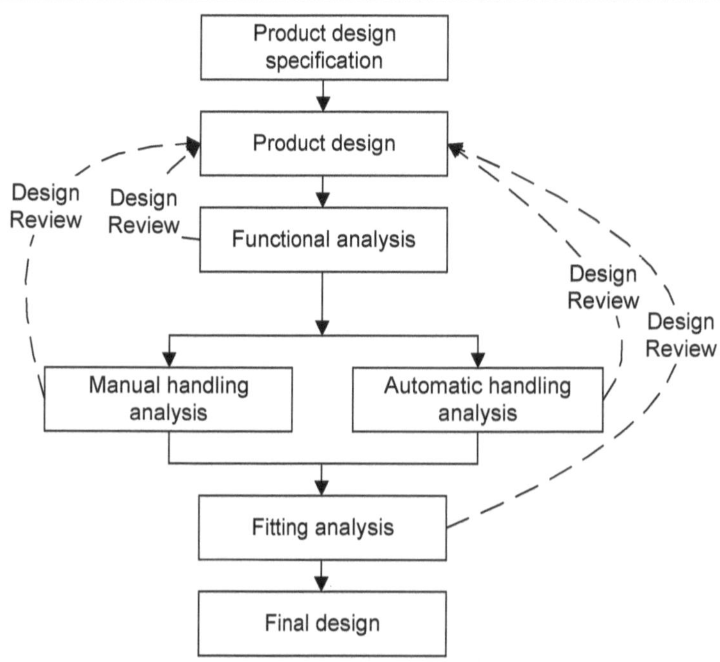


DFA by the Boothroyd/Dewhurst method

The method is a process designed to assess and improve product design for assembly. It will reject less than optimum solutions and will improve designs for manual and automatic assembly.

- Calculate the theoretical minimum number of parts for the assembly by identifying any secondary parts that can be incorporated into the primary part.
- Identify any parts that can be standardised, e.g. if you must use fasteners then reduce the types and sizes needed to the absolute minimum.
- Look for part count efficiency and assess the relative part costs.
- **Tip** - A process flow diagram is a very useful tool to visualise the assembly process (see Section 7.2).
- **Tip** - Threaded fasteners are always a source of quality problems and should be avoided wherever possible.
- **Tip** - Quality is an essential part of the DFA process and designers should use the basic principles to 'mistake-proof' the assembly process.

Functional analysis will almost always result in design modifications to reduce the part count and improve the basic product design.

## Handling analysis

The handling analysis looks at the difficulty of controlling and orienting the part during both manual and automatic assembly.

Handling analysis looks at issues such as:

- Part size.
- Part weight.
- Part flexibility.
- Any other part details that could affect storage and handling, e.g. potential for nesting.
- Handling time and complexity.

All parts and the proposed assembly process should be examined in terms of the DFA basics (see above) to verify that handling is:

- Consistent.
- Easy for both people and machines.
- Capable of delivering high-quality products.

Handling analysis will almost always result in design modifications to improve the ease of orientation in the assembly and to reduce assembly times and costs.

## Fitting analysis

Fitting analysis reviews the fitting methods and looks at issues such as:

- The attachment method used for each part.
- The assembly time needed for each part.
- The elimination of non-value-adding processes.

At each stage of DFA, scores are allocated to the design and the design must achieve a minimum score analysis to go forward to the next stage. DFA is supported by extensive selection tables and software elements to carry out the calculations and assessment.

DFA also helps to decide on the assembly method. Boothroyd and Dewhurst[1] give selection charts to determine the best assembly method (see diagram below) but it should be noted that 90% of production is still best performed by manual or semi-manual methods with only 10% of products being recommended for fully automated assembly of any type.

- **Tip** - DFA needs more than designers, production engineers must be involved at the design level as well.

DFA offers a chance to reduce assembly problems at the design stage.

- 1. Boothroyd, G. Dewhurst, P. and Knight, W., 2010, *Product design for manufacture and assembly.* CRC Press.

- This section is a revised and expanded version of part of Section 2.10 of Kent, R.J. 2012. '*Cost management in* plastics *processing*', Plastics Information Direct.

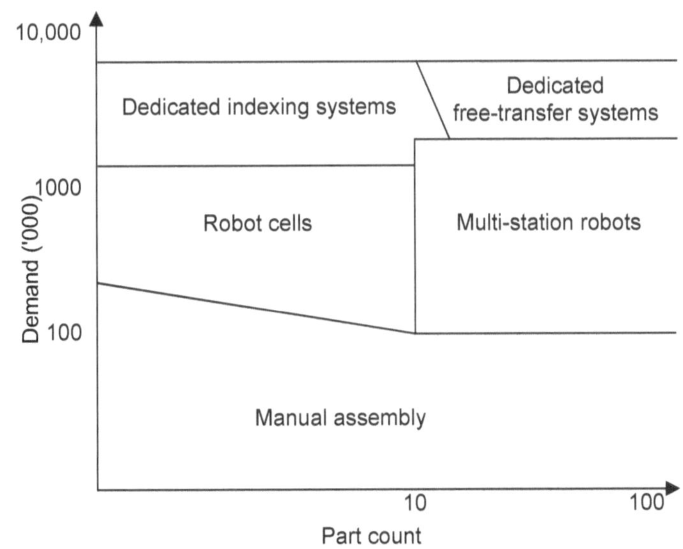


### Choosing assembly methods by DFA

DFA not only improves the product design but assists in assembly method selection. The assembly method is determined by the part count and the demand.

**After Boothroyd & Dewhurst[1]**

# 8.7 Design for recycling and sustainability

## Design for the future

It may not be immediately obvious that quality management includes sustainability but any design that does not consider the complete product life cycle is not a high-quality design. Our focus needs to be on quality for all the phases of life and not simply on the manufacture. The broad aspects of sustainable design in plastics is a complete topic in itself and in this section we will only deal with the key issue of design for recycling.

There is increasing societal and legislative pressure on the users of plastics products to recycle and this needs to be considered at the design stage. The table on the lower right gives the most common disposal methods for plastics and whilst product reuse is the best method, it is not greatly used for plastics products and mechanical recycling is the preferred option for most plastics products.

## Plastics selection and identification

One of the difficulties in mechanical recycling is in identifying the basic raw material (although IR sorting is used by many recyclers). This can be aided by identifying the base material using the standard ISO triangular arrow symbol that includes the internationally recognised abbreviation for the base polymer.

- **Tip -** The material of manufacture should be marked on all plastics parts, using the standard symbols and abbreviations. Standard mould inserts are readily available for marking and should be used to identify the base material. Alternatively, mark the material type with stamping or laser printing.

Whatever material is used, sorting and recycling is easiest if only one basic polymer is used and it is even easier if the number of grades of material used has been reduced.

If it is not possible to reduce the number of polymers then the polymers used should, at the very least, be compatible and capable of forming a polymer blend when reground and recycled. Some polymers are very poor at blending with other materials, e.g. PE has very poor compatibility with most other polymers (except naturally PP) whereas ABS is compatible with most other polymers. Selecting compatible materials can allow bulk regrinding and compounding rather than needing disassembly and sorting.

- **Tip -** Reduce the types and grades of polymer used and use versatile materials with a wide range of applications.
- **Tip -** Select and use compatible materials together to reduce dismantling and sorting.
- **Tip -** Get hold of a table of compatible polymers. Bayer have an excellent table (http://eco3e.eu/materials/plastics/).
- **Tip -** Thermoplastics are better for recycling than thermosetting plastics.

Products that are not designed for recycling can be excellent products but what will the consumer think of them?

Doesn't quality cover the whole of the product life cycle?

## Remove other materials

If an assembly is to be recycled then the presence of any non-plastics materials, e.g. metal inserts, screws and clips, labels adhesives and paints makes recycling difficult and where possible these should be eliminated (this will also reduce cost). If there are metals present then the product will have to be disassembled before the materials can be recycled.

When designing assemblies it is best to

If you want a good laugh then consider the comment I ran across on the Internet that said that 'The identification numbers indicate how many times the plastic can be recycled'.

| Method | Comments | |
|---|---|---|
| Product reuse | Product is designed for more than single use, e.g. returnable bottles. Rare for most plastics products. | Best |
| Mechanical recycling | Products are sorted, cleaned, and reprocessed into new materials for new products. | ↓ |
| Feedstock recycling | Products are broken down to the basic chemicals used to produce new chemicals (plastics or others). | ↓ |
| Energy recovery (incineration) | Products are burned to recover the stored energy (plastics have a higher calorific value than coal). | ↓ |
| Treatment and dumping | Products crushed and treated to reduce volume and pollutants before disposal in landfill. | ↓ |
| Traditional disposal | Sent to landfill. Landfill is being phased out by public opinion and rising costs. | Worst |

avoid using screws and metal parts and to use snap fits, press fits or staking to join parts. If metals must be used then design the joint for easy breakage to separate the metal and plastic components.

- **Tip** - Eliminate or reduce non-plastics parts in an assembly.
- **Tip** - Snap fits are best for disassembly.
- **Tip** - Hot air staking and press fits are best for recycling.
- **Tip** - Welded joints (ultrasonic, hot plate, spin or friction) are good for recycling but can be difficult to dismantle.
- **Tip** - Adhesive bonds can be difficult to dismantle and recycle.

## Make it easy to take apart

Joints are inevitable in assemblies but sometime it is not clear where they are or what they do when it is time to disassemble the product. Design for recycling should clearly signpost where the joints are and make these easily accessible for rapid disassembly.

The first principle is to design assemblies so that joints and connection points are both accessible and easy to recognise. Using snap or press fits (see above) gives joints that are easily seen and broken by hand or by using simple tools such as screwdrivers.

If one-way (irreversible) snap fits are necessary for fit and function then, if possible, access points should be provided to allow the snap fit to be released or broken away using simple tools. This can be assisted by providing designed-in stress-concentrators (break points) to allow the product to be broken apart into separate materials or components.

The best solution for disassembly is to make the process unnecessary. Design for manufacture and assembly (see Sections 8.5 and 8.6) both recommend reducing the part count of an assembly and this can also be a benefit in the area of design for recycling. Integrating multiple functions into a single, well-designed multi-functional part reduces both assembly and disassembly costs as well as having many other benefits.

- **Tip** - Make joints accessible and easy to recognise.
- **Tip** - Use snap and press fits where possible.
- **Tip** - Design in 'break points' for easy disassembly.
- **Tip** - Use 'break points' to separate materials when disassembling.
- **Tip** - Use multi-functional parts to reduce part count and dismantling.

**Note:** In a previous book I stated 'If it can be taken apart then it wasn't put together properly in the first place.' Times have changed and recycling now needs products that can be easily taken apart. Permission to move on and change my mind?

## Reduce material use

Reducing material use (light weighting) is an obvious tactic for reducing environmental impact and product cost although the effect on recycling is debatable.

- **Tip** - Use good structural design to make walls thinner.
- **Tip** - Use good flow design to make walls thinner.
- **Tip** - Use gentle processing conditions to minimise polymer damage.

## The function is still the thing

Quality covers the whole product from design to end-of-life and product designers now need to think about recycling of the product at the design stage rather than after they have produced several million.

Design for recycling is important but designers must never prejudice the design intent to provide for recycling. In the past, recycling was rarely considered at all, now it is simply a normal part of the overall design brief.

- **Tip** - Design for recycling and sustainability, but not at the expense of function or service life.

•1. Kent, R. 2012. *Cost management in plastics processing*. Sections 2.13 to 2.19. PID. This covers the general aspects of sustainable design for plastics processors in more detail.

Design for recycling begins at the start. Leaving it until the end is too late.

Recycling must be part of the standard PDS for all products.

If I can move on and change my mind on the basis of new evidence and market pressure then so should you.

# 8.8 Design fundamentals - where are you now?

## Design fundamentals

The design process is the start of the chain of quality and getting the design right makes all the rest much easier. Good design needs good preparation and the PDS is the foundation for this - it provides an understanding of what the product needs to deliver rather than what it will look like. Rather than focusing on the shape and features, a PDS focuses the design team on the needs of the customer, an essential for the production and delivery of a quality product.

The delivery of a quality product also needs correct selection of the material and process and the matching of the design to the process limitations and design rules. Getting these wrong is often fatal.

The final elements of a good-quality design are the consideration of how the product is going to be manufactured, assembled and lastly, what will happen to the product at the end of the use phase.

Designs that do not consider all of these elements will not be capable of being produced to a high-quality level in high volumes and will end up as the design rejects that we have all seen.

## Completing the chart

This chart is completed and assessed as for the previous charts.

"A bad design is like a rotten egg - it's quality is apparent to everyone after it's busted."

**William T. Bean**

**Design fundamentals**

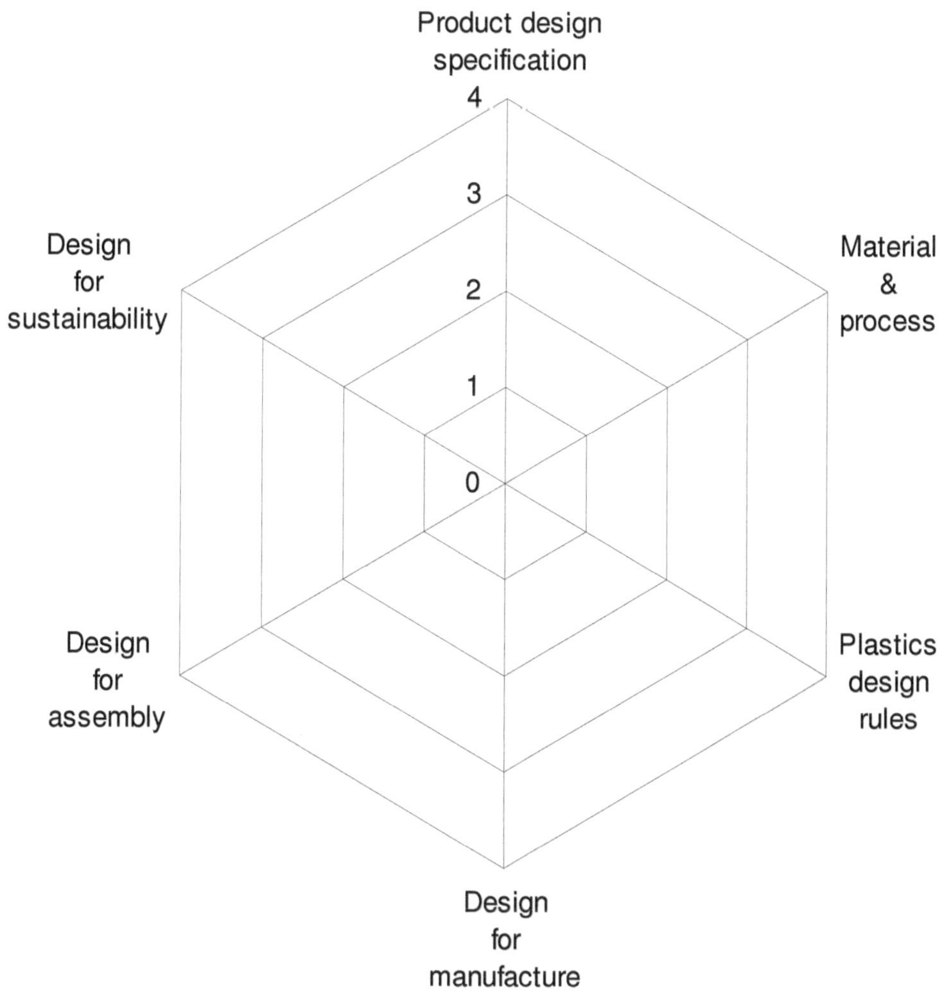


Use the chart to assess where you are in design fundamentals

The numbers from the self-assessment should be transferred to the radar chart for a quick visual guide to where you are in the design fundamentals.

I still remember the first (but not the last) great product 'failure' I had.

We failed to use any of the tools and rushed into the development without really thinking about it.

We paid a heavy price in terms of reject rates, added costs and damaged customer relationships. It was not a pleasant time for anybody and could so easily have been avoided.

## Design fundamentals

| Level | Product design specification | Material & process | Plastics design rules | Design for manufacture | Design for assembly | Design for sustainability |
|---|---|---|---|---|---|---|
| 4 | Excellent PDS before project start & well circulated. PDS gives objectives, requirements & constraints. PDS reviewed & updated regularly. | Designers have excellent material & process knowledge. Training to regularly update knowledge. Wide use of external information sources. | Designers have detailed knowledge of design rules for multiple processes. Designers use rules to produce quality designs. | DFM techniques used widely in all design projects. | Formal DFA used to analyse components & assembly method. | Excellent use of design for sustainability concepts at all stages of product life. Post-use recycling is an essential part of design process. |
| 3 | General PDS before project start but small circulation. PDS does not state objectives, requirements & constraints. PDS rarely reviewed & updated. | Designers have good materials & process knowledge. Some training to update knowledge. Little use of external information sources. | Designers have good knowledge of design rules for multiple processes. Designers use rules to produce quality designs. | DFM techniques known & often used in design projects. | Informal DFA used to analyse components & assembly method. | Good use of design for sustainability concepts at all stage of product life. Post-use recycling is minor part of design process. |
| 2 | Poor PDS generated & lacks vital information, has restricted circulation & is never updated. Projects do not deliver. | Designers have some materials & process knowledge. No training to update knowledge. No use of external information sources. | Designers have good knowledge of design rules for a limited number of processes. Use of design rules is good but limited to few processes. | DFM techniques known but rarely or sporadically used in design projects. | Assembly method formally considered at design stage but no DFA used. | Good use of design for sustainability concepts but only for some stages of product life. Post-use recycling is rarely considered. |
| 1 | No PDS prepared before projects start. Projects evolve & requirements change regularly with little communication of changes. | Designers have poor materials & process knowledge. No training to update knowledge. No use of external information sources. | Designers have some knowledge of design rules for few processes. Use of design rules is sporadic & designs are hard to produce. | DFM techniques known but not used in product. | Assembly method informally considered at design stage. | Poor use of design for sustainability concepts & only for some stages of product life. Post-use recycling not considered relevant. |
| 0 | No PDS prepared before projects start. Projects evolve & requirements change regularly with no communication of changes. | Designers have no knowledge of materials & processes. No training on materials & processes No use of external information sources. | Designers have no knowledge of design rules for processes in use. Designs often break the rules & need substantial revision. | No knowledge of DFM techniques. | No concept of DFA or alternatives in assembly methods. | No use of design for sustainability concepts. No consideration of post-use recycling. |
| **Score** | | | | | | |

# 8.9 The APQP design process

## It is not simply design

Although this chapter concentrates on the design aspects of quality by now it will be clear that it is impossible to segregate design from all the other activities of new product development, especially that of process development. These other activities need coordination and control and the input of more skills than those of simply the design department. Successful product designs are those that make it into production and this needs a team approach.

The Automotive Industry Action Group (AIAG - see sidebar) has developed the Advanced Product Quality Planning (APQP) framework of procedures and techniques for developing and producing new products. This was developed initially for the automotive industry but many of the concepts apply equally well to plastics products. The APQP process also fits well with the PDCA cycle that is the basis of ISO 9001 (see Chapter 4).

The full APQP manual documents the process and also refers to other AIAG documents on:

- Failure Modes Effects Analysis (FMEA - see Section 8.11).
- Statistical Process Control (SPC - see Chapter 5).
- Measurement Systems Analysis (MSA - see Appendices 6 and 7).
- Production Parts Approval Process (PPAP - see Section 10.5).

APQP is focused on planning for quality from the start and retaining that focus through to full production. It is primarily based on the concepts of error prevention and continual improvement rather than simply on error detection. It is also based on using multidisciplinary teams.

The APQP process has the following advantages:

- It reduces development times.
- It encourages simultaneous engineering in design and manufacturing.
- It encourages early procurement of long lead time tooling and facilities.
- It encourages designing within the process capabilities.
- It provides defined objectives that are measured as project stages are achieved.
- It provides defined key events with specified key deliverables.
- It provides significant cost savings through the improved use of resources.
- It provides effective feedback and corrective action throughout the process.

The APQP process consists of 4 stages before series production and these are run concurrently and by the same team.

## Planning and definition

The planning and definition phase of APQP is fundamental to success in the later stages. This stage is where:

- The project is assessed to ensure that it meets the company strategy and business plan.
- The project team is set up and the communications methods defined.
- The project scope is defined.
- The design requirements (the PDS) of the project from the 'voice of the customer' are defined.
- Initial product and process assumptions are assessed and evaluated.
- Initial product and process benchmarking is carried out to check that the proposed product can meet the requirements.

During this stage the team is concerned with defining the customer's needs and planning the quality programme. Typical outputs are shown in the diagram on the right.

## Product design and development

During this stage the team is concerned with developing the design features and characteristics, reviewing the engineering requirements and assessing for potential manufacturing problems. Typical outputs are shown in the diagram on the right.

## Process design and development

During this stage the team is concerned with developing the process and ensuring that this is both comprehensive and effective. A key element of this stage is ensuring that the process is 'capable' of

The AIAG (www.AIAG.org) was set up in 1982 by Chrysler, Ford, and General Motors to develop standards and harmonised business practices. It has grown to include companies such as Toyota, Honda and Nissan as well as retailers, suppliers and manufacturers of other products.

They do great work and their publications and training are highly recommended.

Full details of APQP are available in the AIAG publication that is available from AIAG.

We used a 'modified' simple version of APQP to design and launch products for many years. It changed the way we worked, reduced our costs and never let us down.

PENTAMODE is an alternative general design process developed specifically for injection moulding.

It is covered in detail in Section 2.6 of Kent, R.J. 2012. '*Cost management in* plastics *processing*', Plastics Information Direct.

meeting the customer's requirements. Typical outputs are shown in the diagram below.

## Product and process validation

During this stage the team is concerned with validating the product, the process and the control plan as well as ensuring that the product meets the customer's needs. This validation needs a Production Part Approval Process (PPAP - see Section 10.5). Typical outputs are shown in the diagram below.

After product and process validation is complete, approval is given for series production and planning begins for the next project.

## Feedback, assessment and corrective action

Whilst the four main stages run concurrently, the process of feedback, assessment and corrective action runs through all stages of the model.

•1. AIAG. 2008. '*Advanced Product Quality Planning and Control Plan*'. AIAG.

Good companies:

- Create a culture of designing quality and reliability into their products.
- Use 'product development teams' to include the quality and reliability issues of customers, manufacturing, service and suppliers.

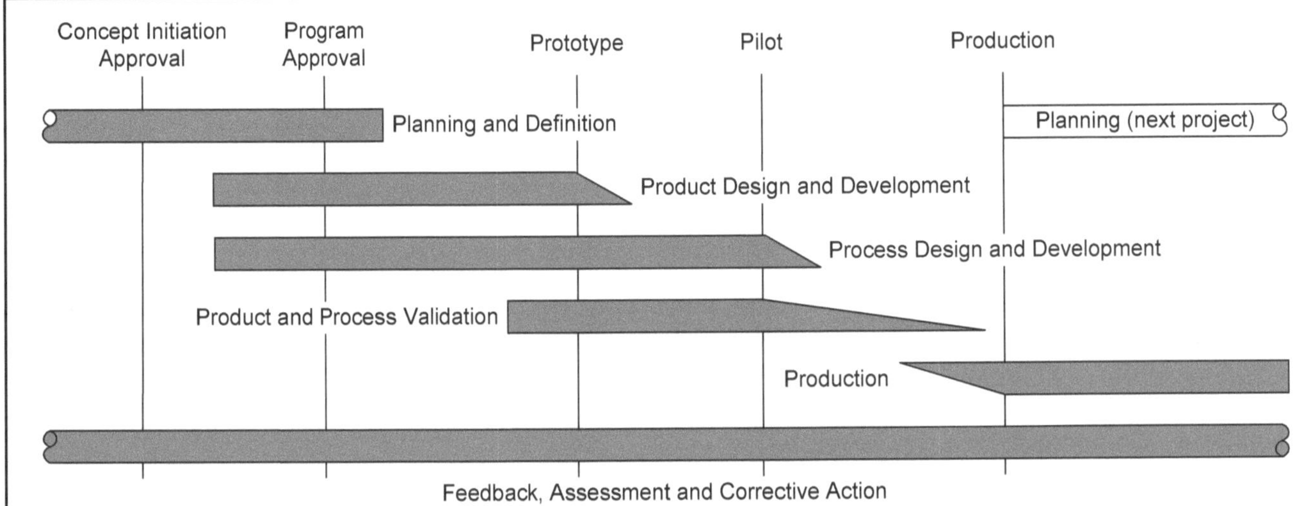


| Planning and definition | Product design and development | Process design and development | Product and process validation |
|---|---|---|---|
| ■Project management.<br>■Design requirements (customer needs).<br>■Reliability and quality requirements.<br>■Draft Bill of Materials (BOM).<br>■Draft process flow chart.<br>■Draft outline of special products and processes.<br>■Draft product quality plan. | ■Design FMEA.<br>■DFM and DFA.<br>■Design verification.<br>■Finite element analysis (FEA).<br>■Engineering drawings and specifications.<br>■Materials specifications.<br>■Tooling specifications.<br>■New equipment specifications.<br>■Draft control plan.<br>■Prototype build.<br>■Design review. | ■Process flow chart.<br>■Process FMEA.<br>■Pre-production control plan.<br>■Process instructions.<br>■Measurement systems analysis plan.<br>■Preliminary process capability study plan<br>■Floor plan layout.<br>■Packaging standards and specifications. | ■Trial production.<br>■Measurement systems evaluation.<br>■Preliminary process capability study.<br>■Production validation.<br>■Packaging evaluation.<br>■Production control plan.<br>■Full quality plan sign-off.<br>■Production part approval. |

Typical outputs of the AIAG model for APQP

The APQP model divides the process of advanced quality planning into 4 separate stages (with series production as the fifth and final stage and with feedback assessment and corrective action as a constant process). Each stage overlaps with the previous and subsequent stages in a continuum of product and process development. APQP is a team process and each stage needs input from a range of people.

# 8.10 Quality function deployment (QFD) - the house of quality

## The house of quality

Quality function deployment (QFD) is a specific use of matrix diagrams (see Section 7.11) and is often referred to as 'The House of Quality' in the product planning stage because of the shape of the resulting matrix diagram which is a combination of the L and Roof-matrices.

QFD is used to translate the 'Voice of the Customer' into the specific technical characteristics and requirements that the design must have to succeed in the market.

QFD is widely used in product planning and in many companies this is the only use of QFD but this misses much of the power of QFD.

## The stages of QFD

When QFD is used throughout product and process development it gives a clear specification of the customer needs and translates these all the way through to the necessary process and quality controls (see box on the upper right). The four stages of QFD are:

### Stage 1 - Product planning

This is the stage that translates the customer needs (the 'want's') into the design requirements (the 'how's'). At this stage the 'customer' can be broadly defined, it does not have to be the end user but can be the production area, the service department or any other area that will be affected by (or influence) the design.

Product planning using QFD gives a better understanding of the underlying customer priorities, improves communication and gives clear design goals for the design area. The output of the QFD process can be explicitly stated (and rated) design requirements with associated target values to be achieved.

This stage can also be used to document and clarify issues such as:

- Warranty data of similar products.
- Competitor analysis of similar products.
- Target values for measurable needs.
- The technical difficulty of meeting the need and achieving the target value.
- The cost of meeting the needs and achieving the target value.
- Comparison of alternative routes to meeting the customer need.
- **Tip -** Good data from the customer is important at this stage and this is often hard to get.

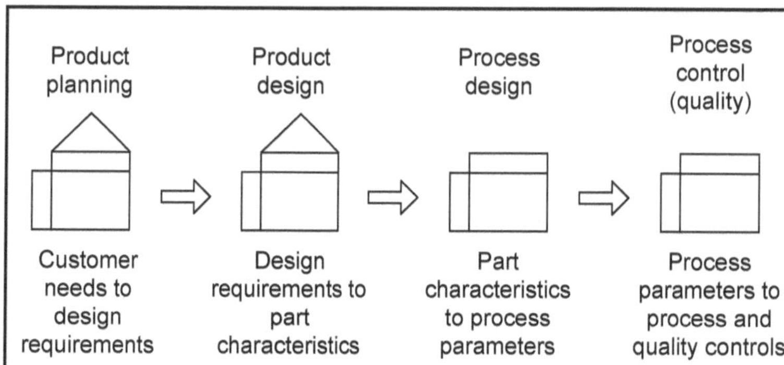


**QFD can be used throughout product development**

- The customer needs are translated into design requirements.
- The design requirements are translated into critical part characteristics.
- The critical part characteristics are translated into critical process parameters
- The critical part and process parameters are translated into process and quality controls for the product.

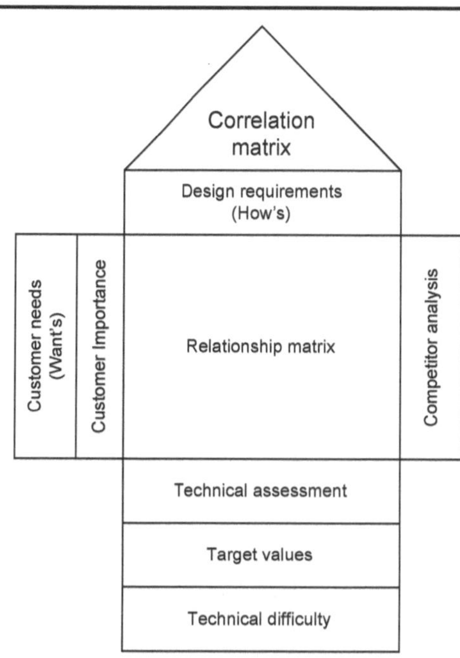


**A basic QFD outline**

The basic QFD sets out the relationship between the customer needs and the how they will be achieved (the design requirements). Additional information (importance, target values, competitor analysis, etc.) can be added as required.

- **Tip -** QFD is a team sport, do not use it alone.
- **Tip -** Stage 1 is a marketing-led process.

## Stage 2 - Product design

This is when the design requirements are translated into the critical part characteristics that the design department can work with, i.e. from 'what it should do' to 'what it will look like'. This is the stage where creativity comes into play. Product concepts are created, reviewed to check that they meet the design requirements and the resulting critical part characteristics are documented for use in Stage 3.

- **Tip -** Stage 2 is a design-led process.

## Stage 3 - Process design

This is when the critical part characteristics are translated into the critical process parameters for the production department to work with, i.e. from 'what it will look like' to 'how we will make it'. The critical manufacturing processes and equipment are identified, the process flow charted and the resulting critical process parameters documented for use in Stage 4.

- **Tip -** Stage 3 is a production-led process.

## Stage 4 - Process control (quality)

This is when the critical part and process parameters are translated into the process controls and parameters for the quality department to work with, i.e. from 'how we will make it' to 'how we will know that we have made it OK'. The critical part and process parameters are identified, the control methods for these parameters established and the inspection and test methods established.

- **Tip -** Stage 4 is a quality- and production-led process.

The four stages of QFD systematically ensure that the customer needs are translated into products that meet the needs, can be economically manufactured and validated as meeting the needs.

# The detail of QFD

Manually generating a QFD can take time to set up and organise the data into the right format. Fortunately, for the lazy ones amongst us, SigmaZone has taken a lot of the work out of the process with SnapSheets XL. This is a relatively inexpensive (<$50) piece of software that is flexible, easy to use and does most of the work for you. The software integrates into Microsoft Excel for easy distribution and use and allows you to focus on the process instead of the software.

The customer requirements are simply fed into the rows and the design requirements fed into the columns. The relationships are recorded in the relationship L-matrix and the interactions recorded in the Roof-matrix. Additional details on competitor and technical details are optional and can be added as desired. The software then automatically creates the ranks and importance of each factor. One of the great features here is that, when moving to Stage 2 for product design, the design requirements are automatically translated into a new matrix for the part characteristics and so on for Stage 3 and Stage 4.

- **Tip -** SnapSheets XL can also be used for Pareto analysis of the design requirements scores, Pugh concept generation and to create FMEAs (see Section 8.12) - all from the same software.

I have used QFD to great effect but I often wonder if the 'voice of the customer' is the only voice you should be listening to. For well-defined products and market needs it works fine but for radical new products the customer may not even know that they want the product until they see it.

Did any customer 'want' an iPad before 2010 or did they only want it after they had seen it?

QFD can be really powerful to get to the heart of the design needs but it does need some perseverance

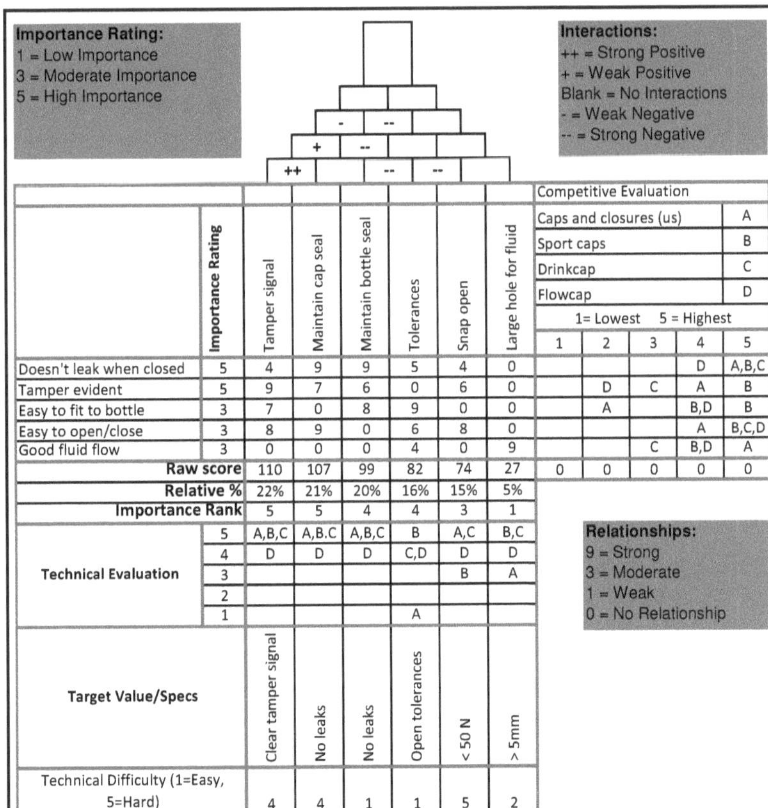


## A simple QFD for product planning of a closure

SnapSheets XL from SigmaZone (www.sigmazone.com) can be used to quickly generate QFDs for product planning and then copy the output of this straight into product design, process design and process control. Highly recommended.

# 8.11 Design FMEA - 1

## Introduction

Failure Modes and Effects Analysis (FMEA) is an essential tool for improving both product and process design. This section covers the use of FMEA for design but it can also be used for system design, process design and machinery design. These uses of FMEA will be covered later in this Workbook.

- **Tip** - Sometimes the Design FMEA is referred to as DFMEA to differentiate it from the Process FMEA (PFMEA). They both use the same process.

FMEA is a structured process of assessing, quantifying and reducing the risks associated with various elements of a design, i.e. it is a risk assessment process. The process assesses what features or failure modes could affect the product quality as perceived by the customer (wherever they are located in the supply chain) and how these risks can be reduced.

The important decisions are what design features are significant and the actions that have been taken (or can be taken) to minimise the risks. The FMEA approach documents these decisions

As with APQP, the AIAG (see Section 8.10) has produced an excellent book on the subject ('Potential Failure Mode & Effects Analysis') that is available from AIAG (www.aiag.org). Equally, SnapSheets XL from SigmaZone (see Section 8.12) can also do most of the hard work for generating a design FMEA (or any other type of FMEA).

- **Tip** - SnapSheets XL can also be used for Pareto analysis of the Risk Priority Numbers to decide on issues for action.

Although a formal FMEA risk assessment should always be part of the design review process, the FMEA should be developed as the design progresses. This allows design changes to be made to reduce risk and improve quality as the product is developed rather than after the design is finalised.

In quality terms, an FMEA provides an overview of the robustness of the product design, makes designs more fault-tolerant and leads to a natural development of the process FMEA and product control plan (see Sections 10.2 and 10.3).

## The FMEA process

FMEA is again a team process where the experience and knowledge of multi-functional teams is used to improve the design. An FMEA is based on asking the following questions:

- How can this product fail to perform as designed?
- What will happen if the product fails to perform as designed?
- How can we reduce the possibility of failure or the severity of the effect?

The process for creating an FMEA is:

### Define the scope

The first step in creating an FMEA is to define the scope of the process. This defines if the FMEA is being developed for a single item, for a sub-system or for a complete system. If you are dealing with a single item then the focus will be on the product and features. If you are dealing with a sub-system or a system then the focus will be on the interfaces and interactions, including the environment and customer.

- **Tip** - You need to understand what you are dealing with.

### Define the customer

The customer can be the next person in the supply chain or the end user, i.e. product quality is not always defined solely by the end user. If the product has to be treated, assembled or in any way handled by someone before the end user then their view of quality counts too.

- **Tip** - Don't forget that regulators are also customers.
- **Tip** - Don't forget that packaging is part of the quality of the complete product.

### Clarify the design intent

The design intent is what the product or system is designed to do and how it is going to do this. The PDS (see Section 8.2) and the output of the QFD process (see Section 8.10) should naturally provide a clear understanding of the function of the product, the requirements and the specifications.

### Identify potential failure modes

A failure mode is any way in which the

A design FMEA can remove many of the small errors in product design that cause poor quality as it is perceived by the customer.

Really, really great product designers probably do FMEA risk assessments informally and in their head.

The rest of us need to do it formally and on paper.

- Beware of trying to make a bad design work by changing the process.

AND EQUALLY

- Beware of trying to make a bad process work by changing the design.

A design FMEA directs the design effort to the critical characteristics and improves design verification to avoid late design changes. It also identifies the characteristics that need to be controlled in manufacturing to maintain product quality.

product can fail to meet the design intent. After the design intent is clarified then the failure modes should become obvious. These should be described in technical terms that are familiar to the design FMEA team and not necessarily in terms of the customer expectation from the PDS or QFD.

- **Tip** - Identifying a failure mode does not assume that it will occur, only that it may occur.

### Identify the potential effects

These are what the customer will see if the failure mode occurs. The potential effects should be described in terms of how the customer (wherever located) would see the failure.

- **Tip** - This can be linked back to the QFD via the target values for the design requirement and the customer needs.

The severity of the effect, as viewed by the customer, can then be assessed.

### Identify the potential causes

This considers each potential failure mode and how it could occur.

- **Tip** - There may be multiple potential causes for a single failure mode and they should be considered separately.

The identified potential causes should be 'root causes' that can be described in terms of design features, process parameters or other items that can be corrected or directly controlled. This is to ensure a direct link between the cause and the failure mode.

- **Tip** - Causes may not only be errors of commission but errors of omission.

The probability of the cause occurring can then be assessed.

### Identify the current controls

These are the planned controls to prevent or detect the causes of each identified failure mode. In a design FMEA this considers the features of the design and how they are implemented to prevent failure. Only if it is not possible to 'design out' the failure mode should process controls be considered.

- **Tip** - Controls can be focused on prevention or correction. Obviously prevention controls are preferable.
- **Tip** - Controls must be 'suitable and sufficient' which is not necessarily the same thing as perfect. It is 'suitable and sufficient' if reasonable precautions are taken so that the remaining risk is low.

The effectiveness of the current controls can then be assessed.

### Identify and prioritise the risks

The Risk Priority Number (RPN) is defined as the product of the severity, occurrence and control assessments (see Section 8.12). The RPN is used to identify and prioritise the risks.

### Recommend and take corrective or preventive actions

The prioritised risks should be assessed and action taken on the significant risks.

- **Tip** - Never set a minimum RPN number for action. Use Pareto analysis to tackle the top 80%.

### Re-assess risks after action

All risks should be re-assessed after action to check that the RPN has actually decreased and that the actions have been effective.

**The magic RPN**

What is the RPN threshold for action?

This question assumes there is a magic RPN number above which action must be taken. The reality is that action is a judgement and the company/ person can decide to take action on any risk. All the RPN does is to prioritise the risks via a defined system.

Process FMEA is covered in Section 10.2.

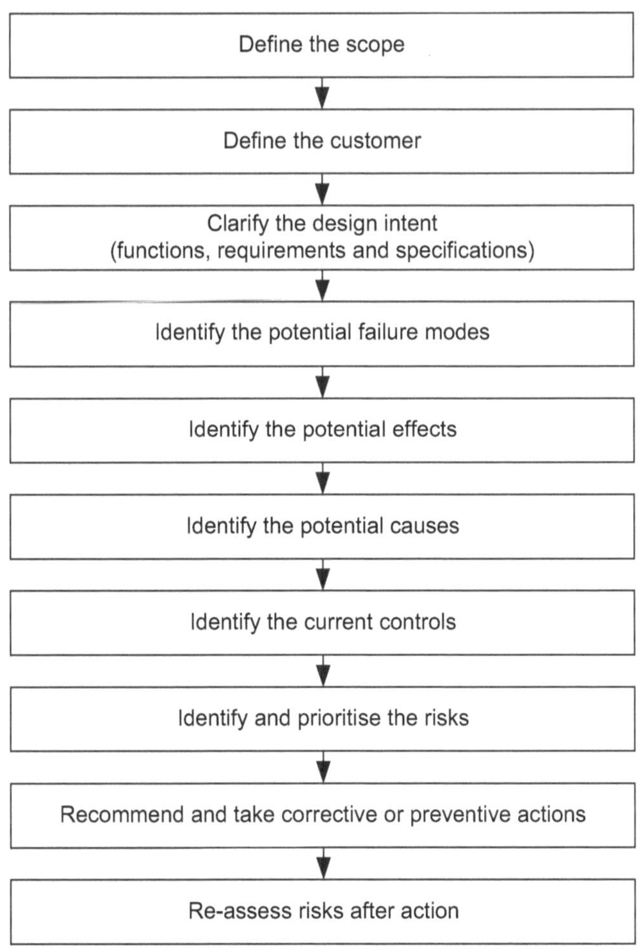


The FMEA process

The process is quite simple but getting the FMEA set up can take some time. Using SnapSheets XL makes it easy and quick to follow the process. The output of the design FMEA (causes) can be fed into the process FMEA to improve the production process.

# 8.12 Design FMEA - 2

## The detail of an FMEA

Section 8.11 reviewed the process of completing a design FMEA but did not consider the actual assessment. The assessment of severity of the effect, the probability of the cause and the current controls are based on simple 1 to 10 ratings.

Some typical definitions of what each rating means are given in the table below. These vary with the industry but those given below can be regarded as being generally applicable.

In some cases the FMEA also includes a 'class' indicator to show if the issue has 'safety/regulatory/legal' implications or 'warranty/field failure' implications. Any issues identified as 'safety/regulatory/legal' must have action taken and issues identified as 'warranty/field failure' are strongly recommended for action.

## Assessing the severity

Severity (S) is rated from 1 (low severity. i.e. little or no impact on the product function) to 10 (a major failure, i.e. failure without warning that has safety, regulatory or legal impact).

## Assessing the probability

Probability (P) is rated from 1 (low probability of failure, i.e. highly unlikely or < 1 in 1,000,000) to 10 (highly likely, i.e. almost inevitable or ≈ 1 in 2).

## Assessing the current controls

Current controls (C) are rated from 1 (high current preventive measures and known design) to 10 (new technology or totally new design with no service history).

## Risk Priority Number (RPN)

For each potential cause, the individual factors are rated and multiplied together (RPN = S x P x C) to produce the RPN.

Action can then be taken and recorded on the significant RPNs to:

- Reduce the severity.
- Reduce the probability.
- Improve the design or temporarily improve the controls.

The RPN is then reassessed after completing the actions.

- **Tip** - Action should be mandatory for any severity rating of 9 or above.
- **Tip** - A design FMEA should be reviewed and revised with time (it is a 'live document') to reflect new equipment, processes and procedures. This will allow the control plan to be reviewed with the experience gained.

The FMEA is a record of the analysis which can be used in future for other new designs.

The design FMEA forms the basis of the draft control plan for the product (see Section 10.3).

The FMEA is a 'live' document and needs to be maintained and kept up-to-date.

| Severity of event (S) | Ranking | Probability of event (P) | Ranking | Current controls (C) | Ranking |
|---|---|---|---|---|---|
| Hazardous: without warning | 10 | Very High: event is inevitable | 10 | Absolute uncertainty of detection | 10 |
| Hazardous with warning | 9 | | 9 | Very remote chance of detection | 9 |
| Very High | 8 | High: Repeated events | 8 | Remote | 8 |
| High | 7 | | 7 | Very low | 7 |
| Moderate | 6 | Moderate: Occasional events | 6 | Low | 6 |
| Low | 5 | | 5 | Moderate | 5 |
| Very low | 4 | | 4 | Moderately high | 4 |
| Minor | 3 | Low: Relatively few events | 3 | High | 3 |
| Very minor | 2 | | 2 | Very high | 2 |
| None | 1 | Remote: Event is unlikely | 1 | Almost certain to detect | 1 |

| | | | |
|---|---|---|---|
| Design | Sash Profile - Compact 121 | Date | 05/10/2015 |
| Prepared By | Robin Kent | | |
| Notes | No consideration of system interactions made at this stage | | |

| | | | | | Current controls | | | | | | Recommended actions | | | Action results | | | | | |
|---|---|---|---|---|---|---|---|---|---|---|---|---|---|---|---|---|---|---|---|
| Function / Requirement | Potential failure mode | Potential effect of failure | Sev (S) | Class | Potential cause | Prob (P) | Prevention | Detection | Controls (C) | R.P.N. | Recommended actions | Responsible | Target completion date | Actions Taken | Sev (S) | Prob (P) | Controls (C) | R.P.N. | Class |
| Seal to outer frame to prevent water ingress | Inadequate gasket compression | Sash leak | 6 | | Profile warping | 3 | Pre-production checking of straightness | QC checks on routine inspection | 3 | 54 | | | | | | | | 0 | WFF |
| | | | | | Inadequate gasket compression on design drawings | 5 | | Check design interference is < 2 mm | 1 | 30 | Gasket compression to be < 1.8 mm | MM | 21/03/2016 | Done | 6 | 2 | 1 | 12 | WFF |
| | | | | | Incorrect gasket selected by fabricator | 4 | Insert clause in system manual | Colour code (stripe) gaskets? | 2 | 48 | Manual updates and colour strip addition | GL | 21/03/2016 | Done | 6 | 1 | 1 | 6 | WFF |
| | | | | | Gasket thickness below specification | 3 | | Gasket control at Inwards Goods | 4 | 72 | | | | | | | | 0 | WFF |
| | | | | | Incorrect handle used with insufficient compression | 6 | Insert approved locking gear list in system manual | | 4 | 144 | Gear list updated | GL | 21/03/2016 | Done | 6 | 2 | 3 | 36 | WFF |
| | | | | | Incorrect keeps used on sash | 4 | Use only keeps specified for locking gear | | 2 | 48 | Keeps specified in system manual | GL | 21/03/2016 | Done | 6 | 1 | 1 | 6 | WFF |
| Hinge location and fitting | Hinge too large for hinge groove | Difficulty fitting hinge | 4 | | Extrusion closing during haul-off | 3 | Increase wall thickness in hinge groove area | Hinge groove check by QC | 4 | 48 | | | | | | | | 0 | WFF |
| | | | | | Extrusion tolerances at lower end of tolerance band | 4 | Increase design width of hinge groove to cope with maximum specification of hinge width | | 2 | 32 | | | | | | | | 0 | WFF |
| | | | | | Hinge width at top limit | 2 | Insert approved hinge list in system manual | | 4 | 32 | | | | | | | | 0 | WFF |
| | | | | | Profile distorting with overtightened hinge screws | 4 | Recommend maximum torque for screw fitting | | 3 | 48 | | | | | | | | 0 | WFF |
| | Hinge too small for hinge groove | Sloppy fit in hinge groove | 1 | | Extrusion tolerances at upper end of tolerance band | 4 | | Hinge groove check by QC | 4 | 16 | | | | | | | | 0 | WFF |
| | | | | | Hinge width at bottom limit | 4 | Insert approved hinge list in system manual | | 4 | 16 | | | | | | | | 0 | WFF |
| | Hinge location | Difficulty opening sash due to clashing | 5 | | Incorrect location of hinge | 3 | Produce and supply (FOC) gauges for hinge fitting | | 3 | 45 | Produce and supply (FOC) gauges for hinge fitting | TM | 21/03/2016 | Drawings done and contract arranged. | 5 | 3 | 1 | 15 | WFF |
| Flatness of extrusion faces | Visible faces are not flat | Visual failure | 2 | | Unbalanced extrusion shrinkage | 6 | Check balance of extrusion flow with die designer at die design stage | | 5 | 60 | | | | | | | | 0 | WFF |
| | | | | | Internal walls too thick for external walls. Breach of design rules | 6 | Design review of all internal wall thicknesses | | 3 | 36 | | | | | | | | 0 | WFF |
| | | | | | Uneven cooling of profile | 4 | Check cooling flow with calibrator designer at calibrator design stage | | 3 | 24 | | | | | | | | 0 | WFF |

Example of a Design FMEA using Snap using SnapSheets XL to generate Risk Priority Numbers

This is not a complete example and the actual design FMEA is much longer and detailed. It is shown only to illustrate the method and the output.

# 8.13 Design and other drawings for plastics products

## Product design drawings

Product design for plastics processing will generally be carried out on a 3D CAD model and any subsequent drawings created from the 3D CAD model. This has many advantages in design terms.

- The 3D CAD model can produce colour-rendered 3D visualisations to allow non-technical people to see what the finished product will look like from a variety of angles. The 3D CAD model can also be exported to a 3D printer for production of a rapid 3D model of the part.
- All drawing views are produced from the same 3D CAD model and are exactly the same in design terms. This can make it easy to look at detailed areas for interactions.
- Product properties, e.g. volume, weight, etc. can be accurately calculated from the 3D CAD model to give better cost and mechanical property calculations.
- 'Fit and function' with mating parts can be easily checked by importing the 3D CAD model of the mating part.
- 3D geometries can be exported to simulation packages for mould filling simulation, extrusion flow simulation and Finite Element Analysis (FEA).
- Product details and drawings can be exported direct to toolmakers for tool design and construction or to other suppliers for 'fit and function' testing.
- Quality and tolerances can be improved by removing any ambiguities in various drawings.
- **Tip -** All drawings should be made in 3D CAD, even for the most basic products to allow them to be used by toolmakers.
- **Tip -** Designers should declare on any (and every) drawing derived from the 3D CAD model that the 3D CAD model is the 'Master' drawing and that any dimensions not shown should be taken from the 3D CAD model.

## Tool design drawings

Despite these advantages, the design of plastic parts is not straightforward and is more difficult than for parts from other materials. The major reason for this is the shrinkage associated with any plastics forming process (see also Section 8.14).

The decision on the exact grade of plastics to use therefore needs to be made very early in the product design process and clearly communicated to the toolmaker, simply specifying 'PP' is not good enough for a toolmaker to do his job.

Designers have to think not only about design but also about production and QC.

Tool design drawings will be used to manufacture the tool and in most cases these will be created by the toolmaker from the detailed product design model. The translation of the finished part drawing (post-shrinkage) into the final tooling (pre-shrinkage) is the responsibility of the toolmaker and, even if presented with a full 3D model, the toolmaker will need to know the specific grade of material being used to account for shrinkage after processing.

Flow simulation for the finished part design and tooling configuration can be carried out by most toolmakers and this is essential for good part and tool design.

- **Tip -** Beware of multiple cavity tooling and flow simulation. This needs careful consideration.
- **Tip -** Always model to the middle of the tolerance range, i.e. the nominal dimension. Do not model to the upper

Drawings should also call up any tests (on-line or off-line) that are to be carried out on the part.

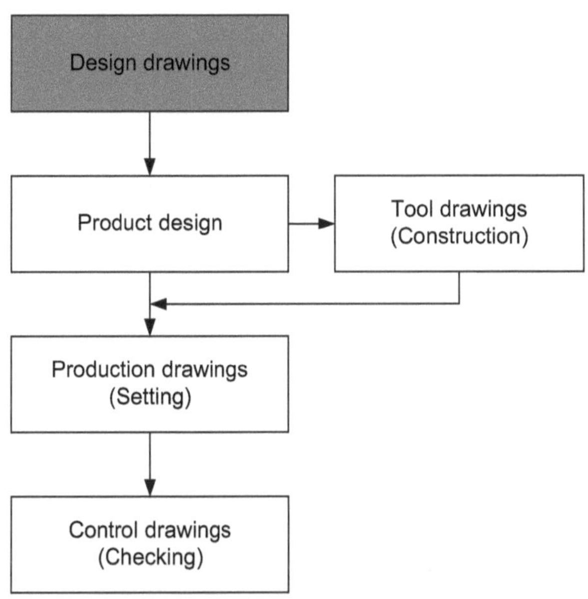


### The layers of drawings for plastic parts

Design drawings will be converted into tool drawings (generally by the tool manufacturer) for tool construction but there is also a need for production drawings (for production approval) and control drawings (for routine QC checking).

limit of the tolerance unless absolutely required and then always note this on the drawing.

- **Tip** - Product designers must work with toolmakers and it is strongly recommended that the toolmaker is selected very early in the project and is part of the design team. It is absolutely essential that the chosen toolmaker is part of the final design review. Their input is vital and can determine the success or failure of a project.
- **Tip** - One of the critical areas for injection-moulded product design is in the area of draft angles for product ejection. Talking to toolmakers early and often about how the tool is going to be made and how much draft they need is essential (see Section 8.14).
- **Tip** - If the toolmaker needs increased draft then always check that the part will meet 'fit and function' requirements with the specified draft angles.

## Production drawings

Product designers must be aware of the need for capable and consistent product measurement at any stage of the production process.

Production drawings need to specify dimensions that are capable of being checked by machine setters and other production staff for initial part approval and for both initial and start-of-run machine setting. These drawings will also be used to validate the tool and the process and this means there will generally be more time available to the person checking the product. At this stage 3D-coordinate measuring using either contact (probes) or non-contact (optical scanners) methods may be used and the number of dimensions listed will be quite high - but obviously not as high as on the 'Master' drawing.

Sites should therefore have a separate set of 'Production Drawings' that call up only the dimensions that will be checked and validated at initial machine start. These will be based on the 3D CAD model but will not give every dimension but only the key dimensions to be checked at this stage.

- **Tip** - It is essential that the production drawings call up dimensions that are physically capable of being measured either by 3D methods or by hand.
- **Tip** - Drawings that are not capable of being measured to check conformance to specification are a waste of time and effort.
- **Tip** - Always use the 3D CAD data and the material properties to specify the approximate weight of the part.

## Control drawings

Even with the best use of process control, it will often be necessary for QC to physically check the final product for conformance to the design drawings. This requires the production of 'Control Drawings' that can be used by QC to physically measure the key dimensions of the product to check that the process is in control. These drawings will generally only have a very few critical dimensions marked, i.e. the dimensions that QC will measure on a regular basis for process control.

Sites should therefore have a separate set of 'Control Drawings' for use by QC personnel that call up only the dimensions that will be checked by QC and that are physically capable of being measured.

- **Tip** - In some cases, design drawings call up dimensions that are physically impossible to measure. It is only after you have tried to validate the length of a part which is specified to the intersection of two projected lines that you really feel frustration! (see box below).
- **Tip** - A 'Control Drawing' should only give dimensions that are capable of being physically measured. QC should have the final decision on what is capable of being measured on the part as produced.

It may sound obvious but getting the drawings right in the first place and getting them physically measurable is a major task.

**Rule for control drawings:**

- Always use A4 size so that it is portable.
- Mark visible surfaces.
- Only mark control dimensions (with tolerances).
- Never put more than 6 dimensions on the control drawing.
- If gauges are used then mark the number of the gauge on the drawing.
- If surface finishes or colours are to be checked then provide samples.
- Always call up any tests (on-line or off-line) that are to be carried out on the part.
- Refer to mating parts (if there are any).

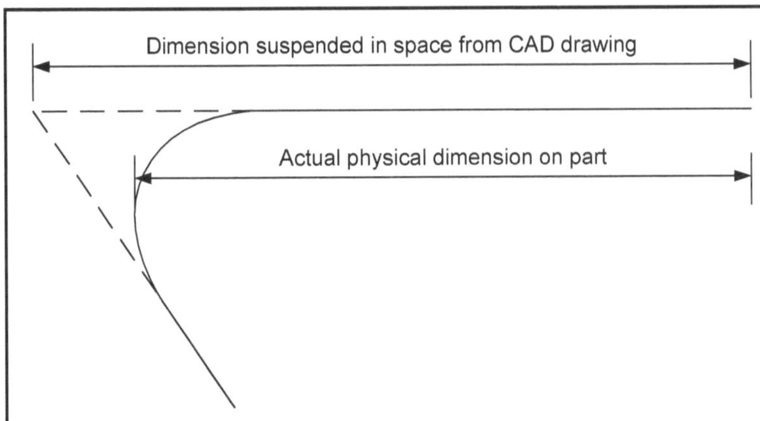


Dimensions must always be capable of physical measurement

Drawings for production and QC checking should always specify dimensions (however derived) that are capable of actually being physically measured - you can draw it but can they measure it?

# 8.14 Plastic parts and tolerances

## What can we achieve?

Selecting and assigning tolerances is not a simple academic exercise carried out by the designer, it has a direct effect on the perceived and actual quality of the product and also a direct effect on the cost of the product. As we learnt in Chapter 5, every process will experience variation and the process output will vary slightly over time. This is true for plastics processes but plastics processing has the additional complication of shrinkage of the plastic which increases product variation and therefore increases the tolerances that are actually achievable.

- **Tip -** Simply taking the tolerances from a metal part and applying them to a plastic part is a recipe for disaster.

It is always necessary to mark tolerances on drawings to indicate the critical dimensions and what will and won't work in the finished article. In most cases the designer will use some type of 'tolerance table' to assign the tolerances based on the type of dimension, the size of the dimension and, for plastics materials, on the type of material. This is all commendable but it is largely irrelevant in most actual operations. I admire the people who set out tolerance tables and work on the arcane principles and methods of geometric tolerancing but the really important point is: 'What is your process capable of doing?' Process capability was discussed in Section 5.22 and is the only thing that counts.

Tolerance tables are based on standard industry practice and are representative of what the broad industry can achieve, they are not necessarily what you can achieve or they may be much worse than what you can achieve. Unfortunately, most companies have no clear idea of what they can actually achieve in practice. If they can perform to the standards of the people who set up the tolerance charts then good. If they cannot then the simple fact of using these tolerances has condemned the production area to a life of 'poor quality' whatever it tries to do.

- **Tip -** It is vital to understand what each specific process can achieve in terms of tolerances so that the appropriate tolerance can be assigned.

## Shrinkage

Shrinkage in plastics products is not uniform and it depends on:

- The exact material being used, i.e. every different material type and grade will have a different shrinkage. This makes early material selection and definition essential. If the selected material has a high and variable shrinkage then some designs may simply not be feasible.
- The length/dimension being considered, i.e. the amount of shrinkage will generally be given in the form 0.XX m/m or '% shrinkage' and longer dimensions will suffer from greater shrinkage and therefore need greater tolerances.
- The direction being dimensioned in relation to the flow in the tool, i.e. for many fibre-filled materials the shrinkage will be greater in the transverse direction than in the flow direction.

Shrinkage makes the tolerancing of plastic parts more difficult than with other materials but in most cases this 'problem' is dealt with by the toolmaker although the designer can assist by considering shrinkage at the design stage.

There is no design that cannot be made worse and more difficult to produce by over-enthusiastic tolerancing at the design stage.

If a process cannot meet the assigned tolerances then, where possible, check the design and change the tolerance to meet the process capability.

This is easier, more reliable and cheaper rather than constantly inspecting and rejecting a product that you cannot reliably make to specification.

### Tolerances

'.........It looked as if some of the tolerances were assigned much closer than should be necessary, and I started to try to find out how they had been fixed. To accomplish this, on every occasion when I met an engineer, I asked him how he decided the tolerances in his branch of the subject: I fear I bored a great many people at this time. I got a variety of answers which sometimes explained things a bit, but often not at all, and I discussed it with quite a number of men, many of them occupying prominent places in different branches of the profession. I came away with the impression that scarcely any of them were really interested in the subject of tolerances. To exaggerate the picture which I got as a result of my inquiry, I concluded that in designing a new machine the chief engineer drew it free hand with dimensions to the nearest inch, and sent it to the draughtsman to work out the details to the nearest thousandth, who then gave it to his junior assistant to mark in the tolerances. Instructions were certainly always given that the tolerances should be as easy as possible, but only lip service was done to them, and the junior assistant, anxious not to get himself into trouble, would, as a general rule, think of the smallest number he knew and then halve it.'

C.G. Darwin - Statistical Control of Production, Nature, vol. 149, pp. 573-575, May 23, 1942.

Things have changed a lot since 1942, haven't they?

- **Tip** - For injection-moulded parts, shrinkage is also affected by the processing conditions. Changing the process parameters will affect shrinkage.
- **Tip** - When using semi-crystalline plastics, e.g. PP or PE then beware of post-moulding shrinkage due to crystallisation.

## The cost of tolerances

Tolerances are not 'cost-free'. They cost nothing to mark on a drawing but tight tolerances will inevitably raise production costs and/or reject rates, both direct costs to the company.

If tolerances are divided into 3 classes, as per many of the standard tolerance tables, then the cost of achieving these tolerances is approximately as follows:

- Normal: Standard cost.
- Medium: 1.5-1.7 x Standard cost.
- Fine: 3 x Standard cost.

Excessively tight tolerances will not only increase the cost of producing the part but will also increase the cost of the basic tooling by approximately the same ratios.

- **Tip** - Designers should only indicate tolerances for dimensions that are critical for fit and function and leave all other dimensions as 'free' or 'normal' tolerances where the toolmaker is free to derive the dimensional data from the 3D CAD model (the master).

## Standard tolerance tables

The standard tolerance tables in most companies are generally derived from a variety of sources:

### Injection moulding

There is a lack of any international standard for injection moulding of plastics. Some standard tolerance tables available are: BS 7010:1988, DIN 16742: 2013 (which replaced DIN 16901), NF T 58-000: 1987 and for those who want to work in the Imperial system there is the invaluable 'Standards and Practices of Plastic Molding' (SPI: 1998). These are generally quite old and do not reflect improved machines and processes. There is a proposal to take DIN 16742 forward for consideration as an ISO standard.

- **Tip** - Draft angles need careful discussion with toolmakers. The draft angle needed depends on the surface finish, if a surface finish is specified then this may affect the necessary draft angle.
- **Tip** - The number of cavities in an injection moulding tool also affects the tolerances. As a 'rule of thumb' increase the tolerances by 5% for each additional cavity, i.e. if you have a 20-cavity tool then the achievable tolerances will be 20 x 5% = 100% those for a single cavity.

### Extrusion

Extrusion is a complex market ranging from film and sheet (which generally only has one effective tolerance) through to profile extrusion for rigid or soft profiles. The main tolerance standard dedicated to plastics extrusion is DIN 16941: 2012-11. Many extrusion companies processing soft profiles use standards such as ISO 3302:2014 or DIN 7715:1979-11 for tolerance tables even though these are standards for rubber products.

- **Tip** - Extruded products will have some degree of 'die-swell' after exit from the die. This can be reduced by drawing down the product by the haul-off but balancing these effects is critical. They are not always uniform across a profile.

## Generate your own tables!

The best tolerance tables are those generated internally by considering the process capability of the main processes. It is relatively simple to generate your own internal tolerance tables and much of the information you need will have already been gathered for other QC reasons.

It is simply a matter of taking historical QC capability study results and converting these into more conventional tolerance tables that can be used by designers to reflect the achievable tolerances at the site.

- **Tip** - Internally generated tolerance tables can also include information on the cost of meeting the given tolerances.
- **Tip** - Processes can be at various levels:
  - Normal running.
  - Normal running and well set.
  - Normal running, well set with no process disturbances.

  The level will vary at every site depending on the setters and the current situation. Designers should understand which level the production site is operating at.
- **Tip** - Beware of using capability study data from a single cavity tool for multiple cavity tools. The variation will be greater for a multiple cavity tool.

We have not considered the build-up of tolerances in this section. This is a complex subject that is beyond the scope of this Workbook.

The number of tolerances marked on a drawing will also affect the process yield.

Increasing the number of specified tolerances decreases the process yield - this decrease is linear for up to about 5 specified tolerances but for > 5 specified tolerances the process yield decreases rapidly.

For a process with a normal distribution, then $C_p$ gives an estimate of the approximate proportion of products that will be out of specification (assuming a centred process), i.e. if $C_p = 1.00$ then 0.27% of parts produced will be out of specification, if $C_p > 1.33$ then it is possible to produce all parts in specification (see Chapter 5).

# 8.15 Designing quality into the product - 'off-line' quality control (DOE)

## Complementary not conflicting

The activities described to date have dealt with 'on-line' quality control, i.e. activities that take place inside the process or on the product. There are, however, complementary techniques of 'off-line' quality control that can be used to improve product design quality. 'Off-line' control methods are used mainly at the design stage for either product or process design - they are not used 'on-line' at the production stage to control the process.

The main 'off-line' quality tool is Design of Experiments (DOE) and this was first developed by R.A. Fisher in his book 'The Design of Experiments' (1935). Conventional experiments are 'one at a time' where one variable is evaluated at a time. This can lead to a high number of experiments that takes a long time, e.g. for 7 variables with 2 possible levels (high/low) each, then you would need to perform $2^7$ or 128 experiments. DOE was developed to allow experiments where multiple variables were changed 'all at once' and then the total effect of each variable evaluated, e.g. for 7 variables with 2 levels (high/low) each then DOE gives the same results with only 8 experiments. This dramatic reduction in the number of experiments needed means that experiments to increase the robustness of designs can be carried out quickly and easily.

These 'traditional' DOE methods have not remained static but have been improved by statisticians such as Bose, Cox and Cochran, Box and Taguchi.

Genichi Taguchi[1] developed a similar approach to classical DOE but was much more focused on implementation and his 'Taguchi Methods' were promoted and widely used in industry. There has been some criticism of the detail of Taguchi's work but it is undoubted that he changed the way industry looked at the design of experiments. Taguchi changed some elements of classical DOE and introduced the concepts of 'loss', screening, targets and 'noise' factors (to estimate variance).

DOE can be used to improve product or process design. In this section we will focus on product design using DOE.

## Why and when to use DOE?

DOE can be used either independently or as a follow-on from a design FMEA where a design element needs further work to make it more robust, e.g. an excessively high RPN with no obvious method of reduction. In contrast to other methods which seek to reduce or remove the 'cause of the effect', DOE seeks to reduce the 'effect of the cause', i.e. it can identify the main sources of variation in a design and to assess how these can be effectively controlled at the design stage. Reducing the effect of the cause makes designs more robust and resistant to outside variations.

Adjust the design not the process.

## Taguchi's concepts

### The stages of 'off-line' QC

As shown in the box below, Taguchi uses 3 stages of design:

- System design: this is the conceptual and non-statistical part of product and process design. It is about selecting the technology and concepts for a prototype design that meets the PDS.
- Parameter design: this is the design and

R.A Fisher developed much of the theory for Design of Experiments at Rothamstead Research where his work was centred on analysing data from crop field experiments.

Rothamstead is about 12 km south of where I sit typing this at the moment.

It is good to feel close to the greats.

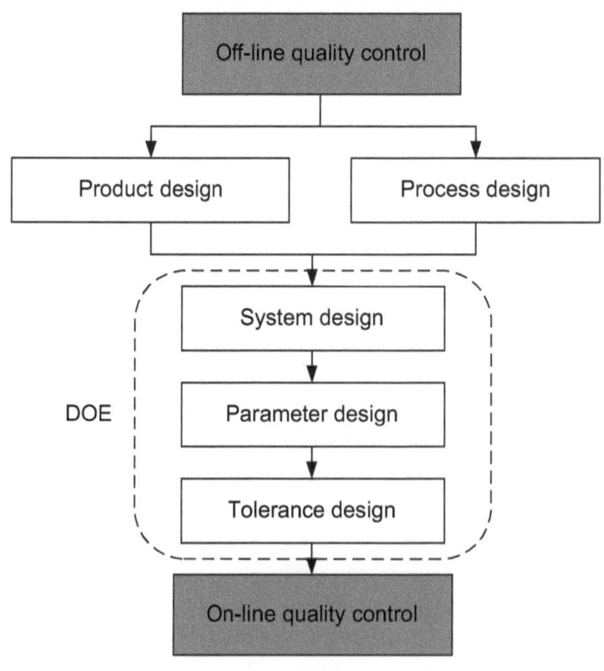


**The relationship between 'off-line' and 'on-line' QC**

'Off-line' QC does not replace traditional 'on-line' QC. It simply makes the design and the process more robust and less likely to vary with inevitable changes in the inputs. Tolerance design is only generally used if parameter optimisation is not successful.

use of experiments to find the optimum level of each system parameter determined in system design, e.g. dimensions, material types.

- Tolerance design: this is where experiments are used to set the tolerances of the parameters defined in the parameter design stage (if necessary). This is only to be used if the parameter design stage does not give sufficiently robust results.

The main objective of the process is to optimise the parameters to give a 'robust product'.

## The signal to noise ratio

In parameter optimisation, we can design the product for one of three responses:

- Type B: Bigger is best, e.g. strength at break.
- Type S: Smaller is best, e.g. deflection under load.
- Type N: Nominal is best, e.g. the nominal value for the characteristic.

Taguchi uses a variable 'signal to noise' ratio for each response, via an 'outer array' to represent uncontrollable noise factors. This is subject to some debate!

## The loss function

Traditionally, designers have set values above or below which a product is unacceptable (the USL and the LSL). All products produced inside these limits are defined as being equally good. And there is no reason to improve beyond getting everything inside the specifications.

For cases where 'nominal is the best', Taguchi introduced the concept of the 'loss function' which states that there exists an ideal or target value for the characteristic (although Fisher was scathing in his response to 'loss functions' in DOE). Any deviation from target value incurs a loss that increases the further away from the target you are. The LSL and USL are regarded as simply points on the loss function (see boxes on the right). This is closer to reality in terms of how customers assess products but is not addressed in conventional tolerances.

## Are all variables equal?

Not all product variables are equal in their effect on the product characteristic to be optimised. Some variables will mainly affect the mean (the location) and some variables will mainly affect the standard deviation (the spread). The results of DOE can be used with analysis of variance (ANOVA) to identify and quantify the relative effect of each parameter on the mean and the standard deviation. It is then possible to analyse the results to show the value of each parameter to optimise location and minimise variation.

## More on DOE

DOE can be extraordinarily powerful in developing robust and optimised designs but is little used (even though there are many impressive case studies). A full treatment of DOE in product design is beyond the scope of this Workbook but designers are urged to consult one of the many longer books on this subject.

Development of DOE did not stop with Taguchi and statisticians advanced DOE into Response Surface Methodology (RSM) to produce experiment designs that reduce experiment complexity even more and give more relevant results.

•1. Taguchi et al. 2004. *Taguchi's Quality Engineering Handbook*. Wiley.

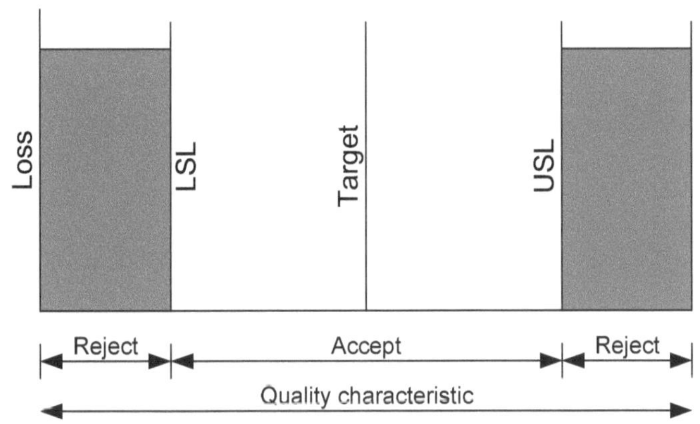


### The conventional view of loss

The conventional view of loss is that the product is 100% satisfactory when it is within the LSL and the USL but then loses 100% of its value when it is less than the LSL or greater than the USL. Whilst useful for production and accounts this is not reality.

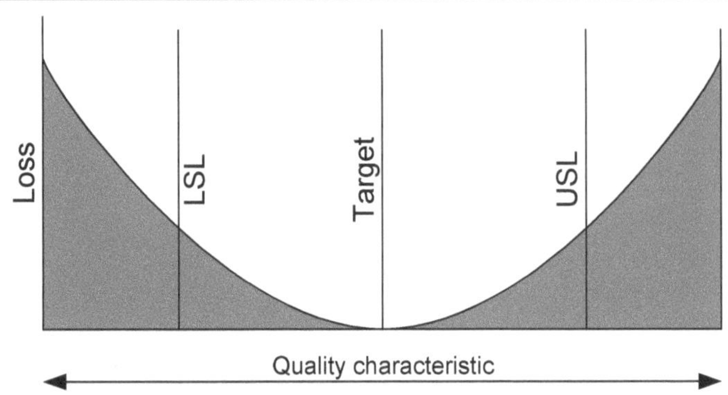


### The Taguchi view of loss

Taguchi takes the view that a product is fully valuable (100%) only when it is at the target value but it then loses value increasingly as the quality characteristic (however defined) moves away from the target value. This is closer to reality.

# 8.16 Design tools - where are you now?

## Tools

The design process is the start of producing a good-quality product. If the design is not correct and suitable for production then no amount of effort by the production or QC functions will produce a high-quality product. Unfortunately, much of the effort in quality management is directed towards the production area and very little effort is spent on ensuring that the quality of the design is suitable. In most companies if the Quality Manager is in the design area then he is either lost or there is an ISO 9001 inspection due!

Getting a high-quality and easily produced product is worth the extra effort in the design stage and the tools to do this are freely available but not often used. Get to know these tools, implement them and ensure that production is easy and trouble-free.

## Completing the chart

This chart is completed and assessed as for the previous charts.

"As tools develop, the content will be what distinguishes the winners/ leaders."

**William Raduchel**

**Design tools**

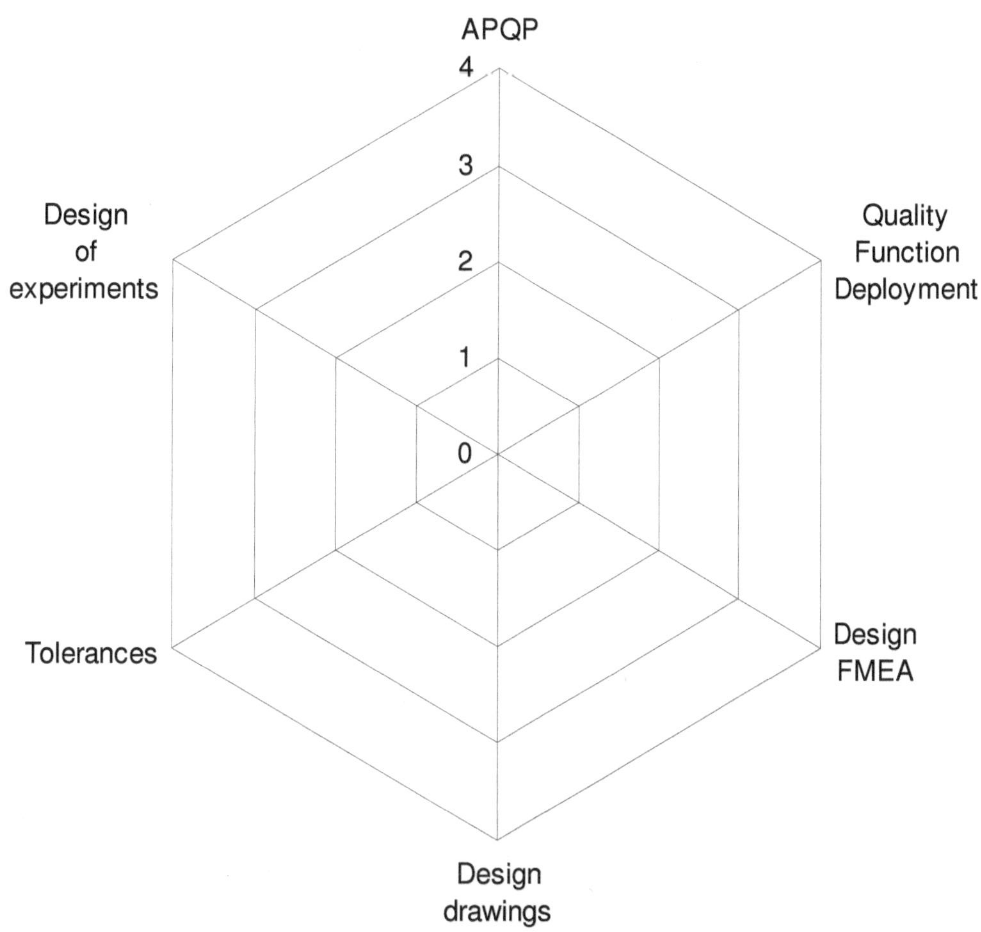


Use the chart to assess where you are in design tools

The numbers from the self-assessment should be transferred to the radar chart for a quick visual guide to where you are in the basics of design tools.

## Design tools

| Level | APQP | Quality Function Deployment | Design FMEA | Design drawings | Tolerances | Design of experiments |
|---|---|---|---|---|---|---|
| **4** | Full APQP process (or similar) used. All documents validated before progress to next stage. Feedback & assessment carried out. | Full customer requirements translated into design requirements. Full QFD analysis (through to process control) carried out. | Full design FMEA done. Effects, causes & controls well identified. Preventive action taken for identified risks & risk re-assessed. | Design carried out by well-trained designer. Production & control drawings produced to match FMEA, control plan & other requirements. | Tolerances analysed for cost of conformance. Based on internal capability studies. | Extensive use of DOE. DOE is used whenever the FMEA indicates a need to improve the robustness of the design. |
| **3** | Full APQP process (or similar) used. All documents validated before progress to next stage. No feedback & assessment carried out. | Most customer requirements translated into design requirements. Partial QFD analysis carried out. | Full design FMEA done. Effects, causes & controls identified but little preventive action taken & poorly documented. | Design carried out by well-trained designer. Production & control drawings produced but do not match FMEA, control plan & other requirements. | Tolerances checked, validated & increased if too tight. Tolerances from standard tolerance tables. Some reference to internal capability. | Some use of DOE but not regular & not always used when FMEA indicates a need to improve the robustness of the design. |
| **2** | Formal design process available & most basic steps carried out. Design process is consistent for most new products. | Only explicit customer requirements translated into design requirements. Some aspects of technical difficulty assessed. | Partial design FMEA done. Not all significant potential effects, causes & controls identified. No documented preventive action taken. | Design carried out by well-trained designer. Production drawings produced but no control drawings produced. | Tolerances checked & some changes made. Tolerances from standard tolerance tables. Applied with no consideration of internal capability. | DOE used sporadically in the past but no longer used even when FMEA indicates a need to improve the robustness of the design. |
| **1** | No formal design process available but some basic steps carried out. Design process is inconsistent for new products. | Poor attempt to translate customer requirements into design requirements. Little consideration of process design or control. | Design FMEA is ad hoc & carried out only by designer using internal knowledge. No documented preventive action. | Design carried out by average designer. No production or control drawings produced. Production & QC use modified copies of master. | Tolerances based on single assessment. Some checked but few changes made. Internal tolerance tables used with no relation to standards or internal capability. | Some knowledge of but no use of DOE to produce robust designs |
| **0** | No formal design process used or available. Design process is ad hoc & variable. | Designers work alone & use their interpretation of customer requirements. No consideration of process design or control. | No design FMEA carried out for any design concept. | Design carried out by poorly trained designer. No production or control drawings produced. Production & QC use copies of master. | Tolerances based on single assessment. No checking or validation. Internal tolerance tables used with no relation to standards or internal capability. | No knowledge or use of DOE to produce robust designs. |
| **Score** | | | | | | |

# 8.17 Design teams and processes - where are you now?

## Teams and processes

Effective management of the design process is covered in greater detail in an earlier book[1]. At this stage we do not intend to go over this information again but the 'where are you now?' assessment is still valid for design quality management and we will include it here for companies to assess their overall management of the design process.

Design is a process and as such it is as manageable as any other process. It may be creative but it can, and must, still be managed.

The best designs are a fusion of the creativity of the designer with the practical knowledge of the manufacturing people and all controlled by effective management of the process.

Good design for plastics processors is not an abstract concept; it is an essential element of quality, productivity and success.

## Completing the chart

This chart is completed and assessed as for the previous charts.

•1. Kent, R. 2012. *Cost management in plastics processing*. PID.

"A good design will fail if improperly processed."

**William T. Bean**

**Design teams and processes**

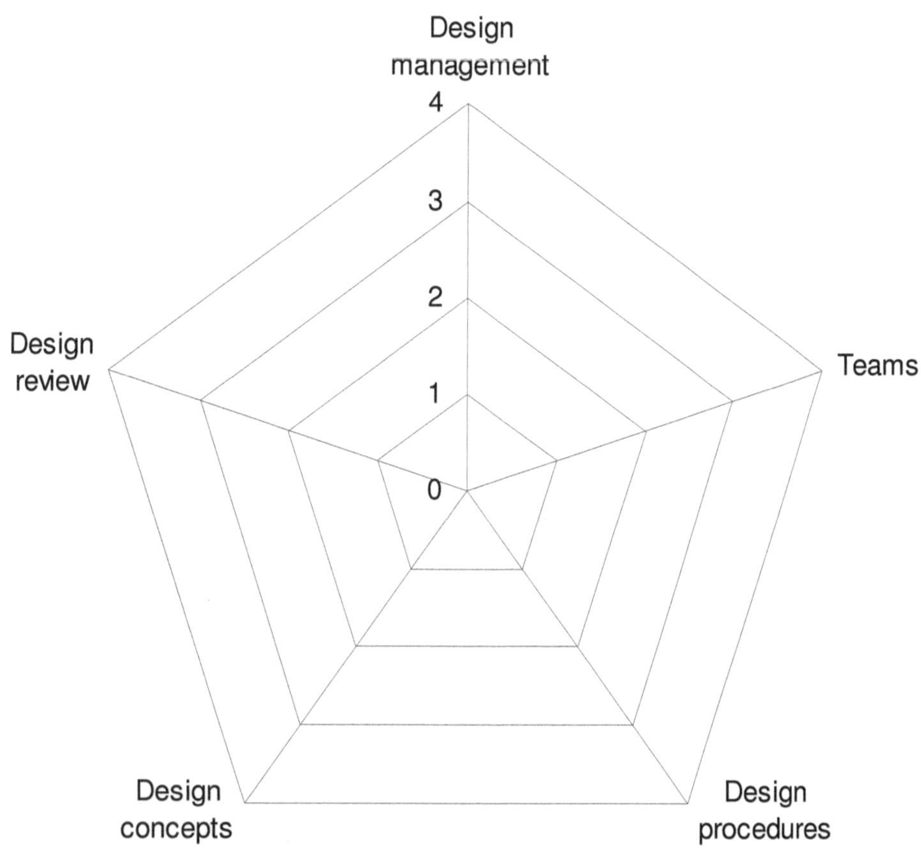


Use the chart to assess where you are in design teams and processes

The numbers from the self-assessment should be transferred to the radar chart for a quick visual guide to where you are in the basics of design teams and processes.

| Design teams and processes | | | | | |
|---|---|---|---|---|---|
| **Level** | **Design management** | **Teams** | **Design procedures** | **Design concepts** | **Design review** |
| **4** | Board Member responsible for design function. Consultants used to extend design capability. Continued training for designers. | Multi-disciplinary teams from all areas work together during the design process. Core team follows project through to completion. | Design procedures clearly written, widely available & rigorously followed. Essential design data are readily available to designers. | Essential & desirable features clearly identified. Concepts evaluated against technical & economic criteria. Design risk assessment carried out. | Formal design reviews regularly held to evaluate design against technical & economic criteria. Clear recording of actions taken. |
| **3** | Board Member responsible for design function. Consultants very rarely used. Design staff well qualified but have no continuing training. | Design by designers with substantial input from other areas. Designers responsible for taking project into production. | Design procedures clearly written, widely available but not always followed. Essential design data are not always available to designers. | Essential & desirable features clearly identified. Concepts not consistently evaluated against technical & economic criteria. | Formal design reviews held to evaluate design against technical & economic criteria. Record of actions taken is often unclear. |
| **2** | High-level manager responsible for design function. Consultants never used ('not invented here') & design staff poorly trained. | Design by designers with considerable input from other areas. Major decisions taken by designers but getting it to work is production's problem. | Design procedures have substantial gaps, not widely circulated or followed. Essential design data often not available to designers. | Some reference made to essential & desirable features but not consistent. Design concepts based primarily on internal ideas. | Informal design reviews infrequently held to evaluate design against technical & economic criteria. Poor record of actions taken. |
| **1** | Mid-level manager responsible for design function. Often overruled by higher managers. Projects poorly managed with little structure. | Design by designers with little input from other areas. When tooling arrives then getting it to work is production's problem. | Design procedures are informal, poorly written & rarely adhered to. Procedures are for show only. No essential data available for designers. | Little reference made to essential & desirable features. Design concepts based on internal ideas only. | Informal design reviews rarely held to evaluate design against technical & economic criteria. No record of actions taken. |
| **0** | Low-level staff with little authority responsible for design function. Projects are not so much managed as evolve with little control. | Design by designers with no input from other areas. Arrival of tooling in Production area is a surprise to production staff. | Design procedures are non-existent & essential data on existing own & competitors' products not available. | No reference made to essential & desirable features. Design concepts based on internal ideas only. | No formal or informal design reviews held during design process. |
| **Score** | | | | | |

# Key tips

- The majority of failures in plastics products are from poor specification, material selection and design. These are entirely avoidable and classed as self-inflicted injuries.
- Aesthetic design is only part of the story, the journey from great aesthetics to great product is a long and difficult one.
- The specification of what the product should do, the Product Design Specification (PDS), is just as important as the actual design itself. It sets out the design intent, i.e. what the product should do.
- The PDS can be quickly created using checklists and standard procedures.
- Material and process selection are interrelated and interdependent. Getting the material and process selection right in the first place makes a lot of other things easier (including the rest of your life).
- Beware of using single-point short-term data to assess variable loadings and long-term results. They are not the same and sometimes bear no relation to one another.
- The basic design rules for most plastics processes are similar.
- Use checklists to make sure that you have covered all of the design basics.
- Ignore the design basics at your own peril.
- Design for X (manufacture, assembly, recycling and sustainability) are becoming essential tools for cost reduction and easy production.
- Design for manufacture is an essential part of the design - if it is pretty but cannot be made then the drawing is art and not a product.
- Most parts made from plastics will eventually be part of an 'assembly' and designing for the assembly operation is an essential part of the design process.
- Most plastics part are designed for short life cycles and failing to design for recycling and sustainability is bad for the product, you and the rest of society too!
- Design is a team sport and a process and needs to be managed as such. There are existing proven systems and processes that can make this easier for designers.
- Advanced Product Quality Planning (APQP) is a proven design management framework/process for many industries.
- The customer cannot be expected to talk in the same terms as the plastics processor. Their view is different and they don't want a plastic part, they want a solution. Quality Function Deployment (QFD) is a method of taking the customer's desire and translating this into a technical specification that the designer can work with.
- QFD can also be used to take the designer's output and translate this into a production-oriented process.
- It is easiest to use custom software to produce a QFD and this is relatively inexpensive.
- Failure Modes and Effects Analysis (FMEA) is one of the keys to a good design by acting as a pre-release fault-finder. It identifies potential faults and provides a process to make reduce the probability of these happening.
- A design FMEA can be done quickly and easily using the same software used to produce a QFD.
- Every product needs drawings. Time spent getting the drawings in shape so that they can be used by the toolmaker, production and quality control is never wasted.
- Tolerances cost money and very tight and unnecessary tolerances are a huge cost to any company.
- Tolerances based on 'standard' tolerances take no account of internal process capabilities.
- The best tolerances are based on internally generated capability studies that show the real capability of the process.
- When it is not possible to reduce the 'cause of the effect' then Design of Experiments (DOE) can be used to reduce the 'effect of the cause'.
- DOE is a versatile technique mainly used in production but it has excellent uses at the design stage to allow the production of robust designs.

# Chapter 9

# Raw materials quality management

The raw materials for plastics processing are primarily the resins but because many plastics parts are assemblies, this chapter will also deal with controlling the quality of any bought-in part for the complete assembly, i.e. springs, screws, or anything else that is incorporated into the final product. Failure or poor quality of any bought in part will lead to failure of the assembly and the quality of these must be controlled.

The starting point for good-quality raw materials is obviously good-quality suppliers. Establishing and maintaining methods for selecting, communicating with and controlling suppliers is a key element of making sure that incoming raw materials do not adversely affect the quality of the finished product. This is not simply about conforming to the ISO 9001 requirements but about adequately defining what you expect from all suppliers and then monitoring that they have delivered to your specification. This is not limited to physical goods but also covers services such as computer support, transport and other non-physical supplies that impact on your ability to deliver a quality product and service to your customers.

After adequately defining the supplier requirements, it is also necessary to check that the suppliers are delivering to the requirements and this generally involves either physical inspection at Inwards Goods level or examination of supplier-provided documents to check/prove conformance with the agreed specification.

It is no longer acceptable to have simple conformance to specification, suppliers must also be able to provide consistency, capability (see Chapter 5) and credibility, i.e. that they are capable of delivering what they say they will.

This needs a good relationship with the supplier to reliably ensure good-quality raw materials and the customer/supplier interface must deal with all of the following:

- Contracts.
- Specifications.
- Test methods.
- Quality data.
- Sales calls.
- Complaints.

Raw materials are one of the keys to quality management in plastics processing and need careful management.

**Note:** We will not discuss the actual tests used for raw materials quality management in detail. There are many texts which cover materials testing and the reader is advised to consult one of these.

**GIGO: Garbage in = Garbage out**

# 9.1 The basics

## Getting it right at the start

As the old phrase GIGO states, Garbage In = Garbage Out. This was originally developed for the computing industry but applies equally well for any industry and is especially apt for the plastics processing industry. If the raw material is not correct then the product will not be correct no matter what production does in terms of machine settings or processing.

If the raw materials quality management process is not correct and all the components are not in place then it is impossible to produce a good product. Controlling the input raw materials (either parts or raw plastic) at the supplier stage is critical to producing good products.

**Note 1:** In this chapter we will treat 'raw materials' as referring to both bought-in parts, e.g. any complete item for incorporation into the finished product, and also to plastic materials for further processing, e.g. granules of compound or polymer resin.

## Get the specification right

The first step is the responsibility of the customer and this is getting the specification right. If the material is inadequately specified then it is very difficult for the supplier to get it right. Crosby (see box on the far right) thinks that most of the concerns identified at inwards goods are the result of 'purchaser faults' and this is mostly due to poor or unclear specification of what the customer actually wants from the supplier. This is the responsibility of the Technical/ Engineering departments and should be finalised as part of the early design decisions (see Chapter 8).

- **Tip -** A clear and unambiguous specification of the raw material (product or material) is absolutely necessary. All specifications should be submitted to suppliers for checking that they are both realistic and achievable at the design stage.

## Get the requirements right

A good specification defines what is required from the supplier but it is not always possible for the inwards goods

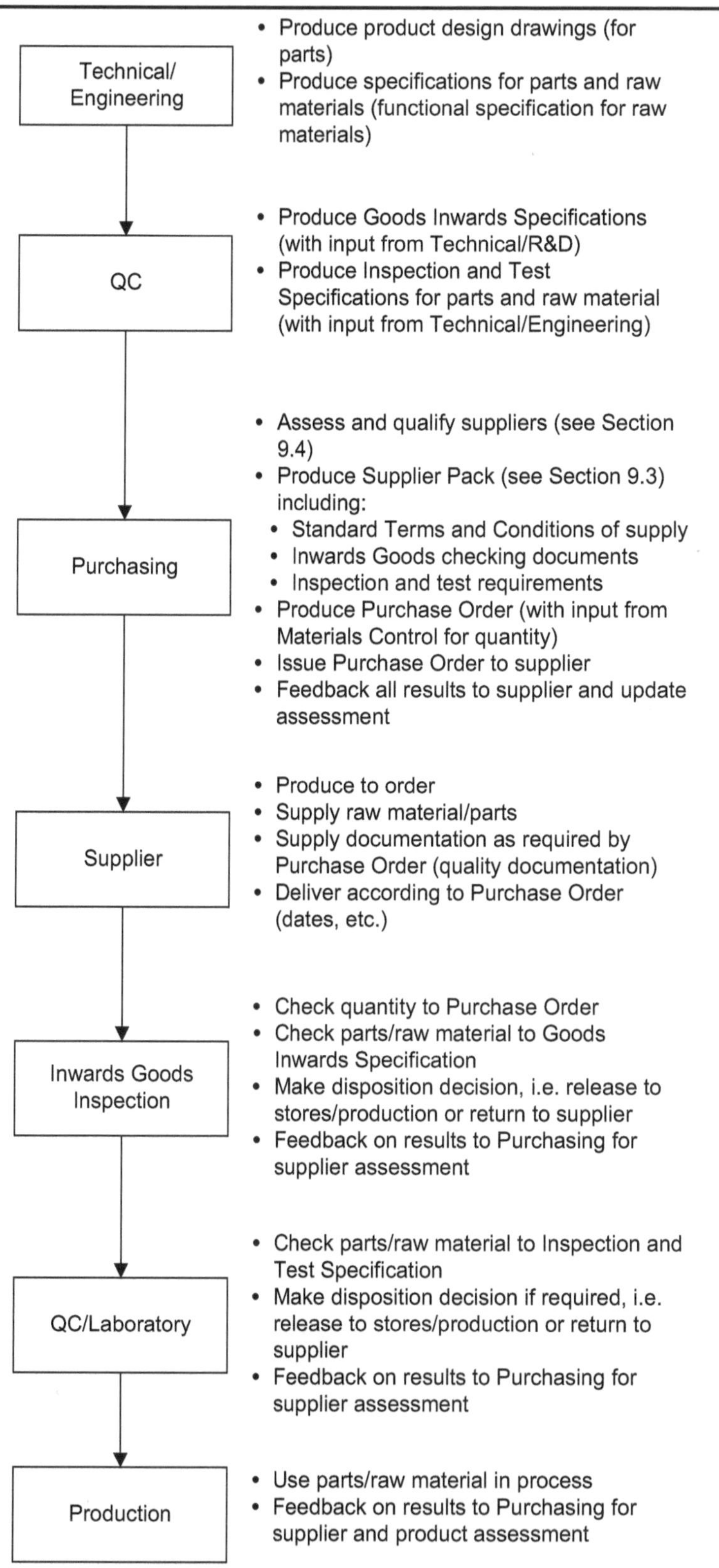


An outline of the raw materials control process (parts and raw materials)

There are many components to controlling the quality of incoming materials (parts or raw materials). If all the components are not effective then poor-quality products will inevitably result.

inspectors to check the raw material to the complete specification and it is also necessary to produce the inwards goods inspection documents and inspection and test requirements. These are a statement of what will be actually checked at inwards goods or in the laboratory and how it will be checked.

If suppliers are required to provide SPC data then this should be specified on the inwards goods inspection documents.

- **Tip** - These requirements should be clearly communicated to the supplier as part of the Supplier Pack (see Section 9.3).
- **Tip** - If SPC data are required then inwards goods inspectors will need training on how to use and understand SPC data.

## Get the right suppliers

Identifying and selecting the right suppliers is the function of purchasing and this involves assessing and qualifying suppliers (see Section 9.4) for quality.

After a supplier is selected then purchasing should produce the supplier pack (see Section 9.3). This is the complete specification of what is required of the supplier. It is not the purchase order but the essential foundation of the purchase order with all the details necessary for the supplier to produce quality raw materials for acceptance into production.

- **Tip** - The supplier pack is the foundation for the order. Take some time getting this right.
- **Tip** - Calculate the total number of suppliers and then the available management time. Can purchasing manage this number of suppliers in this amount of time?

## Check that your suppliers have got it right

Unless all suppliers are qualified for direct delivery then it is necessary to check raw materials at inwards goods before releasing them to production. This is the responsibility of inwards goods inspection and this inspection should be carried out to the inwards goods inspection documents (see above). If laboratory testing is needed before production release then this should be carried out to the inspection and test requirements (see above).

- **Tip** - Acceptance of a raw material at inwards goods does not preclude later rejection. Inwards goods is only the 'first sift'.

## Produce it right

Production is responsible for correct production to the company's specifications.

## Feedback the results to the supplier

The raw materials process is not simply about the customer, it is also about how the supplier and a key element of this is feedback to the supplier. If the raw material fails to meet the requirements at any stage then the results need to be fed back to purchasing and then fed back to the supplier to allow them to take improvement action (see Section 9.6).

Check what you deliver (and are prepared to deliver) to your customers before you specify what you want from your suppliers.

It is good to be consistent.

Do not focus solely on quality problems relating to the actual part or raw material. Quality problems can arise from late deliveries, incorrect invoices and other 'service issues'. It is equally important to manage these aspects of quality.

| Aspect | Crosby | Deming | Juran |
|---|---|---|---|
| Purchasing | Purchasing should simply state the requirements. The supplier is an extension of the company. | It is a team effort. | Quality cannot be improved by internal efforts alone. Suppliers are a key component. |
| Vendor rating | Yes, but purchasing should be rated too. Quality audits are useless. | No, most rating systems do not work. | Yes, but the main aim is to help the supplier to improve. |
| Single source of supply | N/A | Yes | No, it can lead suppliers to be complacent. |
| Inwards goods inspection | Most faults are due to purchaser faults. | Inspection is too late for control. AQL systems allow nonconformities into the system. Statistical evidence is needed to prevent nonconformities entering the system. | Control of raw materials is difficult and the problems are complex. |

**The 'quality gurus' do not agree!**

Unsurprisingly, the 3 main 'quality gurus' do not agree on virtually anything to do with controlling the quality of raw materials. This is a complex subject that has wide-ranging implications to the overall quality of the finished product. I paraphrase somewhat!

# 9.2 The materials team and raw material quality

## The materials team

The purchasing department plays a key role in controlling raw material quality but, as shown in Section 9.1, they are not the only department involved. As with most quality (and other) issues, there is a need for a cross-functional approach to managing quality in raw materials. There needs to be a 'materials team' to look at all the quality issues for all materials used in current products and proposed for new products.

The materials team should include staff from design, quality, production, purchasing, accounts and major suppliers to ensure that raw materials are effectively specified and sourced to improve quality and minimise cost. This is a team effort.

- **Tip** - The materials team should not only be looking at the quality issues, they should also be looking at the cost and design issues, quality is only one of their responsibilities.

The materials team is not a substitute for effective purchasing and the purchasing department must retain the authority and flexibility to negotiate the best deal with suppliers. Purchasing must control all the contractual arrangements for all raw materials. Above all, they must ensure that the delivery and quality standards are clearly communicated to suppliers (see Section 9.3).

## Supplier relations

Approaches to supplier relationships can be either co-operative or competitive.

- Co-operative: This is generally best when the supplier has significant technical resources and can contribute to improved quality, new product development and lower costs as a result of stability or economies of scale. This may involve single sourcing.
- Competitive: This is generally best when there are many suppliers of totally equivalent products or when important raw materials are in danger of having supply interrupted. However, even in these cases the co-operative approach can prove better in terms of additional co-operation from the supplier with regard to quality and delivery.

## Working with suppliers

Reducing the number of suppliers and using 'co-operative single sourcing' is the preferred approach. Experience shows that working co-operatively with suppliers and treating them as an extension of the customer will not only improve raw materials quality but also decrease costs.

- **Tip** - The materials team should always include major suppliers.
- **Tip** - Use Pareto analysis (see Section 7.5) to identify your key suppliers when reducing your supplier base.

Involving key suppliers in the work of the materials team is particularly important at the design stage where suppliers can improve quality and add value through their knowledge of the quality and cost implications of all the design decisions.

Actions to improve the supplier relationship are:

- Start the process of developing the supplier base into a strategic strength for quality improvement.
- Work closely with identified key suppliers to gain the maximum benefit from their expertise.
- Discuss future raw materials requirements at a global level and use

If you have 3 components in a final product and 3 suppliers for each component then you have a total of $3^3 = 27$ variations of the final product.

Enough for mistakes to occur?

Reducing supplier numbers but increasing supplier breadth makes control and relations easier.

Why do you think car manufacturers prefer to deal with Tier 1 suppliers?

Purchasing is not often seen as the profession that it really is. A good purchasing professional, with the support of a materials team, can be a real benefit to the management of quality.

### The 'materials team' approach

Set aggressive targets for the team to stimulate innovative approaches and improve the quality of all raw materials. The materials team should have representatives from all areas of the company. This is not simply a purchasing role.

the increased purchase volumes as a cost and quality lever, i.e. to get the right quality data or levels.

- Use suppliers to reduce the amount of inwards goods inspection.
- Use increased purchase volumes to give greater flexibility in ordering and more frequent delivery of smaller lots.
- Use suppliers to reduce raw materials stocks.
- **Tip** - This should be the start of a process not a single event.

## Existing products

Quality improvement needs an open exchange of information to be effective and to provide the focus for the improvement efforts. The materials team will need all the background quality information to allow them to do their job effectively. They will need detailed reports of any and all issues, either inside the company or inside the supplier, that will affect the quality of incoming materials or services. This allows identification of any areas that can be improved and allows the production of quality improvement plans for the selected raw materials. These plans should also form the basis for an overall quality and cost reduction strategy.

The materials team should:

- Identify the quality implications of every finish, operation and special feature using the quality tools (see Chapter 7).
- Use Pareto analysis (see Section 7.5) to solve the most important issues first.
- Compare competing products, strip them down and look for quality improvements and cost savings - each one may be small but the total will be significant.
- Jointly target quality improvement areas with suppliers so that suppliers can ship direct to the processing area, i.e. minimum stocks with no inspection on delivery.
- **Tip** - If the current supplier doesn't want to do this, then find a new supplier who does.

## New products

It is easier to make changes to designs and raw material specifications at the design stage than at any other time. The materials team is in an ideal position to use suppliers to make changes to reduce potential future quality issues and decrease overall costs. This is the time when suppliers can really improve a company's quality performance with their specialist knowledge and experience.

- **Tip** - Improved working with suppliers can reduce prices but this is not the only aim. Suppliers may need to be reassured that this is not simply a 'cost reduction' exercise but is part of improving quality and reducing the total cost.

## Assessing performance

The effectiveness of the materials team should be assessed by performance measures that reflect the quality and cost improvements that they make.

- **Tip** - Performance assessment always drives improvement.

If confidential information is sought from the supplier then it must be kept confidential and the customer must be prepared to reciprocate with the equivalent information.

Do not become fixated on quality for quality's sake.

Look for the benefits and try to quantify these.

### Measuring the effectiveness of the materials team

Measure the effectiveness of the materials team in improving quality by measuring and recording some or all of the following:

- Number of late deliveries (%).
- Number of times supply shortages have stopped production.
- Number of times nonconforming supplies have stopped production.
- Value of scrap or rework due to nonconforming supplies.
- Reductions in average stock levels.
- Stock turnover of raw materials.
- Average lead time of critical supplies.

### The process for integrating main suppliers into the internal 'materials team'

It is essential that suppliers are integrated into the materials team and see a long-term relationship as preferable to competitive and destructive short-term relationships.

# 9.3 The supplier management process

## Supplier qualification

Working with suppliers as part of the materials team needs suppliers who are capable of delivering value to the materials team and capable of consistently delivering high-quality products direct to production. This type of supplier integration means that more time needs to be spent assessing, qualifying and working with suppliers than in the past where the main activities were placing orders and inspecting what was delivered.

## Initial assessment

Potential suppliers should be subjected to an initial assessment (see Section 9.4).

## The supplier relations manual

If the results of the initial assessment are acceptable then the supplier should be provided with the customer 'supplier relations manual'. This is the basic document for all supplier relations and should consist of:

- The company quality policy.
- A description of the general operations and quality processes.
- An organisation chart (focusing on the purchasing and quality areas).
- An outline of the supplier assessment process, the data that will be collected, how this will be used and how this will be fed back to the supplier (see Sections 9.4 and 9.6).
- A general outline of how company specifications work.
- Data requirements for materials needing SPC data, i.e. $C_p$, $C_{pk}$, control charts.
- AQL procedures and tables for materials where AQL inspection is needed (see Section 9.11).
- Standard terms and conditions for sub-contract processors. These are applied when sub-contract moulders or other processors are used and define:
  - Materials.
  - Tooling.
  - Documentation.
  - Quality control.
  - Price and delivery.
  - Technical assistance.

## Supplier assessment

To fully qualify a supplier it is often necessary to carry out a supplier assessment (see Section 9.4).

## The supplier pack

The supplier relations manual is a general document but the 'supplier pack' is specific to each raw material. The supplier pack should consist of:

- The supplier relations manual (if not

Spend more time on inspecting and qualifying suppliers and far less time on inspecting and sorting incoming goods.

Supplier management is an essential skill for the future.

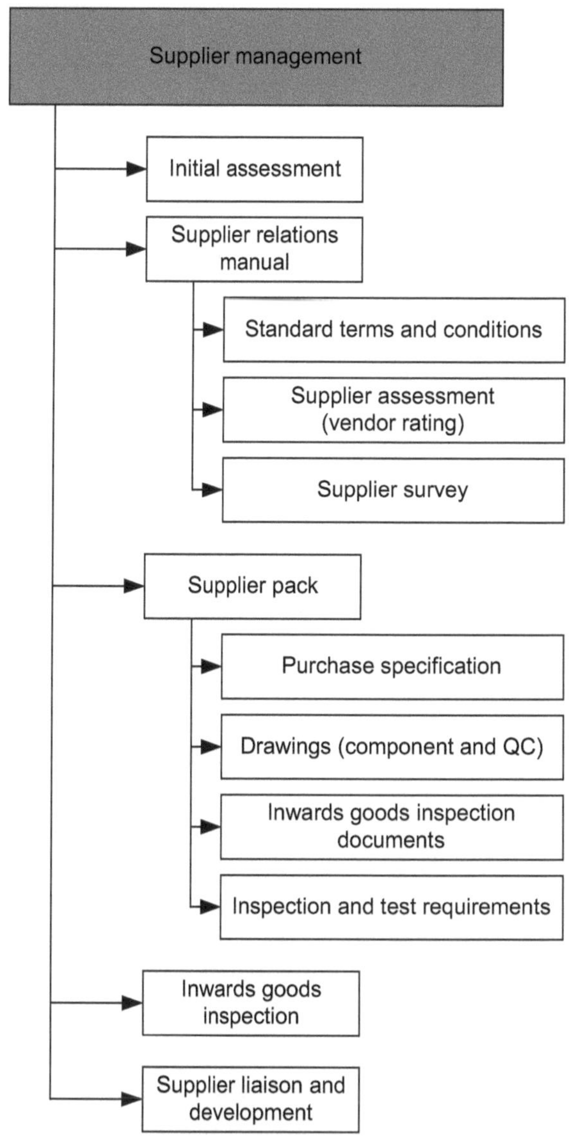


### The essentials of supplier management

Supplier management is a process that needs procedures to both control the supplier and to inform the supplier of what you need. Crosby (the quality guru) thinks that most faults in the process are due to inadequate supplier control or information.

already provided).

- The full purchase specification (see Section 9.7).
- The full component drawing (if applicable).
- The quality control drawing (if applicable).
- The inwards goods inspection schedule for the specified material. This will include specific data requirements for materials needing SPC data and specific AQL requirements for specified issues if AQL is to be used.
- The control plan for the product (see Section 10.3).

The complete supplier pack is provided to qualified suppliers with the enquiry/initial order. The supplier should indicate conformance to these specifications and/or drawings by signing and returning the purchase specification before any order can be placed.

- **Tip -** Purchasing is obviously responsible for the detailed purchasing negotiations with qualified suppliers.

The supplier pack provides the basis for later supplier performance assessment.

- **Tip -** The supplier pack may require the completion of an Initial Sample Inspection Report (ISIR) for new materials, new tooling and first-time supplies. Completion of an ISIR (including capability studies where required) should be mandatory before any production and supply.

## Inwards goods inspection

After goods are delivered by the supplier, they are inspected to the inwards goods inspection documents and inspection and test requirements (see Section 9.10).

- **Tip -** Quality does not come from inspection but from improving what the suppliers and you do. Inspection ties up people, space, handling costs and produces paperwork (even if most of it is done on computer).
- **Tip -** Inwards goods inspection may not only be inspection of materials but also relevant documents, e.g. SPC data.

## Supplier liaison and development

The supplier management process should not stop at the delivery and inspection of the material. Good supplier management always involves supplier development but this is a two-way street and sometimes the supplier can educate the customer, e.g. I have been involved in teaching customers how to run SPC so that they see the benefits and implement SPC themselves.

Quality improvement actions between customers and suppliers could involve:

### Joint improvement plans and working groups

These can set targets for product or process improvement and provide a mechanism for supplier quality improvement (if only by modifying expectations). Working together will require changes at both the supplier and at the customer but will always result in better quality, less inwards goods inspection, lower costs and increased customer satisfaction (see Section 9.6).

### Joint design processes

Integrating suppliers into the design process (see Chapter 8) can result in lower-cost improved designs - the supplier always knows more about his product than you do. Learn to use their knowledge at the design stage and if possible use their standard products to improve quality and reduce cost.

Use Pareto to understand the supplier base:

- 80% of your purchase spend will be with 20% of your suppliers - these are the ones to concentrate on when building relationships.
- 80% of your raw materials problems will come from 20% of your suppliers - these are the ones to concentrate on when managing the process. They are not necessarily the ones you spend the most money with!

Purchasing and suppliers are a two-way street. Good suppliers need good purchasers and if the purchasing process is poor then the supplied goods will also probably be poor.

Supplier management should not simply be at the sales-purchasing interface. Good supplier management works at all levels.

Getting the supplier's production people talking to your production people can be a great way to improve quality by improving communication between the two.

# 9.4 Supplier assessment - the basics

## Are they any good?

Whilst the quality gurus disagree about the value of supplier assessment, most companies will carry out some type of supplier assessment even if it is informal and unstructured.

A well-structured and comprehensive supplier assessment is not simply about the quality system and whether they have ISO 9001 or not. It is about whether the supplier can meet all of the customer's requirements.

## ISO 9001 and suppliers

Many companies assume that if the supplier has ISO 9001 then all is well with their quality but nothing could be further from the truth. ISO 9001 certification is about the quality management system and does not say anything about the product (see Chapter 4). The producer can choose the specifications that their product should meet and as long as their products meet these specifications then the system can and will be certified - these may not be the same as the customer's specifications or requirements. The recent developments in ISO 9001 certainly push for continual improvement but they do not specify what standards, if any, the product should achieve.

ISO 9001 tells you:

- That the basics of a good-quality management system are in place and saves you the effort of having to assess the basic quality system.
- That the quality management system has met the essential ISO 9001 requirements but it does not tell you how well it has met these, i.e. it is essentially a pass/fail assessment and is not the same thing as meeting your quality standards.
- **Tip** - Would you want to be operated on by a surgeon who scored 50% in their surgery examination? Maybe you would ask for something more than a simple pass/fail assessment.

ISO 9001 does not tell you:

- That the supplier is capable of producing the product to customer's or any other standards, i.e. the supplier may be able to make a sewer pipe and have ISO 9001 but that does not mean that they know how to make a window profile. These are different extrusion technologies and have different specifications.
- That the supplier is capable of meeting delivery schedule requirements.
- That you will be able to work with the supplier.
- That the supplier is financially stable.

ISO 9001 is only the start of supplier assessment, it is not the finish and supplier assessment is still a vital part of managing a supplier's quality.

## Initial assessment

Potential suppliers should be subject to an initial assessment. This should not be based simply on the first-time cost (see Section 9.8) but also on quality, reliability, technology and financial stability. This assessment is based on the following type of information:

- What is the supplier's financial position and credit rating?
- Who are the key contacts in design, production and finance?
- What are the current and projected levels of business?
- What credit terms are offered and how do these compare with other suppliers?
- Is the initial price competitive and reasonable?
- How is production scheduled and can they meet current and future schedule requirements?
- What are the cost implications of overdue deliveries against the schedule?
- How is special tooling purchased (if required)?
- What are the design and development procedures for customer-specific products (including change processes)?
- Is the supplier's quality system externally certified to ISO 9001?
- What is the supplier's history of non-conformities and corrective action and are there any outstanding corrective actions?

The initial assessment is mainly a desk exercise carried out by Purchasing and should be carried out using a checklist approach.

Some suppliers will never make the grade but don't worry, they will select (or de-select) themselves through the supplier assessment process.

**Control of sub-contractors**

One aspect that is often overlooked is how a customer (or a supplier) controls sub-contract moulders and processors. For a sample set of 'Standard terms and conditions for sub-contractors' see www.tangram.co.uk/quality.

These can be downloaded and modified as required.

## New suppliers

Assuming a supplier is satisfactory at the initial assessment stage, then it will normally be necessary for a supplier survey. In the past, this mainly considered the quality management system but if the supplier has ISO 9001 then it is possible to look more at the capabilities and processes than at the simple quality system. The supplier survey should be carried out by the Purchasing and Quality Managers. The supplier survey should form an important part in the decision whether to use suppliers or not.

- **Tip** - Use a standard form and focus on the capability to produce materials or services to the relevant quality levels on time and at the agreed price.

## Existing suppliers

Supplier assessment is not a single event, it is an ongoing process and some of the possible measures of supplier quality are given in the box on the right.

Whichever measures are chosen, these should be regularly measured to give a regular formal vendor rating.

A regular vendor rating gives:

- A benchmark that allows measurement of improvement (or not).
- A method of telling suppliers how well they are doing (and an opportunity to improve).
- A target (from the best suppliers) for all suppliers to aim for.
- A metric for use in future supplier selection (or de-selection in the case of bad suppliers).
- **Tip** - Vendor rating should be designed to help make decisions on which suppliers to choose. If it doesn't do this then it isn't working.

Vendor ratings should be published at least annually and distributed to the suppliers. This distribution should be not simply to the sales person but to the CEO, it is his business to know about this. The ratings should not simply be a value but should include as much raw detail as possible and an indication of their performance relative to similar suppliers and overall.

- **Tip** - Be careful when publishing vendor ratings. This could be sensitive information for your suppliers. Those rating well will be pleased, those rating poorly will be less pleased!

Where suppliers show good or improving performance then they should be recognised and congratulated, preferably in the most positive way, by offering them more business. These quality-assured suppliers can provide products that go straight to production with minimal inspection or sampling at reduced levels.

Quantifying quality for services is more difficult than for products but it should still be done if services are a large part of the 'sales package'.

Where suppliers show poor or worsening performance then the rating process is an opportunity to ask for and implement improvement plans (see Section 9.6) or take other action.

- **Tip** - If a supplier is unwilling to implement improvement plans then the action is clear, i.e. find a new supplier.

A vendor rating system, when objectively and consistently applied, is a management tool that will concentrate attention on supplier performance. Simply eliminating the bad suppliers will improve the average supplier performance.

More suppliers are de-selected for quality than for price - so negative and so much wasted opportunity.

### Supplier quality measures

The quality measures for supplier assessment that are easily measured include:

- On-time delivery performance.
- Quantity delivered versus quantity ordered.
- Quality of data submitted, e.g. $C_p$, $C_{pk}$, control charts, etc.
- Quality level at inwards goods inspection.
- Rejects on delivery (% of lots rejected).
- Rejects in process relating to supplied product.
- Number of customer complaints relating to supplied product (internal and external).
- Number of times supply shortages have stopped production.
- Number of times nonconforming supplies have stopped production.
- Value of scrap or rework due to nonconforming supplied product.

These measures are quantifiable and can be scaled to be applicable across the supplier base.

The quality measures that are more difficult to measure include:

- Flexibility in meeting changes to requirements (specification changes or volume changes).
- After-sales service.
- Enthusiasm for joint product development and cost reduction.
- Good communication processes.
- Future investment programme.
- Effective invoicing and accounts processing.

These measures are not easily quantified. They are largely qualitative but a judgement still needs to be made and recorded.

# 9.5 Supplier assessment - the extras

## It is more than internal

Supplier assessment naturally focuses on the ability of the supplier to provide quality goods and services but supplier assessment should also look outside the actual quality of the product. This means looking at the broader aspects of the product and company such as the legislative, social, economic and environmental practices of the supplier as part of the assessment. This is not simply about the quality of the product, it is about the overall quality of the supplier company and how they meet their duties and obligations (which will vary with the country).

## Materials data

Suppliers must ensure that raw materials do not contain any prohibited product, material or substance. In Europe, this is controlled by the REACH (Registration, Evaluation, Authorisation and restriction of Chemicals) system. In the USA, it is controlled by the American Toxic Substance Control Act (TSCA).

Suppliers should ensure that their products do not contain any material that is a 'critical or hazardous' substance and that suitable controls are in place.

### Materials Safety Data Sheets (MSDS)

These have various names around the world but all serve as a method of proving compliance with REACH (or equivalent). Suppliers should be prepared to supply MSDS for any substance supplied.

An MSDS will provide vital information on:

- Identification of the substance/mixture and of the company/undertaking.
- Hazard identification.
- Composition/information on ingredients.
- First-aid measures.
- Fire-fighting measures.
- Accidental release measures.
- Handling and storage.
- Exposure controls/personal protection.
- Physical and chemical properties.
- Stability and reactivity.
- Toxicological information.
- Ecological information.
- Disposal considerations.
- Transport information.
- Regulatory information.
- **Tip** - Suppliers to the automotive industry also need to comply with the requirements of the European End of Life Vehicles (ELV) directive.
- **Tip** - Suppliers to the automotive industry may need to use the International Material Data System, which is the automobile industry's material data system.

Do not dismiss the extras as a 'nice to have'.

They can make a real difference in the long term.

## Health and safety

All suppliers should ensure health and safety risks are reduced and that operations are carried out to the relevant regulations, approved codes of practice and industry best practice. All suppliers

**This Bangladeshi company is proud**

The need for checking compliance may be higher in some countries than in others but this Bangladeshi company proudly states their compliance and displays it at their site. They also actually do all of the things that they say they do. We checked!

should show a commitment to health and safety and maintain effective policies and procedures.

- **Tip** - Ask about training (manual handling, fire safety, emergency response, etc.), risk assessment, personal protective equipment, improvement plans, accident incidence and their monitoring.
- **Tip** - Health and safety issues can shut down a supplier and prevent them from delivering to you so this is not simply about them. After you have had a fire at a tool maker and lost 4 months on a project you take this sort of thing seriously.

## Social

The social activities of a supplier are also important and all suppliers should:

- Obey the laws and regulations in all of the countries in which they operate and/or sell.
- Not use child labour, i.e. those under the age of 16 or the country's legal minimum age, whichever is higher, and comply with the International Labour Organisation's provisions for the health, safety and morality of young people aged between 15 and 18.
- Not use forced or compulsory labour, i.e. any work which is forced upon any person under the threat of a penalty and which the person has not entered into of his or her own free will.
- Ensure that employees understand their rights to payment of fair wages, overtime and retention of identity documents, etc.
- Ensure that working hours comply with the country's laws & regulations and international conventions, e.g. overtime work should be voluntary and paid.
- Comply with all applicable laws and regulations, including those relating to minimum wages, overtime hours and benefits as well as paying employees in a fair and timely manner.
- Not discriminate against any worker based on race, colour, age, gender, sexual orientation, ethnicity, disability, religion, political affiliation, union membership, national origin, social origin, or marital status.
- Respect the right of workers to associate freely, form and join worker's organisations, seek representation and bargain collectively, as allowed by the applicable laws and regulations.

## Economic practices

Fair business practices are an essential to building a relationship with suppliers and are a good indication of how they will treat not only their customers but also their suppliers. All suppliers should:

- Have processes and procedures in place to prevent corruption, bribery and extortion. They should not, directly or indirectly, offer, give, demand or accept any bribe to obtain or retain business.
- Have internal controls, ethics and compliance programmes to prevent and detect bribery, e.g. financial and accounting procedures to ensure transparent and accurate accounts.
- **Tip** - In days gone by many suppliers offered substantial 'gifts' or 'reward programmes' to purchasing departments for business. This is no longer acceptable and everybody should know this.
- Have processes and procedures in place to ensure fair competition, e.g. no cartels.
- Have processes and procedures in place to identify, reveal and avoid conflicts of interest.

## Environment

Environmental issues are of growing importance for any company and all suppliers should:

- Comply with all relevant environmental laws and regulations.
- Have and publish an environmental policy.
- Take all possible measures to protect the environment and minimise the effect of their products on the environment during the whole of the product life cycle.
- **Tip** - Good suppliers will have an Environmental Management System that meets ISO 14001.
- Calculate their carbon footprint and take all possible measures to reduce this.
- **Tip** - Good suppliers will have an Energy Management System that meets ISO 50001.
- **Tip** - Suppliers to the food industry may already have met this in the Walmart™ 'Sustainability Index' where suppliers have to answer 16 questions on sustainability. They are serious about the sustainability efforts of their suppliers.

The 'extras' make a real difference in terms of knowing whether you are dealing with a supplier who takes it seriously or not.

Asking the questions will tell you a lot about a supplier's attitudes.

# 9.6 Quality improvement with suppliers

## Part of the process

Working with suppliers is not simply a matter of assessing the supplier, measuring the incoming quality and criticising the supplier - although some companies do not even do this. Raw materials and supplier quality management is not simply a method of driving down the price. It is a two-way street with the objectives of improving the raw material quality and driving down the Total Acquisition Cost (see Section 9.8) as well as building and increasing business with good suppliers.

Suppliers need feedback on their performance so that they can get better (see Section 9.4) and to do this we need to not only measure and communicate supplier performance but also to actively work with suppliers to improve quality.

Whatever the supplier performance, joint quality improvement projects are an opportunity to improve.

- **Tip** - Good suppliers are generally the easiest to work with for quality improvement but poor suppliers are often those who will benefit most from a joint quality improvement process.
- **Tip** - Rate the purchasing department as well as the suppliers. Good suppliers need good customers.

## What do we want from suppliers?

Obviously, supplier improvement plans are needed for specific persistent raw materials concerns but they should also be used for general process and product improvement to allow delivery straight to production. This means improving the process and quality improvement. When viewed via SPC (see Chapter 5 and box on the right), this means that we want our suppliers to prove that they are:

- Capable - suppliers should have a proven ability to produce the raw material to specification. This means the provision of capability study data for the process and a $C_p$ of > 1.33.
- Centred - suppliers should not only have capability but this should be centred on the mid-specification area, i.e. $C_{pk} > 1.33$.
- Convergent - suppliers should be able to demonstrate that the process is improving to reduce the spread and improve the capability.
- Consistent - suppliers should be able to demonstrate that processes and products are consistent from batch to batch.
- Credible - suppliers should be able to produce credible and trustworthy information on the process and products.

This type of improvement is less urgent than urgent issues with raw materials but can have a greater and more beneficial long-term effect on quality.

## The quality improvement process

The basic operation of a supplier quality improvement programme is shown in the box on the far right. The initial stages of the process involve establishing trust between the supplier and the customer and building a good supplier/customer relationship. The 'materials team' (see Section 9.2) is an essential part of this process but trust will only be built on the

Supplier engagement can also be improved by methods such as:

- 'Supplier of the year/month' awards.
- Supplier newsletters giving details of supplier quality progress.
- Supplier 'open days' giving details of future plans and products.

Purchasing is not a one-way street.

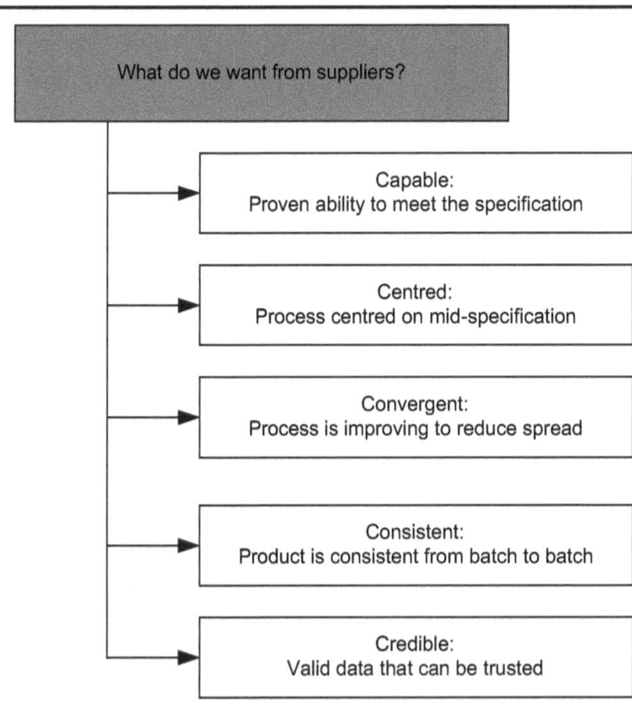


### What do we want from suppliers?

Suppliers need to demonstrate, via good data and information, not only that they are capable of delivering consistently good materials but also that they are improving the quality of their products.

two foundations of communication and data. These will take time to establish and initially inwards goods testing will be as per standard products (see Section 9.10).

Communications should be developed by mutual visits at the operational level, it is not enough for only high-level visits (although these are useful). The most useful visits are when the operational people can meet and exchange views on the product and issues with the raw material.

Data should be developed from inwards goods inspection reports, production reports and other sources. Typical data are:

- Product data, e.g. try carrying out a capability study on the inwards goods to get an idea of the supplier's capability.
- Product non-conformances (from inwards goods checking).
- Process non-conformances (from production reports).
- Procedure non-conformances (from accounting and other sources).

This type of data allows concerns to be identified and a joint quality improvement programme with the supplier to be set up for specific items and features. The process steps in a supplier quality improvement programme are very similar to the process steps in any internal quality improvement programme with the additional difficulty or benefit that the work is not totally internal.

- **Tip** - Supplier quality improvement programmes need management just as internal quality improvement (see Chapter 2).

One important part of supplier improvement is that it is also a 'never-ending' process, i.e. after the most important concern has been identified and solved then the next concern on the list is identified and solved.

## Benefits for the customer

A supplier improvement programme can help customers by:

- Using suppliers to drive quality improvement.
- Providing extra resources and skills from the supplier to improve quality.
- Providing an external reference on internal quality improvement.
- Decreasing costs by reducing inwards good inspection costs.

The main benefit is that improving raw materials quality will also improve outgoing process and product quality and help you to demonstrate quality improvement to your customer.

## Benefits for the supplier

A supplier improvement programme can help suppliers by:

- Developing positive communication with the customer rather than simply hearing about concerns.
- Developing long-term relationships based on capability rather than price.
- Improving customer confidence at both the production and sales levels.
- Raising the supplier to 'preferred supplier' status.
- Increasing the amount of work given to the supplier.
- **Tip** - Supplier quality improvement programmes must be confidential.
- **Tip** - You are a supplier too so get started on working together with your customers on a joint quality improvement programme. It can work wonders for you too.

Limit the variety of suppliers used in the business (despite what engineering says - I have been there too).

Use national or preferably international standards and avoid specifying by brand name.

I've done that to force purchasing to buy from my preferred supplier and to avoid having to do the engineering work necessary to get the standard material to work.

I know I was lazy but there really was no excuse.

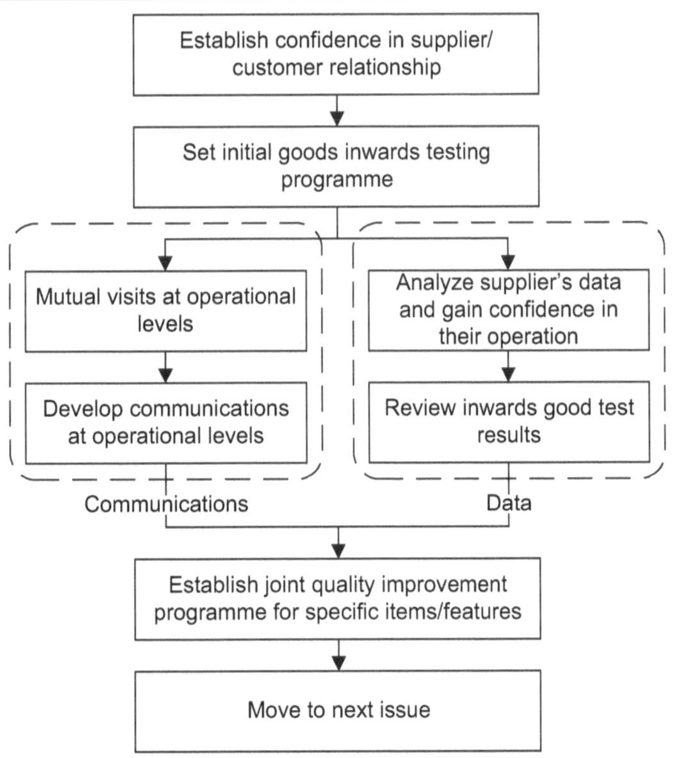


### Supplier quality improvement needs trust

The main requirement of a supplier quality improvement programme is that the supplier and the customer trust one another. Quality improvement is data-driven and as Deming once said: 'In God we trust, all others bring data'.

# 9.7 Purchase specifications

## Specifications

Ultimately, it is the Purchasing Department's responsibility to ensure that only raw materials suitable for use in the company's products are purchased. To ensure this, they must be provided with a purchase specification that gives measurable properties and requirements of the raw material, e.g. size, weight, performance parameters, safety requirements and quality levels or requirements.

This purchase specification is used as the basis for supplier assessment and the purchase contract.

- **Tip -** Items that do not conform to the purchase specification should never be purchased.
- **Tip -** If sub-contract moulded or other parts are purchased then a set of standard terms and conditions for sub-contract moulders that should be used is applied in addition to the normal purchase conditions. A sample set of these is available at www.tangram.co.uk/quality.

## What does a purchase specification include?

A purchase specification should give the following details:

### Heading

#### Specification number

Each specification should have a unique reference number and it is recommended that these should vary depending on the type of raw material, e.g.

- 01-XXXX = Raw materials.
- 02-XXXX = Sub-components.
- 03-XXXX = Finished articles.
- 04-XXXX = Packaging.
- 05-XXXX = Capital purchases.

Index letters A, B, C etc., should be used for revisions.

#### Issuing authority and date

Purchase specifications require authorising signatures on preparation, approval and release. The release date of the revision should also be noted on the specification.

### Materials

Materials should be specified by properties without giving the supplier's name, unless absolutely necessary. This allows purchasing to perform their function.

If a material colour is specified then this should be by a RAL or similar colour reference or by a colour sample.

- **Tip -** Do not just attach or copy a supplier's data sheet to the purchase specification. This is laziness.

### Dimensions

If the specification is for a product then the relevant drawings (both full component and quality control drawings) should be referenced in and attached to the specification.

If the product has 'mating parts' then the mating parts, and their drawings, should be referenced in and attached to the specification. If at all possible, reference samples of the mating parts should also be attached as part of the specification.

Trade names will lock you into a single supplier.

If a specification includes physical characteristics or properties then include tolerances for these.

Specifications that are too tight or too loose will cost money and if it isn't specified then the supplier can provide whatever he likes. If you are using AQL then specify the level for acceptance and the requirements.

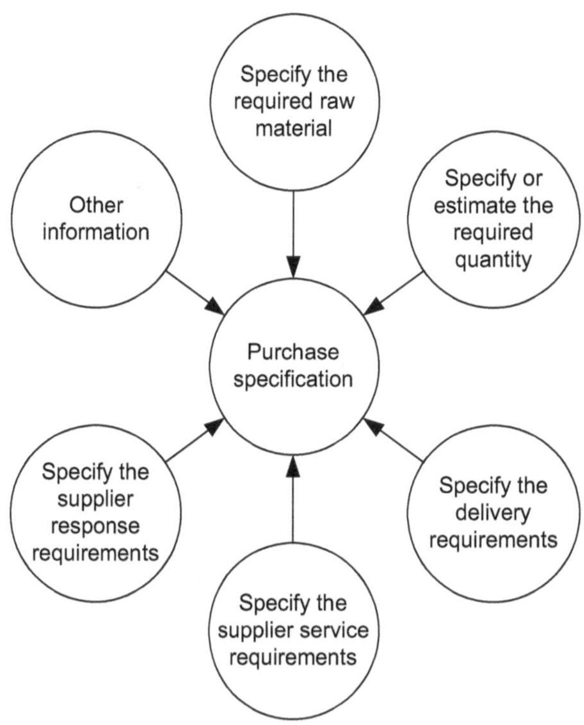


Factors in the purchase specification

Purchase specifications must do more than simply specify the required raw material. They must also specify service levels and other issues that will affect the quality of the product as seen by the customer.

## Construction

If the specification is for a capital purchase or similar items then notes on the construction standards or requirements should be included, e.g. CE marking requirements, etc.

## Testing

This section should contain the inwards goods inspection schedule. This will include:

- Details of any tests to be carried out and their initial frequency.
- Specific AQL requirements for specified features/issues if AQL is to be used.

If the specification is for a product and gauges are to be used for inwards goods inspection then this section should also include details and drawings of the gauges to be used.

This section should also include the laboratory testing schedule. This will include:

- Details of tests to be carried out and their initial frequency.
- Acceptable tolerances for all laboratory test results.
- **Tip** - All testing should be to national or international standards.

## Documentation required

This section should contain details of the technical documentation to accompany the material on delivery (this does not include delivery notes, etc.). This will include:

- Details of laboratory testing carried out at the supplier.
- Statistical evidence of the supplier's control procedures such as SPC control charts and/or other data.
- **Tip** - Do not ask for 'Certificates of Conformity', these are worse than useless (see sidebar in Section 9.10).

## Packaging

Packaging is needed to protect and deliver raw materials and the specification should give details of:

- The method of packaging.
- The quality of packaging, e.g. grade of card, etc. to be used.
- The information or labelling required on the packaging.

## Statutory information

If statutory information is required then it should be requested in this section. Typical information would be:

- Information required for conformance with product liability legislation.
- Information required for conformance with health and safety legislation.

## Control of purchase specifications

Control of purchase specifications is generally the responsibility of Technical/Engineering and they should be responsible for issuing and releasing all purchase specifications. This will be carried out in conjunction with other departments, e.g. purchase specifications for plastics materials will be written after liaison with production and trialling of materials.

- **Tip** - If you do not have purchase specifications for all raw materials then roll them out slowly by introducing purchase specifications for raw materials as they are ordered from suppliers.
- **Tip** - The purchase specification is NOT the purchase contract. It is an essential part of the purchase contract but they are not the same thing.
- **Tip** - The purchase specification is to allow purchasing to seek competitive quotations by forwarding the purchase specification to prospective suppliers.
- **Tip** - Never design and make a product if there is an 'off-the-shelf' part available. These will be cheaper, more reliable and easier to source. Give purchasing the discretion to suggest using an 'off-the-shelf' product and be prepared to modify the design to use 'off-the-shelf' parts.

Suppliers should be told what the function is and asked for their recommendations to meet the function.

They may be able to suggest cheaper but equally acceptable products or more expensive but far better materials that can add value to your customer.

Some years ago in the motor industry our buyer purchased 7 million O-rings to a detailed purchase specification. The supplier met the specification and supplied the goods to this.

It was only after 2 years that we accidentally discovered that the supplier could easily supply an alternative. This was less than half the price and did the job more than adequately.

The specification was changed.

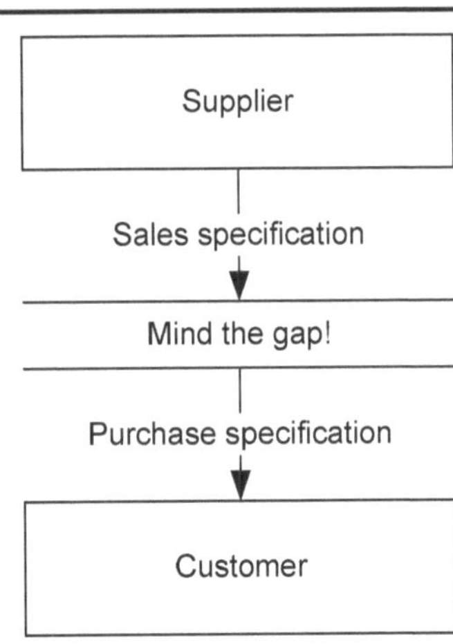


### The gap between the sales and purchase specifications

For standard raw materials, most suppliers will have a 'sales' specification. This should be compared with the purchasing specification to identify the inevitable gap between the two.

# 9.8 Price and cost at suppliers

## The Total Acquisition Cost

The total cost of any product is not simply the price and most good purchasing departments have already realised this but the message does not seem to have got through to everybody in the organisation.

The price of the raw material or part is almost certainly the largest element of the total cost but this does not include any of the other direct and indirect costs. The total of the direct and indirect costs is called the Total Acquisition Cost (TAC). This is rarely calculated and in most companies is only vaguely assessed.

It is not necessary to calculate the TAC to great detail for all products. Select the largest suppliers (by spend) and quickly calculate the TAC for these suppliers.

- **Tip** - It is impossible to accurately calculate the TAC and approximations are good enough in almost all cases.

## Quality aspects of TAC

The basic elements of the TAC are shown in the diagram on the right. For the purposes of this Workbook, we will naturally concentrate on the quality aspects of the TAC.

### The price

This is the most obvious cost and is easy to assess. If all the other factors in the TAC were equal then the decision would be simple, i.e. lowest price. Unfortunately, this is rarely the case and, as with quality costs (see Chapter 3), the ramifications of selecting poor-quality suppliers is more than simply the price of the initial material or part. If things go wrong then the initial price is the least of your costs.

### Purchasing and administration costs

These will be similar for most suppliers but poor-quality suppliers will always generate more costs in terms of dealing with rejections, trying to obtain credits, dealing with dissatisfied customers (if the materials have made it into your system) and the extra effort in developing the supplier to a suitable level (if you decide they are worth the effort). The administration costs of poor quality are always substantial.

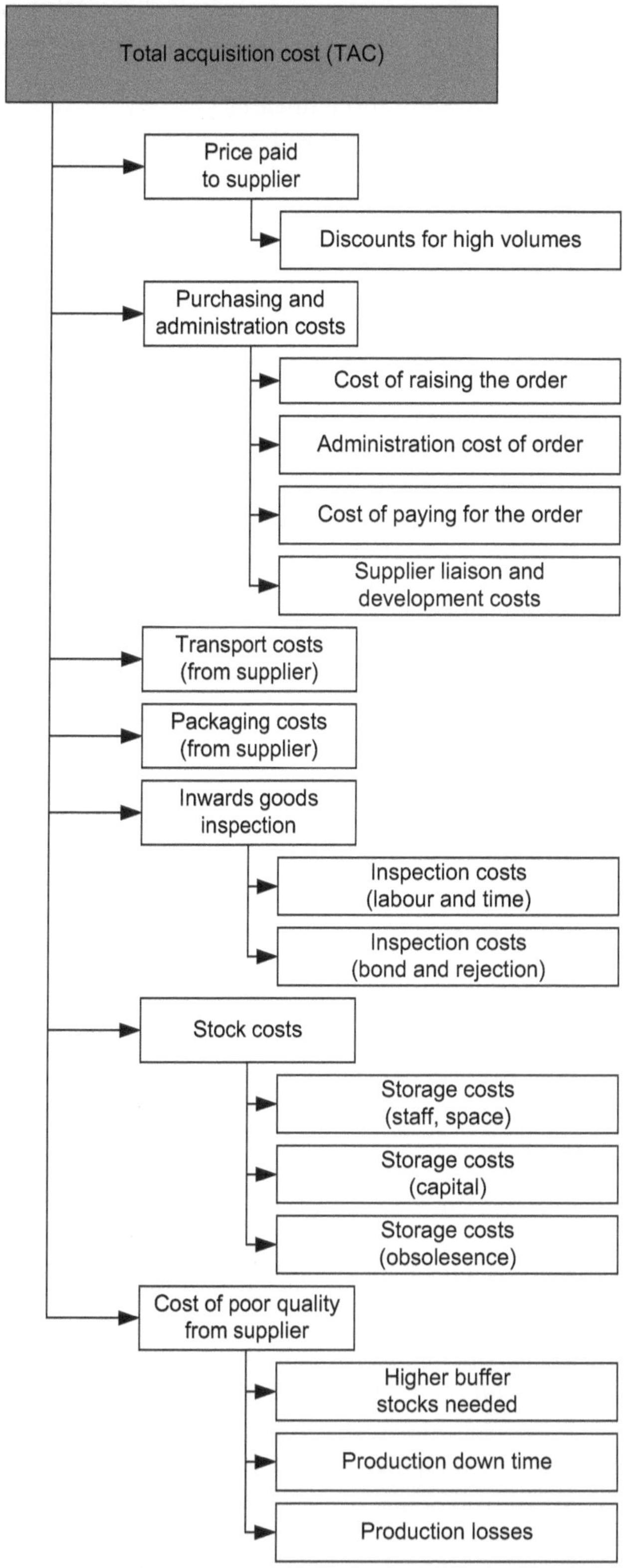


**The price paid to the supplier is only part of the cost**

The Total Cost of Acquisition (TAC) is far greater than simply the price paid to the supplier. All of the other costs must be considered. Quality and the effect that poor quality has on the company is an essential consideration.

### Transport and packaging costs

Again, these will be similar for most suppliers but poor-quality suppliers may cause more costs in transporting materials, especially if faulty materials need to be returned.

### Inwards goods inspection

After they are identified, poor-quality suppliers will need a higher level of inwards goods inspection (see Section 9.10) and this will add to all the associated inwards goods inspection costs (labour, time, bond areas and the problems of rejected goods). This is a major cost that is rarely identified if the company is fixated by the initial price.

### Stock costs

Inventory is never free and the cost of simply holding stock is estimated to be in the region of 20% of the total value of the stock.

High-quality suppliers can be trusted to deliver straight to production with minimal inventory and no inspection but poor-quality suppliers will always need buffer stocks and inspection to cope with any quality issues.

- **Tip** - Assess the value of the total raw materials inventory, calculate the cost of holding the stock and inspecting it. Compare this with the cost of using higher-quality suppliers.
- **Tip** - Increased buffer stocks will increase the possibility of increased costs if products become obsolescent.

### Cost of poor quality from the supplier

Poor-quality raw materials (including late deliveries) that make it into the production system will raise costs on many levels such as:

- Increased production disruption, e.g. production delays whilst problems with the material are resolved.
- Increased production down-time and stoppage costs.
- Increased production losses from wasted materials.
- Increased opportunity losses from idle machinery and labour.
- Increased operating costs by operating machinery with no productive output.
- Increased costs for re-working, repairs or scrap.
- Increased customer dissatisfaction from late deliveries.
- Increased warranty and product liability costs.

## Non-quality aspects of TAC

There are many non-quality aspects and the most important of these are:

### Cost of lead-time

Lead-time costs relate to how long it is between when the order is placed and when the product is delivered. If payment is made at the order stage then the money for this is tied up in goods that are not saleable. This increases the risk of product or process obsolescence and increased costs.

It is often hard to judge the effect of development lead-times and the time to market but in FMCG markets delays in product development can be critical for product success. The average 'life' of a mobile phone is around 2 years but most of the sales will be made in the first 6-9 months from release. If it is 3 months late to market then it loses around 40% of the potential sales.

- **Tip** - To see how important it is:
  - Charles Lindbergh was the first man to fly the Atlantic solo and we have all heard of him. Who was the second?
  - Neil Armstrong was the first man to walk on the Moon. Who was second?

### Cost of new suppliers

Finding and assessing a new supplier involves all of the activities listed in Section 9.4 and this work is not cost-free. Poor-quality suppliers will force new supplier assessments and increase costs.

## Assessing the TAC

Whilst the full calculation of TAC is almost impossible, purchasing professionals should be able to put together an approximate TAC calculation quickly and use this to assess the effect of poor quality on the TAC.

- **Tip** - Set up a spreadsheet based on 3 to 4 levels of supplier quality and use this as an approximation of the effect of raw materials quality on the total TAC.

Price is but one element of total costs - and sometimes the smallest. Ignore the other elements at your peril.

Capital equipment also needs a TAC approach to costing and this should include a life-cycle cost approach (including energy costs).

Quality is one of the criteria to be applied for a given item and, depending on the product, may be more important than or less important than the actual price paid for the goods.

In each individual case, purchasing is responsible for deciding on the relevant ranking for the criteria such as quality, price, delivery and quantity.

Quality is not measured by price alone.

# 9.9 Purchasing quality - where are you now?

## It is an art and a science

Purchasing is a key skill in improving the quality of raw materials. Purchasing professionals need to use their skills to select quality suppliers through supplier assessment and rating and to integrate the selected suppliers into the company's activities. This all needs to be done whilst at the same time ensuring the actual price paid is both fair and reasonable.

However, purchasing is not simply about prices, it is about defining the specification so that the supplier has a chance to reduce prices and increase quality whilst using their skills to the best of their ability. Establishing good supplier partnerships and developing the skills of the supplier base allows suppliers to provide a valuable input into the process of improving quality and reducing the total cost of acquisition.

## Completing the chart

This chart is completed and assessed as for the previous charts.

Purchasing should never be confrontational or driven by a desire to win at all costs.

Capitalism may sometimes be short-sighted but it should not be blind.

Every customer needs good suppliers and must be prepared to pay for their continued existence.

**Purchasing quality**

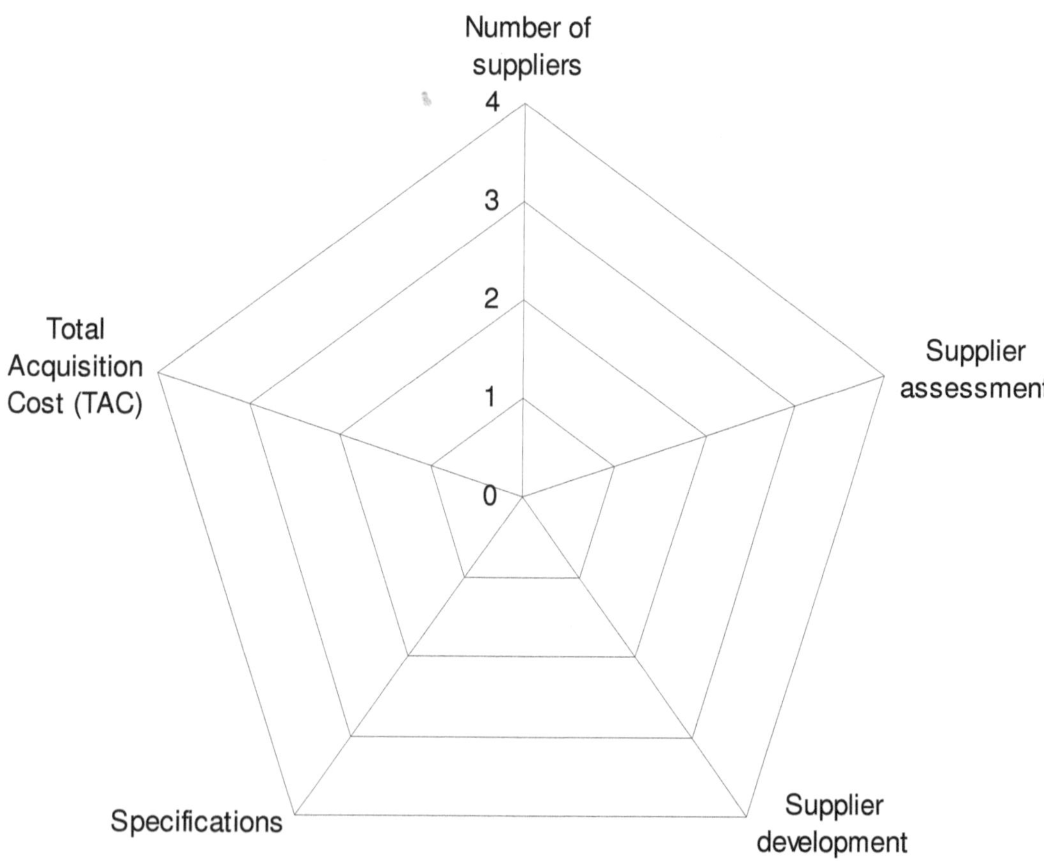


Use the chart to assess where you are in purchasing quality

The numbers from the self-assessment should be transferred to the radar chart for a quick visual guide to where you are in the basics of purchasing quality.

Suppliers know their products better than you will ever know them. Learn to ask for advice from suppliers and heed their advice.

Invest in the development of the 'soft' purchasing skills to add value to the process rather than to simply drive prices lower.

This is a skilled profession and should be treated as such. If you want simple 'buyers' then be prepared to pay the eventual cost for them.

## Purchasing quality

| Level | Number of suppliers | Supplier assessment | Supplier development | Specifications | Total Acquisition Cost (TAC) |
|---|---|---|---|---|---|
| 4 | Minimal number of suppliers. Single sourcing encouraged & alliances/partnerships established. | High-quality local certified suppliers preferred & used extensively. Feedback of certification & rating results to supplier & open discussion of these. | Extensive supplier development program in place to bring suppliers into the process. | Clear & concise delivery & quality specifications set in conjunction with supplier to use supplier's skills to best advantage. | TAC known, measured regularly & decreasing. |
| 3 | Low number of suppliers & actively being reduced. Dual sourcing is discouraged. | Supplier rating carried out to high level but little feedback of results or discussion with supplier. Suppliers rated & high ratings preferred. | Moderate supplier development program in place for selected large suppliers. | Good delivery & quality specifications but little use of supplier's skills to best advantage. | TAC known, measured regularly & increasing. |
| 2 | Very high number of suppliers. Action planned to reduce dual sourcing & number of suppliers. | Supplier rating carried out to moderate level. Some recording of performance but historical personal relationships greatly influence supplier choice. | Small supplier development program in place for very limited number of suppliers. | Delivery & quality specifications exist but considerable room for improvement in use of supplier's skills. | TAC known but rarely quantified & increasing. |
| 1 | Large number of suppliers with significant amount of dual sourcing common on large-volume products. | Supplier rating carried out at basic level. No recording & primarily anecdotal evidence. Little action taken on adverse rating & no preference for high rating. | No supplier development program in place or in development. | Poor & ambiguous delivery & quality specifications arbitrarily imposed on suppliers. | TAC only vaguely known & not quantified. |
| 0 | No controls over number of suppliers. Dual sourcing common & encouraged. | No supplier rating carried out. | Suppliers seen as not worthy of development. | Delivery & quality specifications are non-existent or vague. Frequent disputes with suppliers over quality standards. | TAC not known or measured. |
| Score | | | | | |

# 9.10 Assessing inwards goods

## Why are we doing this?

Inwards goods inspection is to ensure that only products suitable for use in the company's products are accepted into the production process and that no incoming material is used or processed until it is verified as conforming to requirements.

## The inspection choices

Inwards goods inspection can be by:

- No incoming inspection: This is a high-risk strategy unless the supplier is very reliable, highly qualified and supplies SPC data.
- 100% incoming inspection: This is a high-cost strategy with no guarantee of success because inspection is not always reliable (see Section 9.11).
- Spot checking on random materials with no statistical basis: This may stop some nonconforming materials getting through but is generally the same as no incoming inspection, i.e. high risk.
- Sampling a constant % of the material: Sampling is the best compromise between 100% and no inspection but sampling on a constant % is not ideal. This type of sampling is under inspects for small batch quantities and over inspects for large lots. As noted in Section 6.1, sampling at 10% for a batch of 10,000 is far more stringent than sampling at 10% for a batch of 50. This may appear counter intuitive but it is statistically correct.
- Statistical sampling methods: This is the Acceptance Quality Level (AQL) method (see Chapter 6). AQL can distinguish between critical and non-critical characteristics, has statistically known risks and provides the same inspection level for small and large batches.

For inwards goods inspection, statistical sampling techniques based on AQL are strongly preferred unless the supplier is very good and provides full statistical data.

**Note:** We will only consider inspection by attributes as this is the most common type of inspection system used at inwards goods inspection.

- **Tip -** Full details of AQL systems are given in Chapter 6.
- **Tip -** For any inwards goods inspection system to work then the supplier must know what is expected of them, i.e. they must have the full supplier pack (see Section 9.3).

## SPC data

If high-quality and reliable SPC data are provided by suppliers (see Section 9.6) then it is possible to remove inwards goods inspection entirely and deliver raw materials straight to production. This reduces costs but inwards goods inspectors need to be trained so that they can satisfy themselves that the necessary data have been supplied either with the goods or direct to purchasing.

Inwards goods inspection does not relieve a supplier of their responsibility to ensure that all delivered products conform to the specification.

Acceptance of a material by goods inward inspection does not eliminate the possibility of later rejection.

## Get it in the Ts & Cs

The supplier pack and purchase contract (see Section 9.3) should include details of the terms and conditions regarding inwards goods inspection for the raw material, e.g. SPC data requirements and AQL procedures. The pack should also include details of what happens in the

It is essential to check that the customer and supplier are using the same standards, especially for any sensory data, e.g. visual standards.

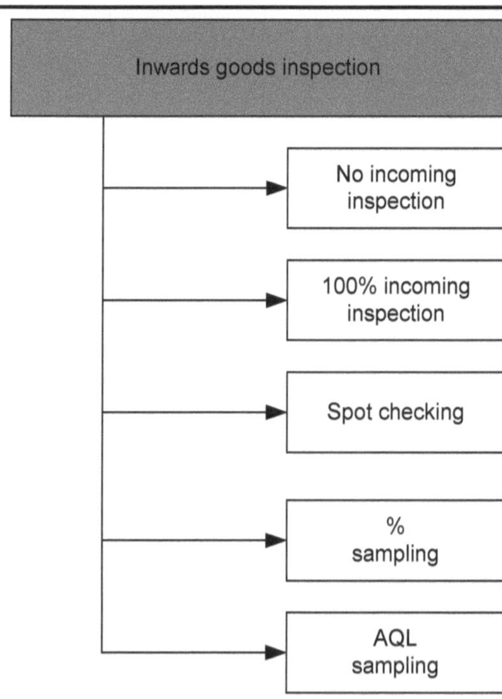


### The choices for inwards good inspection

Inwards goods inspection can use a variety of methods. All have risks and costs but the essential thing is that incoming raw materials need either inspection or very good suppliers who have good performance and provide reliable data with every delivery.

event of raw materials failing the inwards goods inspection process, such as:

- The supplier agrees to either debit rejected and returned batches or to pay for sorting and reworking if the batch is urgently needed.
- The supplier agrees to pay any transportation costs involved in returning nonconforming materials.
- The supplier agrees that the customer can return for credit and claim production costs on any nonconforming items from accepted batches that are found during subsequent production.
- The supplier agrees that AQLs may change for subsequent deliveries after a nonconforming batch is found.
- The supplier agrees that nonconforming deliveries will be recorded in the supplier rating process.

## Inwards goods specifications

The purchase specification and the supplier pack define the requirements of suppliers but it is possible that these will require tests or measurements which cannot be carried out at inwards goods. Therefore, an 'inwards goods specification' is needed for inwards goods checking. The inwards goods specification is an abstract of the purchase specification which details those tests which can be reasonably carried out during the inwards goods check, e.g. it may request a specified Melt Flow Index (MFI) of the plastic that cannot be checked at inwards goods but must be checked in the laboratory. The inwards goods specification defines those tests to be carried out at inwards goods inspection and gives the relevant AQL for the particular feature/measurement. For a sample of an inwards goods specification see Section 9.11.

- **Tip** - Inwards goods specifications can use a check sheet for simplicity and speed but must refer to the relevant AQL level for the feature.

## Inwards goods test specifications

For tests and measurements that must be carried out in a laboratory, it is necessary to have an 'inwards goods test specification'. This is, again, an abstract of the full purchase specification which details those tests to be carried out in the laboratory. Inwards goods test specifications should always refer to national or international standards for the specific test method.

- **Tip** - Laboratory testing is mainly used to measure batch-to-batch variations and is rarely the basis for specification of the material or the product.

## Lot and component data?

Inwards goods inspection can be designed to provide data on the lot or on the individual components.

Lot data are related to the complete lot inspected. These data are used to determine what happens to the material, i.e. accept, sort or reject.

Component data are related only to the component inspected or tested and are generally provided by testing (in-house or external laboratories). These data are generally related to determine whether to specify the component or not.

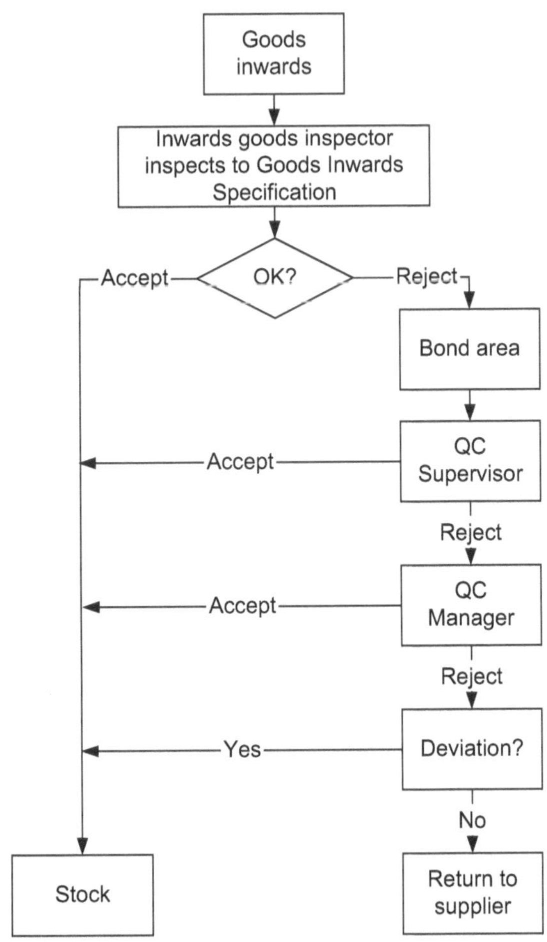


### The inwards goods process flow

Raw materials should be checked to a defined inwards goods specification. If there is doubt about the acceptability of the raw material then it should be held in a 'bond' area to prevent potentially faulty material being delivered to production.

**Do 'Certificates of Conformity' work?**

Try this simple test for how effective CoC's are:

How many of your suppliers are stupid enough to send you a form that says 'This part is faulty'?

Would you ever send a form like that to your customers?

They may provide limited reassurance for product liability but that is about all.

# 9.11 AQL/SPC at inwards goods

## Speeding up the process

Inwards goods inspection can be tedious and time consuming for staff as well as tying up material before it can be used. It is therefore in everybody's interests to get it done as swiftly and as accurately as possible.

The inspection process can be speeded up considerably if goods inwards inspectors are provided with good data and a typical goods inwards specification sheet for a component as shown on the right. This gives a sketch of the component, the method of checking, the gauges or test equipment to be used and the relevant AQL for the feature.

- **Tip** - Resin control needs different information (see Section 9.12).
- **Tip** - This sheet uses nonconformities to assess the product but can just as easily be used to assess non-conforming items (see Section 6.2 for the difference between the two).

## AQL level specification

For most features, it is sufficient to use one of the three AQLs recommended in Section 6.4, these are:

- AQL = 0.15: For components or features that are safety and functionally critical.
- AQL = 1.0: For components or features that are not safety critical but are functional or visually critical.
- AQL = 4.0: For components or features that are non-critical but are still functionally or visually important.

Restricting the number of AQLs used makes it easier for inspectors to be given simple AQL tables and to learn to work with these.

Initial inspection (for a new supplier or component) should start out with single sampling at the normal level and the switching rules followed if the results show a need to change between normal, tightened and reduced levels.

- **Tip** - The AQL level in the goods inwards specification should not be changed unless the switching rules indicate a change and approval has been obtained from the production and purchasing departments.
- **Tip** - The results of inwards goods

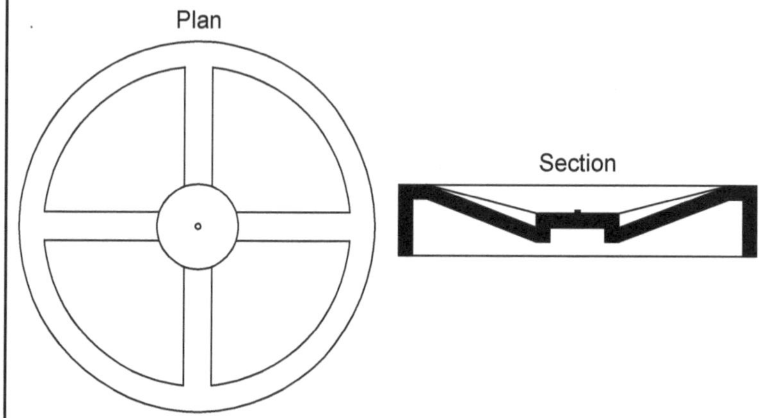

Product sketch

| No. | Item/Feature | Method | AQL |
|---|---|---|---|
| 1 | Short shots | Visual | 0.15 |
| 2 | Flash around rim | Visual | 0.15 |
| 3 | Distortion | Visual and test | 0.15 |
| 4 | Ejector marks | Visual | 1.0 |
| 5 | Large gate | Standard | 0.15 |
| 6 | Colour density | Standard | 4.0 |
| 7 | Ovality | Gauge 001 | 1.0 |
| 8 | Depth of centre | Gauge 002 | 0.15 |
| 9 | Contamination | Visual | 4.0 |
| 10 | Damage | Visual | 0.15 |
| 11 | Stress crack test | Test rig 001 | 0.15 |
| 12 | Taper | Visual and test | 0.15 |
| 13 | Gassing | Visual | 1.0 |
| 14 | Flow lines | Visual | 1.0 |
| 15 | Trimmings in box | Visual | 4.0 |

**Gauges:**

Gauge 001

Gauge 002

**Test equipment:**

Test rig 001

Initial inspection at normal level and switching rules apply.

All results to be recorded and copied to purchasing for incorporation in the supplier rating.

Typical inwards goods specification sheet giving AQL values for various features

The inwards goods specification sheet tells the inspector what to check, the method of checking, gauges or test equipment to be used and the relevant AQL for the feature.

inspection should always be copied to purchasing for incorporation in the supplier's rating.

## Selecting samples

It is always difficult for inwards goods inspectors to select a truly random sample from a batch because of simple logistics and access issues. For trusted suppliers, it is possible to ask the supplier to select the samples to be inspected and to submit these separately from the batch and labelled as 'random samples'. These samples can then be checked by inwards goods inspectors without the need to try to select random samples.

This type of system needs to be run in parallel with conventional sampling for some time to verify that the supplier's 'random samples' are truly representative of the batch. The advantages of this are:

- Faster inwards goods inspection with no need for inspectors to select samples.
- Reduced materials handling at inwards goods because most of the material can go direct to the bond area until the inspection is complete.
- Reduced inwards goods inspection area.

## Audit of decisions

If suppliers submit truly random samples (see above) then it is possible to get the supplier to carry out the inspection and testing to the specified AQL and to self-audit their own production output. This means the supplier using the same inwards goods specifications and gauges as the inwards goods inspectors.

The data from the supplier inspection can then be compared with the data found at inwards goods inspection. If they are the same, then the customer can have confidence that the supplier is making the right decisions and reduce or even eliminate inwards goods inspection.

- **Tip** - The supplier/customer results will not always match exactly but should be in the same area, i.e. rejecting bad batches and accept good batches. This is sampling we are dealing with.
- **Tip** - This process is more concerned with checking that the supplier makes the same decisions than with checking the raw material. It is auditing the supplier's decisions rather than auditing the raw material itself. Information on the product is secondary to information on the decision.

## Using SPC at inwards goods

Using AQL at inwards goods is useful because it assesses the risk of accepting poor-quality materials and needs less time and product handling than other methods. However, it only detects poor quality rather than seeking to improve it, it does not add any value to the process and provides little feedback to the supplier for process improvement. For suppliers with stable and 'in control' processes (see Chapter 5), SPC can be considered for use at inwards goods inspection.

SPC is not designed to assess products for lot by lot acceptance, it is designed to assess and improve processes. It is possible to use SPC using attributes (see Sections 5.19) to assess inwards goods to provide feedback on process changes. This requires a change of emphasis, i.e. instead of focusing on lot-by-lot acceptance, the data from inwards goods are used to create a control chart as follows:

- Each delivery is inspected for attributes using the sample size, i.e. the AQL sample size, and items as on the inwards goods specification.
- The sample is treated as a sub-group and the inspection data are plotted on the relevant attributes chart (p, np, c or u-chart).
- The centre line, UCL and LCL are calculated and plotted on the control chart (see Section 5.19 for details).
- The control chart is used to identify special causes and to feed back performance to the supplier.

**Note:** This process will need historical data to establish the control limits and validity but can be very effective in the long-term.

**This is only sampling.**

Do not forget that any use of AQL is still based on a sample. It is possible that nonconformities will still make it into the system.

Don't hang the messenger if nonconformities make it into the system.

Inspection is not a method of assuring quality. It should be used to understand and improve processes that produce products and services.

Combining AQL and SPC at inwards goods inspection can give both effective inspection and long-term supplier quality improvement.

However, the best strategy is to work with suppliers to get them to implement effective quality management, to improve their quality and then to deliver straight to production with no inwards goods inspection at all.

Mayfair "Number One"
cigarettes are made from the finest
rich matured Virginia leaf,
grown in districts famous for the smoothness
and flavour of their tobaccos.
THE HOUSE OF MAYFAIR, LONDON

### How good are you at inwards goods inspection?

Try this simple test to see if you would make the grade as an inwards goods inspector. Count the number of 'f's' in the main text above within 20 seconds. The correct answer is given in Section 9.14. Did you pass as an inwards goods inspector?

# 9.12 Resin control - as delivered

## Production quality starts here

The inwards goods quality control of resin is essential for quality control in plastics processing but the inspection and control processes are very different from those used for products. Resin control is mainly based on laboratory testing using relatively few samples and also involves relatively sophisticated equipment compared to normal inwards goods inspection.

Resin control is an essential part of processing but some companies carry out little testing. They prefer to trust their supplier and sometimes pay the price later for unacceptable materials variations.

## Raw material selection

Selecting the right material is the responsibility of the design and production areas (see Section 8.3) and purchasing is responsible for ensuring that the specification is correct (see Section 9.7).

- **Tip -** It is assumed that the material as specified runs successfully in production and produces acceptable products.
- **Tip -** The material specification can be taken from a supplier's data sheet but should not refer to the supplier, i.e. keep the specification 'open' for alternative suppliers who can meet the specification.

## The raw material properties

For resin control, the function of inwards goods control is primarily to validate that batch-to-batch variations are minimised and acceptable to the production area, i.e. the material delivered is the same as was previously delivered and ran correctly. This is not always easy because the full purchase specification will include properties that are difficult to quantify and do not assist in deciding if the material is the same as previously delivered.

Detecting batch-to-batch variations at inwards goods needs testing that:

- Measures the relevant properties (or preferably spectra) to validate the material.
- Is quick to operate and low-cost.
- Uses simple equipment.
- Produces unambiguous results.
- Is accurate enough to distinguish between acceptable and unacceptable variations in the material.

This means that testing at inwards goods will not often always match the methods and values on the purchase specification. At inwards goods, the idea is not to fully characterise the material (that should have been done already), simply to check for variations that may adversely affect production or the end product.

- **Tip -** It is rare that inwards goods will be able to check resin to the full purchase specification and order.

If you are using a manufacturer's data sheet as part of the purchase specification then be sure to ask for the tolerances on the properties.

The main decision at inwards goods is:

Is the delivered material the same as was delivered before?

| Typical material tests (as delivered) | |
|---|---|
| **Simple tests** | |
| Test type | Property measured |
| Density | Density |
| Melt Flow Index (MFI) | Viscosity (related to physical structure and average molecular weight) |
| Tensile test | Mechanical properties |
| HDT (Heat Deflection Test) | Thermal and mechanical properties |
| **Complex tests** | |
| Test type | Property measured |
| DSC (Differential Scanning Calorimetry) | Physical structure via thermal transitions ($T_g$, $T_m$ and % crystallinity) |
| TGA (Thermo-gravimetric analysis) | Physical and chemical structure via thermal decompositions |
| TMA (Thermo-mechanical Analysis) | Coefficient of thermal expansion and thermal transitions (rarely used) |
| DMA (Dynamic Mechanical Analysis) | Physical structure via visco-elastic mechanical properties ($T_g$ and $T_m$) |
| FT-IR (Fourier Transform Infrared) | Chemical structure |
| GPC (Gel Permeation Chromatography) | Molecular weight distribution |

**Main test methods for resin control**

The simple tests are low-cost and easy to carry out but do not provide much information at the detailed level. The complex tests use more expensive equipment but can provide much more information on whether the material has changed or not.

## Test decisions

Inwards goods control of materials must decide:

- What measurements are to be made and are they going to be values or Pass/Fail ?
- How often measurements are made?
- Are parallel measurements needed?
- What are the tolerances on the results?
- **Tip** - For some highly regulated products, e.g. pipes and medical products, the test type and minimum frequency is specified by the regulator or the standard. These cannot be ignored!

## Test methods

The main tests used at inwards goods are listed in the table on the left with the property that is measured by the relevant test. These will not be fully discussed as this is available in a variety of good textbooks. At this stage we will simply discuss their suitability for inwards goods testing.

### Simple methods

Simple test methods may be low-cost but they will always give far less information on whether the material is consistent with previous deliveries. Generally the output will be a single number that will not reveal anything about the material performance over the processing and service temperature range. This means that they do not easily discriminate between materials, particularly when the materials are very similar (although they may be different enough to affect processing or the final product properties).

- **Tip** - Single point data whether at inwards goods or at specification can be misleading.

### More complex methods

The more complex methods give far more information about the material, generally over a wide range of temperatures or strain rates, and are particularly good at providing information on whether the material varies between batches or not. These tests do not generally give a single data point but a curve of results that can be used to both characterise the material and to detect if the resin in a delivery is the same as previous deliveries that have both been processed successfully and have produced acceptable products.

The problem with all the more complex methods is that the test equipment is expensive and often difficult to operate.

This has been improved for the new-generation machines which have reduced the need for high operator skill. They have replaced this with simpler 'rule-based' decision making, which allows a pass/fail decision to be made by the machine.

For comparative inwards good testing, DSC probably gives the most useful information provided the user has built up a sufficiently large database of previously acceptable materials.

### New technologies

A new development for inwards goods testing is the Identipol range of machines. These are relatively low-cost and use a combination of DSC and DMA in a single system that can confirm if a delivery is acceptable and can also check how an alternative material compares to that currently used.

- **Tip** - Visit www.identipol.com.
- **Tip** - If you are using (or buying) high quantities of recycled material then the simple methods will not generally give enough information to see if the material is good enough.

## Data analysis

If inwards goods testing generates 'single point data' then the data can be used with an 'Individuals and Moving Range chart' (see Section 5.18) to detect changes and drive improvement.

The more complex methods generate complex curves and make it difficult to use any SPC method.

Higher quality demands on final products increases the importance of quality assurance of raw materials.

Working with resin suppliers to improve the quality and consistency of the incoming raw material is one of the most valuable things a processor can do.

The real problem is that most of the raw materials suppliers are much bigger than any individual processor and can adopt a 'take it or leave it' approach.

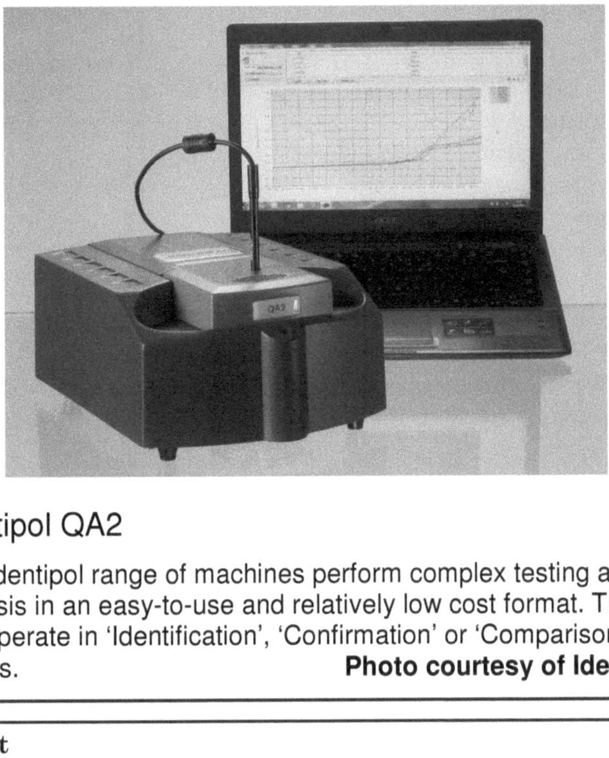


Identipol QA2

The Identipol range of machines perform complex testing and analysis in an easy-to-use and relatively low cost format. They can operate in 'Identification', 'Confirmation' or 'Comparison' modes. **Photo courtesy of Identipol**

# 9.13 Resin control - test pieces and products

## After processing

As discussed in Section 9.12, the inwards goods control of resins is mainly to check that the delivered material matches previously delivered good material. There are, however, raw material properties that cannot be checked simply from the resin as delivered and which need either testing on test pieces or on the final product to verify conformance with the full purchase or regulatory specification.

**Note:** Some product standards (rather than purchase specifications) require raw material testing by batch for simple tests such as density and MFI and on a regular basis, e.g. quarterly, for more complex tests such as impact and tensile strength. It will then be necessary to carry out these tests to conform to the product standard.

- **Tip** - Test piece and product testing is generally more suitable for materials validation at the selection stage (see Section 8.3) than for inwards goods checking. This is because of the care needed for sample preparation and the time taken for testing.
- **Tip** - Test piece testing is generally not suitable for production control of products.

Some of the typical tests that are made using test pieces or products are shown in the tables on the right.

## Test pieces

Most data sheets generated by resin suppliers (and any purchase specifications based on these) will refer to properties that are measured by standard tests using standard test pieces. This will almost always give 'single point data' and will not be representative of the material performance over the wide range of temperatures or loading rates that will be experienced in use.

A selection of typical tests using test pieces is shown in the upper box on the right.

- **Tip** - Tests should always be to national or preferably international standards to allow for comparison of results.
- **Tip** - Single point data are of limited use but you may be forced to provide this or test for it. That is OK for conformance to specification but do not expect it to represent the true performance of the material during use under different temperatures and strain rates.

The result of a standard test will almost certainly not be representative of the behaviour of the material in the final product. This is standard test piece performance and will not describe the performance of the final product. This

Testing at this stage is to validate conformance of the material to specification not the conformance of the product to specification.

**Test piece**

| Test | Measurement |
|---|---|
| HDT - (Heat Distortion Test) | Temperature properties |
| Linear coefficient of thermal expansion | Temperature properties |
| Yield stress | Mechanical properties |
| Tensile modulus | Mechanical properties |
| Izod impact strength | Impact properties |
| Flammability - Oxygen Index | Fire resistance properties |
| Dielectric strength | Electrical properties |
| Light transmission | Transparency |
| Dynamic friction coefficient | Mechanical properties |

**Main test-piece based test methods for resin control**

All of these tests need careful sample preparation, significant equipment and testing to national or international standards but they are still designed to validate the material and not the product.

**Product**

| Test | Measurement |
|---|---|
| Falling weight impact | Impact |
| Solvent stress cracking | Brittle failure |
| Shrinkage | Distortion |
| Dimensions | Weight |
| Weatherability | Exposure |

**Main product-based test methods for resin control**

These are tests that are carried out on the product to account for the effect of processing on the raw material. They can tell us about the raw material but often also tell us something about the processing and the final product.

testing is to validate conformance of the material to specification, not conformance of the product to specification (which should have been already validated).

Even the results for relatively well-controlled test piece testing can vary considerably. This is true for tests such as fracture, impact, fatigue and other mechanical properties where the test result is highly sensitive to the exact shape of the test piece, the exact loading conditions and the processing history of the material.

- **Tip** - Very careful sample preparation is necessary for reproducible results in test piece testing.
- **Tip** - Fatigue test results will generally vary more than for other types of mechanical testing because of the fatigue process.
- **Tip** - Section 8.3 showed that the majority of product failures were due to the cumulative effects of synergies between creep, fatigue and other environmental factors. Single tests will not reveal these synergies or how the material will perform in the real world.

Even when test piece results show low variability then in-service results on the actual product will inevitably show higher variability due to the real and more complex product geometry and loading conditions.

## Products

Apart from the different geometry, the major disadvantage with testing on test pieces is that the test pieces do not replicate the processing history of the resin on real machines. To account for this, some tests to validate the conformance of the material to specification are carried out on actual products rather than on test pieces, e.g. impact testing of rotational mouldings.

This is particularly true for weathering tests where you need to replicate the processing conditions to assess the performance of the material.

A selection of typical tests using product-based testing is shown in the lower box on the left.

- **Tip** - This is, again, testing to validate conformance of the raw material to specification and not to validate conformance of the product to specification - although sometimes, just to confuse you, it can do both.

## Correlating testing and final product performance

Whilst the testing we have discussed so far is inwards goods testing and is concerned with validating the delivered material, it is always useful to try to establish a correlation between the test results and the final product performance. If a correlation can be established then this will provide extra information on the quality of the final product. In the first instance, collect the results of any resin control testing and final product testing and plot these in a simple scatter diagram (see Section 7.7).

When looking for correlations, start by comparing similar properties, e.g. try comparing Izod impact testing on test pieces with falling weight impact testing on products, and move on to comparing properties that have a less direct relationship, e.g. comparing HDT with product shrinkage.

- **Tip** - If the testing is being done then always seek to get extra information from it.
- **Tip** - Establishing a correlation between testing and actual product properties can be easier for product testing where the same test can often tell us something about the raw material and about the final product.
- **Tip** - Looking for correlations adds value and can save later product testing.

## Data analysis

As with direct inwards goods testing (see Section 9.12), test pieces and product testing will generate 'single point data' and the data can be used with an 'Individuals and Moving Range chart' (see Section 5.18) to detect changes that are likely to later have an affect on processing or on the final product.

Increase the value of resin control testing. Get as much information as you can from the testing and do not simply look for a Pass/Fail result.

Colour measurement and conformance to specification of delivered resins can be difficult to assess from test pieces and should generally be assessed using final products.

Compounders can now use in-line colour measurement on compounding lines to validate colour conformance of resins as they are produced. This provides a reference to previously good compounds and an assurance that the delivered resin will perform consistently.

# 9.14 Raw materials control - where are you now?

## It is all in the inputs

Good raw materials control is a fundamental for good production. If the material does not meet specification then it may be impossible for the production area to correctly process the material to give a high-quality product.

For plastics processors, the most important raw material to be controlled is the basic resin from the supplier but many processors either have poor controls or measure only the most basic features of the resin. This will inevitably lead to 'insoluble' issues in the production area.

The keys to raw materials control are:

- Good supplier management and assessment.
- Well-defined specifications for purchased products.
- Well-defined inwards goods inspection and test specifications with appropriate checking levels.

## Completing the chart

This chart is completed and assessed as for the previous charts.

Inwards goods inspection often does not get the investment that is appropriate to the value of the activity.

This is an area that is worth spending some money on to get the material correct at the start.

**Raw materials control**

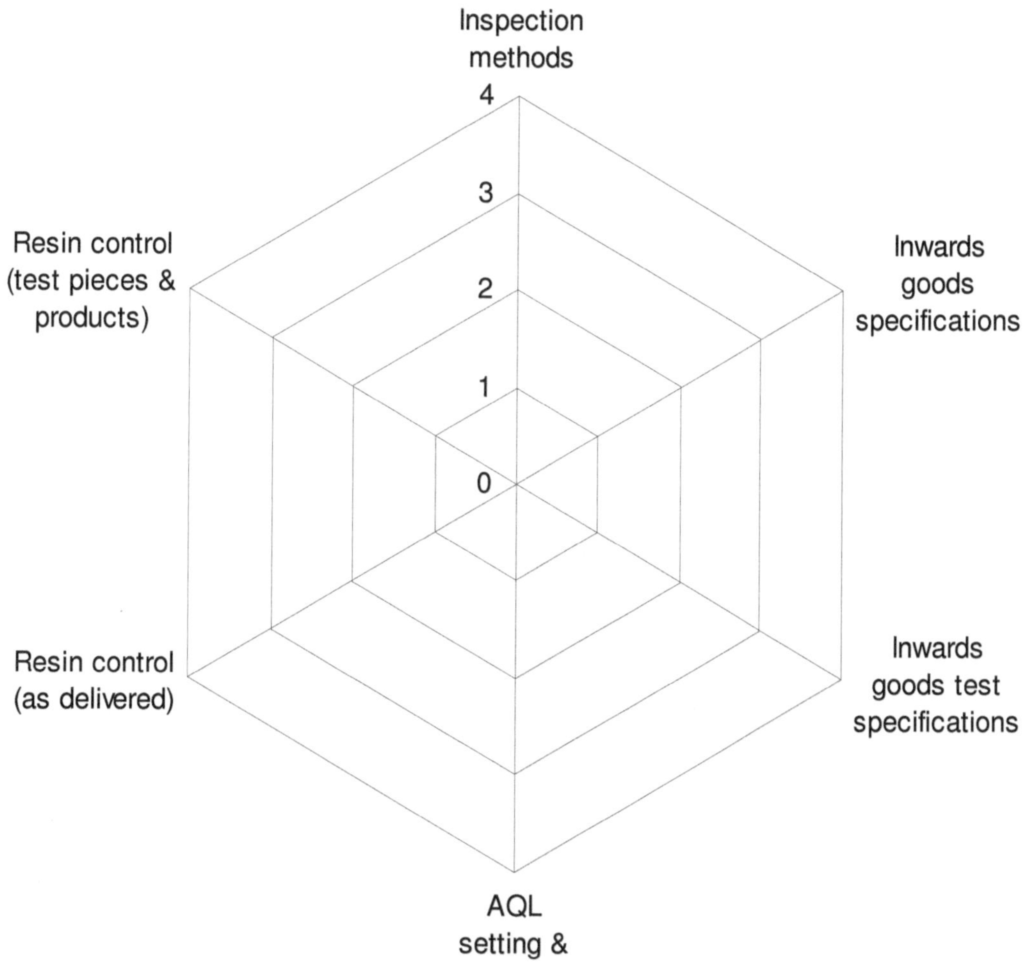


Use the scoring chart to assess where you are in raw materials control

The numbers from the scoring chart can be transferred to the radar chart for a quick visual assessment of where you are in terms of raw materials control.

Answer to Section 9.11:

There are 10 'f's' in the text.

Now go back, count them again and remember that you just let sub-standard material into the production area.

## Raw materials control

| Level | Inspection methods | Inwards goods specifications | Inwards goods test specifications | AQL setting & control | Resin control (as delivered) | Resin control (test pieces & products) |
|---|---|---|---|---|---|---|
| 4 | Excellent definition of inspection method. AQL sampling used for all inwards goods. SPC data accepted from qualified suppliers. | Inwards goods specifications available for all products. Specifications clearly define checks at inwards goods. | Inwards goods test specifications available for all products. | Suppliers select & audit samples to appropriate AQL level for all features/items in the inwards goods specification. Supplier audit results validated & accepted with no further inspection. | Blend of simple & complex test methods used to show material is the same as previously delivered. SPC tracking of results. Excellent material control. | Test piece & product testing used for resin control by delivery. SPC tracking of results. Excellent control of material. |
| 3 | Good definition of inspection method. Inspectors use AQL sampling for all inwards goods. SPC data not accepted. | Inwards goods specifications available for most products at inwards goods. Other products use full purchase specification & decision on checks made by inspector. | Inwards goods test specifications available for most products. | Suppliers select samples for customer inspection to appropriate AQL levels for all features/items in the inwards goods specification. | Simple test methods regularly used. Complex testing if results show issues or on regular basis. SPC tracking of results. Good material control. | Test piece methods used for resin control by delivery. Product testing infrequently used. SPC tracking of results. Good material control. |
| 2 | Moderate definition of inspection method. Inspectors use AQL sampling for some inwards goods. SPC data not accepted. | Inwards goods specifications available for some products at inwards goods. Other products use full purchase specification & decision on checks made by inspector. | Inwards goods test specifications available for some products. | All sample selection & inspection by customer. Good AQL levels set for all features/items in the inwards goods specification. | Simple test methods used for resin control. No SPC tracking of results. Complex testing only on regular basis, e.g. 3-monthly. Moderate material control. | Test piece methods used for resin control by delivery. No product testing. No SPC tracking of results. Moderate material control. |
| 1 | Poor definition of inspection method. Inspectors use constant % sampling or spot checking. No use of AQL & SPC data not accepted. | Only full purchase specification available at inwards goods. Checked items are chosen from full purchase specification by inspector. | Inwards goods test specifications available for very few products. | Poor selection of AQL levels for many features/items in the inwards goods specification. | Simple test methods used for resin control. No use of more complex testing. Little material control. | Test piece methods used for resin control but only on a regular basis, e.g. 3-monthly, to meet standards. No product testing. Little material control. |
| 0 | No defined inspection method. Inspectors use own inspection & sampling system. No use of AQL & SPC data not accepted. | No specifications available at inwards goods. Checked items are decided by inspector. | No inwards goods test specifications available for any product. | No AQL system or inwards goods specification at inwards goods. | No resin control at inwards goods inspection. Material delivered direct to production & used without testing. Poor material control. | No checks on resin through test piece or product testing. Poor material control. |
| Score | | | | | | |

# Key tips

- A clear and unambiguous specification is a prime requirement for inwards goods control.
- The specification should be used to select only suppliers who are capable of delivering to the requirements.
- Define the raw materials control process to ensure that only good materials enter the production system - it is too late afterwards.
- Suppliers need to be integrated into the 'materials team' so that they fully understand the requirements of the product and can reduce both quality issues and cost.
- Regularly assess the performance of the materials team to drive improvement.
- Supplier relations can be co-operative or competitive but the co-operative approach works better.
- Supplier management is the key to good supplier relations and performance. This needs:
  - An initial assessment of the supplier.
  - A full supplier relations manual that clearly defines the customer requirements.
  - A supplier pack of all the relevant documents for the product being delivered.
  - A clear statement of what is expected of the supplier at inwards goods inspection and test.
  - A commitment to supplier liaison and development.
- Supplier assessment is not simply 'do they have ISO 9001 or not' - it is about far more than simply their quality management system.
- Supplier assessment is about how well the potential supplier and their products can meet your needs as a customer.
- Supplier assessment is not a 'one-time' event. It is a continuous process of vendor rating based on clearly defined and well-understood performance measures.
- Supplier assessment also needs to include areas such as full materials data, health and safety data and social, economic and environmental practices.
- The feedback from vendor rating can be used for quality improvement with suppliers but this is a process that needs careful and consistent management.
- Good purchasing specifications will help to define and maintain the quality of raw materials.
- The first 'price' of a product is not the total cost. Customers need to calculate and control the Total Acquisition Cost to improve quality and reduce costs.
- Assessment of inwards goods is a decision process.
- If the supplier is very good then no inspection may be necessary.
- If the supplier has a poor record then inspection is needed but it must be the right inspection and have a good statistical basis.
- Acceptance sampling by attributes using the appropriate AQL for defined features provides an economic and quick method of assessing the quality of inwards goods but the features need careful definition.
- SPC can be used for inwards goods control but needs some historical data to work properly.
- Control of resin is the key issue for plastics processors.
- Resin control at inwards goods is not to fully characterise the material. It is to check that the material is essentially the same as previous deliveries that have been processed successfully.
- Many of the simple resin tests give little information on the acceptability of the resin.
- More complex testing methods give good information and the prices are dropping for this type of equipment.
- Testing of test pieces and products can produce good information on the batch-to-batch consistency of materials but be careful about using 'single point data' to extrapolate to real performance.
- When using test pieces or product testing to validate resins then the consistency of the sample preparation and test method are very important.

# Chapter 10

# Processing quality management

Processing is where the difficult stuff starts and quality processing is a key to success in plastics processing. In the preceding chapters we have covered the basics of quality management but this section goes into the detail of managing plastics processing to produce high quality products to the customer specification. This is what plastics processors do every day and is at the heart of why they exist. If you are not producing high-quality products then no matter how low your price, you will not be in business for long.

In this chapter we will cover most of the main plastics processing methods and consider some of the unique issues that need to be considered to consistently deliver quality products cost-effectively from each process.

Initially, we will deal with some problems that are common to many industries, e.g. setting process standards, creating process FMEAs, establishing control plans, deciding between product and process control and testing, the best ways to set machines and the basics of off-line quality control. These are all vital issues whichever process you are using and help to set the scene for controlling quality in processing.

We then deal with the unique issues for several of the major processes used with plastics. The main processes used by the industry are extrusion and injection moulding (by volume) and naturally we will spend longer in the discussion of these processes. The other, smaller-volume, processes will be dealt with but simply not to the extent of extrusion and injection moulding. In any case, many of the issues, such as shrinkage, are unique to the plastics industry but are shared by all of the common processes.

We will also cover Engineering Process Control (EPC) a development in the control of processing equipment to deliver quality automatically and the process for initial sample approval (PPAP).

In many ways, this chapter brings together the threads that have been developed earlier in this Workbook and readers are advised not to jump in here without reading much of the preceding sections - still, you won't listen to me anyway.

"If you have always done it that way, it is probably wrong."

**Charles Kettering**

# 10.1 Process design and development

## Taking product design to process design

As the product is being designed (see Chapter 8) the production process itself needs to be designed to ensure that the system is capable of delivering to the customer requirements. Process design must be carried out in parallel with product design.

The product design is validated at the prototype stage and the outputs of the product design and development review are shown in the left column of the table on the lower right. All of these outputs are necessary to complete an effective process design. However, process design and development will always continue for some time after pilot production to accommodate changes in product design.

During process design and development the team develops the process to ensure that this is both comprehensive and effective, i.e. that the process is 'capable' of meeting the customer's requirements. Process design and development in the context of the AIAG APQP process is shown on the far right and the outputs of the process design and development phase are shown in right column of the table on the lower right.

## Process flow chart

The process flow chart is one of the most important outputs from process design and development. This is a complete flow chart of the process flow and can be used to identify sources of variation using cause and effect charts) for the complete process. A process flow chart means that the team can look at the complete process rather than on the individual steps in the process and this is invaluable when completing the Process FMEA (see Section 10.2) and the Control Plan (see Section 10.3).

The process flow chart is completed as per a standard flow chart (see Section 7.2). It must:

- Show the sequence and linkage between production and inspection/control areas.
- Include links to the Process FMEA and the Control Plan.
- Describe, in detail, how the product will move during the process.
- Consider 'push/pull' production decisions.
- Identify where any reworked products (if any) will be inspected before re-entering the production process.
- Identify and correct any potential quality issues with handling or external processes.

If you are inspecting lots of products then you have got it wrong further up the chain.

## Process FMEA

A process FMEA is essential before production trials are started. A process FMEA is similar to a design FMEA but covers the possible failure modes during production rather than design failure modes. The methodology is similar but the boundaries are very different.

The process FMEA is covered in more detail in Section 10.2.

## Pre-production control plan

A pre-production control plan is essential before production is started. This is similar to the site-wide control plan discussed in Section 4.18 but is much

| Inputs from product design and development (see Section 8.9) | Outputs to product and process validation (see Section 10.5) |
|---|---|
| ■Design FMEA.<br>■DFM and DFA.<br>■Design verification.<br>■Finite element analysis (FEA).<br>■Engineering drawings and specifications.<br>■Materials specifications.<br>■Tooling specifications.<br>■New equipment specifications.<br>■Draft control plan.<br>■Prototype build.<br>■Design review. | ■Process flow chart.<br>■Process FMEA.<br>■Pre-production control plan.<br>■Process instructions.<br>■Measurement systems analysis plan.<br>■Preliminary process capability study plan<br>■Floor plan layout.<br>■Packaging standards and specifications. |

**The inputs and outputs of process design and development**

Initial process design is carried out in parallel with the product design but after the product design is complete, process design and development continues.

more product-specific.

The pre-production control plan is covered in more detail in Section 10.3.

- **Tip** - A pre-production control plan will inevitably contain more product and process controls than the final production control plan.

## Process instructions

The process instructions are at the heart of successful plastics processing but many companies neither set machines in a logical manner, record these settings accurately nor enforce the use of these settings.

Getting the process instructions correct is covered in more detail in Section 10.4.

## Measurement systems analysis

The pre-production control plan gives details of the controls and measurements to be made during production but analysis of the measurements to ensure that they are valid is rarely carried out in the plastics industry. In fact, it appears to be almost unknown except for those companies dealing with the automotive industry.

The use of measurement systems analysis and a discussion of this is given in Appendices 6 and 7.

## Preliminary process capability study plan

The characteristics identified in the pre-launch control plan (see above and Section 10.3) will need assessment for process capability during product and process validation. This will necessarily be a machine capability assessment (see Section 5.4) because full process capability (see Section 5.22) can only be carried out after long-term production to allow other external factors to be included.

At this stage the output is the plan for machine capability assessment and not the actual assessment.

## Floor plan layout

The floor plan layout describes, in detail, the layout of the production area. The floor plan layout must show the location of all machinery, tools, process instructions, inspection areas, control charts, visual aids and storage areas for materials (raw materials, work-in-progress, suspect materials, rejected materials and finished goods). Material flow in the floor plan layout must be linked to the process flow chart and to the pre-launch control plan.

- **Tip** - Check that all process and inspection areas are of the correct size and adequately lit.
- **Tip** - Check that inspection areas are located to prevent the shipment of non-conforming products.
- **Tip** - Check that the layout and/or controls prevent the mixing of finished goods with other similar products.
- **Tip** - Check that the layout and/or controls prevent environmental contamination from external sources.

## Packaging standards and specifications

It is usual that the customer will have specified the packaging for delivery and use but if this is not available then packaging standards and specifications must be established and documented so that the product characteristics are maintained during storage and transport to the customer.

## Product and process validation

After completion of the process design and development, the next step is validation of the product and process via pilot production and a production part approval process (PPAP). This is covered in Section 10.5.

A site cannot operate at more than one quality level.

The quality level for all of your products will effectively be the same whether you want it to be or not.

•1. AIAG. 2008. *Advanced Product Quality Planning and Control Plan.* AIAG.

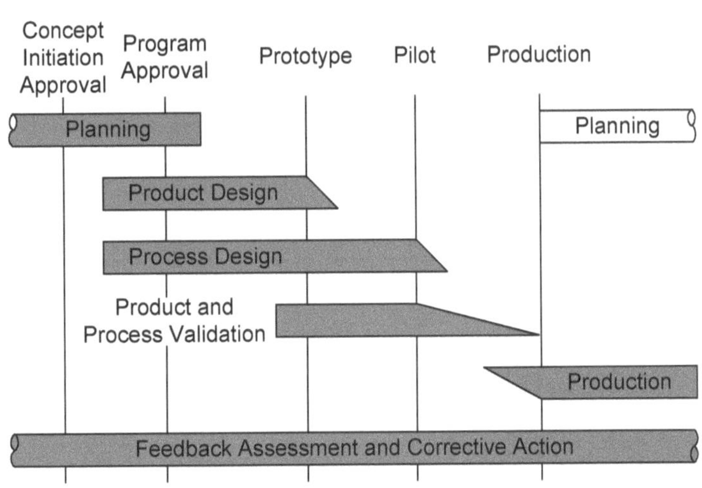


### Process design and development in APQP

Process design is essentially complete after pilot production when the product and the process are both validated to ensure that they produce a product that meets the customer's needs (see Section 10.5).

# 10.2 Process and machinery FMEA

## It is all coming together now

A process and machinery FMEA is very similar to a design FMEA (see Sections 8.11 and 8.12) and uses the same basic risk assessment methodology. In this case, the assessment is concerned with the production process and machinery.

A process and machinery FMEA should be part of the process design to assess potential risks to quality and the action that can be taken to reduce these risks. This will lead to more fault-tolerant production processes and a reduction in production-related quality issues.

The process and machinery are investigated to assess potential areas where poor performance or failure could impact on product quality or safety and each potential risk is assigned a rating after considering the following:

- Severity (S) - 1 is low severity and 10 is high severity.
- Probability (P) - 1 is low probability and 10 is high probability.
- Current controls (C) - 1 is high current controls and 10 is low current controls.

Values for S, P and C for each risk/event are given in the tables below. For consistency, these are the same tables that are used for a design FMEA.

## Risk Priority Number (RPN)

For each potential cause, the individual factors are rated and multiplied together (RPN = S x P x C) to produce the relevant RPN rating.

Action can then be taken and recorded on the significant RPNs to:

- Reduce the severity.
- Reduce the probability.
- Improve the process or temporarily improve the controls.

The RPN is the reassessed after completing the actions.

- **Tip** - Action should be mandatory for any severity rating of 9 or above.
- **Tip** - A process and machine FMEA should be reviewed and revised with time (it is a 'live document') to reflect new equipment, processes and procedures. This will allow the control plan to be reviewed with the experience gained.
- **Tip** - A process and machinery FMEA should be linked to each part of the process flow chart.
- **Tip** - As noted in Section 8.11, SnapSheets XL from SigmaZone can do most of the hard work in generating a process FMEA.

Process and machinery risk assessment can remove many of the small errors in process design that cause poor quality as it is perceived by the customer.

There is no 'magic' RPN number above which you must take action. This is simply a method of highlighting and prioritising risks.

You can take action on whatever you like but it does makes sense to reduce the highest risks first.

The best FMEAs are often the shortest. Get the basics of the process right and the rest follows.

| Severity of event (S) | Ranking | Probability of event (P) | Ranking | Current controls (C) | Ranking |
|---|---|---|---|---|---|
| Hazardous: without warning | 10 | Very High: event is inevitable | 10 | Absolute uncertainty of detection | 10 |
| Hazardous with warning | 9 | | 9 | Very remote chance of detection | 9 |
| Very High | 8 | High: Repeated events | 8 | Remote | 8 |
| High | 7 | | 7 | Very low | 7 |
| Moderate | 6 | Moderate: Occasional events | 6 | Low | 6 |
| Low | 5 | | 5 | Moderate | 5 |
| Very low | 4 | | 4 | Moderately high | 4 |
| Minor | 3 | Low: Relatively few events | 3 | High | 3 |
| Very minor | 2 | | 2 | Very high | 2 |
| None | 1 | Remote: Event is unlikely | 1 | Almost certain to detect | 1 |

| | | | |
|---|---|---|---|
| Process | Sash Profile - Compact 121 | Date | 05/10/2015 |
| Prepared By | Robin Kent | | |
| Notes | | | |

| Function / Requirement | Potential failure mode | Potential effect of failure | Sev (S) | Class | Current process controls: Potential cause | Prob (P) | Prevention | Detection | Controls (C) | R.P.N. | Recommended actions: Recommended actions | Responsible | Target completion date | Action results: Actions Taken | Sev (S) | Prob (P) | Controls (C) | R.P.N. | Class |
|---|---|---|---|---|---|---|---|---|---|---|---|---|---|---|---|---|---|---|---|
| Extruder and upstream | see previous sheet | | | | | | | | | | | | | | | | | | |
| Die heating | Die heat failure (complete) | Extrusion difficult Profile out of tolerance Potential damage to thrust bearing | 8 | | Controller/thermostat failure | 2 | Regular maintenance check for operation | QC on output | 5 | 80 | Start-up check by operator/setter. Add visual check start-up checklist | Production control for checklist Operator/setter for regular checks | 21/03/2016 | Added to checklist | 8 | 2 | 2 | 32 | |
| | | | | | Wiring failure | 2 | Regular maintenance check for operation | QC on output | 5 | 80 | Start-up check by operator/settter. Add visual check to start-up checklist | Production control for checklist Operator/setter for regular checks | 21/03/2016 | Added to checklist | 8 | 2 | 2 | 32 | |
| | | | | | Heater failure | 2 | Regular maintenance check for operation | QC on output | 5 | 80 | Start-up check by operator/settter. Add visual check to start-up checklist | Production control for checklist Operator/setter for regular checks | 21/03/2016 | Added to checklist | 8 | 2 | 2 | 32 | |
| | Die heat inaccuracy (too high) | Profile out of tolerance | 6 | | Controller/thermostat failure | 3 | Maintenance check for calibration | QC on output | 5 | 90 | Add separate thermostat and warning light to system | Maintenance | 21/03/2016 | To be completed | 6 | 3 | 3 | 54 | |
| | Die heat inaccuracy (too low) | Extrusion difficult Profile out of tolerance Potential damage to thrust bearing | 8 | | Controller/thermostat failure | 3 | Regular maintenance check for calibration | QC on output | 5 | 120 | Add separate thermostat and warning light to system | Maintenance | 21/03/2016 | To be completed | 6 | 3 | 3 | 54 | |
| Profile calibration | Water bath temperature (too high) | Profile not fully cooled and not calibrated properly | 6 | | Chiller settings for complete chiller system | 2 | Regular maintenance check for settings | LED indicator of water flow temperature visible in production area | 1 | 12 | No action | | | | 6 | 2 | 1 | 12 | |
| | | | | | Chiller failure (general system) | 1 | Regular maintenance check for settings | LED indicator of water flow temperature visible in production area | 1 | 6 | No action | | | | 6 | 1 | 1 | 6 | WFF |
| | | | | | Pump failure (general system) | 2 | Regular maintenance check for settings | LED indicator of water flow temperature visible in production area | 1 | 12 | No action | | | | 6 | 1 | 1 | 6 | WFF |
| | | | | | Hose restrictions (for specific machine) | 4 | Regular maintenance check for hose condition | | 4 | 96 | Start-up check by operator/settter. Add visual check to start-up checklist. | Production control for checklist Operator/setter for regular checks | 21/03/2016 | Added to checklist | 6 | 4 | 2 | 48 | WFF |
| | Water bath temperature (too low) | Profile cooled too quickly before calibration effective | 2 | | Chiller settings for complete chiller system | 2 | Regular maintenance check for settings | LED indicator of water flow temperature visible in production area | 1 | 4 | No action. | | | | 2 | 2 | 1 | 4 | WFF |

Continued on next sheet

Example of a Process FMEA using Snap using SnapSheets XL to generate Risk Priority Numbers

This is not a complete example and the actual process FMEA is much longer and detailed. It is shown only to illustrate the method and the output.

# 10.3 Control planning and control plans

## A documented description of systems and processes for controlling the product

Control planning and the control plan are essential elements of advanced quality planning and are some of the building blocks by which quality is planned into the product.

Control planning for production is similar to the global quality plan concept shown in Section 4.18. In the global case, the plan is used to gain an overview of the quality system for the complete company. In the production case, control plans are product-specific and deal with the checks and balances used during production using specific processes.

The control plan:

- Identifies the controls required to ensure product quality with a focus on any special characteristics defined by the customer.
- Defines the reaction plans to be implemented where nonconformities are identified, e.g. product containment and 100% inspection to ensure that the process becomes stable and capable.
- Is an output from the FMEA process.

The control plan provides information on the process and product controls during production and for finished products. It also provides the essential guidance on defect prevention at the operator level.

Overall, the control plan is a systematically compiled document which summarises the quality planning and the requirements for all significant characteristics of the product. It can be used not only for control of internal production but also as a reference document by sub-contract moulders for the quality standards to be applied.

- **Tip -** If external suppliers are being used then the control plan must be supplied as part of the supplier package (see Section 9.3).

The control plan is a 'live' document and must reflect any significant changes made in the sourcing, processing or requirements of the product.

- **Tip -** Control plans can be made at the system, sub-system or component level.

## Stages of control planning

A control plan should be produced and revised at 3 separate stages during the journey to full production:

- The prototype control plan focuses mainly on the product design and conformance to requirements rather than on the process.
- The pre-production control plan is a full description of the dimensional, material and functional checks that will take place after prototyping and before full production. This will include the additional checks needed before full production validation. These may include:
  - Increased inspection levels (SPC or otherwise).
  - Increased points for in-process checks.
  - Assessment of machine and/or process capability.
- The production control plan will often be the shortest of all the control plans. This is because the process will have been validated (see Section 10.5) and the checks that remain will be to ensure control rather than capability.
- **Tip -** All the stages of control planning use the same basic format. The only difference is in the detail and the type of control plan should always be noted.

## Completing the control plan

The control plan should use the process flow chart and the FMEA data (design, process and machinery) to provide the basic information for the control plan and should cover:

- Incoming raw materials inspection points and requirements.
- Production inspection points (both in-process and final) and requirements.
- Off-line inspection points, e.g. laboratory tests, and requirements.
- Fit and function testing for mating parts.
- Assembly inspection points and requirements.
- Packaging requirements and inspection points.
- Final inspection points and requirements.

Every part must have a control plan, but family control plans can be used, e.g. an extrusion site producing many different profiles for similar uses could devise a 'family' control plan' for the all production.

The control plan is a 'living document'.

It should be used to identify and communicate any changes to the process or the product and the associated changes to the controls.

- **Tip -** Standards requirements (both customer and regulatory/legal) should be clearly identified on the control plan.
- **Tip -** If specific customer issues have been identified these should be identified in the control plan and be signed off by the customer.
- **Tip -** Any special characteristics (product or process) should be identified using the following key:
  - CTQ: Critical to Quality.
  - KPOV: Key Process Output Variable.
  - KPIV: Key Process Input Variable.
  - USL: Upper Specification Limit.
  - LSL: Lower Specification Limit.
- **Tip -** Where gauge testing, e.g. go/no-go or similar, then gauges must be compatible with any used by the customer.
- **Tip -** The control plan is the ideal opportunity to review the control methods for both process and product so that they are robust and defensible.
- **Tip -** If process improvements are made, and shown to be so through measurement of the process capability, then it is appropriate to reduce inspection and monitoring. Equally, if the process capability decreases then increased inspection and monitoring may be needed.

## Reaction planning

One of the key benefits of creating a control plan is that it forces consideration of reaction planning, i.e. what you are going to do when something goes wrong. Reaction planning should include all of the following cases:

- Action when incoming materials are not to specification.
- Action when processes are out of control either through special or common causes.
- Action when products are out of specification but the cause is not known.

The actions to be taken must be planned and documented for all the staff so that suspect or nonconforming product is quarantined to prevent further processing or delivery. The reaction plan is the essential reference for risk-based thinking.

Any inspection or control activity costs money. The control plan allows an assessment of the overall cost of control (and equally the risk and cost of failure to control).

Error proofing is more effective (and cheaper in the long term) than any other type of control method. The control plan provides the opportunity to look for opportunities to error-proof the process.

**Control Plan**

☐ Prototype ☑ Pre-Launch ☐ Production — Page: 2 of 5

| | | | |
|---|---|---|---|
| Control Plan Number: CP 010 153 | Key Contact/Phone: Robin Kent (034 146 457 697) | Date (Orig.): 12/03/2015 | Date (Latest Rev.): 28/10/2015 |
| Part Number/Latest Change Level: C60-1101, Drawing Issue 5 (12/03/2015) | Core Team: MA, TM, MMcM, GL, DS. | Customer Engineering Approval/Date: Not required. Internal production for stock supply. | |
| Part Name/Description: Main Frame: Compact 60 | ~~Supplier~~/Plant Approval/Date: Michael Adams, 30/10/2015 | Customer Quality Approval/Date: Not required. Internal production for stock supply. | |
| ~~Supplier~~/Plant: Shuttleworth Extrusion; Supplier Code: N/A | Other Approval/Date: N/A | Other Approval/Date: To meet EN 12608. | |

| Part/ Process Number | Process Name/ Operation Description | Machine, device, jig or tool | Characteristics: No. | Characteristics: Product | Characteristics: Process | Special Char. Class | Methods: Specification or tolerance | Methods: Evaluation or measurement technique | Methods: Sample Size | Methods: Sample Freq. | Methods: Control method | Reaction plan |
|---|---|---|---|---|---|---|---|---|---|---|---|---|
| 3 | Calibration of extrusion | Calibration bath (water temperature) | 36 (Process FMEA) | - | Calibration bath water temperature | CTQ | 12°C ± 1°C | Temperature sensor and alarm | N/A | N/A | Automated alarm for temperature variation | Cease production and investigate. Check calibration bath water temperature. Contact maintenance to resolve. |
| 4 | Calibration of extrusion | Calibration bath (vacuum level) | 37 (Process FMEA) | - | Calibration bath vacuum levels | CTQ | Calibrators 1 to 3: 25 -kPa ± 5 -kPA Calibrators 4 to 6: 20 -kPa ± 5 -kPA | Vacuum sensor (6x) | N/A | N/A | Automated alarm for vacuum variation | Cease production and investigate. Check calibration bath vacuum levels. Contact maintenance to resolve. |
| 7 | Final extrudate dimensions | Operator measurement | 45 | Profile Depth (D) | - | USL and LSL | 60 ± 0.3 mm (EN 12608) | Micrometer and/or Shadowgraph | 5 | Hourly | Mean and range control chart | If chart indicates that the process is out of control then take local action to resolve. Check production since previous check for conformance to specification. |
| | | Operator measurement | 46 | Profile width (W) | - | USL and LSL | 75 ± 0.5 mm (EN 12608) | Micrometer and/or Shadowgraph | 5 | Hourly | Mean and range control chart | |

Control planning gives all the details necessary to show compliance with the customer requirements

The control plan is a method of getting all the product and process requirements in one place to review the level and extent of control in one place. It must include both the product and process controls as well as the full reaction plan in the event that either of these indicate the potential for the production of nonconforming product.

# 10.4 Process setting

## Getting the best from the start

The initial process settings selected at process design and development are key to getting the best results from the process but many companies do not realise the effect that variability has on process setting and the need to explore the complete 'process space' when deciding on these settings.

## The effect of process variability

The classic experiment to illustrate the effect of process variability and 'tampering' on running production is the Deming 'Funnel Experiment'. This describes what happens when processes are constantly adjusted in response to the natural process variations.

The same type of effect is seen when initial process setting during process design and development is based on very few samples and, in the worst case, on a single sample. Setters are often desperate to get a process approved (the project started late and they are trying to make up time) and they may base the initial process settings on very few samples.

Consider the following series of events:

- The setter selects one sample (which happens to be at the top of the process distribution) and finds that it is 'high'. This is despite the fact that the process is actually totally fine (see upper right).
- The process is adjusted downwards by the same amount that it was running high - therefore pushing the process outside the set limits so that it produces parts outside specification. The process is measured again but this time the selected sample is at the bottom of the process distribution. The setter obviously finds that it is 'low' (see middle right).
- The process is again adjusted and this time overshoots so that it is running high and parts are again produced outside of specification (see lower right).
- The cycle continues and the process see-saws in and out of specification whilst the setter continues to adjust the process in a futile attempt to find settings that actually work.

This situation happens day after day in industry as setters try to find settings that

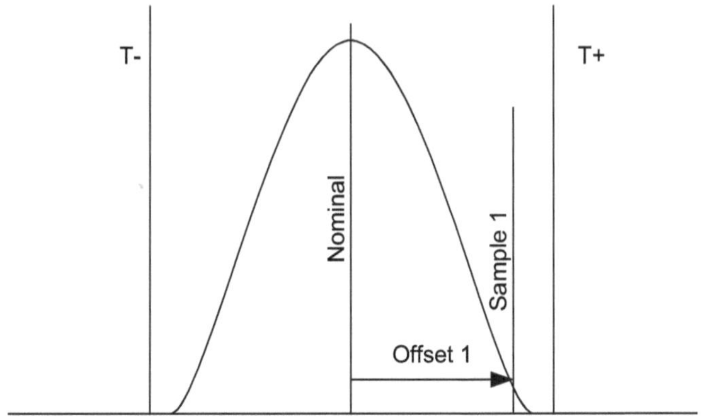


The process is set with a single reading. If this is located at the top of the distribution then it will appear that the process is 'high', even though the process is actually centred and in control.

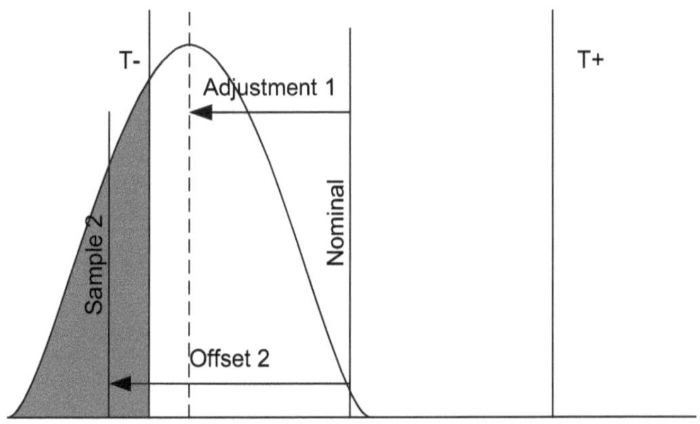


The process is adjusted down by the amount that it was 'high' - pushing the process so that it produces outside specification. The process is measured and is 'low'.

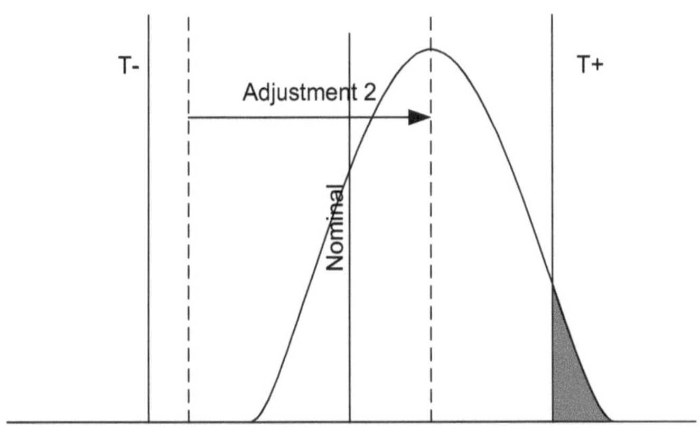


The process is again adjusted and this time overshoots so that it is running 'high' and parts are again produced out of specification.

### Do not set processes from a single sample!

Process variability is similar to tampering and introduces excessive variation into the system. Setting processes from a single sample will always result in excessive variation and difficulty in finding stable process settings.

will allow them to sign off the process.

Even worse (perhaps) is when the setter selects a single sample that shows the process to be OK when it is mostly outside the tolerance limits - the process is then approved with settings that will inevitably result in nonconforming parts being produced when the process goes into full production.

- **Tip** - To set a process you should never rely on a single sample result. Always make at least 5 measurements and set the process on the basis of the average of the 5 measurements.
- **Tip** - If you really are in a hurry then you can always try using 'pre-control' to account for process variability (see Section 10.6).

## Searching for the local maximum

At the process design stage, processes are often set using a first 'guess', i.e. some traditional settings that worked for a similar product. The setters then play with the settings for a while until they get the 'best' settings for the process. This first 'best' setting is assumed to be the overall best setting and this is recorded as the full production settings. This may work if the 'best' setting is actually the overall 'best' setting but, in most cases, it will be a local maximum rather than the system maximum (which can be very different).

The figure on the right shows an imaginary response surface with a local maximum (A), and a system maximum (B). If the first 'guess' based on traditional settings is in the area of 'A' then changing the settings, (often via the 'blind man's walk' approach) to increase the response will result in the optimum settings being those at which 'peak' performance is encountered, i.e. A.

Any changes from these settings lead to a decreased response and the process is regarded as optimised at the 'A' settings and the process is set to run under these conditions.

This method both ignores and fails to discover the higher 'global' maximum in the area of 'B' where the settings could be better for consistency and speed.

Setting for local maxima on a complex response surface, e.g. most plastics processing methods, is always misleading. It can prevent progress not only in plastics processing but also in many other areas such as evolution where the attempt to achieve a 'local maximum' can prevent further progress to the 'global maximum' because any change results in a lowering of performance (for a more complete discussion of local maxima in evolution see 'Climbing Mount Improbable' by Richard Dawkins, 1996, Penguin, 978-0141026176).

- **Tip** - Processes must be set using correctly designed experiments and studies - Taguchi methods and DOE and scientific moulding studies are the best for this (see Section 8.15).
- **Tip** - Always ensure that the material used is the same as the production material and that the machine used is the same model that will be used for subsequent production.
- **Tip** - The initial parameters should be used to calculate the control chart parameters to allow SPC charts to be set up for future runs.

Setting sheets are a valuable tool to reduce process variation. Failing to use them is a mortal sin in terms of quality management.

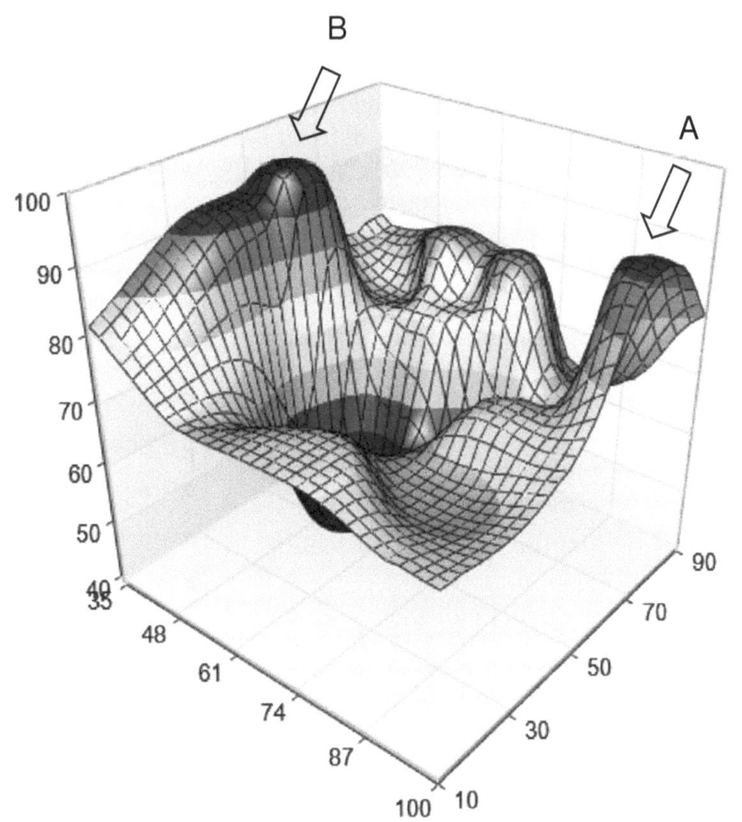


Process setting by the 'blind man's walk' method

Having a process set on the parameters for the local maximum 'A' means that any small process change will involve a decrease in response and a subsequent return to 'A' as the 'best' setting. The global maximum of 'B' will remain undiscovered.

# 10.5 Product and process validation

## Checking that you got it right

Product and process validation is to ensure that products made from the full production tools and processes meet the relevant engineering requirements or specifications, that the processes are capable (using $C_p$, $C_{pk}$ and other SPC data) and can produce acceptable product consistently over time.

- **Tip -** Product and process validation should also check that all the production documentation (control plan and process flow chart) is valid and being followed.

The product and process validation procedure is called PPAP (Production Part Approval Process) in the automotive industry but in the medical or pharmaceutical industries it is often called IQ-OQ-PQ (Installation Qualification-Operational Qualification-Process Qualification). Despite the different names, these are designed to give much the same output, i.e. process and process validation. We will use 'PPAP' to represent the complete validation process.

The format of the output will vary with the industry and is often closely defined by the industry, e.g. the AIAG PPAP process. Whatever format is used, the end result should be a documented process that can reliably deliver the final product to meet the customer's requirements.

Successful completion and sign-off of the PPAP process is a key event in the journey to full production and can act as the signal for several events:

- Approval for full production and to ship production to meet customer orders.
- Payments from the customer for tooling and/or machinery.
- Payments to suppliers for tooling and/or machinery.
- Revision of supplier assessments by the customer.

The financial implications of PPAP can be very significant and getting it right is a crucial milestone in process and product development. The main actions that are needed for PPAP are listed below.

## Trial production

Trial production is obviously one of the main parts of product and process validation and should be a 'significant production run'. This is a production run of between 1 and 8 hours (longer is better) with a minimum of 300 consecutive parts. This will obviously increase for multi-cavity tooling.

The significant production run should:

- Be at the final production site.
- Use the proposed production rates/cycle times.
- Use full production tooling.
- Use production gauges and test methods from the pre-production control plan.
- Use the proposed production materials.
- Use the production operators.

Trial production should be carried out on every unique production process in the process flow chart. This includes all downstream operations, e.g. assembly and decoration operations. Only by looking at the complete process can the effectiveness and efficiency be assessed.

- **Tip -** There is often pressure to make trial production short to make up for time lost in other parts of the process. Resist this temptation at all costs. This is the time to find out if the process is

PPAP provides a logical and thorough process for assessing the readiness of a process and product for full production.

Some companies (mainly German and working to VDA 2) require an ISIR (Initial Sample Inspection Report) for acceptance of production.

All of the data for an ISIR will already have been completed as a result of the PPAP process (and more).

| Inputs from process design and development (see Section 10.1) | Outputs to production and feedback |
|---|---|
| ■Process flow chart.<br>■Process FMEA.<br>■Pre-production control plan.<br>■Process instructions.<br>■Measurement systems analysis plan.<br>■Preliminary process capability study plan<br>■Floor plan layout.<br>■Packaging standards and specifications. | ■Trial production.<br>■Measurement systems evaluation.<br>■Preliminary process capability study.<br>■Production validation.<br>■Packaging evaluation.<br>■Production control plan.<br>■Full-quality plan sign-off.<br>■Production part approval. |

**The inputs and outputs of process and product validation**

After completion of the product and process design, the complete system needs to be validated to ensure that it delivers the required level of customer satisfaction.

robust. After PPAP is signed off you are into production orders and fixing something up is much more difficult.

- **Tip** - Pay attention to the ability of linked processes to match production rates. If decoration cannot keep up with production there is pain in the future.

## Measurement systems evaluation

Trial production should be used to carry out measurement systems evaluation (see Appendices 6 and 7).

## Initial process capability study

Machine capability data (see Section 5.4) may have been carried out earlier in the process development but trial production provides the extended production run that is necessary to assess the overall process capability (see Section 5.22). Initial process capability studies should be carried out on all of the characteristics identified on the pre-production control plan.

- **Tip** - These characteristics will have also been identified on the control drawings (see Section 8.13).

The initial process capability study will reveal areas where the process is capable but may also reveal areas where changes may need to be made to drawings, tooling, mating parts or downstream process equipment.

- **Tip** - In many cases, process capability studies are carried out on too many dimensions. To reduce the workload, make sure that the control plan is focused on those dimensions that are important and 'control' the product.
- **Tip** - There will inevitably be non-critical dimensions that do not require a capability study. These values can be simply reported (without capability studies).

## Production validation

In addition to the process capability studies, there will be other tests, e.g. material tests, mechanical tests and 'fit and function' tests that must be carried out away from the process area. This testing must all be carried out on the trial production to fully validate the process and product.

- **Tip** - PPAP is concerned with the whole product and not simply with the dimensions and appearance.

## Packaging evaluation

Most products will be packaged for delivery to the customer, i.e. in 'bulk' packaging or 'display' packaging and the trial production should be used to evaluate the 'fit and function' of the packaging that will be used to protect the product.

This is the last chance to get it right.

Do not rush the product and process validation stage. After this, you are into series production and any changes are not only difficult but painful.

## Production control plan

The results of the initial process capability studies and production validation are used to produce the production control plan.

- **Tip** - The initial production control plan will typically contain less detail than the pre-production control plan.

## Full quality plan sign-off

After the control plans and process flow chart have been validated then the overall quality plan can be signed-off and issued.

## Production part approval

Final production part approval will be given by the customer (if the product is an external development) or by internal release (if the product is an internal development). The status of production part approval can be one of:

- Approved for production.
- Interim approval for production (subject to further studies).
- Rejected for production.

The output of PPAP should be complete documentation of the process and product. Use a checklist to make sure that all the components have been completed.

- 1. AIAG. 2008. *Advanced Product Quality Planning and Control Plan.* AIAG.

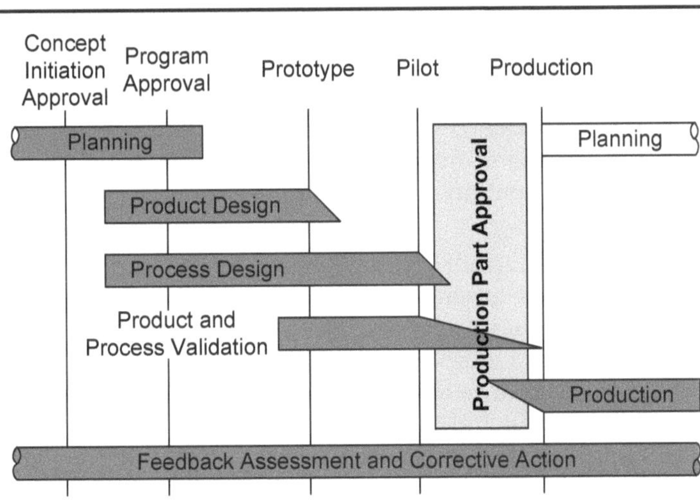


### Where production part approval fits into APQP

Production part approval is the final link in the APQP process. It is a significant production run at the production site using the production facilities (including operators). This process is designed to validate the process at volume production levels.

# 10.6 Pre-control (zone charts)

## Start-up control

Pre-control (sometimes called 'zone charting' or 'stop-light' control) is a statistical process control technique that differs from the conventional SPC methods discussed in Chapter 5. In conventional SPC, the control limits are based on the performance of the process (the voice of the process). In pre-control, the control limits are set based on the specified tolerances (the voice of the customer). There is a great deal of controversy over the use of pre-control (see Section 10.7) but at this stage we will simply describe the method and the significant uses.

Pre-control is very useful for working with short run processes where there may never be time to collect all the relevant SPC data to set control limits and for starting up for a process that has already been qualified. It gives immediate feedback on the state of the process and guidance on the next steps, i.e. when to adjust or seek help, when it is safe to use sampling, when the process needs examination and when to do nothing.

## Pre-control limits

The limits for pre-control (for bilateral tolerances) are set by dividing the total tolerance zone into three zones as shown on the lower right. These are:

- The middle half of the tolerance band is the 'green' zone. For a process with a $C_{pk}$ of 1.00, i.e. the tolerance band is equal to the process spread and the process mean is the same as the nominal value, this green zone will contain ≈ 86% of the population.
- The two outer quarters of the tolerance band between the green zone and the upper and lower set limits are the 'yellow' zones. For a process with a $C_{pk}$ of 1.00, i.e. the tolerance band is equal to the process spread and the process mean is the same as the nominal value, these two yellow zones will contain ≈ 14% of the population, i.e. 7% each.

  On this basis the probability of getting two consecutive readings in the yellow zone will be 0.0049 or less than 1 in 200.
- The areas outside the upper and lower set limits are the 'red' zones.

## The pre-control rules

### Start-up rules

At start-up, the operator checks five consecutive samples for the required tolerance (dimension, weight, etc.) and then applies the following rules:

- If the five consecutive samples are in the green zone, then the set-up is OK to run and the operator can move to the 'running rules'.
- If one of the five samples is in the yellow zone then restart counting and sampling.
- If two consecutive samples are in the yellow zone then adjust the process (if trained) or call for help to adjust the process.
- If one sample is in the red zone then adjust the process (if trained) or call for help to adjust the process.
- **Tip -** There are several slight variations to these rules but those listed above are reasonably well accepted.

This means that a start-up cannot be validated and cannot take place until five

Pre-control was developed by Satterthwaite at Rath and Strong in 1954 and has been controversial ever since.

It is with some trepidation that I include it in this Workbook (and risk the wrath of the control chart purists) but this is a Workbook for industry.

As a technique it is regarded as statistically valid and can be a useful tool.

"Talk to the parts; they are smarter than the engineers."

**Dorian Shainin**

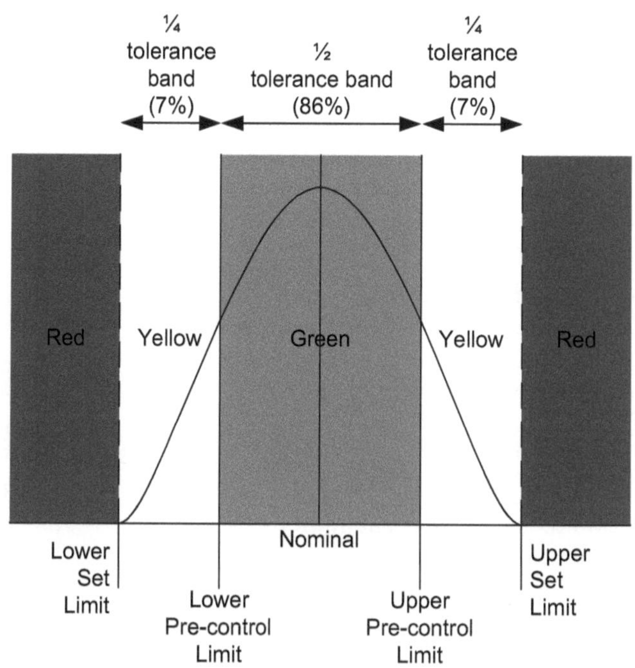


### The basics of pre-control

The complete tolerance band is divided into three areas: The green zone (OK), the yellow zone (caution) and the red zone (nonconformity). These zones are used to assess the process status at start-up and during running.

samples in a row are in the green zone. If this is not the case, then the process is incapable of producing parts reliably within specification (for whatever reason) and process improvements are required to improve the process capability ($C_p$).

If the operator attempts to continue production without meeting the 'five greens in a row' then every part will need to be checked.

## Running rules

Running under pre-control is equally simple. The operator checks two consecutive samples for the required tolerance (dimension, weight, etc.) and then applies the following rules:

- If both samples are in the green zone or one is in the yellow zone (upper or lower) and the other is in the green zone then continue production.
- If both samples are in the yellow zone (on the same side) then adjust the process (if trained) or call for help to investigate and adjust the process.
- If both samples are in the yellow zone (on different sides) then 'stop' and call for help immediately. This indicates a potential change in the process capability and requires more investigations and action than can be taken by an operator.
- If any sample is in the red zone then stop and call for help to investigate and adjust the process (if trained) or call for help to investigate and adjust the process. Inspect all parts since the last check.
- **Tip** - The original pre-control rules call for adjustment but there is legitimate concern that this may lead to excessive tampering with the process. Plastics processing rarely has a single 'dial' that you can change to adjust the process and in most cases there will be a need for more process investigation rather than a simple adjustment.

## One-sided tolerances

If the tolerance is not bilateral then the rules for one-sided tolerances are:

- Lower tolerance is zero, e.g. ovality, flatness, then:
  - 'Green' is from zero to 1/2 of the tolerance band.
  - 'Yellow' is from 1/2 of the tolerance band to the full tolerance.
  - 'Red' is greater than the full tolerance.
- Upper tolerance is maximum, e.g. tensile strength, burst pressure, then:
  - 'Green' is from the best result recorded down to 3/4 of the tolerance band.
  - 'Yellow' is from 3/4 of the tolerance band down to the lower tolerance limit.
- 'Red' is less than the lower tolerance limit.

Pre-control charts are useful to quickly centre and qualify a process. It should not be used for significant process adjustment without a full process investigation.

## Product not process

Pre-control differs significantly from conventional SPC in that it looks more at the product than at the process and sets the process limits based on the 'voice of the customer' (the tolerances) rather than the 'voice of the process' (the actual process capability). This makes it more closely aligned to conventional product inspection (with all the defects that this has) than to SPC but it is sometimes more compatible with operations in most plastics processors.

I think that most readers will have understood the concept a lot quicker than the concept of control charts.

That is why it is in this Workbook.

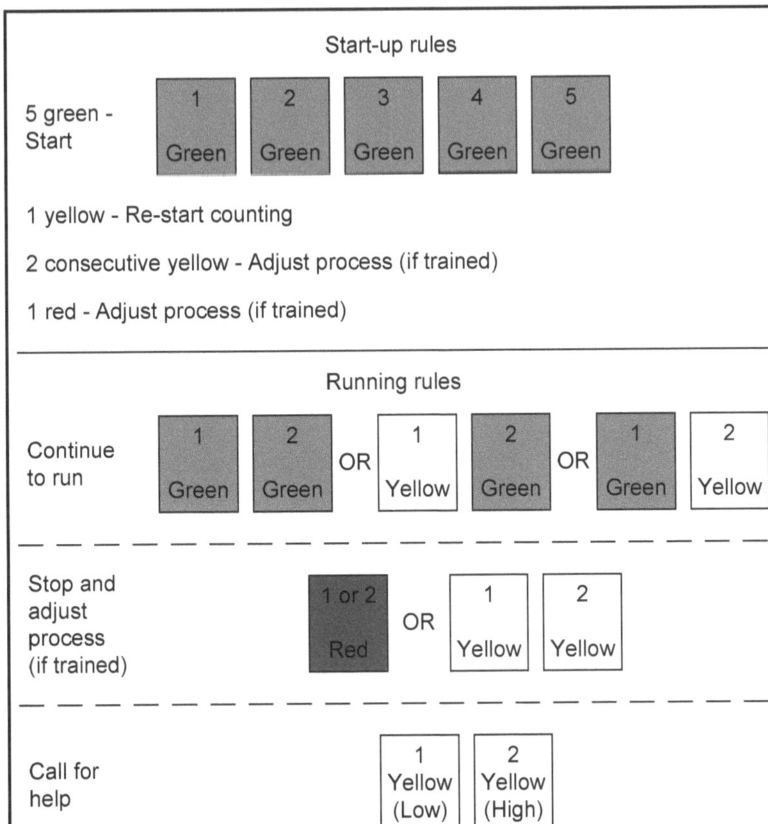


**The start-up and running rules for pre-control**

The rules of start-up and running using pre-control are very simple and easily understood. It operates by checking that the product is being produced in the green zone and meeting the requirements of the customer.

# 10.7 Inspection, pre-control or control charts?

## What do we want to control?

In the introduction to SPC (see Section 5.1) we discussed the difference between simple product inspection (defect detection and after the event) and process control (defect prevention and before the event). The conclusion was that defect prevention was a far preferable strategy and this remains the case.

Pre-control is somewhere between these two extremes of simple product inspection and the more complex control charting of SPC. It compares individual product values against the specification limits but uses these to make a judgement on the capability and setting of the process rather than simply accepting or rejecting the inspected products.

The traditional process of inspecting either single or multiple samples (without a sampling system - see Chapter 6) for simple conformance to specification is very limited in effectiveness and efficiency.

## Inspection (product control)

### The advantages

- Inspection is easy to carry out.
- Inspection is initially inexpensive to carry out.
- Inspection is easy to understand.
- Inspection is easy to schedule.
- Inspection does not need calculations or software.

### The disadvantages

- Inspection tells you about the inspected product but not about the other products being produced or about the process.
- Inspection gives no information about the process spread or location.
- Inspection does not separate common and special causes.
- Inspection provides no reliable guidance for process adjustment. In fact, process adjustment based on limited inspection can be detrimental to outgoing product quality (see Section 10.4).
- Inspection can easily lead to excessive tampering with the process.
- Inspection does not provide any information or driving force for process improvement.

## Pre-control (product and process control)

### The advantages

- Pre-control is useful in 'short-run' processes where the batch size is small and control charts are difficult to set up.
- Pre-control can be used to rapidly check start-up for the correct location of the process.
- Pre-control gives clear guidance on when to adjust a process or make tool changes.
- Pre-control is easily understood and used by operators.
- Pre-control is simple to operate, i.e. the

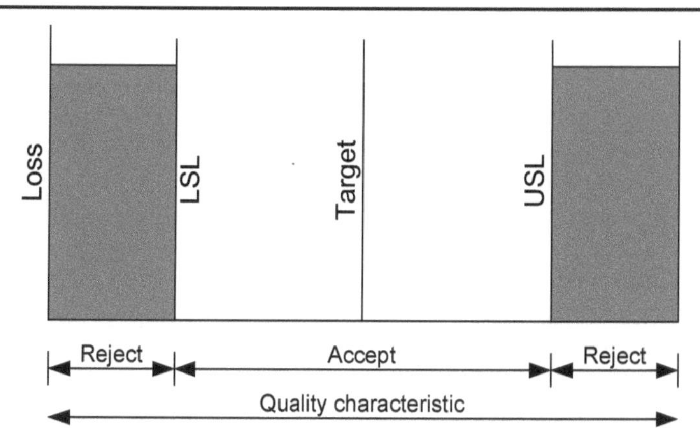


**The inspection and pre-control view of loss**

Inspection and pre-control treat all parts that meet specification as 100% 'good' and parts outside the specification as 100% 'bad'. This is the conventional view of loss and is useful for production and accounts but is not reality (see Section 8.15).

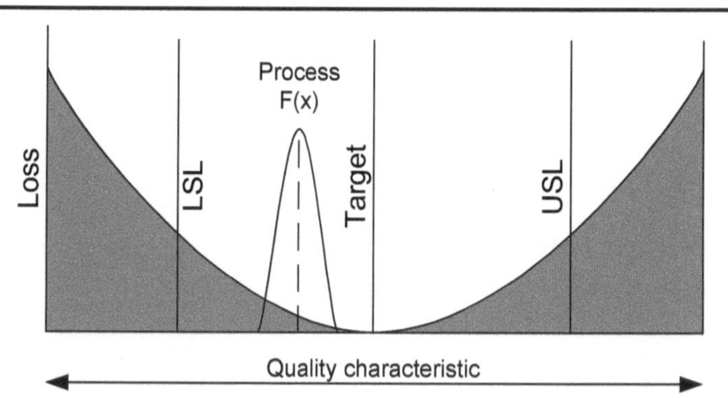


**The control chart view of loss**

Conventional control charts are similar to the Taguchi Loss function (see Section 8.15), i.e. the further away from the target then the 'worse' the product is and the greater the loss. The aim is to get on the target with minimum variation.

target is to stay in the middle of the specification.

- Pre-control is inexpensive to operate, i.e. it does not require expensive calculations or charting software.
- Pre-control can be used with attributes (gauging) to start a process.
- Pre-control has a low risk of rejecting parts or processes, i.e. <2% in the worst conditions and 0% if $C_{pk} > 1.66$.
- Pre-control works from the specification limits ('the voice of the customer') rather than the process limits ('the voice of the process').

### The disadvantages

- Pre-control needs good and realistic specification limits.
- Pre-control needs a stable process that can operate well within the specification limits, i.e. a process with good capability.
- Pre-control does not distinguish between common and special causes and is therefore not useful for process improvement, i.e. it provides no incentive to improve the process because all good parts are treated as equally good.
- Pre-control is not useful where the process has a poor capability and pre-control can signal process changes when these are not really needed.
- Pre-control is best used for processes with low part-to-part variation and can increase process adjustments if part-to-part variation is high, i.e. there is some danger of false signals.
- Pre-control does not provide a record of the process, i.e. there is no requirement for a written record or chart.
- Pre-control is not accepted by some major automotive companies as being acceptable for process control.

## Control chart (process control)

### The advantages

- Control charts allow process variation to be separated into common and special causes and the causes to be treated separately.
- Control charts use common causes to set the control limits.
- Control charts give clear guidance on when to adjust a process and when to leave it alone.
- Control charts drive process improvement (if used properly).
- Control charts are accepted by all types of customers.
- Control charts provide a written record or chart.
- Control charts provide a common language for all levels of staff.

As with anything else, you need to choose your weapons and tools carefully.

### The disadvantages

- Control charts are based on the process limits ('the voice of the process') rather than the specification limits ('the voice of the customer') but the process should never take precedence over the customer.
- Control charts work best when the process capability is high.
- Companies can suffer from control chart 'overload' where control charts multiply and take over the production area.
- Control charts need trained operators.
- Control charts often use expensive software or time-consuming calculations to operate.
- Control charts are best in long-runs where the process has a chance to become stable for short runs they can be difficult to operate.

## Summary

Having described and discussed inspection, pre-control and control charts, it should be evident that inspection is not adequate for modern plastics processing but that both pre-control and control charts have strengths and weaknesses.

Unfortunately, many plastics processors have not got beyond product control (inspection) - the least useful technique of all and need to move forward significantly and quickly.

Those plastics processors who have taken the initiative and moved into process control have generally done so as a result of customer pressure and have been forced into control charting by the same forces that forced them into ISO 9001, i.e. 'if you don't comply then we won't buy'. They have rarely critically examined control charting or even considered pre-control. Despite this, pre-control is a technique that can offer, almost, the best of both worlds and a technique that should be seen as complementary to control charts rather than an enemy.

- **Tip** - If you have never used pre-control then try it on some products that have good capability and see if it works for you.

# 10.8 Engineering process control (EPC)

## Dynamic changes

To date we have focused on a variety of SPC techniques to reduce process output variability by detecting and eliminating assignable causes of variation (special causes). In SPC, unless the process shows a warning then it is left alone to avoid 'tampering'.

Engineering Process Control (EPC) takes an alternative view and seeks to reduce process output variability by regularly measuring a variable and then continuously adjusting one or more input variables to compensate for any variation. In EPC the process is being continuously changed in response to the measured process output.

SPC was developed largely by statisticians and quality engineers but EPC was developed largely by process and control engineers and for many years these two different groups approached variability very differently.

The fundamental difference between the two approaches is shown in the flow diagrams on the right.

This section introduces EPC as a concept and Section 10.9 shows how the two concepts can potentially be integrated to reduce process variability and improve quality at the same time.

## The basics of EPC

In the simplest form, EPC makes dynamic changes to an input variable to control the resulting measured output variable. EPC therefore needs a model of the dynamic behaviour of the process, i.e. which knob to turn to adjust the process.

The control used by EPC can be:

- Open-loop control - this is where the adjustment to the input variable is by an operator taking action based on the difference between the output result and the desired target. If the operator takes no action then the system can run out-of-control. This is the least used of the EPC control systems for obvious reasons.
- Closed-loop control with feedback - this is where there is no operator intervention and the process control system automatically takes action based on the difference between the output result and the desired target (or set point). This type of control uses a 'summing junction' to calculate the difference between the output result and the target and make the required change to the input variable, e.g. a thermocouple measuring barrel temperatures can be used to control barrel heaters. Closed loop control with feedback works very effectively for processes where there is

SPC can detect special causes quickly but EPC can keep a process on target. They are very different things.

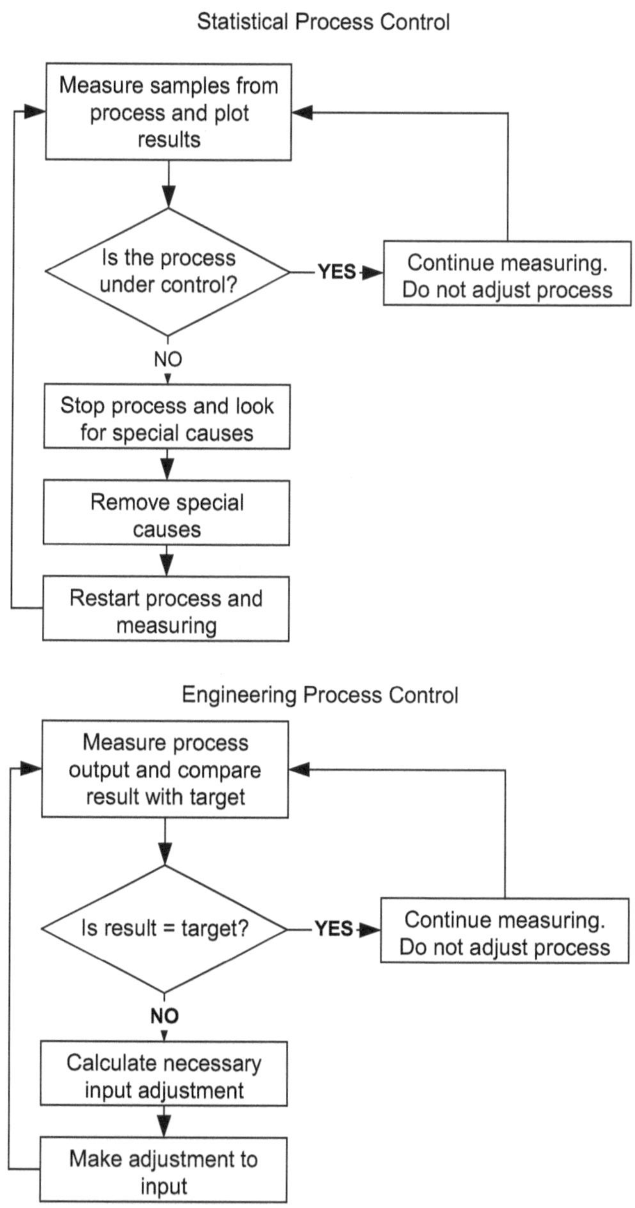


### The difference between SPC and EPC

SPC and EPC are different in their approach to process control because they come from different developers. The downside of EPC is that it does not look for special causes (it adjusts the process) and therefore does not fit the concept of improvement.

only a small time delay for the change in the input variable to take effect

- Closed loop control with feedforward - there is, again, no operator intervention and the process control system automatically takes action. However, in this case the action is based not on the output result but on the disturbances to the system. These disturbances are measured (before the process output) and changes are made to the input before the output is affected or even measured, e.g. measuring the moisture content of a plastic and adjusting the processing conditions before processing to prevent moulding concerns. Closed loop with feedforward is most useful where there is a long time delay for the change in the input variable to take effect and where it is possible that faulty product could be produced for some time before the correction was effective.
- Cascade control - this is where multiple control loops, either feedback or feedforward, are used to control complex processes.

These control options are shown in the diagram on the right. In practice, most current systems for plastics processing will use closed-loop control with feedback because of the difficulty in setting up the necessary model for effective control using feedforward systems.

The controller used in EPC will affect the response of the system and controllers can use a variety of techniques such as:

- Proportional control (P) - where the applied correction is proportional to the measured difference between the output result and the desired target.
- Integral control (I) - where the applied correction is proportional to the time integral of the measured difference between the output result and the desired target.
- Derivative control (D) - where the applied correction is proportional to the rate of change of the measured difference between the output result and the desired target.

In practice, a combination of these control methods is generally used and one of the most common is the PID (Proportional/Integral/Derivative) control.

## Problems in plastics

One of the major problems with EPC in plastics processing is that, except in simple cases, there is rarely a single 'knob' to turn to change the process output, i.e. changing any single process input variable can often have multiple and sometimes unpredictable effects on the output result. This is because most plastics-forming processes are 'multivariate', i.e. there is more than one variable involved in the process. Multivariate systems are much more difficult to control, even with SPC where it is the output that is being measured and manually controlled. With EPC it is the system model, however established, that is trying to make multivariate adjustments and this is a more complex task.

Previously, this was extremely difficult but advances in both computing power and in polymer rheology now make EPC a serious contender for controlling plastics-forming processes to ensure that processes remain in control to produce high-quality products.

EPC has been described by some (harsh) critics as a 'bandage to cover a wound not to cure it'.

One of the major disadvantages of EPC is that it corrects the process to reduce the effect rather than seeking to remove the underlying cause of the problem.

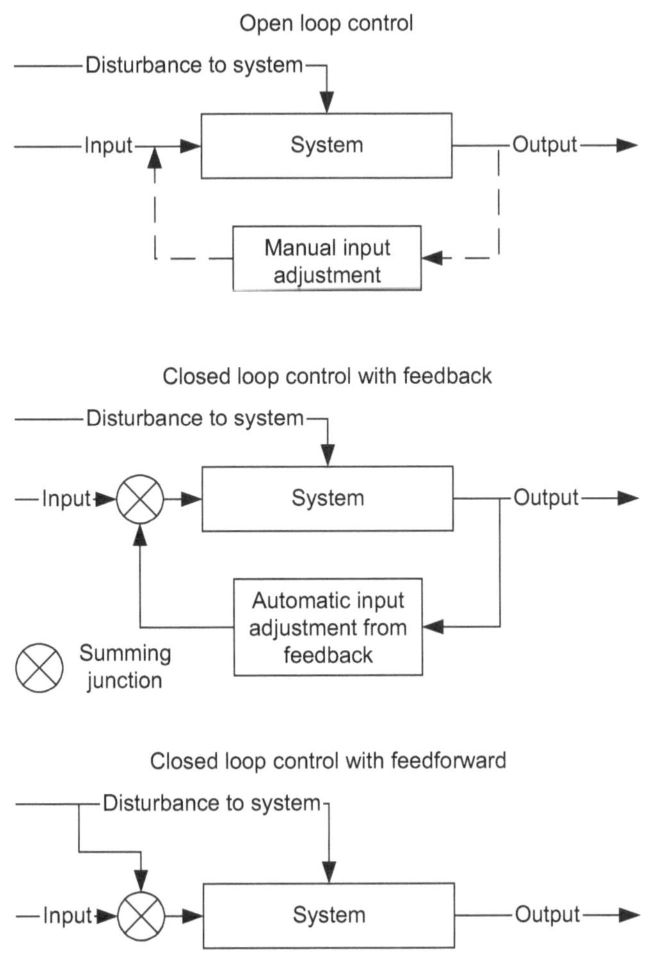


### The control options for EPC

Control under EPC can be open loop, closed loop with feedback or closed loop with feedforward. Closed loop with feedback is the most common system due to the difficulties of establishing the model necessary for closed loop with feedforward.

# 10.9 Integrating SPC and EPC

## Can we use both?

SPC and EPC were developed separately and in separate industries and the main differences between the two approaches are shown in the table on the lower right. Despite this, they are largely complementary and look at different aspects of controlling the quality of the output. In the past, the lack of computing power, the lack of reliable sensors and a poor understanding of polymer rheology restricted the development of EPC for plastics processing but these issues have largely been removed. The concept of integrating SPC and EPC to improve quality is therefore rapidly gaining ground. Integration of the two techniques offers the potential to use SPC to monitor the process for special causes and EPC to reduce the effect of measurable process variations. This idea has gathered much support and research with very good results but implementation in industry is currently very poor, i.e. most plastics processors are still at the inspection stage let alone contemplating the integration of SPC and EPC.

## The approaches to integration

EPC was first introduced to statisticians and quality engineers by MacGregor[1] in 1988. There was initially considerable controversy about the idea of 'continuous tampering' with a process that was in control even if this approach had been used by process and control engineers for many years and had a long history of success. EPC has historically been used extensively to control processes, e.g. a simple temperature controller on a barrel heater typically uses closed loop control with feedback and a PID controller and the EPC ideas should not come as a surprise to any engineer. The issue was really how to convince the statisticians and quality engineers that these ideas could be used to control the complete process and product and be successfully integrated with SPC. Apart from simple process control, e.g. barrel heaters, progress has been slow in the plastics industry where this approach has rarely been trialled.

There are several basic ways to approach integration and these are:

## Conventional SPC

The simplest approach is to use SPC to verify that the process/product is in control and, if any of the traditional SPC 'out of control' signals are seen, then to adjust the process based on a set of EPC rules. Successful corrective actions are stored as the rules for action in the future if SPC signals a special cause. This is only a slight variation on standard SPC where corrective action is not explicitly stored to provide a basis for future action.

EPC is rapidly developing into APC (Automatic Process Control) where the system controls itself to monitor and control the process.

| | SPC | EPC |
|---|---|---|
| Developers | Statisticians and quality engineers. | Process and control engineers. |
| Industry basis | Product industry. | Process industry. |
| Function | Process monitoring to control product. | Process adjustment to control process. |
| View of process | Process is either in control or not. | Process is continuously adjusted to stay at target value. |
| Process changes | None unless output warning signal seen. | Continuous process changes made in response to output signal. |
| Process model | Not necessary except when process is out of control. | Good process model needed to make adjustments. |
| Expects | Process to be stationary (centred). | Continuous process drift from centre line. |
| Measures | Quality features. | Process features. |
| Timescale | Medium- to long-term control. | Short-term control. |
| Technology | Low. | High. |
| Results | Process improvement. | Process optimisation. |

### The differences between SPC and EPC

SPC and EPC attack the 'quality problem' from different but complementary angles and there are obvious benefits to both approaches. This has lead to attempts to integrate the two approaches to give both improved quality and control.

## Algorithmic SPC (ASPC)

This approach combines an algorithm (model)-based controller and SPC techniques. It uses EPC to reduce predictable variations and then monitors the system using SPC. Instead of using SPC to monitor and making adjustments only when necessary, ASPC turns this around by automatically making adjustments and then monitoring to ensure that they are effective.

There are some concerns about ASPC and the main issues are:

- What are the variables that are monitored to assess control? This is not only about the actual variables but about whether they should be output variables (quality features) or input variables (process features).
- How accurate does the process model need to be for effective control?
- Should the controller be automatic or manual, i.e. open loop or closed loop?

This type of approach has been used successfully for multivariate systems using multivariate SPC and shows a great deal of promise for plastics processing.

**Note:** When SPC is applied in conjunction with EPC it is most common to use an Exponentially Weighted Moving Average chart (EWMA chart). EWMA charts are beyond the scope of this Workbook.

## Active SPC

SPC is traditionally an 'off-line' process, i.e. samples are taken and measured off-line. Active SPC takes SPC 'on-line' and uses a process model to define the control limits and the necessary control actions. The on-line SPC results are used to decide if action is required, i.e. if any of the traditional SPC 'out of control' signals are seen, and then takes action following the rules of the process model.

This type of approach minimises the effects of 'continuous tampering' and has been used for many continuous processes although applications in plastics processing are rare.

## Run-to-run (or run-by-run)

This approach has mainly been used in the semi-conductor industry where a 'run' is a group of products, e.g. a batch or a group of products produced under the same conditions. In this approach, changes to the process conditions can only be made at the end of a run and the need for changes is detected/signalled by conventional SPC methods. The process changes are made by EPC controllers.

This type of approach minimises the effects of 'continuous tampering' but still needs a good process model to define the control actions.

## Integrating SPC and EPC in plastics processing

Integrating these complementary approaches in plastics processing needs:

- A good process algorithm to control the process - these are now available for most processing methods.
- A good correlation between the input variables and the output variables - multivariate approaches to plastics processing now give much better correlations for plastics processing.

This progress in understanding the basics of plastics processing means that the time has come to start this integration to improve process control and quality.

This is shown by the fact that in September 2015 KraussMaffei released Adaptive Process Control (APC) to control IMMs (www.kraussmaffei.com/en/new-adaptive-process-control-apc.html). This uses data from a series of sensors to continuously monitor and adjust the machine changeover point and holding pressure (even during the shot) to dramatically reduce the shot weight spread and improve part quality. This is just the beginning.

## How effective is integration?

There is limited reporting of integrating SPC and EPC in plastics processing but Montgomery et al.[3] showed that the proper use of an integrated SPC and EPC control system 'can always outperform the use of either alone'. This was a purely theoretical paper using simulation results but validates the use of EPC for control and the use of SPC to search for special causes.


•1. MacGregor, J.F. 1988, On-Line Statistical Process Control, *Chemical Engineering Process*, Vol. 84 (No. 10), pp 21-31.

•2. Box, G.E.P. All of Box's work in this area is valuable and worth reading, even if it is very high level. See www.stat.wisc.edu/people/george_box/publications for a list of these.

•3. Montgomery, D.C.; Keats, J.B.; Runger, G.C. and Messina, W.S. 1994, Integrating SPC and EPC, *Journal of Quality Technology*, Vol. 26 (No. 2), pp 79-87.


Integrating SPC and EPC is still at the very early stages in plastics processing but offers the 'best of both worlds' with a future of self-adjusting machines automatically producing quality products.

That is why it is important.

"All models are wrong but some are useful."

**George Box**

**Machine and plastic variables?**

John Bozelli, in an excellent article (Plastics Technology, September, 2015), makes the vital point that we should consider the plastic variables and not the machine variables for successful moulding.

Machine variables will depend on valves, voltages, temperatures and set points. These will also suffer from variability and this will vary the experience of the plastic in the process and the results.

EPC mainly works on the machine variables so maybe there is some extra work to do here?

# 10.10 Injection moulding - scientific moulding

## Taking the art out and putting the science in

Design of experiments is one of the key tools for finding the system maximum (see Sections 8.15 and 10.4) and there has been some excellent work carried out in this field but it is not the only way to improve quality in injection moulding. One of the greatest advances in injection moulding in the past 20 years has been the use of what is termed by some as 'scientific' moulding where the injection moulding process is effectively broken down into three stages:

- The filling stage of ≈ 95% of the volume of the mould which is controlled by the ram/screw velocity.
- The packing stage of the remaining ≈ 5% of the volume of the mould which is controlled by the packing pressure.
- The cooling stage after gate seal where no more material can enter the mould and the packing pressure can be released as the product is simply cooling before ejection.

The recognition that these were separate stages allows separate optimisation of the stages to give better process control. An idealised graph of the cavity pressure and the hydraulic pressure (for a hydraulic machine) is shown on the lower right. The curves are not aligned because of time delays in signalling and valve switching.

**Note:** The curve will be different for amorphous and semi-crystalline plastics.

## The stages of moulding

### Velocity-controlled filling

This controls the majority of the cavity filling and sets the basic shape of the product. Changes to the fill speed can be used to control orientation, crystallinity, appearance and other aspects at the surface of the product. Excessive speeds can result in jetting (in the early stages), flash formation and even mould damage.

After the mould is nearly filled, control is changed from velocity control to pressure control. This changeover point will be set using a closed-loop system based on a signal from screw position, hydraulic pressure, cavity pressure or cavity temperature.

- **Tip** - Cavity pressure is also used to control valve gates for multi-cavity tools.

### Pressure-controlled packing

After changeover, the packing pressure is used to fully compress the plastic as it cools and shrinks in the mould. Changes to the packing pressure will control the compression of the melt along the flow path. Insufficient packing pressure can result in voids, short shots, excessive shrinkage on ejection, warping and other filling defects.

### Cooling

The cooling stage, whilst one of the longest stages, is simply waiting for the material to cool sufficiently for successful ejection without significant post-ejection warping.

- **Tip** - If the cavity is fully filled then the experience of the melt during cavity filling (temperature and pressure) is a major factor in product quality.
- **Tip** - If a cavity pressure sensor shows a significant drop at ejection then it is likely that the gate has not yet fully sealed.

The theoretical and practical setting of IMMs has advanced hugely in the past 20 years. 'Scientific moulding' is a general term for using a defined set of tests to fully determine the moulding settings in a logical manner.

Scientific moulding is sometimes referred to as 'decoupled moulding' - this term is a service mark of RJG Inc. The company set up by Rod Groleau in 1985 and now a world leader in sensor and moulding technology.

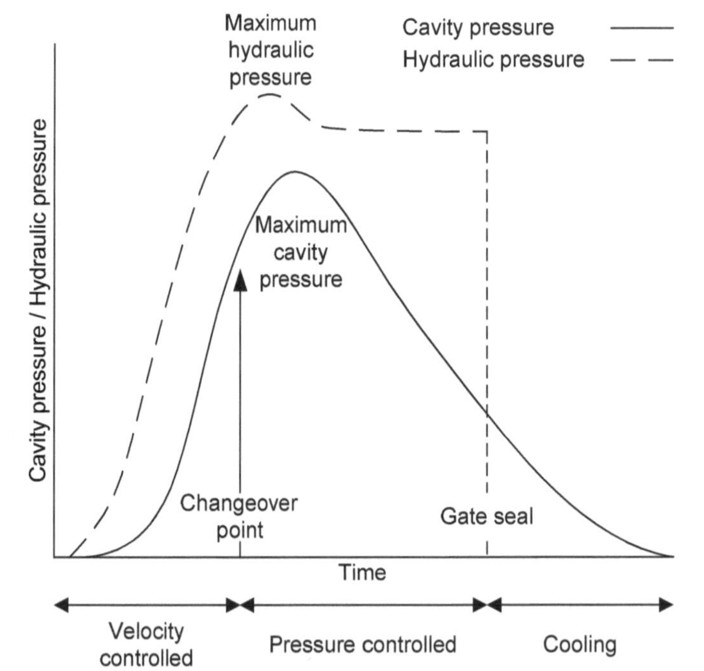


An idealised cavity and hydraulic pressure curve for injection moulding (not to scale)

The cavity and hydraulic pressures will change throughout the cycle, most noticeably near the changeover point and as the gate seals. After gate seal, the hydraulic pressure can be removed.

## The scientific setting process

Scientific moulding uses the separate stages to set the moulding parameters to fall within the Moulding Area Diagram (see right). This is also sometimes termed the 'process window'[5] as it defines the basic process conditions for successful processing. Scientific moulding looks at variations in process and material properties by studying:

- The effect of processing speed (shear rate) on material viscosity - this is to select an injection speed which gives consistency of the melt viscosity with changes in injection speed.
- The effect of pressure drops on the process - this is to avoid a 'pressure-limited process' where the machine is not able to deliver the material at the required injection speed.
- The time to gate seal - product weight versus time shows when the product weight stops increasing and if the packing time is long enough.

Additional studies can also be carried out on:

- The cavity balance of the specific mould - this is to validate cavity-to-cavity fill consistency in multi-cavity moulds.
- The effect of processing conditions on the product cosmetics - this is to set the conditions to maximise the process robustness in terms of cosmetics.
- The cooling time - this is to determine the shortest cooling time to produce an acceptable part.

Scientific moulding has many quality advantages for processors:

- It provides a robust process that is more consistent in production - this improves product consistency and therefore quality.
- It provides a disciplined and scientifically valid approach to solving injection moulding quality issues.
- It introduces science to what was, for many, a 'black art' based on experience.

## Setting sheets

The output of a scientific moulding process is a setting sheet which fully defines a robust set of moulding parameters but they have to be used to be successful. Settings should:

- Be recorded on machine and mould setting sheets.
- Always be used to set up the machine and mould for all runs.
- Never be an optional practice - failure to follow the setting sheet should be a disciplinary offence (it is money).
- Not stop at the settings of the main machine but should include all checks and settings for all ancillary equipment.
- **Tip** - If settings can be recorded directly on the machine then they should be set (and locked out) to prevent unauthorised changes.
- **Tip** - Setting sheets should include a 'Special Notes' section for details of safety concerns or notes, start-up and shut-down procedures and any details of purging to be carried out.
- **Tip** - Setting sheets reduce set-up times by allowing machine parameters to be set rapidly.


- 1. Goff, J.R. 2013. '*Moulding Masterclass*'. PID.
- 2. Routsis, A. 2015. '*Injection Molding Reference Guide*'. An excellent general guide that is available free from www.traininteractive.com.
- 3. Bozelli, J. 2016. Various publications free from www.scientificmolding.com.
- 4. Kulkani, S. 2010. '*Robust process development and scientific molding*'. Hanser.
- 5. Kazmer, D. 2009. '*Plastics Manufacturing Systems Engineering*'. Hanser.


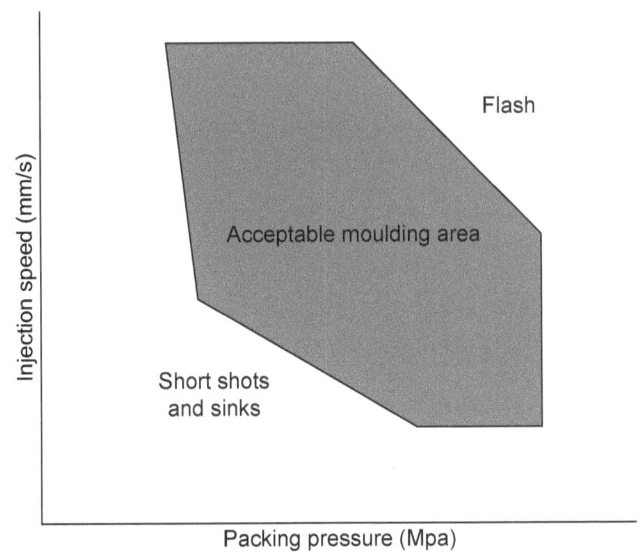


A Moulding Area Diagram (MAD) or process window

The MAD summarises the conditions that will give acceptable mouldings.

**After Kazmer[5]**

Scientific moulding is a complex subject and the reader is referred to one of the references given for further details.

This Workbook concentrates on the quality aspects of plastics processing and scientific moulding is a valuable tool to improve injection moulding quality.

### Machine and plastic variables (again)?

An article by John Bozelli (see Section 10.9) shows the effect of both the machine and plastic variables on a part.

Any setup sheet will list the machine variables, e.g. injection speed. These are machine parameters and not the plastic parameters which are the ones that really count.

# 10.11 Injection moulding - multivariate analysis

## A multivariate process

Whether you use scientific moulding or not (and we would always recommend that you do), at some stage you have to determine and fix the settings of the IMM. This also means that you have to determine the critical parameters to control during the process.

One of the main issues with controlling injection moulding is that the process is a multivariate process, i.e. there is no single 'knob' to turn to manage the process and the variables are related. This issue can be reduced by using simple DOE (see Section 8.15) to improve the robustness of the process, i.e. to make it less sensitive to changes in the variables. However, simple DOE often does not fully consider interactions between the factors being considered and the issue of a multivariate process still remains.

The best work in multivariate analysis (MVA) for injection moulding is being carried out by Kazmer and his group at UMass Lowell and they have published many excellent papers on this subject in the last 20 years.

- **Tip** - Search the internet for any of the papers on injection moulding by Kazmer. There is some excellent information on the moulding process and particularly the work that has been carried out with Hazen at MKS Instruments.

## Dependent or independent variables?

The simple case of measuring a single variable using SPC is shown on the top right. In this case, the control limits are set at $\overline{X} \pm 3\sigma$ and these limits will contain 99.73% of the population, i.e. there is a single set of control limits for the process. If the variable is within these limits then the process is regarded as being in control.

If two independent variables are involved then this can be considered as shown on the middle right. In this case, the control limits for the independent variables are both set at $\overline{X} \pm 3\sigma$ and these limits effectively define a square or rectangular control space. If both variables are within these limits then the process is regarded as being in control.

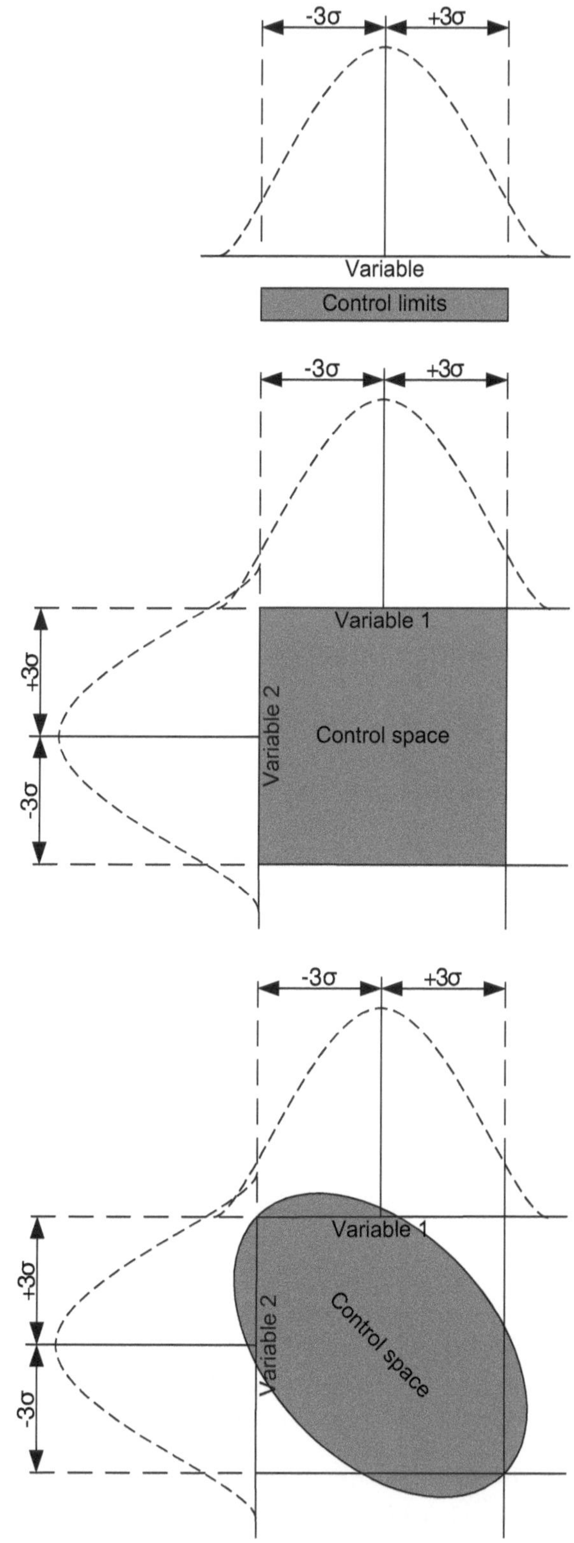


The effect of dependent variables on control limits

With a single variable there is a single set of control limits. For 2 independent variables there is a rectangular control space but for dependent variables the control space is an ellipse that is smaller than the rectangle.

**After Hazen[1]**

If two dependent variables are involved then this can be considered as shown on the lower left. In this case, the control limits for the two dependent variables are both set at $\overline{X} \pm 3\sigma$ but, because of the interdependency of the variables, the control space is no longer a square or rectangle but an ellipse. This reduces the allowable control space and it is possible for both variables to be nominally 'in control' but for the actual process to be 'out-of-control'.

This is because the control space is defined not simply by the values of the two variables but also by the interaction and correlation between the two variables.

A moulder is then left with three choices:

- Set the process window so that it lies entirely within the elliptical multivariate control space. This is the cautious option and will reduce the control limits but will also cause multiple false alarms showing that the process is 'out-of-control'.
- Set the process window as for the independent variables control space. This is the standard option for many companies but will allow the process to continue producing when it is actually 'out-of-control'.
- Use multivariate analysis (MVA) with DOE to set the process limits to those which reduce false alarms, reliably indicate if the process is 'out-of-control' and produce good products.

  This will typically use Principal Components Analysis (PCA) to reduce the size of the data set and to develop a multivariate model of the process.

In reality, for injection moulding, there are more than two dependent variables and traditional variables such as injection speed, packing pressure and melt temperature will all depend on each and interact, i.e. it is even more complicated than shown in the diagram on the left.

The use of MVA, PCA and DOE to optimise the injection moulding process is outside the scope of this Workbook and is best described in publications that are available from MKS Instruments [2,3] which describe this process in detail.

This process has shown excellent results in practice, e.g. decreasing cycle times by up to 10% and reducing rejects by up to 5%, but the disappointing aspect is the lack of real utilisation by industry of a proven process.

- **Tip** - Find out about using MVA with DOE to set and run moulding machines to get the best out of the process. The information is out there and waiting to be used.

Validation of processes using MVA and DOE works but the investment in staff, training and equipment has resulted in very limited take-up by industry.


•1. Hazen, D. 2009. *'How better fault detection methods improve molding quality'.* Injection Moulding magazine, UBM, December 2009.

•2. Hazen, D. and Kazmer, D. 2007. *'Using multivariate process analysis to predict injection molded part quality'.* Injection Moulding magazine, UBM, October 2007. Available from MKS Instruments, www.mksinst.com.

•3. Hazen, D. 2012. *'The Optimization of Injection Molding Processes Using Design of Experiments'.* MKS Instruments Application Note 06/12-7/12,USA), www.mksinst.com.


### Taking time to save time - traditional but it works.

Whilst working with an in-house moulder some years ago we had a 'problem' tool with an average scrap rate in the region of 25%. The mould was never capable of running for more than 2 hours without having to be seen by the setters and the output was always inconsistent and of poor quality. An ideal candidate for improvement!

We spent some time on the basics:

- Finding out the best settings (using capability studies and a set of properly designed experiments).
- Recording the settings on the setting sheet.
- Training the operators in SPC and 'when to do nothing'.
- Establishing the SPC limits and checking protocols.

The simple control was to measure the weight of the output and use this for a control chart. Operators were warned that tampering with the controls whilst the chart showed the process to be in control was a disciplinary offence.

The product ran for around 56 hours with absolutely everything running fine and no setting changes by anybody. This was regarded as a true miracle by all concerned but we knew it couldn't last.

Finally the control chart indicated 'out-of-control' and we all panicked. We checked the settings, the material, the control charts and everything we could think of. We were just about to give up and modify the settings when the Senior Tool Setter found the problem. The cleaner had moved a rubbish bin at the back of the machine and accidentally kinked a cooling water hose. This restricted the flow and the cooling of the mould. We straightened the hose and waited. After 20 minutes the chart indicated that the process was 'in control' once more and the run continued.

The process does work but sometimes thought is needed before action.

# 10.12 Injection moulding - managing variables

## Too many variables

Most injection moulders will not use MVA with DOE for process control and will use some other control system, e.g. SPC or simple random inspection. SPC whilst valuable for many applications, measures and monitors a single variable and for multivariate processes can be misleading as it does not consider dependent variables (see Section 10.11). The other disadvantage of SPC is that the variable to be monitored must be specified and in injection moulding there are simply too many variables. We need a way to make this simpler.

## The cavity is the thing

The major controller of product quality in injection moulding is the mould itself. The mould not only defines the product dimensions but acts as a 'heat exchanger' to control cooling. Fill the mould under consistent conditions with a consistent amount of plastic and you are more than halfway towards a quality moulding.

- **Tip** - Machine conditions may appear constant but it is the cavity conditions, i.e. what the plastic experiences, that counts.

In the past, it was difficult to control the process using real-time variables due to both their rapid changes and their inter-relationships but it is now possible to measure real-time cavity pressure and temperature data. This can be used to both validate each moulding and to control machine parameters for consistent mould filling based on the p-V-T relationship for the plastic being moulded.

Cavity pressure and temperature are both sensitive to and proportional to the resulting part weight and dimensions (assuming shrinkage is allowed for) and provide the information necessary for process and thus part control.

Cavity pressure sensors, e.g. Kistler, Priamus and RJG, can be mounted near the gate to show and control:

- When the flow front has passed the sensor.
- The rate of fill of the cavity - this should be as uniform as possible to avoid warping and the cavity pressure should increase linearly during fill.
- The changeover point from speed to pressure control - cavity pressure is a sensitive indicator of when changeover should take place (or has taken place).
- The peak pressure in the cavity.
- The gate freeze point.

Temperature sensors can be mounted at the end of the flow path to show and control:

- The presence of the flow front - this can be used to control changeover, i.e. to indicate filling of the cavity.
- The cooling of the cavity during the cooling phase.

Real-time cavity pressure and temperature measurements provide the essential information for process control because these measures are very sensitive to changes in either the machine conditions or the material. Consistent cavity pressure and temperature curves through the cycle, as measured by the integral of the curves, will allow consistent moulding and part production from:

- Run to run.
- Cavity to cavity.
- Machine to machine.

If good products are produced for defined, maintained and consistent cavity pressure and temperature curves then moulders are assured that the final product will also be consistent, even if the mould is transferred to another machine or if the material changes, e.g. an increase in regrind percentage. These curves show what the plastic is experiencing rather than simply the machine variables.

- **Tip** - Cavity pressure alarm limits can be set to alarm when not in tolerance to indicate poor-quality parts.
- **Tip** - Cavity pressure and temperature are much more sensitive than screw position for determining shot size and cushion.
- **Tip** - In the past, thermocouples were relatively slow in indicating temperature changes but new thermocouples are much faster in registering temperature changes.
- **Tip** - Fast-acting servo valves allow better control of changeover and machine control.

Critical process parameters such as the integral of the cavity pressure curve and the melt temperature curve are very sensitive in detecting process variations.

Quality measures from the finished moulding still need to be correlated to the process parameters, e.g. it needs to be shown that the integral of the cavity pressure does indicate an optimised process. However, after this, the effort should be focused on monitoring the process and not on monitoring the moulding.

Changing materials, e.g. increasing the regrind percentage, should be treated as a change in the machine settings. Varying the MFI will change the cavity pressure and change the moulding parameters needed for a good moulding.

Unlike extrusion, measuring the current (amps) drawn by an IMM does not provide much useable information.

The current drawn varies widely during the cycle and much of the energy used is due to machine losses and not due to processing loads.

- **Tip** - For basic quality control, the maximum values of cavity pressure and temperature can be monitored instead of the integral of the curves.

## Temperatures

The mould temperature and melt temperature are two other measures that have a large effect on moulding quality.

### Mould temperature

The mould temperature is probably the most important temperature in terms of quality and this affects quality in many ways. The mould temperature is not the same thing as the cooling water temperature, it is the temperature of the actual mould cavity after it has stabilised. The mould temperature determines:

- How fast the product will cool and how much stress is locked into the product.
- The surface finish of the moulded product. A higher mould temperature will increase gloss levels.
- How much crystallisation take place in the mould.
- **Tip** - Getting a stable and controlled mould temperature is essential.
- **Tip** - The gate should always be the first place to freeze.

### Melt temperature

Good control of and the correct melt temperature is vital for any plastics process but this is not often well controlled or measured. The main controller of the melt temperature is the shear heating that takes place in the screw/barrel area and the barrel heaters are primarily to adjust and maintain the temperature. A complication for injection moulding is that the screw changes purpose during the cycle, initially it is used to move and condition the melt and then it is used as a ram to fill the mould.

These dual functions make controlling the melt temperature complex. Barrel temperature settings, which measure the barrel temperature and not the melt temperature, are not the same as the melt temperature because they do not allow for shear heating. Other factors that affect the melt temperature are injection rate, i.e. increased shear heating, back pressure, cycle time and residence time.

- **Tip** - Getting a stable and controlled melt temperature is essential.

### We don't have all that stuff!

Although we are strongly in favour of using cavity pressure and temperature to control the injection moulding process, many processors simply do not have the time, money or staff to invest in and operate process control tools such as cavity pressure or temperature measurement or to set up and use MVA/DOE for establishing robust process settings. In cases such as these then what are the options?

- The first option is to manually measure the key dimensions identified on the control drawings (see Section 8.13) and to use standard SPC variables charts (see Section 5.18) based on these dimensions.
  This approach needs a lot of operator time, relies on the operators measuring dimensions correctly and can end up with a lot of charts to control the process. In cases such as this, it is normal to assign one or two dimensions as the 'control dimensions', i.e. they are the dimensions that effectively control or measure the other dimensions of the product. The other dimensions are not measured as they are assumed to be dependent on the control dimension.
  This approach also relies on manual intervention and adjustment if the chart indicates a need for process changes and often the necessary process changes to be made are not clearly defined.
  However, for plastics processing the main difficulty with measuring product dimensions is that they do not stabilise for some time after moulding. For most polymers this can be some hours due to thermal shrinkage, i.e. until the moulding has fully cooled, but for polyolefins this can be some days due to crystallisation shrinkage, i.e. until crystallisation is complete. This means that measurements must be made using 'proxy' dimensions or 'hot' gauges to allow for thermal and crystallisation shrinkage and it is not always easy to predict how the shrinkage will change with moulding conditions and time.
- The second option is to use a single surrogate measurement for all of the process parameters. The most common surrogate measurement is the product weight. If the tooling has been dimensionally validated and the process has been set correctly before series production then the product weight will be a measure of continued process control, i.e. if the cavities are not fully filled and packed then this will show as a weight variation. This is again a manual approach that requires operator measurement but weighing the product needs less time and operator skill than measuring a dimension. Weighing the product also relies on manual intervention and adjustment if the control chart indicates a need for process changes and as for dimension measurement, the necessary process changes to be made are often not clearly defined .
  It is reasonably simple to set up and use a variables control chart based on product weight (see Section 5.18) and then to use the product weight as a surrogate measure for process control. I have used this with good results in the past but it must be recognised that product weight is not as sensitive to the inevitable process and material variations as the cavity pressure and melt temperature curve.
- **Note:** If using product weight as a surrogate measurement then the control limits may need to be reduced to be sure of reliably identifying process changes, i.e. try using $\overline{X} \pm 2\sigma$ instead of $\overline{X} \pm 3\sigma$ to set the control limits for the variables chart.

# 10.13 Injection moulding - managing attributes

## Visual properties

Good process setting and cavity pressure control allows moulding within the process window and this should also allow consistent processing within the 'cosmetic window', i.e. the processing conditions necessary for good part cosmetics. However, variation is always present in any process, and processes (material, mould or machine) can change during a run. Attributes (visual features) are not only extremely important to customers but are often an early indicator of process changes or excessive variability in the process.

- **Tip -** Define a 'cosmetic window' as well as defining a 'process window'. They may not be the same thing!

Attributes in injection moulding are also multivariate, i.e. they depend on more than one variable, but in the case of attributes we are concerned with what we see rather than what we measure and this does make it simpler to measure attribute data (although resolving any non-conformities can be just as difficult).

## Have we got a concern?

The easiest way to monitor attributes in injection moulding is to use SPC charts for attributes (see Sections 5.19 and 5.20). Attributes control charts are under-used in plastics processing but provide an excellent early warning method to determine if there is a real issue with the visual properties and if action needs to be taken.

Attributes control charts are easy to set up (see Section 5.19 ) and analysis is easier than for variables control charts as there is only one chart to look at. The control limits for attribute charts will give warnings of any out-of-control conditions (the rules are the same as for variables control charts) and the need to take action.

- **Tip -** Accurate assessment of attributes needs consistent application of standards and judgements (see Section 5.20).
- **Tip -** Use a constant sample size to remove any issues with changing sample size. This makes sampling easier for operators (and everybody else) to understand.

## Follow a process

If the rules for attribute charts indicate a non-conforming attribute then the process is 'out-of-control' and action needs to be taken. The process is as follows:

### What is the concern?

There are many visual non-conformities that can arise in injection moulding and the terminology varies not only from company to company but also from country to country. A list of some of the terms for the most common defects is shown below.

The first step in solving attributes issues is to define what the actual concern is and to get everybody to agree on what you are dealing with. Without agreement on what the non-conformity actually is it will be impossible to resolve the issue.

- **Tip -** The range of terms used to describe visual defects is very wide. Visual samples or photographs of each type of concern will help people to agree on what the issue is. Start a collection for the future now - I have my own samples.

Attributes are generally the visual aspects of the moulding but can include dimensions if these are monitored by 'go/no -go' gauges. A fail on a 'go/no-go gauge' is then treated as an attribute defect.

If the mould is the cause of attribute failure then fix the mould.

Do not try to adjust the process. You might get the mould working but this will only push the 'good' cavities close to the edge of the processing window and reduce the robustness of the complete process.

### Surface and visual (attribute) concerns

There are many different types of concerns that can be tracked via attributes. The terminology varies throughout the industry and one man's 'silver streak' is another man's 'splay', some of the more common terms are:

- Black specks or streaks.
- Blisters.
- Burn marks.
- Brittleness
- Bubbles.
- Colour variations.
- Contamination.
- Cracking.
- Crazing.
- Delamination.
- Discolouration.
- Ejector pin marks.
- Flash.
- Flow lines.
- Flow marks.
- Gate blush.
- Gas marks.
- Gloss marks.
- Jetting.
- Knit lines.
- Matt marks.
- Short shots.
- Shrinkage.
- Silver streak.
- Sink marks.
- Splay/splash.
- Voids.
- Warping/bowing.
- Weld lines.

## When is the concern?

When an attribute concern appears is vital to finding a solution and the options are shown on the upper right. Concerns can be:

- Mould and machine trial related - this type of concern must be resolved before the process is accepted for production.
- Start-up related - this type of concern is at start-up with a mould, machine and material that previously produced good product. In this case the focus is on determining what has changed between the runs. Start-up-related concerns must be corrected by the setter before the machine is handed over to production.
- **Tip** - The main reason for start-up concerns is generally changes in machine settings made on the previous run. Restore the machine to the 'good' settings for production.
- **Tip** - If the mould is being used on a different machine then settings will obviously need to be modified.
- **Tip** - Check that the mould has not been modified since the last good run.
- **Tip** - Check if the material is from a batch that previously produced good products. If a new batch/delivery has been used then try a different batch to eliminate material issues.
- **Tip** - Retain a sample of 'good' material that can be used to quickly check for material issues if start-up is a problem.
- Quality control-related - this type of concern occurs on a running process when the chart shows that the process is 'out-of-control'. In this case the focus is on determining what has changed during the run and any changes in material, machine or mould should be checked.
- **Tip** - For QC-related concerns look at the material and machine first.

## Where is the concern?

The location and frequency of the concern, especially in multi-cavity moulds, can give excellent information on the source of an attribute concern and the options are shown on the lower right

- **Tip** - The important thing is to use all of the available information.

## How do we solve the concern?

The Internet has many 'troubleshooting guides' for visual defects but my favourites are from Andy Routsis[1], 4Plas[2], and Kenplas[3].

- 1. Routsis, A. 2015. '*Injection Molding Reference Guide*'. Pg 46-47. Available free from www.traininteractive.com.
- 2. 4Plas. 2014. '*Injection Moulding Troubleshooting Guide*', www.4plas.com/download.php?id=111.
- 3. Kenplas. 2016. '*Trouble shooting for the injection molding process*', www.kenplas.com/service/imtroubleshooting.aspx.

Analyse everything and use the Quality Tools (see Chapter 7) to isolate the root cause and then solve it permanently.

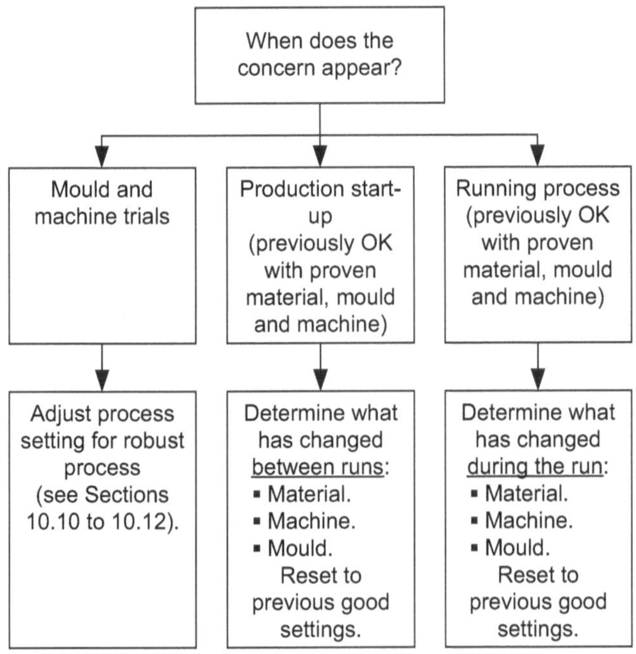


### When is the concern?

QC is mainly concerned with when a process fails to start-up correctly or when a running process starts to produce poor-quality parts. The first step is to return to the previous good settings to see if this resolves the concern.

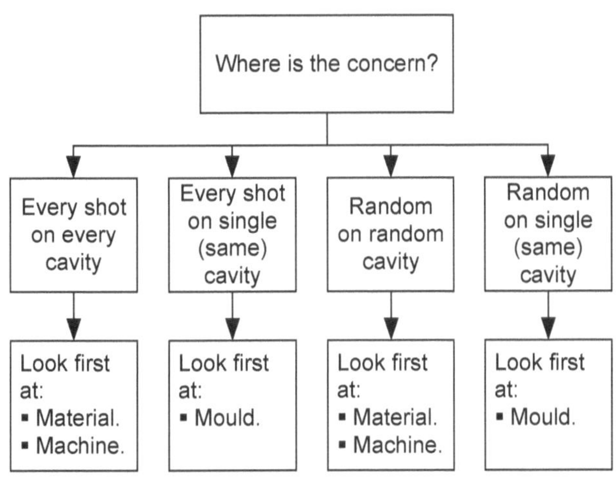


### Where is the concern?

Multi-cavity moulds give good information on the possible cause of the attribute concerns. The important thing is to use all of the available information to determine the cause of the concern and resolve it.

# 10.14 Extrusion - managing variables

## The die is the thing

The basic shape or outline of an extrusion is set at the exit from the die lips whether it is a profile, pipe, sheet or film. The extrudate wall thickness or film thickness is set by a combination of the die lip opening and the haul-off speed. Depending on the product, these can be adjusted whilst running by changing the die lip opening or by changing the haul-off speed to draw down the extrudate to the required wall or film thickness.

- **Tip** - Haul-off speed is the main method of adjusting wall thickness for profiles but can be limited by the need for cooling of the extrudate before handling.
- **Tip** - For profiles, the shape can be adjusted slightly during calibration but the main shape is still set at the die lips.

Film and sheet are the simplest extrusion processes in terms of dimensional tolerances and product control focuses mainly on the film or sheet thickness. Pipe and tube is slightly more complicated and has two critical product tolerances (wall thickness and ovality). Profile extrusion can be much more complicated and has many product tolerances to control. It is possible to assign one or two dimensions as the 'control dimensions' (see Section 10.12), i.e. they are the dimensions that effectively control or measure the other dimensions of the product. The other dimensions are not measured as they will be dependent on the control dimensions (assuming a properly built and sized die and consistent calibration and draw down).

- **Tip** - Control dimensions can be used with standard SPC variables charts (see Section 5.18) for basic control.
- **Tip** - Always note the control dimensions on control drawings (see Section 8.13).

Manual measurement of 'control dimensions' can provide a basic quality control strategy but this needs a lot of operator time, relies on the dimensions being correctly measured and often needs several charts to control the process. This also needs manual intervention and adjustment if a chart indicates a need for process changes. There is a clear need for process measurements and controls similar to those for injection moulding.

## Key parameters

Extrusion is, in some ways, simpler than injection moulding because there are fewer process variables. The motor driving the screw is set at a nominally constant rotational speed to deliver a constant volume of well-conditioned plastic to the die. The consistency of the melt delivered to the die is a major factor in process control for extrusion and process measurements can be either direct, i.e. they measure the plastic properties, or indirect, i.e. they measure the machine properties.

### Direct process measurements

Direct process measurements relate to the condition of the melt and are:

- Melt pressure - despite the nominal 'steady state' of the extrusion process, the melt pressure will always fluctuate and can lead to surging and dimensional variations. Melt pressure measured with pressure sensors at the die is a sensitive measure of the overall process stability.
- **Tip** - The best method of providing a consistent melt pressure is to use a gear pump. This will remove any pressure variations and deliver a constant material flow at a constant pressure.
- **Tip** - Melt pressure can be measured by pressure transducers at the end of the barrel, at barrel vents or in the die.
- Melt temperature - melt temperature is a vital part of a well-conditioned melt. It can be measured by thermocouples at various points in the melt flow but control is more difficult. This is because melt temperature is mainly affected by shear (frictional) heating in the interior of the melt and heater bands or cooling will only control heating or cooling of the melt at the barrel wall (these can be very different). Shear heating is related to the square of the shear rate and even small increases in the motor speed can lead to large increases in the melt temperature that cannot be quickly or easily controlled by adjusting the heating or cooling at the barrel wall as this can take some time to affect the overall melt temperature.
- **Tip** - Temperature sensors on the barrel do not indicate the melt temperature.

Most extruded products will be cut to length whilst still warm and are likely to contract after cutting. There will often be a cutting tolerance applied to ensure that the customer gets the correct length of product.

Try monitoring the shrinkage and using SPC with variables to minimise the excess length needed and increase quality and profits.

We once worked with a BOPP film producer whose process showed a $C_p$ = 1.407 with $Z_{upper}$ = 0.712 and $Z_{lower}$ = 2.103, i.e. $C_{pk}$ = 0.712.

This was a process that was capable in statistical terms but the location was 'high' and the product thickness would often exceed the upper tolerance.

We suggested moving the process to reduce the thickness and still stay within the tolerance. This would reduce the amount of material used for the same $m^2$ value.

However the company did not sell their film by the $m^2$ (what their customer wanted to buy) but by the kg (what the company wanted to sell).

Nothing was changed: There is no accounting for a company's strategy.

- **Tip** - Adjustments to heater band controls will suffer from a time delay in terms of adjusting the melt temperature.

### Indirect process measurements

Indirect process measurements mainly relate to the screw speed. The extruder motor is set for a constant speed to deliver a constant flow of melt to the die but this will vary with any variations in the melt. Common measures are:

- Motor speed - the effect of shear means that drive systems must maintain a constant screw speed to maintain a constant melt temperature. Accurate control of the motor speed is therefore essential for melt temperature control.
- **Tip** - Good control of motor speed will result in good melt temperature control.
- **Tip** - Overheating of the melt will reduce the viscosity and, if the motor speed is poorly controlled, this can lead to an increase in motor speed, an increase in shear heating and a 'runaway' process. This is an issue because of the difficulty of cooling the melt close to the screw via fans on the exterior of the barrel.
- Amps drawn by the main motor - this is a measure of the power drawn by the motor (and therefore indirectly of the power input to the melt in the form of shear heating). This makes it a good measure of the melt condition.
- **Tip** - A change in the amps drawn is a good indicator of a change in the melt viscosity (for whatever reason).
- **Tip** - The melt pressure at the die will also change with any change in screw speed and this will cause dimensional variations (see above).

Measurement and control of the direct and indirect process measures through closed-loop controllers (see Section 10.8) will give stable process conditions and good process and product control.

- **Tip** - Minor fluctuations in amps and motor speed are inevitable but using an SPC variables chart can provide an early indication of potential process problems such as material variations, etc.

## Profile extrusion

Product thickness is generally the main measurement used for process control. Assuming a stable process, product thickness can be controlled by adjusting the haul-off speed relative the die exit speed to control the dimensions of the finished product.

- **Tip** - An SPC variables control chart of thickness can provide an early indication of potential process problems such as blocked melt filters, etc.
- **Tip** - Select the control dimensions carefully to control the complete product.
- **Tip** - If the extruder is a twin-screw processing powder, e.g. PVC, then material feed control is essential for good mixing and process control.
- **Tip** - If the product is an extruded foam then it is not generally possible to make any significant wall thickness adjustment via haul-off control.

## Pipe and tube extrusion

Thickness, diameter and ovality are generally measured using ultrasonic, laser or other measurement systems. Depending on the system, the gauges can either rotate to measure the complete product or be stationary using two-, three- or four-axis thickness, diameter and ovality measurement.

Adjustment of thickness and ovality can be carried out by adjusting the position of the mandrel in the die either automatically or manually.

- **Tip** - Unlike four-axis systems, stationary two- or three-axis systems do not provide full coverage for pipes.

## Film

Film thickness can be measured using a variety of non-contact methods, e.g. lasers, X-rays, beta, gamma or infrared, to measure the overall film thickness and also to measure the layer thickness in multi-layer films. It is also possible to use mass measurement as an indicator of film thickness.

Adjustment of film thickness can be carried out by adjusting either the haul-off speed or the die lips and this can be automatic or manual.

- **Tip** - Film requires not only control of variations in the machine direction but also in the transverse direction.

## Sheet

Sheet thickness can be measured either manually (see Section 10.15) or by using a laser callipers.

Adjustment of the die bolts to control the sheet thickness can be carried out by adjusting the die bolts to change their length and this can be automatic or manual.

**Cooling control**

For profiles, pipes and tubes the ability to cool the product defines the production rate, i.e. an extrudate cannot be removed from the haul-off until it is sufficiently cooled to be dimensionally stable. Cooling and heat transfer therefore define the production rate, i.e. slow the line down or make it longer.

For blown film and sheet the production rate is also limited by heat transfer and most systems use PLC-based line controllers to manage not only the extruder and all downstream equipment but also the material feed and blending systems.

Close control of thickness in film extrusion allows down-gauging whilst still meeting the specification.

Shrinkage will affect extruded products just as it does any plastic product and needs to be considered in any product dimensional measurement.

Changes in factory conditions can have a large effect on extruded products, e.g. temperature changes due to draughts from open doors, because it is an 'open' process.

Extrusion needs good environmental controls for consistent products.

# 10.15 Extrusion - using variables data

## A case study

An extrusion site was producing extruded sheet for later thermoforming into food trays. The site regranulators were stacked with 'reject' product from the extruders on the basis of thickness variations.

## Capability study

A rapid process capability study (see Section 5.22) was carried out on 50 consecutive historical data points extracted from the QC records for sheet thickness. These data were taken over both the day and the night shifts.

These data were used to produce the capability study histogram shown on the lower right. This does not look exactly like a normal distribution but this is explained when it is considered that the operators are making manual measurements with micrometers. The operators are reluctant to measure in '5s' and prefer to measure in '10s'. This is easy to do with a manual micrometer and the operators will tend to 'fudge' the results. This is normal human behaviour and is not a criticism of the operators. This behaviour is shown quite clearly in the histogram.

The specification for the product is 575 μm ± 6% (± 34.5 μ, ). This gives:

- USL = 609.5 μm (Upper Set Limit or Upper Acceptable Tolerance).
- LSL = 540.5 μm (Lower Set Limit or Lower Acceptable Tolerance).

The nominal thickness, the USL and the LSL are all marked on the capability study histogram.

The distribution was considered to be approximately a normal distribution (subject to the operator adjustment) for the purpose of analysis.

The results of the capability study showed that the average ($\overline{X}$) of the results was 574.02 μm and the standard deviation ($\hat{\sigma}$) of the results was 11.96 μm.

## Process spread

The process spread (relative to the specified tolerances) is described by the process capability index, $C_p$ (see Section 5.22). For the process studied, $C_p$ = 0.96.

When $C_p$ is < 1.00 then the normal distribution width is greater than the tolerance band width. This means that defects will always be produced at the top or bottom of the tolerance band and it will not be possible for the process to produce parts within tolerance. For this process ($C_p$ = 0.96), out-of-tolerance products will be produced even if the process is exactly centred and does not vary.

## Process location

$C_p$ describes the process spread but does not describe the process location. This is described by $C_{pk}$ (see Section 5.22). For the process studied, $C_{pk}$ = 0.93.

When $C_{pk}$ is < 1.00, as in this case, then it will not be possible for the process to produce parts within tolerance. The process will always be located such that out-of-tolerance products are produced.

This is a process which is not capable of reliably producing products within specification, i.e. nonconformities will always be produced.

In cases such as this, where the process is not capable, then the process:

- Should be examined for improvement to reduce the variation.

  AND/OR

- The tolerances adjusted to more realistic levels, in consultation with the customer, to increase the process capability.

The data are out there, but they are rarely converted into information that can be used to drive process improvement, improve quality and reduce the cost of quality.

Simple techniques and tools can be used to examine processes to gain an insight into how good they are compared to the specifications.

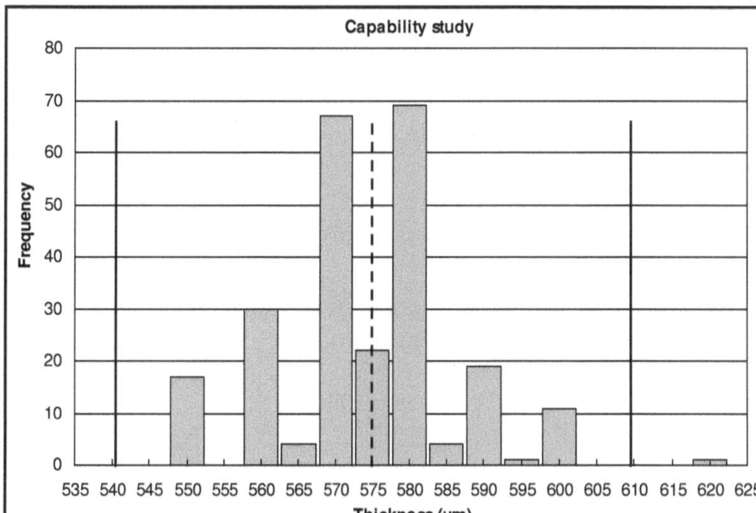


**An initial process capability study for sheet extrusion**

The process capability study shows that $C_p$ = 0.96 and $C_{pk}$ = 0.93. This process is not capable in terms of either spread or location. Out-of-tolerance products will always be produced whatever the process settings.

If this is not done then the process will continue to produce excessive out-of-tolerance products and reject material will continue to be sent for regranulation.

- **Tip** - In reality, the situation will actually be worse than this because this is more of a machine capability study than a process capability study and will not show the full variation of the process.

## Process control

The QC records were further analysed to produce average and range ($\overline{X}$ and R) charts for the running process. The 5 values measured at each inspection were taken as the five samples for SPC by variables (see Section 5.8).

The control limits were calculated using standard SPC formulae and are:

### Range chart:

$UCL_R = 24.9$ µm.

$\overline{R} = 11.65$ µm.

$LCL_R = 0$ µm.

### Average chart:

$UCL_{\overline{X}} = 567.27$ µm.

$\overline{\overline{X}} = 574.02$ µm.

$LCL_{\overline{X}} = 580.78$ µm.

The control limits and process data are shown on the right for the range (upper right) and the average (lower right).

The range chart shows that the range chart was broadly in control for the whole period with the exception of two short periods between 9 and 42 (the alarm signals for range charts are given in Sections 5.10 to 5.13). The rapid changes in the later section of the range chart (37 to 50) may indicate that more than one person was making the measurements or 'tampering' with the machine when it should not have been adjusted.

- **Tip** - The resolution of the measuring equipment for this case is suspect due to the operator tendency to measure in 10s and not in 5s.

The average chart shows that the process was out of control almost all of the time (the alarm signals for average charts are given in Sections 5.14 to 5.16). This is thought to be due to the operators 'hunting' for a stable process to meet a specification that it simply could not meet (see above). This behaviour explains the excessive rejects and the excessive use of the regranulators in the extrusion process.

## Using the numbers

The available QC records show that this is a process that is fundamentally not capable of producing to the set specification. The management were not aware of this and this information prompted a review and improvement of the complete process (and a reduction in the scrap product being produced).

- **Tip** - This type of data is available in many companies but is rarely used for process management.

This is not high-tech, it is simple statistics based on the natural variation of processes.

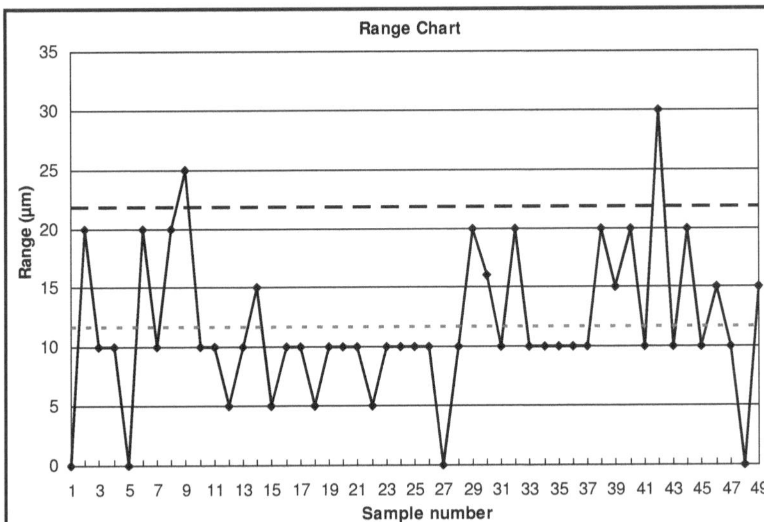


A range chart for sheet extrusion

The range chart (the first chart to look at) shows striation due to the measuring instrument resolution and also several times where alarms were indicated for process control. These were not seen by the operators and scrap was produced.

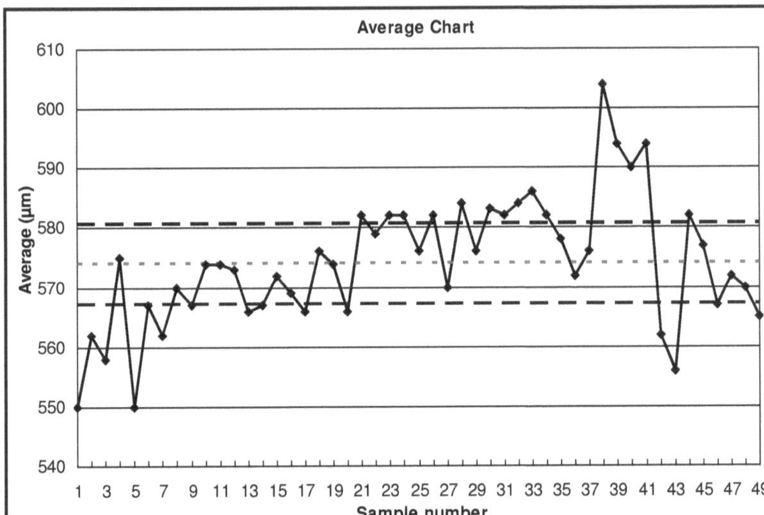


An average chart for sheet extrusion

The average chart (the second chart to look at) shows poor control of the process due to either a moving process, i.e. the average appears to be increasing, or to poor process control. There is also evidence of 'hunting' or tampering with the process.

# 10.16 Extrusion - managing attributes

## A range of processes

Extrusion is used for a broad range of process but the basic principle of screw extrusion remains: a rotating screw (or screws) feeds, pressurises, conditions and moves the polymer melt through a die (profile, sheet, film or wire coating) to form the final product. This seemingly simple process is the basis for most plastics forming methods, e.g. injection moulding uses a rotating screw to condition and move the melt into the cavity, but the simplicity should not be underestimated, this is a complex process. We will not cover the basics of extrusion process as these are covered in many standard texts. By far the best general text for all the necessary theory and practice is Rauwendaal[1] and the best for troubleshooting the process is Noriega and Rauwendaal[2]. They are essential.

## Screw selection

Getting the correct screw for the material being processed is a basic requirement. 'General-purpose' screws are sometimes used but these are more accurately defined as 'no-purpose' screws because they requires so many compromises in terms of screw design that they become unsuitable for most applications.

- **Tip -** Get the right screw for the material and application.

## Die care

The die is the heart of extrusion and it is the die that determines the final shape of the product as well as many of the visual concerns that occur in extrusion. Despite the seeming robustness of extrusion dies, they are delicate, susceptible to damage and must be handled with care (as with any tooling). Damage to any die will show in the final product and not only create dimensional concerns but also visual or attribute concerns.

- **Tip -** Always use soft tools when cleaning dies and never use steel tools to dig plastic out of a die.
- **Tip -** Disassembled dies are very vulnerable. Take extra care and always use wooden surfaces.
- **Tip -** If you have 'rusty fingers' that rust dies then use gloves when handling dies.

## Quality concerns in extrusion

Whichever extrusion process is being managed, it is essential to follow a process to identify and remove the concern (see Section 10.13). The steps are:

### Have we got a concern?

Attributes control charts are under-used in plastics processing but provide an excellent early warning method to see if there is a real issue with the visual properties and if action needs to be taken.

Attributes control charts are easy to set up (see Section 5.19 ) and the control limits for attribute charts will give warning of any out-of-control conditions and the need to take action.

- **Tip -** Do not over-react. Unless the control chart shows a need to adjust the process then leave it alone.
- **Tip -** Accurate assessment of attributes needs consistent application of standards and judgements (see Section 5.20).
- **Tip -** Use a constant sample size to remove any issues with changing sample size. This makes sampling easier for operators (and everybody else) to understand.

### What is the concern?

The range of visual non-conformities found in extrusion is as wide as those found in injection moulding and the

Manage the attributes (surface concerns) found in any of the extrusion processes using control charts for attributes (see Sections 5.19 and 5.20).

Attributes charting and the control limits tell you when to take action and when to leave the process alone

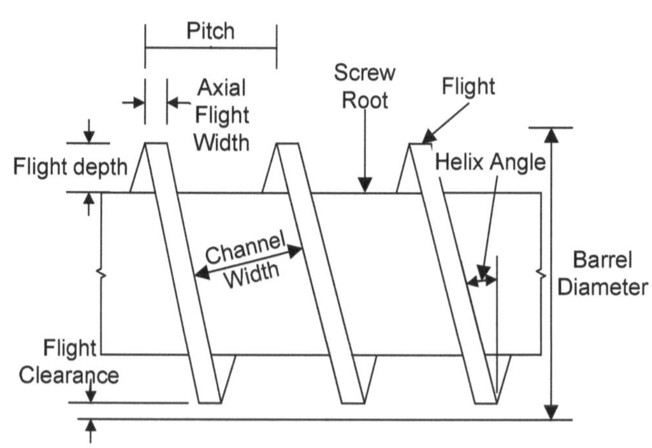


The key screw dimensions in extrusion

Good screw selection is important in getting good quality extrusion. The screw should always be specific for the material. As the saying has it: 'A *general purpose* extrusion screw is actually a *no-purpose* extrusion screw'.

terminology also varies. Without a clear definition and agreement on what the concern is then it will be impossible to resolve the concern.

- **Tip** - The range of terms used to describe visual defects is very wide. Visual samples or photographs of each type of concern will help people to agree on what the issue is. Start a collection today.

## When is the concern?

When an attribute concern appears is vital in troubleshooting. Concerns can be:

- Die and machine trial related - this type of concern should be resolved before the process is accepted for production.
- Start-up related - this type of concern is at start-up with a die, machine and material that previously produced good product. In this case the focus is on determining what has changed between the runs. Start-up related concerns must be corrected by the setter before the machine is handed over to production.
- **Tip** - Start-up concerns are generally due to changes in machine settings made on the previous run. Restore the machine to the 'good' settings for production.
- **Tip** - If the die is being used on a different machine then settings will obviously need to be modified.
- **Tip** - Check that the die has not been modified since the last good run.
- **Tip** - Check if the material is from a batch that previously produced good products. If a new batch/delivery has been used then try a different batch to eliminate material issues.
- **Tip** - Retain a sample of 'good' material that can be used to quickly check for material issues if start-up is a problem.
- Quality control related - this type of concern occurs on a running process when the chart shows that the process is 'out-of-control'. In this case the focus is on determining what has changed during the run and any changes in material, machine or mould should be checked.
- **Tip** - For QC related concerns look at the material and machine first.

Although extrusion is seen as a continuous process, it is actually a cyclic process where many concerns can be identified by relating their frequency to the crew rotation frequency. Rauwendaal[1] gives an excellent chart relating the frequency of concerns to the potential cause.

## Where is the concern?

The location of the concern in extrusion does not give as much information as in injection moulding but it is always good practice to look for regularity in location to find patterns and possible solutions.

## How do we solve the concern?

Cause and effect charts (see Section 7.4) are invaluable in resolving extrusion concerns and can be used in conjunction with troubleshooting guides to provide a guide to the possible root cause of the concern.

- **Tip** - For an excellent trouble-shooting guide to film extrusion see the Qenos publication[3].

Extrusion attribute concerns will generally relate to one of the following:

- Material quality and preparation - these can be due to either poor incoming material , e.g. some gels in film extrusion, or to excessive gas in the material which is not vented sufficiently.
- Process instabilities at the extruder - these can be high frequency, e.g. shark skin, melt fracture or draw resonance, or low frequency, e.g. surging, pressure fluctuations.
- **Tip** - Looking for a relationship between the extruder screw frequency and the concern frequency can sometimes reveal the cause of the concern.
- Die concerns - these can be due to poor die maintenance or a build up of material in the die, e.g. continuous die marks along the length of a profile or lines across the profile.
- Process instabilities downstream from the extruder - these are generally product specific, e.g. wrinkles in the machine or transverse direction for film production that are caused by the downstream handling of the product.

If attributes charting indicates a need for process adjustment then it is essential to follow a process to define the root cause and not to jump to conclusions.


•1. Rauwendaal, C. 2014. '*Polymer Extrusion*'. Hanser.

•2. Noriega, M and Rauwendaal, C. 2010. '*Troubleshooting the Extrusion Process*'. Hanser.

•3. Qenos Pty. Ltd. 2015. 'Film extrusion and conversion'. Free from www.qenos.com/internet/home.nsf/web/PolyethyleneTechGuides.


Never underestimate the power of the Mark 1 'Human Eyeball' in assessing attributes information. Be warned that it can also be misleading, particularly in the area of colour assessment.

Improve the power of the Mark 1 'Human Eyeball' by investing in a simple portable microscope. I have several, including one that plugs into the USB port of my laptop and displays the view on screen and can also capture this as a file.

Invaluable tool for detailed cosmetic and general product assessment.

You will never be able to carry out effective 100% inspection of your production.

The end user (potentially your customer) will always carry out 100% inspection of the final product.

# 10.17 Blow moulding

## Even more processes

Blow moulding is actually not a single process. Although the principles are much the same, i.e. a parison or pre-form is heated and blown with compressed air to form the final product, there are actually 3 distinct processes and these will be discussed separately.

## Extrusion blow moulding (EBM)

EBM is widely used for the production of containers in PE-HD or PE-LD. As with any processing method, quality starts at the design stage and an important but often neglected issue is the location of vents in the mould. Vent areas are generally the last areas of the mould to take the shape of the mould and if venting is inadequate then this can prevent full formation of surface details.

### Parison formation

Parison control is used to control and vary the wall thickness of the parison and hence the wall thickness of the final product. Parison control varies the flow from the die head through the length of the parison (see diagram on the right) to provide a consistent parison and consistent production. The important points in achieving a good parison are:

- Consistent pressure at the gap head - this is needed to allow parison control to be effective.
- Adequate land length - as with any extrusion process, die swell at the die exit can be minimised by providing an adequate land length.
- Consistent polymer viscosity - parison control works by varying the die annulus and this is only effective if the polymer viscosity is consistent.

Parison control also allows parison droop to be controlled.

### Blowing the parison

After the parison is formed it is trapped and compressed air is blown into the parison to form it into the shape of the mould. Adequate compressed air pressure and flow is needed to form the parison to the mould shape before it cools too much to take up the surface details of the mould.

- **Tip** - Venting is important.

### Cooling the parison

The formed parison is finally cooled by the chilled mould to form the finished product.

### Concerns

EBM can suffer from a range of concerns and the main concerns are:

- Gels and black specks from extrusion - treated as for standard extrusion.
- Extrusion die lines - treated as for standard extrusion.
- Die swell - treated as for standard extrusion and corrected by parison control.
- Axial thickness variations in the parison - unless the parison is accurately centred then the thickness will vary around the extrusion axis and create weak areas.
- Parison pleating (parison folds in on itself as the length increases) - this will be more pronounced if there are weak areas around the extrusion axis.
- Parison droop (elongation of the parison due to own weight) - this can be reduced by reducing the parison length or by using a polymer with a higher zero shear viscosity.
- Mottled surface due to water or uneven cooling.
- Pinch scarring from poor mould closing.

Manage dimensions in any of the blow moulding processes using control charts for variables (see Section 5.9).

Manage the attributes (surface concerns) found in any of the blow moulding processes using control charts for attributes (see Sections 5.19 and 5.20).

The crystallinity, density of the material and the melt strength define which material can be used for which blow moulding process, e.g. PE-HD and PE-LD are used for EBM because of their high melt strength whereas PP is not often used because of the lower melt strength.

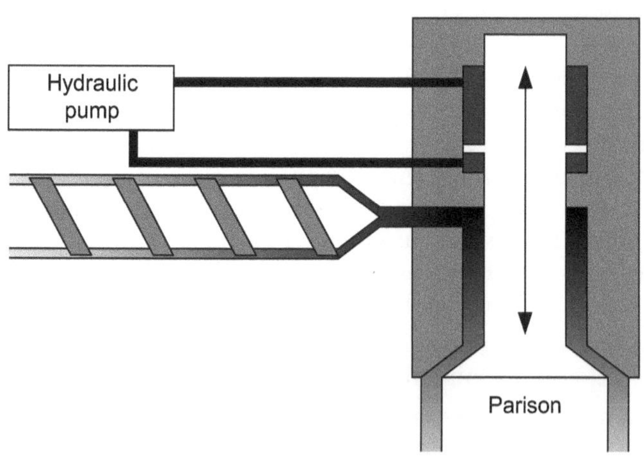


The principle of parison control in EBM

Good parison control is the key to reducing product weight by changing material distribution in the part. Not all areas are blown to the same extent and the wall thickness can vary widely across the part. Parison control puts the material where it is needed.

- Trimming marks from poor trimming.
- Tops and tails folding - where excessive tops or tails fold back onto the main product and stick to it.

### Testing

As most EBM products are designed to contain liquids, it is common to carry out a pressure test on every product.

- **Tip** - For an excellent trouble-shooting guide to extrusion blow moulding see the Qenos publication[1].

## Injection blow moulding (IBM)

IBM is an integrated process (see diagram on the right) which does not suffer from many of the issues of EBM but is limited to smaller products. In IBM the pre-form is injection moulded and, whilst still hot, transferred to the blow stage before being transferred again for cooling and ejection.

IBM gives better thickness control than EBM, has no tops and tails, i.e. greater process efficiency, and has a reduced tendency for surface defects due to the closed nature of the process. It is also capable of using PP as there is no concern with the tensile viscosity of the melt.

### Testing

As many IBM products are designed to contain liquids, it is common to carry out a pressure test on every product except where it is carrying solids, e.g. pills.

- **Tip** - For an excellent trouble-shooting guide to injection blow moulding see the Qenos publication[1].

## Injection stretch blow moulding (ISBM)

ISBM is largely used to produce beverage bottles and one great advantage is that the process creates biaxial orientation which greatly increases the mechanical properties of the product.

### Moulding the pre-form

Pre-forms are moulded using standard injection moulding techniques (see Section 10.10) and stored before blowing.

- **Tip** - Good control of the neck area in the pre-form is needed as this is the final form for the blown bottle.

### Blowing the pre-form

After storage, the pre-forms are clamped at the neck area and the body area is reheated with IR heaters for blowing with high-pressure air (35 bar). The blowing process produces a highly biaxial oriented material which greatly increases the mechanical properties.

### Concerns

- PET moisture content before moulding - the presence of moisture will reduce product strength and give a poor surface finish to the pre-form (and product).
- Heating distribution during blowing - uneven heat distribution will result in preferential blowing and uneven wall thickness.
- Pearlescence (stress-whitening) on the inside of the bottle is due to over-stretching in the blowing stage.
- Haze or whitening of the outside of the bottle is due over-heating and crystallisation of the PET at any stage.
- Low load strength is due to low wall thickness and general bottle design.
- Impact failure is due to inadequate orientation strengthening.

### Testing

Almost all ISBM products are designed to contain pressurised liquids and it is common to carry out a pressure test on every product.

- 1. Qenos Pty. Ltd. 2015. '*Polyethylene Blow Moulding*'. Free from www.qenos.com/internet/home.nsf/web/PolyethyleneTechGuides.

Injection stretch blow moulding of PET bottles is one of the leading users of vision systems for control of dimensions. These allow the inspection of up to 100,000 bottles/hour for a range of dimensions and all automatically.

Vision systems and automatic inspection are a new frontier in inspection.

Infrared cameras can be used to see the temperature distribution in the parison or pre-form to check for uniform temperature distribution.

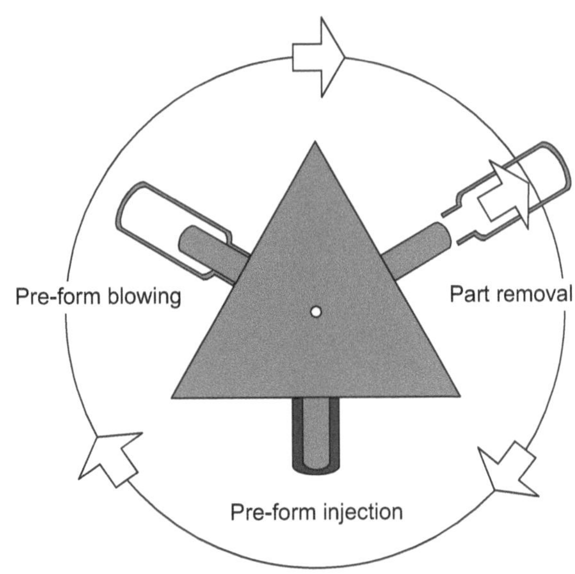


**The IBM process**

IBM is often carried out on a 3-station rotary turntable. The turntable can have multiple stations (only one is shown for clarity). After pre-form injection, the turntable is rotated for the blow operation and then rotated again for part removal.

# 10.18 Thermoforming

## A secondary process

Thermoforming can be considered to be a secondary process because it uses the output of the extrusion process to produce products. However, it is such an important process that it will be considered in it's own right.

The most common type of thermoforming is roll fed, plug-assisted thermoforming and this is the process that is considered but many of the points here are common to other thermoforming processes.

## Sheet orientation and quality

Variations in the orientation of the extruded sheet may show as poor or uneven sheet performance during thermoforming and a simple check (see diagram on the upper right) can provide an early detection and recording method for uneven sheet orientation.

## Sheet heating (reheat)

Sheet heating is a critical phase of thermoforming and affects all the following phases. For most methods of sheet heating, and particularly for thick sheet, the temperature of the sheet will be uneven through the thickness due to the poor thermal conductivity of the sheet. Thermoforming has a 'processing window' (see diagram on the lower right) for good production and the centre-line temperature must be inside the processing window before the surface temperature goes outside the process window. If the surface sheet temperature is higher than the processing window then degradation will occur and the product will show surface blisters and bubbles. If the sheet temperature is lower than the processing window then the sheet will show incomplete forming or tearing.

- **Tip** - Understanding and controlling the processing window is essential for good thermoforming.
- **Tip** - The processing window defines the process speed as well as quality.
- **Tip** - Using infrared heating can minimize temperature variations through the sheet by matching the infrared heater emissivity to the absorption characteristics of the material being processed. Infrared heaters also

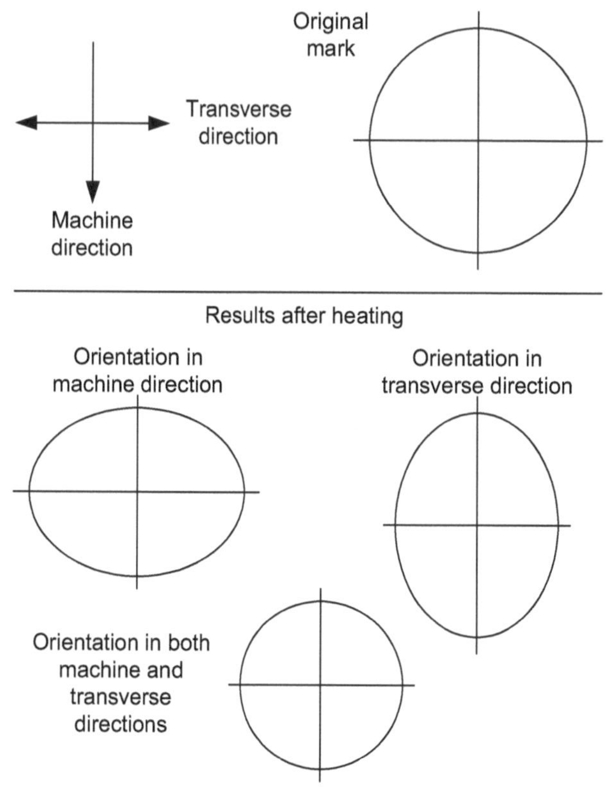


### The 'quick and easy' test for sheet orientation

Sheet orientation can affect thermoforming quality. The 'quick and easy' test is to mark the sheet (see 'Original mark' above), heat the sheet with unconstrained edges and check the resulting shape of the mark. The test will reveal the sheet orientation.

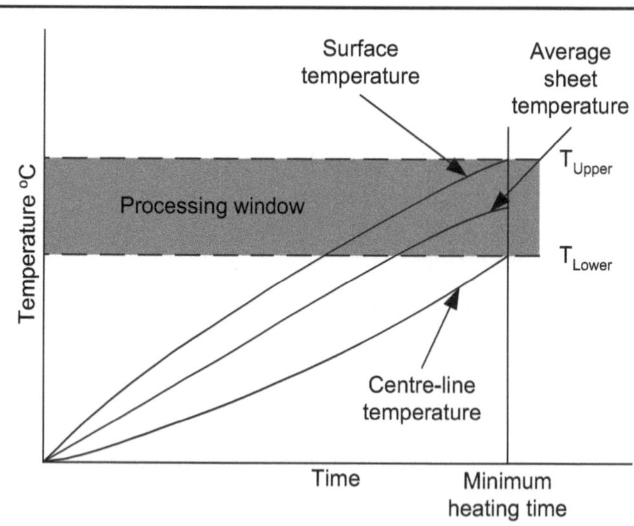


### Sheet heating and temperature are critical parameters

Close control of sheet temperature is essential for good thermoforming but the poor thermal conductivity of plastics means that the temperature of the sheet will vary through the thickness.

have a better response time and can reduce energy costs.

Directional effects can also give uneven sheet heating across the sheet. If standard radiation heating is used then the centre areas will 'see' more radiation than the edges and this can complicate getting the sheet temperature correct even more although in this case the temperature variation will be across the sheet rather than through the sheet.

- **Tip** - Sheet temperature is a key process and quality variable. Get the sheet temperature inside the processing window through and across the sheet for consistent processing and quality products.
- **Tip** - Temperature control is more important for semi-crystalline materials such as PP than for amorphous materials such as PS because the processing window is smaller.
- **Tip** - As a simple process control it is possible to use IR temperature sensors to measure the sheet surface temperature and alert the operator if this goes outside specified limits.

The sheet heating phase of thermoforming is ripe for improved process control (see Section 10.8) and whilst some attempts have been made, these do not appear to have made it into production machines to improve quality and the majority of machines are still poorly controlled.

- **Tip** - As with all plastics forming processes, the thermoforming process can take a little time to 'settle down'.

## Sheet forming

Sheet heating is not the only process variable in plug-assisted thermoforming and other variables are:

- Plug speed - this controls the strain rate that the material experiences during forming. Increasing the plug speed will increase the force on the plug and an excessive plug speed can lead to sheet tearing due to the strain rate sensitivity of polymers. Decreasing the plug speed should decrease the plug force but the sheet will be cooling whilst it is clamped and the plug force can actually increase.
- Plug shape - this controls how the material deforms and is stretched and consequently thickness variations in the product.
- Air pressure - this will provide the final driver to deform the plug assisted sheet into the final shape of the product.
- Vacuum assist - this can be applied to assist in forming the final shape.

Monitoring the plug location and force, the air pressure and the vacuum through the cycle provides excellent process control information and allows quality monitoring by setting acceptable limits for each of these parameters.

- **Tip** - Toolvision (www.toolvision.de) uses this type of curve to provide excellent information for process control, process development, quality control (to detect quality issues in real time) and maintenance.
- **Tip** - As with the cavity pressure curve for injection moulding (see Section 10.10), the plug force versus time curve through the cycle is a sensitive measure of process consistency.

Manage dimensions in thermoforming using control charts for variables (see Section 5.9).

Manage the attributes (surface concerns) found in thermoforming using control charts for attributes (see Sections 5.19 and 5.20).

## Quality controls

In the absence of good process control, simple attribute control charts for the most common concerns (see below) can provide excellent process control information for the heating phase - link the identified attribute concerns to the possible causes and then solve the underlying cause.

At the design stage the draw ratio, i.e. total surface area of the formed part/the footprint of the part before forming, is a key value.

Given the desired final wall thickness then the draw ratio will provisionally define the starting thickness of sheet needed for the part.

This assumes a perfect material distribution and can be affected by local features such as sharp corners.

### Surface and visual (attribute) concerns

Thermoforming has a range of concerns that can be tracked via attributes. As with injection moulding, the terminology varies throughout the industry. Some of the more common terms are:

- Blisters/bubbles.
- Incomplete forming
- Poor details.
- Blushing.
- Color intensity change.
- Webbing.
- Bridging
- Wrinkling.
- Excessive sag.
- Sag variations.
- Chill marks/striations.
- Shiny streaks.
- Post-forming shrinkage.
- Distortion.
- Poor material allocation.
- Thin corners.
- Part sticks in mould.
- Sheet sticks to plug.
- Sheet tearing.

# 10.19 Rotational moulding

## Leaving the 'Dark Ages'.

As with most plastics processing methods, the control methods used for rotational moulding have improved dramatically in the past 30 years. The start of this was the pioneering work of Crawford[1] and the team at Queens University in Belfast who first introduced science to the process and developed entirely new process control systems based on 'what the polymer saw' as opposed to 'what the machine saw'. As we saw in injection moulding, this is what really counts.

These developments in both the theory and process control provide the basis for huge improvements in process control and part quality.

## The basics

Despite the developments in process control, there are still some basics that need to be correct to produce quality products in rotational moulding.

### Get the material right

Raw materials quality management in rotational moulding is potentially more complex than in other methods. In addition to all the other controls that are required (see Chapter 9), rotational moulding needs the correct size of powder. This is because rotational moulding uses fundamentally different materials to other plastics processing. All other processes use either granules or mixed powder (twin-screw extrusion) but rotational moulding generally uses a mechanically milled powder of between 74 μm and 2,000 μm.

As well as checking the material and grade, controlling the polymer powder size is a key input variable.

- **Tip -** A powder that is too coarse will not only increase cycle times but will potentially give 'under-fused parts'.

### Get the right amount of material

Although rotational moulding is capable of controlling wall thickness within ± 10% this is only true if the correct amount of material is both weighed out and put into the mould.

- **Tip -** Use auto-weighing systems if possible.
- **Tip -** Control initial weighing and use 'check weighing' at the machine to validate the shot weight.
- **Tip -** Control material and shot weight though the process.

### Avoid contamination

The fine powders used for rotational moulding and the 'open' process makes material contamination a higher probability than with other processes and good contamination controls are needed.

- **Tip -** Keep the production area clean. Don't forget to look up and down as well as around for potential contamination sources.

### Protect the parting line

Parting lines on rotational moulds are sensitive to damage from mould opening and from cleaning.

- **Tip -** Protect the parting line area by building in features for mould opening.
- **Tip -** Always use the correct tools for mould opening and parting line scraping.
- **Tip -** Never, ever use compressed air to open a stuck mould - it can kill!

## Process control

Internal air temperature is the most important process control parameter for rotational moulding and a typical trace of oven temperature and internal air temperature is shown on the lower right.

This shows that recording or controlling the oven temperature tells us very little about what the bulk of the polymer is seeing, i.e. the internal air temperature.

Manage raw materials (mainly particle size) in rotational moulding using control charts for individual measurements (see Section 5.18).

Manage dimensions in rotational moulding using control charts for variables (see Section 5.9).

Manage the attributes (surface concerns) found in rotational moulding using control charts for attributes (see Section 5.19).

Standard process control tests for rotational moulding are:

- Impact strength.
- Bubble content.
- Internal surface appearance.
- External surface appearance.
- Dart impact.
- Bubble content - the bubble content is inversely related to cure, i.e. high bubble content indicates low cure.

### Surface and visual (attribute) concerns

Rotational moulding has a range of concerns that can be tracked via attributes. As with injection moulding, the terminology varies throughout the industry. Some of the more common terms are:

- Under-fused parts.
- Over-cured parts.
- 'Ringworm' on outer surface.
- Mould sticking.
- Non-uniform wall thickness.
- Part and parting line bubbles.
- Blow holes.
- Parting line flash.
- Warped parts.
- Powder bridging.
- Inside surface discolouration.
- 'Lightning' colours.
- Low impact resistance.
- Low stiffness.
- Long-term part failure.
- Stress cracking.

Initially the oven quickly reaches the set-point but the air takes longer to heat up until it reaches the point at which the material starts to stick to the mould. The rate increase of the internal air temperature then slows as the material absorbs the input energy.

After the mould is fully coated with polymer, i.e. all the material is effectively molten, then the internal air temperature will increase again as the sintering and densifying process begins. The oven heating is then turned off but the internal air temperature continues to rise and drive coalescence of the polymer powder.

This sintering and densifying continues and is largely complete when a specified peak internal air temperature is reached. The peak internal air temperature defines the point at which the material is fully coalesced and cooling can start.

As the mould cools, the material cools and eventually starts to solidify. During solidification the rate of change of the internal air temperature decreases until all the material is solid and it will then increase again until cooling is complete.

## Peak internal air temperature and 'degree of cure'

Initial work[1] showed that it was essential for the internal air temperature to reach a specified level to give sufficient time for sintering and densifying. If this temperature was reached then moulders could reliably produce good mouldings. Later work[3] has shown that it is not simply the peak internal air temperature that counts but also the time at this temperature and this was called the 'degree of cure'. This is unsurprising as coalescence will be affected not only by the temperature but by the time at the temperature, i.e. the amount of input energy. Degree of cure is defined in units of (°C x min) where the °C is measured as being above the melting point and the time is also that above the melting point. The concept of the 'degree of cure' is shown on the right.

- **Tip** - Controlling the degree of cure gives good control of the process as seen by the polymer.

The 'degree of cure' can be considered analogous to the cavity pressure curve (see Section 10.10) for process control. A controlled and consistent 'degree of cure' will give a controlled and consistent process.

## Cooling

Cooling is also important in rotational moulding. During cooling, the product will shrink due to thermal shrinkage and, if it is a semi-crystalline material, due to crystallisation shrinkage. The cooling rate affects the degree of crystallisation and the shrinkage. If the material is cooled too rapidly during cooling then warping will increase.

Warping is a major issue and one of the main causes of warping is non-uniform cooling.

## Quality and troubleshooting

Quality is linked to general troubleshooting and both Crawford and Throne[1] and LyondellBasell[4] have excellent charts for process troubleshooting and quality improvement.

Under-fused parts will show many small bubbles in the wall and a rough slightly powdery internal surface.


- 1. Crawford, R.J. and Throne, J.L. 2002. '*Rotational Molding Technology*'. William Andrew.
- 2. Beall, G. 1998. '*Rotational Molding*'. Hanser.
- 3. Spence, A. and Scott, R. 2003. '*Rotational Molding Process Control*', ANTEC. Available from http://www.centroinc.com/Content/Our-Processes/Case-Studies.aspx.
- 4. LyondellBasell. Undated '*A Guide to Rotational Molding*'. Free from www.lyondellbasell.com.


The degree of cure can be related to impact strength and bubble content and is a sensitive process control measure.

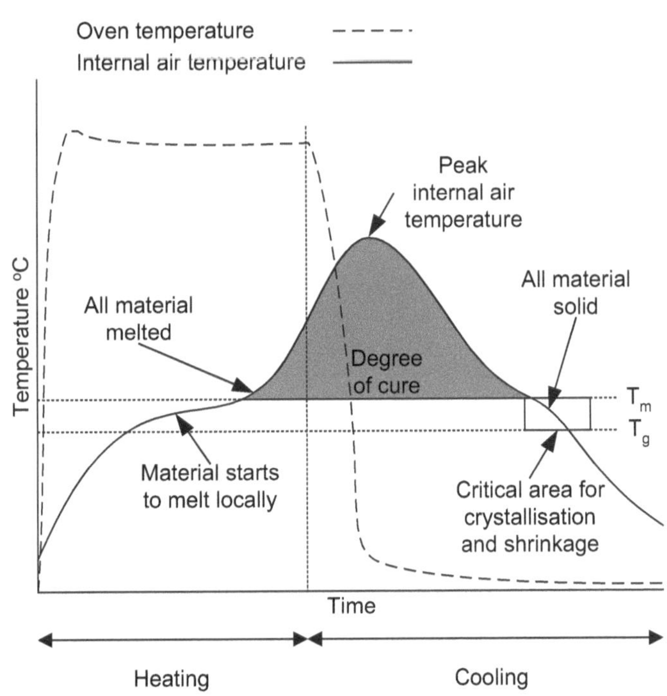


### Temperature changes in rotational moulding

The oven temperature is not the same as the temperature that the polymer sees. What the polymer sees is the important thing. Understanding and controlling the internal air temperature rather than oven temperature is vital to good quality parts.

# 10.20 Processing systems - where are you now?

## Get the systems right

High quality design and development work counts for nothing if the transfer to production is not managed consistently and well to deliver the required results.

The process models, e.g. APQP, for this transfer are widely available but poorly or little used in many plastics processors. These models provide both the tools and good practices to set up processing correctly from the start and for clearly specifying what should happen in the event of a process not performing correctly.

The keys to good processing systems are:

- Planning for process design and development.
- Creation and completion of FMEAs for the complete process and machinery.
- Control planning for the process, including the necessary reaction planning.
- Good machine setting at the transfer stage.
- Full product and process validation before series production.

## Completing the chart

This chart is completed and assessed as for the previous charts.

Good plastics processing needs good systems to transfer processes from design and development into series production.

Get the systems right to get the production right.

**Processing systems**

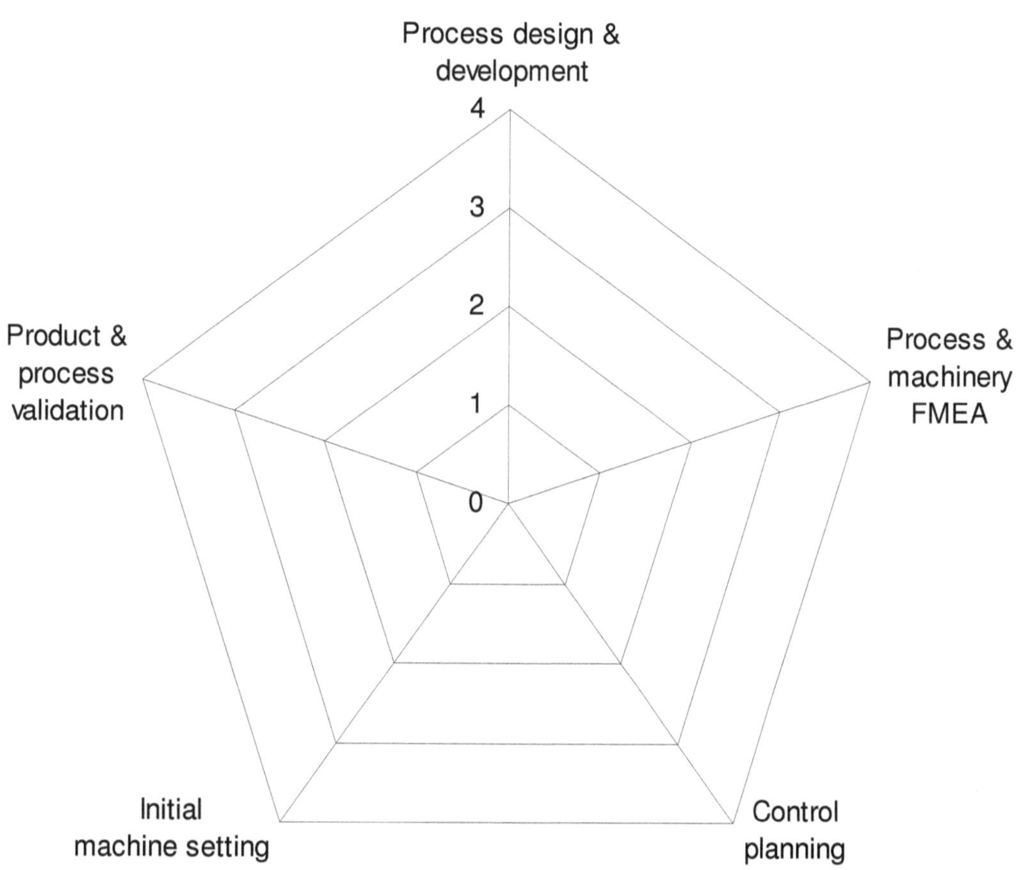


Use the scoring chart to assess where you are in processing systems

The numbers from the scoring chart can be transferred to the radar chart for a quick visual assessment of where you are in terms of processing systems.

A systematic approach to process and machinery planning and control will always give reduced quality issues in series production.

Failing to plan is the same as planning to fail.

| Processing systems | | | | | |
|---|---|---|---|---|---|
| **Level** | **Process design & development** | **Process & machinery FMEA** | **Control planning** | **Initial machine setting** | **Product & process validation** |
| **4** | Full process design & development carried out using logical steps. Records show all required inputs considered & all outputs produced. | Full process & machinery FMEA carried out & available. | Full & detailed control plan available with all process & customer requirements specified. Reaction plans in the event of failure specified. | Initial machine setting carried out in logical manner taking account of process variability. Process setting always based on multiple samples. | Full product & process validation carried out using logical steps. Records show all required inputs considered & all outputs produced. |
| **3** | Majority of process design & development carried out using logical steps. Records show most required inputs considered & most outputs produced. | Partial process & machinery FMEA carried out. Identifiable gaps in FMEA coverage. Written records available. | Partial control plan available. Identifiable gaps in process & customer requirements. Partial reaction plans in the event of failure specified. | Initial machine setting carried out in a logical manner but little consideration of process variability. Process setting mostly based on multiple samples. | Majority of product & process validation carried out using logical steps. Records show most required inputs considered & most outputs produced. |
| **2** | Some process design & development carried out using logical steps. Records show some required inputs considered & some outputs produced. | Poor process & machinery FMEA carried out. Large gaps in FMEA coverage. Written records available. | Poor control plan produced. Large gaps in process & customer requirements. Poor specification of reaction plans in the event of failure. | Initial machine setting based primarily on previous 'best' settings with no consideration of process variability. | Some product & process validation carried out using logical steps. Records show some required inputs considered & some outputs produced. |
| **1** | Informal & poorly documented process design & development procedure. Poor written records available. | Informal process & machinery FMEA carried out. Poor written records available. | Informal & poorly documented control plan. Little reference to process & customer requirements. No reaction plans available. | Initial machine setting carried out with little regard to process variability. | Informal & poorly documented product & process validation procedure. Poor written records available. |
| **0** | No discernable process design & development procedure. No written records available. | No process & machinery FMEA carried out. No written records available. | No control plan prepared. | Initial machine setting is poorly controlled & managed. | No discernable product & process validation procedure. No written records available. |
| **Score** | | | | | |

# 10.21 Processing - where are you now?

## Get the systems right

The variety of plastics processing methods and products means that the controls applicable and those used vary widely with the process and also throughout the industry. Despite these process and product differences, the standard methods of control charting for process control of variables and attributes can be applied for most of the processes (see Chapter 5). This is more difficult when processes are effectively multi-variate (see Section 10.11) but any attempt at process control by controlling the key process variables will always give better results than traditional simple end-of-line inspection.

The key to controlling any process is to identify and understand the effects of the main process variables affecting the quality of the output product. These variables are different for every plastics processing method but controlling the process will always be more effective than simply trying to control the product.

## Completing the chart

This chart is completed and assessed as for the previous charts.

**Processing**

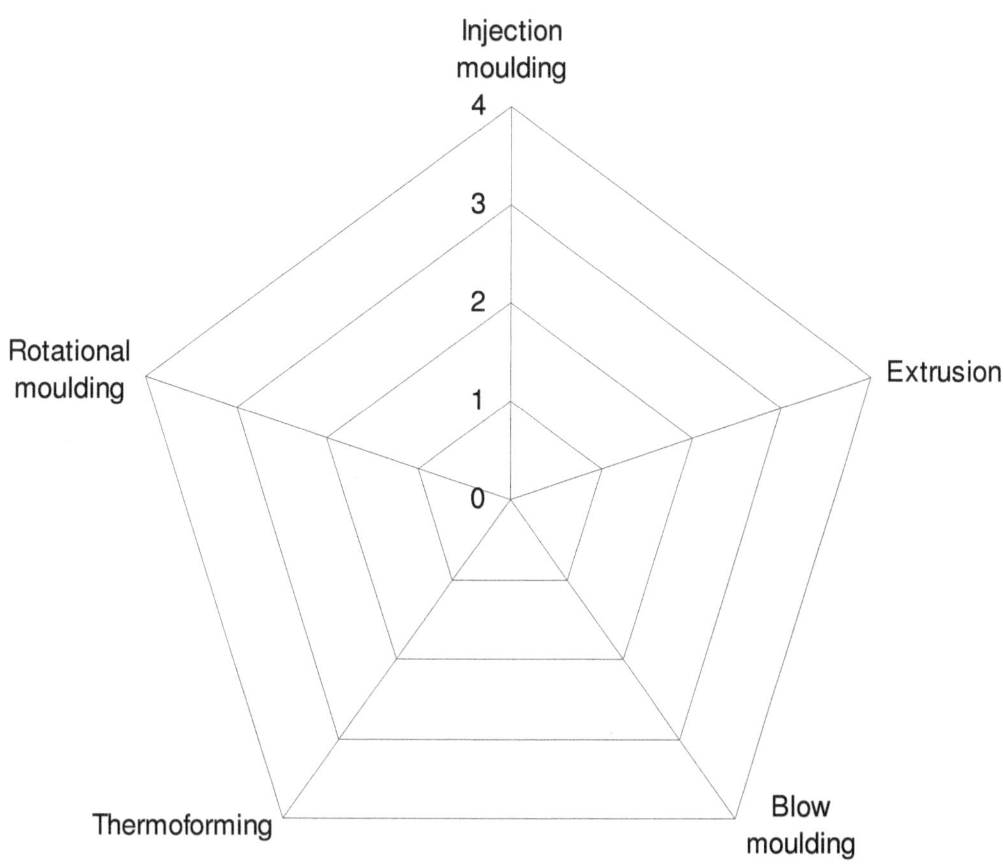


Use the scoring chart to assess where you are in processing

The numbers from the scoring chart can be transferred to the radar chart for a quick visual assessment of where you are in terms of processing.

Using attributes data for process and product control is sometimes seen as being easier than using variables data.

Variables data gives much more information that can be used to control the process and reduce the possibility of producing non-conforming product.

Where possible, my preference is to convert attributes data to variables data for better control, e.g. a colour attribute can be OK or not OK but an actual colour measurement gives much better opportunity for control.

"The disciplines we've created don't correspond to real-world problems.

One of my friends says, '*The world has problems, and we have departments.*'

The fact is, there is no electrical engineering. There is no mechanical engineering. There are problems, and we solve them with whatever tools we have available."

**Jim Melsa**

Think of this when you next approach a quality concern and equip yourself with the best tools you can possibly find - from wherever you can find them.

## Processing

| Level | Injection moulding | Extrusion | Blow moulding | Thermoforming | Rotational moulding |
|---|---|---|---|---|---|
| 4 | Scientific moulding used for all setting. Cavity pressure curves used to control process variables Attribute charts used to control surface features. | Key direct process & indirect process variables monitored for process control. Product dimensions automatically measured & monitored for control. Attributes monitored for process control. | Key direct process & indirect process variables monitored for process control. Product dimensions automatically measured & monitored for control. Attributes monitored for process control. | Key direct process & indirect process variables monitored for process control. Product dimensions automatically measured & monitored for control. Attributes monitored for process control. | Key direct process & indirect process variables monitored for process control. Product dimensions automatically measured & monitored for control. Attributes monitored for process control. |
| 3 | Scientific moulding used for majority of setting. Cavity pressure sometimes used to control process. Attribute charts sometimes used to control surface features. | Some direct process & indirect process variables monitored for process control. Product dimensions manually measured & monitored for control. Attributes monitored for process control. | Some direct process & indirect process variables monitored for process control. Product dimensions manually measured & monitored for control. Attributes monitored for process control. | Some direct process & indirect process variables monitored for process control. Product dimensions manually measured & monitored for control. Attributes monitored for process control. | Some direct process & indirect process variables monitored for process control. Product dimensions manually measured & monitored for control. Attributes monitored for process control. |
| 2 | Scientific moulding used for some setting. Cavity pressure rarely used to control process. Attribute charts rarely used to control surface features. | Some direct process & indirect process variables measured but not for control. Product dimensions manually measured as part of inspection. Attributes assessed as part of inspection. | Some direct process & indirect process variables measured but not for control. Product dimensions manually measured as part of inspection. Attributes assessed as part of inspection. | Some direct process & indirect process variables measured but not for control. Product dimensions manually measured as part of inspection. Attributes assessed as part of inspection. | Some direct process & indirect process variables measured but not for control. Product dimensions manually measured as part of inspection. Attributes assessed as part of inspection. |
| 1 | Product weight is main control variable monitored. No use of attribute controls apart from inspection. | No process variables monitored. Product dimensions manually measured for inspection. Attributes assessed for inspection only. | No process variables monitored. Product dimensions manually measured for inspection. Attributes assessed for inspection only. | No process variables monitored. Product dimensions manually measured for inspection. Attributes assessed for inspection only. | No process variables monitored. Product dimensions manually measured for inspection. Attributes assessed for inspection only. |
| 0 | No use of variables or attributes charts. Control is by regular, or irregular, inspection of final product. | Control is by regular, or irregular, inspection of final product only. | Control is by regular, or irregular, inspection of final product only. | Control is by regular, or irregular, inspection of final product only. | Control is by regular, or irregular, inspection of final product only. |
| Score | | | | | |

# Key tips

- Taking product design to the required process design needs a process and APQP provides an excellent model for this.
- APQP provides a template and '*aide memoire*' for all the things that you need to do to make the transition from the design to the product.
- Carrying out a process and machinery FMEA can not only help make the transition easier but also help with proving that you carry out risk-based thinking for ISO 9001.
- Control planning helps everyone to understand what is needed to control quality, when and where the controls should take place and, most importantly, what to do when things don't go to plan (which is always the case).
- Process variability is always present and process setting must take this into account.
- Process setting should always use multiple samples to account for variability.
- After a process is fully designed and implemented then it must be validated to make sure that it delivers what it was designed to deliver.
- APQP provides a template and '*aide memoire*' for all the things that you need to do to validate the product and process.
- Pre-control (zone charts) offers a rapid method of assessing process control at start-up and during running.
- Pre-control is an alternative to either inspection or full control charts.
- Engineering process control (EPC) uses established process control methods to dynamically change the input variables to control the output variables.
- EPC and SPC both try to control quality but approach the issue from different directions.
- Scientific moulding provides a rational approach to setting injection moulding machines by understanding what the process is actually doing.
- Injection moulding is a multi-variate process and this must be considered when making any process change.
- Cavity pressure control in injection moulding provides a method of understanding 'what the plastics sees'.
- Managing the quality of attributes in injection moulding needs a defined process to get the most information from the process.
- Extrusion can use direct or indirect process measurement of variables data for process control.
- Managing the quality of attributes in extrusion needs a defined process to get the most information from the process.
- Blow moulding is a range of processes but both variables and attributes can be controlled using standard process control techniques.
- The critical process control in thermoforming is sheet heating and getting a the complete sheet into the process window is key to production quality.
- The critical process control in rotational moulding is the 'degree of cure' which defines the temperature and time needed for coalescence of the polymer powder.

# Chapter 11

# Operations quality management

This chapter on 'operations' is designed to look at some of the aspects of a company that affect quality other than in the production area.

These areas are all part of the 'quality' of the product as seen from the customer viewpoint. 'Overselling' of a product at the sales stage creates a customer expectation that can never be reached and leads to disappointed customers who then see the product as being less than what they were sold and therefore of 'poor quality'. Equally a perfect product that is followed by an incorrect invoice and subsequent accounting problems creates a poor image of the 'product quality'.

These are not theoretical issues and most of us have suffered from them at one time or another. The overall impression that we are left with is of a poor-quality product and all the effort of the production people in making the actual product is wasted.

Therefore this chapter will look at quality management in sales and how this can be achieved to ensure that the customer's expectations are managed and how performance can be measured and improved. It will also look at accounting processes and then how they can be measured and improved to raise the perceived quality.

Finally we will look at some of the more common office processes and how simple techniques can reduce the frequency and severity of errors to improve the customer experience.

Quality management in office processes has a significant bearing on the overall perception of quality and readers are urged not to skip this section simply because they are production oriented.

"By a small sample we may judge the whole piece."

**Miguel de Cervantes**

**Don Quixote de la Mancha (1605)**

# 11.1 Measuring sales quality

## It is a process

Improving the product quality is not a guarantee of success when the customer assesses the company in terms of the complete process and it is just as important to improve the sales process. The old saying 'build a better mousetrap and the world will beat a path to your door' is no longer true (if it ever was).

Sales is a process and will suffer from the same variations as manufacturing processes. Despite this, the proven quality management tools and techniques discussed for manufacturing are rarely used in the sales process and in many cases the sales process suffers from poor data gathering and handling.

Most companies have sales performance measurement systems in place but how useful are these and do they give information rather than simply data? Sales is more than lead generation systems, customer relationship management (CRM) systems and systems for setting sales targets and recording their achievement (or not). The whole of the sales process, from lead generation to invoicing, needs to be considered.

- **Tip -** If you want to improve the complete process then you should measure the complete process.

The statistical tools of quality management need good data and this is often not available for the sales process but, in many cases, it is better to try for relevant 'good enough' data than to try for highly accurate but non-existent data.

- **Tip -** The real question is 'Are the data good enough for the purpose'?

There are many methods for assessing the suitability and capability of specific people for sales jobs and some of these show good results in selecting and defining the right people for the right job. The conventional interview process is now regarded as a poor predictor of performance, especially when you are trying to recruit for specialist positions, i.e. the requirements for 'goal scorers' or 'hunters' are very different to those for 'goal keepers' or 'farmers'. In this Workbook we will not consider how to assess people for the job, simply how to assess if the process is succeeding or not.

- **Tip -** Getting the right people in the sales process is critical - spend some time on this and get it right.

## Measuring the process

Sales performance measurements for external salespeople can suffer from several issues:

- Setting measures will lead salespeople to perform to those measures, e.g. if the measures do not include long-term customer development then this will not be done and the focus will be on short-term sales, i.e. it is easy to measure 'new sales' but more difficult to measure 'sales retained'.
- Data used for sales performance measures will rarely be totally objective and quantifiable.
- Setting unrealistically high targets can demotivate sales staff.
- Measures are generally more complicated than simply 'meeting specification'. If you measure conversion rate then this may not capture the quality of the conversion. If you measure contact rate, e.g. 6 calls/day, then how is the quality of the call measured?
- Setting sales measures linked to rewards will always lead people to 'game' the system.

Most companies will have measurement systems for the performance of external sales people but the unfortunate thing is that they are mainly aimed at control and reward rather than understanding and improvement.

- **Tip -** Measuring simple sales values seeks to control the product, i.e. it is similar to inspection, rather than seeking to control the process. Measurement should not be simply for control and reward but also for understanding and improvement.

Relatively few companies have any equivalent measurement systems for performance of the internal sales service and support staff. These are equally important measurement systems and in most cases, because they are under your control, these parts of the process can be accurately measured and assessed using conventional quality methods and improved using conventional quality

The objective of a sales person is not to make sales, but to make customers.

They should be supported in this by the whole of the sales function.

**Sell to the specification**

A major cause of 'quality problems' is overselling of the product in the first place. If the customer is promised features or services that you cannot deliver then they will feel dissatisfied with your product and service. Sales staff desperate to get an order can set you up to fail by overselling the product in the first place.

Over-promise and under-deliver and you will always have problems.

Under-promise and over-deliver and you will always have satisfied customers.

Measurement is about improving the process.

Do not simply concentrate on the product, i.e. the sales values.

It takes less effort to keep an old customer satisfied than to get a new customer interested.

improvement techniques.

- **Tip** - Too many companies (and their fancy dashboards) simply track the external sales people and their results and ignore the essentials of the complete process.
- **Tip** - Do not forget that measurement of the sales process should include all stages of the process and not simply the external actions. Assessing external performance if they are let down by the 'back-office' will lead to resentment from the external people. Assess the whole process.

Whichever part of the process is being measured any performance measurement must be assessed in terms of the cost of making the measurement, the value it provides and if the information from the measurement is actually productively used inside the company to improve the performance or the process.

- **Tip** - If a measurement does not add value then don't make it.

## Measurements

A selection of possible measurements for the sales process are shown in the box on the right. Notice that there are as many, if not more, for the 'back office' processes as for the external processes. Whilst this might initially appear a daunting list then spare a thought for the production people who are assessed against a multitude of different measures every day (and expected to meet them).

- **Tip** - Initially select only a few measures to gain an understanding of the sales process and to identify areas for possible improvements (see Section 11.2). Do not be too ambitious at the start.
- **Tip** - Speed of response to customer queries is important. Are data input forms designed for the company or the customer?

## It is a quality issue

One issue that always appears when the quality department starts to get involved in measuring the performance of the sales process is the objection that 'This is not a quality issue, this is a sales issue'. This is because of the assumption that quality is only concerned with the product which is far from the truth from the customer's viewpoint. The customer is concerned with the whole package, not simply what is in the box.

**Quality measures in sales**

| Measurement | What is measured | Typical specification |
|---|---|---|
| **Strategic measures** | | |
| Customer satisfaction | Market share | Target market share achieved |
| Sales performance | Budget achievement | 100% of budget |
| **Operational measures** | | |
| Lead generation | Number of new leads generated | 30% of budget from new customers |
| Lead list | Lead quality | 80% of leads validated |
| Lead qualification | Lead quality | 80% of leads categorised for value |
| Sales performance | Sales value by sales person | Target achievement |
| Product information | Product description accuracy | 100% product description accuracy |
| Incoming phone calls | Response time | 98% answered before three rings |
| Sales enquiries | Response time | Quotation within 24 hours |
| Sales orders | Response time | Order acknowledged within 24 hours |
| Shipment | Response time | Order shipment within 24 hours |
| Invoicing (see Section 11.3) | Invoice accuracy | 100% of invoices accurate at line item level |
| Invoicing (see Section 11.3) | Invoice speed | 100% of invoices issued within 24 hours of despatch |
| Customer satisfaction | Repeat customers | 70% of sales from repeat customers |

**Some quality measures for the sales process**

There are many measures to assess the quality of the sales process. Most companies focus on the simple sales performance numbers but the 'back office' service and support are just as important in delivering a quality sales operation.

# 11.2 Improving sales quality

## Map the process

Nobody goes to work to make mistakes and in most cases the reason for something going wrong is a process failure. The first step in improving the quality of the sales process is to map the complete process using process maps (see Section 4.6) and flow charts (see Section 7.2). These help to understand:

- What the process is doing (in detail).
- Who is doing each part of the process.
- Where measurements should be made.
- What the critical measurements are.
- How often the measurements should be made.
- Who should make the measurements.
- What you are going to do with the data and how you will show it.

## Improving the process

Creating the process map/flow chart will start to reveal areas of process failures and the potential for quality improvements such as:

- Reducing the number of process steps to reduce the process complexity.
- Reducing the number of interfaces in the process. Interfaces between functions are always a source of errors (and friction) and reducing the number of interfaces will lead to quicker and more accurate processes.
- Reducing redundant information, i.e. record information once only and at the point of first contact (but get it right).
- Improving staff understanding of the process.

The standard quality tools (see Chapter 7) and process improvement methods can then be used to improve the process using the process measurements and data, e.g. Pareto analysis, check sheets, etc.

- **Tip** - One of the key issues in the sales process is how time is spent and whether the activity actually adds value to the process or simply adds cost. Look carefully at all the activities in the process.

Improvements in process efficiency, by concentrating on the core activities of the process, will always improve the quality of the process as seen by the customer.

## Control charts in sales

The standard control chart method (see Chapter 5) was originally developed for manufacturing processes but there have been many attempts to use it in service processes.

- **Tip** - The focus should always be on finding and preventing the causes of process deviation rather than inspecting the output and correcting poor output.

Some of the issues with using SPC for service processes are:

- Deciding on the measures: In sales, deciding on the specification to meet is

Any analysis of data gathered should focus on the process rather than on the output.

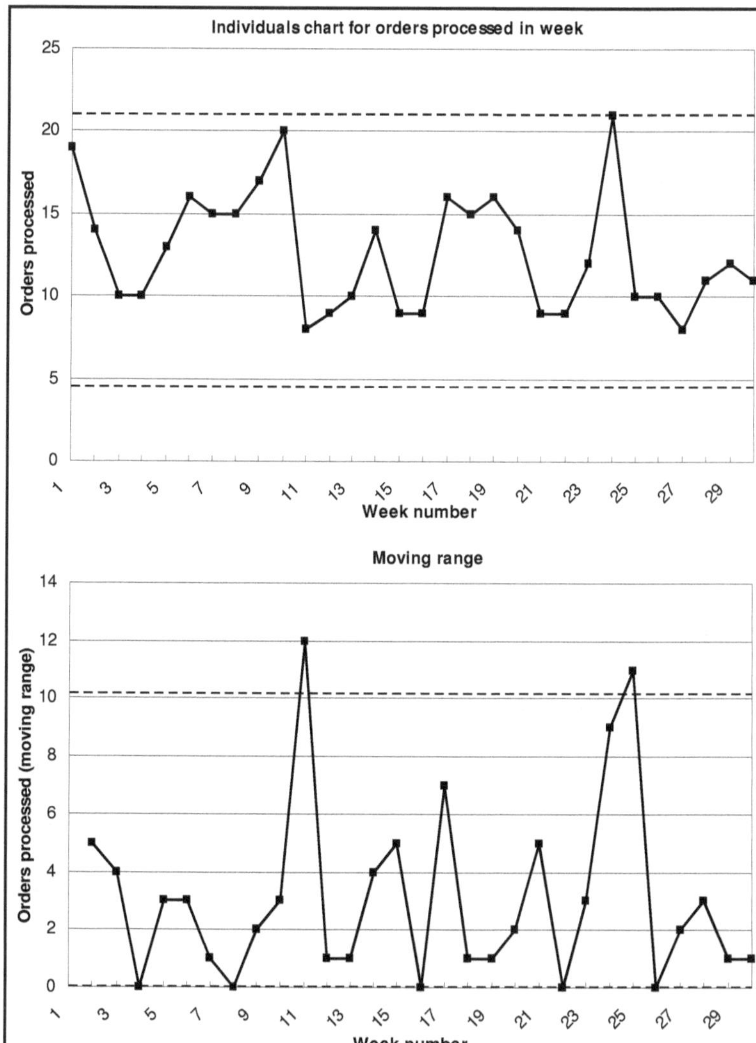


Individuals chart for orders processed

Individuals and moving range charts are useful for when there is only one data point available for each measurement. They can show not only when orders are made (top) but also when orders exceed normal volumes (the points outside the control limits).

more subjective. The measures chosen need to be sensitive to process variation and detailed enough to allow action to be taken. They should also relate directly to the customer experience.

- Linking the measures to reward: There is a tendency when measuring service processes for measures to be used for reward as well as prevention. This can contaminate the results.
- The type of data: Variables in manufacturing are generally centred around a specified value. In service processes, variables most often need to be minimised or maximised, e.g. the minimum number of rings of the phone.

This does not mean that control charts cannot be used for service processes, simply that they need to be chosen and used with care.

The process measures chosen will naturally vary and, as with any process, these can be divided into 'common' and 'special' causes (see Section 5.2):

- Common causes are predictable variations due to natural random causes, they can only be reduced by process improvements at the management level. Common causes are generally process issues such as:
  - Poor process design.
  - Poor interface management.
  - Poor staffing levels.
- Special causes are significant identifiable events which can be removed by local action. Special causes are generally people issues such as:
  - Poor skills or aptitude.
  - Poor training or demotivated staff.

The most useful control charts to use for sales are:

- The individuals and moving range chart for variables data (see left).
- The p-chart for proportion nonconforming for attributes data, e.g. invoice accuracy.

In all cases, the aim is to determine if the process is capable of delivering the required output, to find 'out-of-control' points and improve the process.

- **Tip** - If the process is 'in control' then any variations are probably due to common causes, i.e. with the process.

## Other charts in sales

Simply recording the output does not improve the process but it can be useful to show progress and the use of run, deviation and CUSUM charts to record sales progress (the output) is shown below.

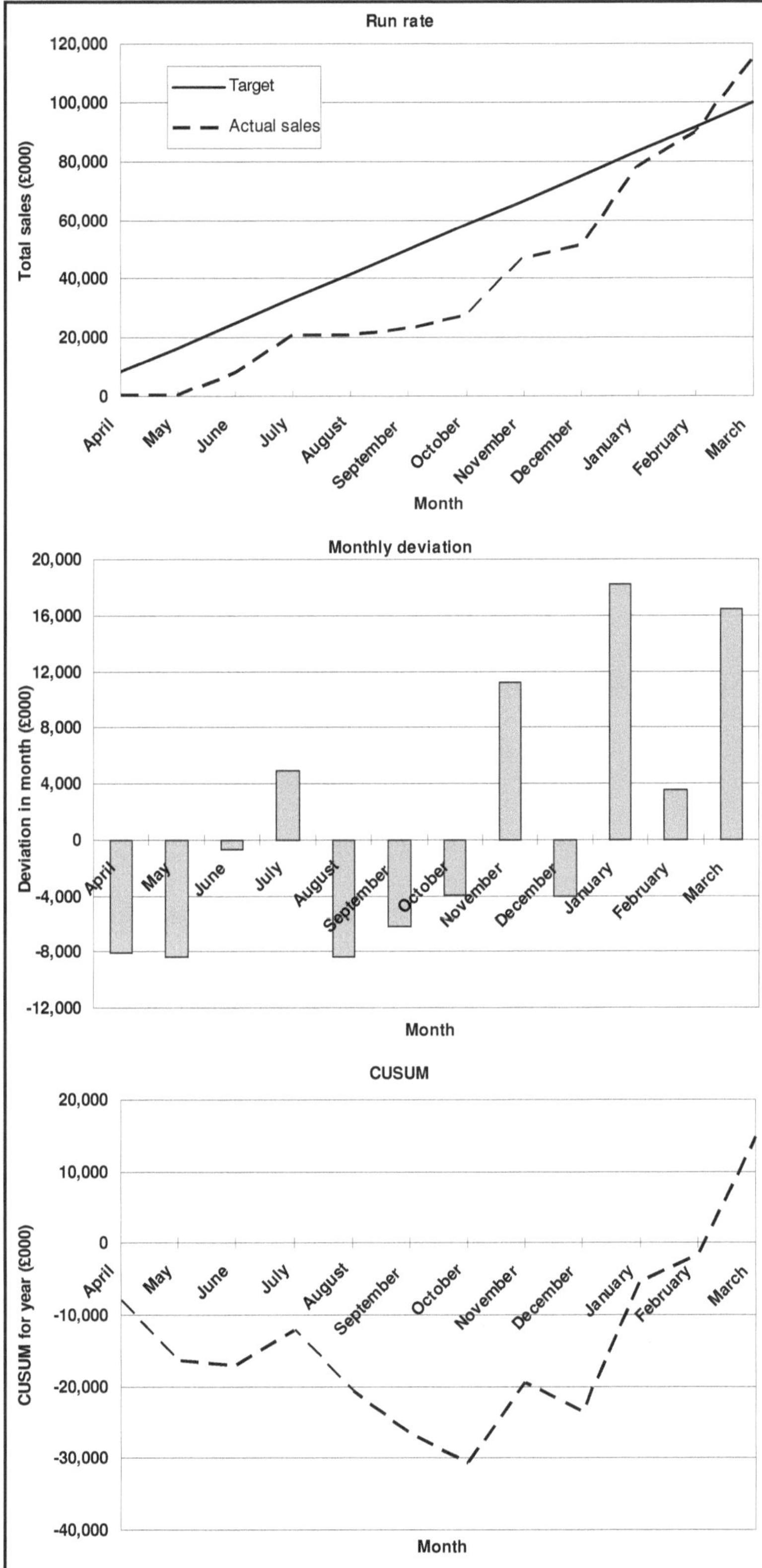


### Tracking sales volumes

Simple run charts (top) show performance relative to budget on a time basis, deviation charts (middle) show the monthly deviation from budget and CUSUM charts show the year-to-date deviation from budget. Sales started slow but caught up in the end.

# 11.3 Accounting quality

## Quality in accounting

Financial accounting is the external side of accounting and is subject to many standards and regulations that define accounting reporting, e.g. the International Financial Reporting Standards define the reporting standards for the EU and similar standards define the reporting standards in other countries. These standards define the output and format of financial reporting and the audit process is designed to check that the output is accurate.

- **Tip** - A creative accountant can make the numbers show anything he likes. The rules are there to stop them doing that.

Management accounting is the internal side of accounting and, whilst subject to many excellent 'good practice guides', there are no legal requirements for output, format or auditing. These are decided by the company's operational needs.

The process of generating and controlling accounting information is rarely examined for quality management and improvement. This is despite the fact that accounting information is one of the key methods of monitoring other processes (see Chapter 3). It often appears that nobody 'watches the watchers'.

- **Tip** - Quality improvement in accounting can be managed using standard quality improvement techniques.
- **Tip** - The payroll section must get it right every time. If the payroll is late or incorrect then feedback from staff is both immediate and strong. What would you accept as the error in the activities of the payroll section with your wages?
- **Tip** - Good cash flow is fundamental to any company and this needs correct and timely invoicing. Invoice delays or errors will always slow down cash flow, getting it right is critical.

## The accounting process

Accounting has many different and distinct functions and processes, e.g. payroll, invoicing and supplier payment, as well as the more obvious functions of preparing the financial and management accounting data.

Before the obvious step of process mapping and flow charting, it is worthwhile creating a global process map (see Section 4.5) to map the overall work of the accounting function. This will identify the main processes and their outputs. Each process can then be mapped and flow-charted (see Section 4.6 and Section 7.2).

- **Tip** - Mapping accounting processes can often be most effectively carried out by a 'paper chase exercise'. This involves tracking the 'paper work' or 'information flow' through the process. Look for areas where information is input or manipulated for potential process and

Why do managers have to sign off invoices? They have already signed the purchase order and goods received note.

If these match then what is added except time to pay the supplier?

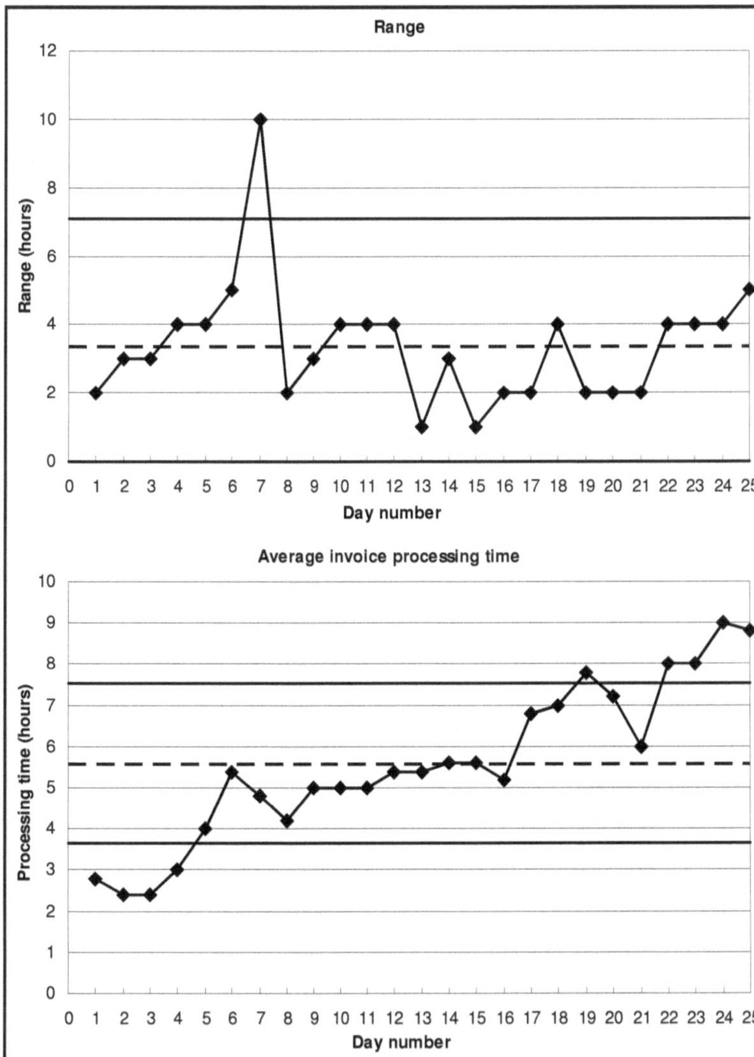


### Simple control charts for invoice processing time

Taking a sample of 5 invoices for each day and running a simple variables control chart (see Section 5.8) shows that the system is 'out-of-control'. There is an out-of-control point on Day 7 and the process shows a drift upwards from Day 8 onwards.

quality improvements.

- **Tip** - In many companies, a large amount of the information input or manipulation is via spreadsheets maintained by individuals. These spreadsheets are rarely compatible with each other, rarely consistently formatted, rarely fully error-checked and often riddled with inconsistencies. They create 'islands of data' that may be very valuable but are not integrated into an MIS (Management Information System) and cannot be used by other areas.

## Improving the process

Accounting processes have the same potential areas for quality improvement as the sales process (see Section 11.2) and the standard quality tools (see Chapter 7) and process improvement methods can be used to improve the process.

## Control charts

Speed of response is important in any accounting process and standard control charts can be used to assess process performance.

An example of a standard $\overline{X}$ and R chart for invoice processing time is shown on the left. This shows a single out of control point on Day 7 but more worryingly shows a process drift upwards from Day 8 onwards. Something has obviously changed with the invoicing process and action needs to be taken. Similar charts can easily be generated to monitor and control many accounting processes.

## Run charts

Run charts can be used to look at the output of a process over time. They do not have control limits but can still be used to find trends in processes.

An example of a run chart for invoice processing time is shown on the upper right. This shows an obvious process disturbance around Day 11 that needs to be investigated in greater detail - is this a temporary issue, a system change or is it a sign of system failure? Further investigation using the quality tools should reveal the cause and potential improvement action.

- **Tip** - These are 'people processes', there will always be common causes for variations in times.
- **Tip** - Run charts can be used in many of the accounting functions such as payroll, invoicing and supplier payment.

## Check sheets

Check sheets (see Section 7.3) are another useful quality tool for accounting processes to record events in a structured manner. They are especially useful for tracking issues where the frequency and type of error is important so that it can be traced to a particular process step, e.g. inventory errors.

Monitoring credit notes can be a quick way of assessing the invoicing process.

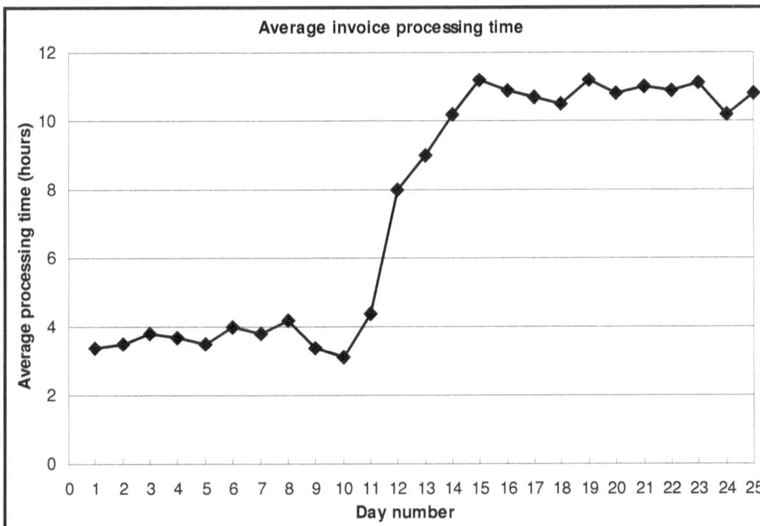


### A run chart to assess invoice processing time

A simple run chart of the average invoice processing time shows that the system has been 'disturbed' around Day 11 but has stabilised by Day 15. Does this show a lack of holiday cover, a new employee or more serious system issues?

### Auditing the output - payment of suppliers

A long time ago (in the days when all payments were by cheque) I was a 'cheque signatory' and I used to sign all of the company cheques to suppliers. This was probably a punishment for something I had done wrong because I used to get around 200 cheques each week to sign.

As my name was on the cheques I was concerned about their accuracy. I started a process of checking about 1 in 10 cheques for accuracy to the supplier's invoice. The general result was that 1 in 10 of the payments I checked was incorrect for various reasons.

My response was to refuse to sign any further cheques after I found a error and to send the complete box of cheques back to accounts for a full review - they did not like it but imagine how our suppliers (the customer of the accounts department) would have felt. They were getting cheques that did not match their invoice and was either an under-payment (bad for them) or an over-payment (bad for us). Not surprisingly, the accounts department starting improving the process and checking the cheques before they came to me to eliminate the errors.

This would not work now with direct transfers but how is the payment process controlled at your company?

**Note:** The one thing I did not do (to my great regret) was to start an audit of the sales invoice process. The same people were involved and this was about the money owed to the company.

# 11.4 General office quality

## It is all about processes

Sales and accounting are probably the two main office processes but there are many other office processes and they all need quality management. This section covers the more general office processes and methods to both monitor and improve them. We will not cover all the processes in detail but instead will give more general guidance on how to improve the quality of these processes. This may seem strange in a Workbook on plastics processing but quality is not simply about the main product, it is about all the company processes, but most importantly about the interfaces between the customer and the company. As an example, I use some very expensive testing equipment and it is excellent in build quality and what it does, but if you want to get it serviced (and there is only one place) then the service arm has systems that are 'as user-friendly as a cornered rat'. They give dates that are inaccurate, they don't meet their inaccurate dates and they do not communicate with you. This is so much so that my next purchase will be from their competitor even thought the actual product is not quite as good.

Office processes all have the same basic characteristics and we will deal with these rather than the specifics for any process.

## Work cells

Just as the time taken for a product to make it through a process can be predicted by the distance the product has to travel, so the time taken for a process to be carried out can be reliably predicted by the number of 'touches'. The number of people touching, handling or otherwise involved in the process is a good indicator of both the time the process will take, the number of errors that will be introduced and the difficult in improving the process.

Manufacturing has met this challenge by cell-based manufacturing[1] and office processes can be similarly improved. Instead of using a 'production line'-based model, as did 'old' manufacturing, the time is ripe for office-based processes to adopt cell-based manufacturing where a 'cell' of office staff carry out complete processes.

Office processes based on cells gives:

- Clearer process function and responsibilities.
- Improved process quality.
- Reduced 'touches' and time to complete the process.
- Reduced waiting time for processing.
- Reduced 'lost' paperwork.
- Increased fulfillment for staff.
- Greater customer satisfaction.

The greatest benefit in work cells is the empowerment of the staff to improve the process. Workers in a process will always be able to see quality and process improvements but they will only be able to improve the process if they are empowered to do so. Cell-based manufacturing empowers the staff and allows them to monitor their process using standard charts and to improve the process using the standard quality tools.

- **Tip -** A retailer in the UK has a policy that everybody (and that means everybody) in the company has the power to give a refund on a product for any reason. No questions, no exceptions, no paperwork and no hassle. If it was a mail order purchase then they don't even want you to send the old product back - they just refund your money, BUT they keep their customers forever!

Office processes are not exempt from the drive to improve quality. This is about the whole company and not simply about manufacturing.

Manufacturing is measured to within an inch of it's life but there is always resistance when measurement of office processes starts.

We all do this or nobody does it.

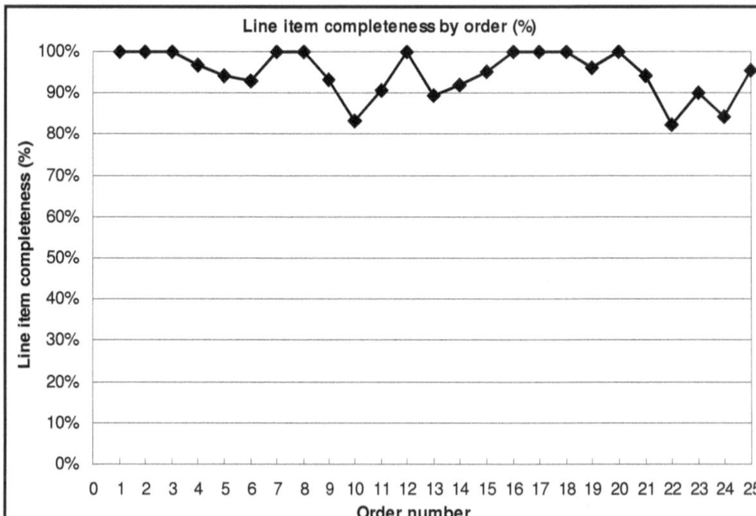


Line item completeness is an important indicator

Office processes need good measurements to assess the quality of their services. Despatch processes are best monitored using line item data - it is no good getting 99% of the order value when the 1% missing prevents the rest being used.

Cell-based processes can be more difficult for smaller companies where people multi-task. In these cases, the processes are almost always already cell-based, in that one person carries out the complete task, i.e. they are a one-person cell.

- **Tip -** The concept of 'touches' applies for your own work. Dealing with something immediately rather than 'touching' it multiple times gives greater throughput, provided you deal with the important things first.

## 5S for offices

Office processes often have large amounts of paperwork and this not only makes processes slower but also allows errors to be introduced. Manufacturing has also attacked this issue with the 5S concept. 5S is a method of both cleaning out the manufacturing area and maintaining the cleanliness to improve process quality. The 5S process is based on:

### Sort (Seiri)

This is sorting and removal of unnecessary items in the work area. Sorting is designed to make the work area tidy, organised and arranged so that necessary items can be found easily. If it is not necessary for the process then it should not be in the work area.

- **Tip -** Everything has a place and should be in it - even if it is the rubbish bin!

### Simplify or Set in order (Seiton)

This is organising, arranging and identifying everything in the work area so that they can be easily selected for use. Frequent use items should be close by and infrequent use items can be further away in a central area. All storage areas should be clearly labeled to allow items to be put in the correct place, e.g. where did I leave the stapler again?

### Shine (Seiso)

After sorting and simplifying the work area, it is necessary to keep it that way and shine is the process where the area and any equipment are regularly cleaned, kept orderly and kept safe. Shining is also an inspection process for the area, i.e. is everything in good condition.

### Standardise (Seiketsu)

A clean and tidy work area allows the process to be standardised (using flow charts and checklists) and examined for quality or process improvements. Best practices are documented and rolled out across the work area, standards and process measures are established and displayed in the work area.

### Sustain (Shitsuke)

This is the process of 'holding the gains' made by the previous steps. This can only be achieved by auditing work areas and processes to make sure that the 5S standards are maintained.

An additional 3 S's are sometimes added and these are:

- Safety.
- Security.
- Satisfaction.

5S in offices streamlines the workflow to reduce errors as well as improving process times and employee satisfaction.

## Keystroke errors

Whenever office processes are considered, a major source of errors is simple keystroke errors. Many keystroke errors in text can be found with spell-checking but this will ignore errors such as using 'two' instead of 'too' because both are legitimate words.

The real issue is with errors in inputting alphanumeric information, e.g. SG5 2DF is more difficult and error-prone than 245 289. They may be easier to remember but they give higher keystroke error rates.

- **Tip -** Use fully numeric information wherever possible on warehouse locations, product numbers, transactions codes and all paperwork to reduce keystroke areas.

- 1. Kent, R.J. 2012. '*Cost management in plastics processing*', Section 5.8. Plastics Information Direct.

"Excellence is in the details. Give attention to the details and excellence will come."

**Perry Paxton**

### Is it necessary? - the paper chase pays off

Many years ago I started a new job and found that my secretary was keeping records of capital expenditure for the whole company. She was very efficient in doing this and could locate any record immediately. The only problem was that the company had computerised the whole system some years previously and this automatically tracked the expenditure - but nobody had told her. All the records were already being kept in the finance department and it involved considerable effort to send the copies to her for filing and recording, all of which duplicated the records already kept. After doing the paper chase and analyzing the process we decided that there was no need for her to keep them and her workload was reduced by 30%. She was being efficient but not 'effective'. To improve quality we need to be effective in what we do rather than simply concentrating on the efficiency of the process.

# 11.5 Operations - where are you now?

## Operations quality counts

Operations is about the non-production activities and staff. Although the focus of most quality texts is on the product, there is more to it than this.

Manufacturing has undergone a revolution in the past 20 years and yet, apart from mass computerisation, the 'back office' functions have remained largely unaffected. It is time for the administrative functions to learn the lessons of manufacturing in terms of measurement, organisation and techniques. There is has been too little action to date to apply manufacturing process improvement and quality knowledge on the office tasks. Operations are ripe for transformation in terms of quality and process design. This transformation should not to be feared but embraced as a way of making meaningful improvement to the quality of office processes.

## Completing the chart

This chart is completed and assessed as for the previous charts.

Quality management does not stop at the door to the manufacturing area. It is applicable throughout the company and can improve all aspects of a company's operations.

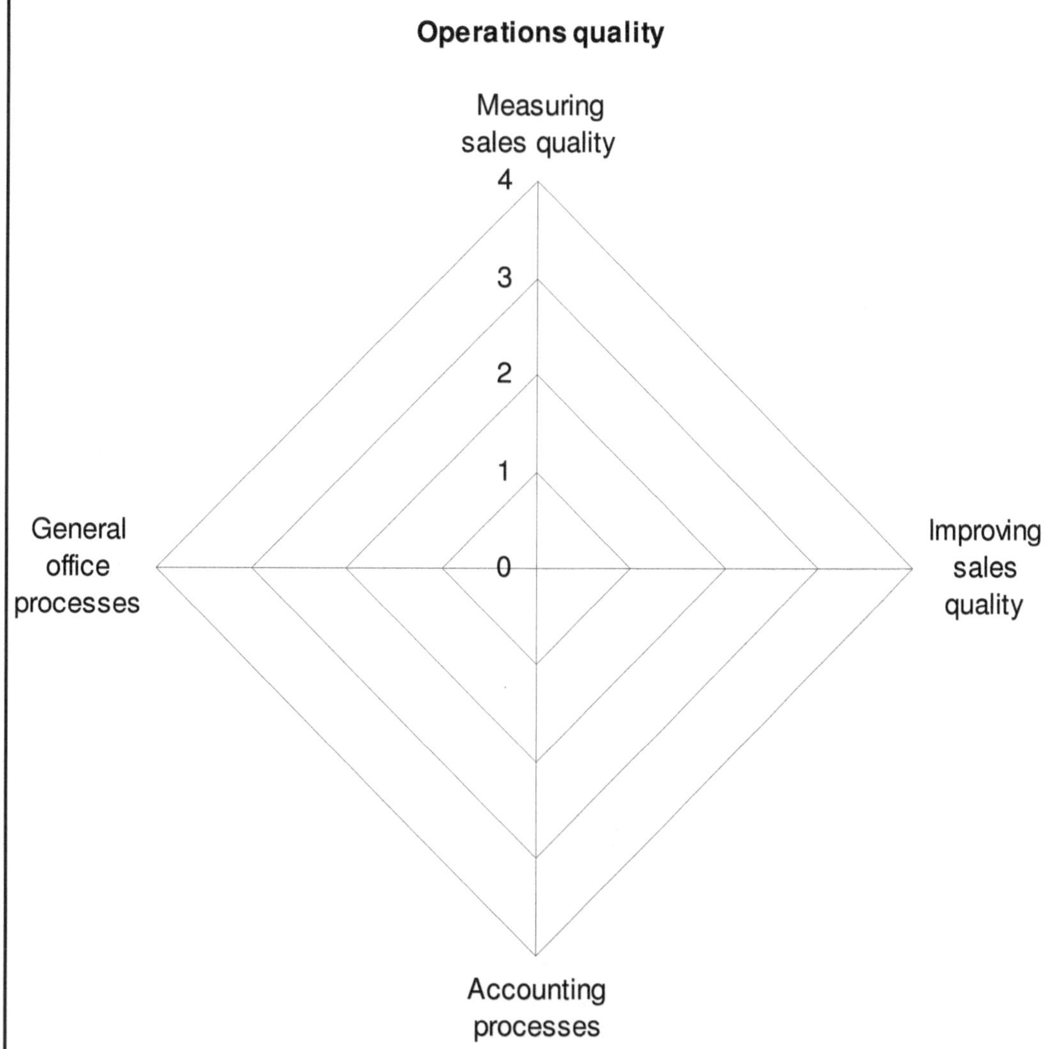


Use the scoring chart to assess where you are in operations quality

The numbers from the scoring chart can be transferred to the radar chart for a quick visual assessment of where you are in terms of operations quality.

Simply measuring the output of the process for reward (or punishment) is not enough. Measurements should to be made to improve the process and the output will then take care of itself.

| Operations quality management | | | | |
|---|---|---|---|---|
| **Level** | **Measuring sales quality** | **Improving sales quality** | **Accounting processes** | **General office processes** |
| **4** | Sales process (external & internal) evaluated & measured for process control. Process control tracking used. Measurements made of internal processes as well as external outputs. | All sales processes mapped & flow-charted. Processes changed to improve customer experience. All processes evaluated using good control charts. | All accounting processes mapped & flow charts available. Paper chase carried out & redundant process steps, forms & information eliminated. All processes evaluated using good control charts. | All general processes mapped & office processes divided into work cells with clear work-flow & responsibilities. Processes have clear process measures & standards. 5S used to manage office areas. |
| **3** | Sales process (external & internal) evaluated & measured for process control. Measurements are mainly the outputs of internal & external processes. | Major sales processes mapped & flow-charted. Major processes changed to improve customer experience. Major processes evaluated using control charts. | Major accounting processes mapped & flow charts available. Limited paper chase carried out to identify redundant process steps, forms & information but no action taken. Major processes evaluated using control charts. | Major general processes mapped & most office processes divided into work cells with clear work-flow & responsibilities. Most processes have clear process measures & standards. 5S used to manage some office areas. |
| **2** | Sales process (external & internal) measured but no attempt at process control. Only measurements are of the outputs of the processes (internal & external). | Some sales processes mapped & flow-charted. Some processes changed to improve customer experience. Few processes evaluated but company has good data. | Some accounting processes mapped & flow charts available. No paper chase carried out but no visible redundant process steps. Some processes evaluated using good measures but mainly for output assessment. | Some general processes mapped. Complex work-flow but some responsibilities allocated. Few processes have measures or standards. 5S used to manage few office areas. |
| **1** | External sales process measured. Measurements made of external outputs only & displayed through sales dashboard. No consideration of internal parts of the sales process. | Few sales processes mapped. Processes are good for company but no consideration of customer experience. Very limited process evaluation using poor data. | Few accounting processes mapped & few flow charts available. No paper chase carried out but no visible redundant process steps. Processes evaluated using unsuitable measures & mainly for output assessment. | Few general processes mapped. Complex work-flow & poor responsibility allocation. No process measures or standards. No use of 5S for office areas. |
| **0** | Only recording of sales process is the results/output from external sales. No plotting or control. No consideration of internal parts of the sales process. | No sales processes mapped or flow charts available. Processes are good for company but no consideration of customer experience. No process evaluation. | No processes mapped or flow charted. No paper chase carried out. Visible redundant process steps, forms & information present. No process measures in place. | No mapping of any office process. Processes carried out using a variety of methods. No process measures or standards. No concept of 5S for office areas. |
| **Score** | | | | |

# Key tips

- Operations quality management requires looking at the processes and identifying how to improve these.
- When measuring operational processes remember the rules for manufacturing processes - you are trying to control the process and not inspect the output.
- Measuring sales processes for quality improvement should focus on improving the process not simply on reward or punishment.
- There are many measures possible of the sales process. Only select a few of these at the start to gain experience.
- Map the sales process to understand what the process is doing.
- Streamline the sales process to reduce errors and time taken.
- Use control charts to measure, control and improve the process.
- Common and special causes exist in sales processes as well as in manufacturing processes.
- Accounting processes can affect the survival and profitability of a company.
- Quality improvement is possible in accounting processes.
- Use a paper chase (or an information chase for computer systems) to identify where errors can be introduced, where there is too much information and areas for improvement.
- Avoid 'islands of information' created by unlinked and uncontrolled spreadsheets. Try to use an integrated MIS for consistency.
- Run charts can quickly show trends in the accounting process (as well as in the accounting information).
- Use cell-based working for the major office processes.
- Implement 5S in office processes. If in doubt, ask the manufacturing people about this.
- Reduce keystroke errors by selecting easy input values.

# Appendices

Quality management involves a great deal of statistics and rather than interrupt the flow of this Workbook we have taken the choice to put many of the detailed statistical tables and charts in Appendices.

These are for reference only and most readers will not need the detail given here.

"By nature, all men long to know."

**Aristotle - Metaphysics**

# Appendix 1 - Standard SPC formulae for variables and attributes charts

### $\overline{X}$ and R chart

**R chart**

- Centre line:

$$\overline{R} = \frac{\text{Sum of R samples}}{\text{Number of R samples}}$$

- Upper Control Limit:

$$UCL_R = D_4 \times \overline{R}$$

- Lower Control Limit:

$$LCL_R = D_3 \times \overline{R}$$

**$\overline{X}$ chart**

- Centre line:

$$\overline{\overline{X}} = \frac{\text{Sum of } \overline{X} \text{ samples}}{\text{Number of samples}}$$

- Upper Control Limit:

$$UCL_{\overline{X}} = \overline{\overline{X}} + A_2 \times \overline{R}$$

- Lower Control Limit:

$$LCL_{\overline{X}} = \overline{\overline{X}} - A_2 \times \overline{R}$$

**Estimate of σ:**

$$\sigma_X = \frac{\overline{R}}{d_2}$$

### Median chart ($\tilde{X}$ and R)

**R chart**

- Centre line:

$$\overline{R} = \frac{\text{Sum of R samples}}{\text{Number of R samples}}$$

- Upper Control Limit:

$$UCL_R = D_4 \times \overline{R}$$

- Lower Control Limit:

$$LCL_R = D_3 \times \overline{R}$$

**$\tilde{X}$ chart**

- Centre line:

$$\overline{\tilde{X}} = \frac{\text{Sum of } \tilde{X} \text{ samples}}{\text{Number of samples}}$$

- Upper Control Limit

$$UCL_{\tilde{X}} = \overline{\tilde{X}} + \tilde{A}_2 \times \overline{R}$$

- Lower Control Limit:

$$LCL_{\tilde{X}} = \overline{\tilde{X}} - \tilde{A}_2 \times \overline{R}$$

**Estimate of σ:**

$$\sigma_X = \frac{\overline{R}}{d_2}$$

### $\overline{X}$ and s chart

**s chart**

- Centre line:

$$\overline{s} = \frac{\text{Sum of s values}}{\text{Number of samples}}$$

- Upper Control Limit:

$$UCL_s = B_4 \times \overline{s}$$

- Lower Control Limit:

$$LCL_s = B_3 \times \overline{s}$$

**$\overline{X}$ chart**

- Centre line:

$$\overline{\overline{X}} = \frac{\text{Sum of } \overline{X} \text{ samples}}{\text{Number of samples}}$$

- Upper Control Limit:

$$UCL_{\overline{X}} = \overline{\overline{X}} + A_3 \times \overline{s}$$

- Lower Control Limit:

$$LCL_{\overline{X}} = \overline{\overline{X}} - A_3 \times \overline{s}$$

**Estimate of σ:**

$$\sigma_X = \frac{\overline{s}}{c_4}$$

### Individuals chart (X and MR)

**MR chart**

- Centre line:

$$\overline{R} = \frac{\text{Sum of R samples}}{\text{Number of R samples}}$$

- Upper Control Limit:

$$UCL_R = D_4 \times \overline{R}$$

- Lower Control Limit:

$$LCL_R = D_3 \times \overline{R}$$

**X chart**

- Centre line:

$$\overline{X} = \frac{\text{Sum of X samples}}{\text{Number of samples}}$$

- Upper Control Limit

$$UCL_X = \overline{X} + E_2 \times \overline{R}$$

- Lower Control Limit:

$$LCL_X = \overline{X} - E_2 \times \overline{R}$$

**Estimate of σ:**

$$\sigma_X = \frac{\overline{R}}{d_2}$$

## Proportion nonconforming (p chart)

- Centre line:

$$\bar{p} = \frac{\text{Total number of nonconforming items}}{\text{Total number of products inspected}}$$

- Upper Control Limit:

$$UCL_p = \bar{p} + 3 \times \sqrt{\frac{\bar{p}(1-\bar{p})}{\bar{n}}}$$

$\bar{n}$ = Average sample size

- Lower Control Limit:

$$LCL_p = \bar{p} - 3 \times \sqrt{\frac{\bar{p}(1-\bar{p})}{\bar{n}}}$$

## Number nonconforming (np chart)

- Centre line:

$n \times \bar{p}$ = Average number of nonconforming items in a sample of constant size 'n'.

- Upper Control Limit:

$$UCL_{np} = n\bar{p} + 3 \times \sqrt{n\bar{p}\left(1 - n\bar{p}/n\right)}$$

- Lower Control Limit:

$$LCL_{np} = n\bar{p} - 3 \times \sqrt{n\bar{p}\left(1 - n\bar{p}/n\right)}$$

## Nonconformities per unit (u chart)

- Centre line:

$$\bar{u} = \frac{\text{Total number of nonconformities}}{\text{Total number of products inspected}}$$

- Upper Control Limit:

$$UCL_u = \bar{u} + 3 \times \sqrt{\frac{\bar{u}}{\bar{n}}}$$

$\bar{n}$ = Average sample size

- Lower Control Limit:

$$LCL_u = \bar{u} - \sqrt{\frac{\bar{u}}{\bar{n}}}$$

## Number of nonconformities (c chart)

- Centre line:

$\bar{c}$ = Average number of nonconformities found in a set of samples.

- Upper Control Limit:

$$UCL_c = \bar{c} + 3 \times \sqrt{\bar{c}}$$

- Lower Control Limit:

$$LCL_c = \bar{c} - 3\sqrt{\bar{c}}$$

**Standard formulae for all the types of variables and attributes control charts**

The constants referred to ($D_4$, $D_3$, $A_2$, $d_2$ etc.) in the formulae are given in Appendix 2 for various sample sizes.

# Appendix 2 - Standard SPC constants for variables charts

$\overline{X}$ **and R chart**

| Sample size | $A_2$ | $D_3$ | $D_4$ | $d_2$ |
|---|---|---|---|---|
| 2 | 1.880 | - | 3.267 | 1.128 |
| 3 | 1.023 | - | 2.574 | 1.693 |
| 4 | 0.729 | - | 2.282 | 2.059 |
| 5 | 0.577 | - | 2.114 | 2.326 |
| 6 | 0.483 | - | 2.004 | 2.534 |
| 7 | 0.419 | 0.076 | 1.924 | 2.704 |
| 8 | 0.373 | 0.136 | 1.864 | 2.847 |
| 9 | 0.337 | 0.184 | 1.816 | 2.970 |
| 10 | 0.308 | 0.223 | 1.777 | 3.078 |

$\overline{X}$ **and s chart**

| Sample size | $A_3$ | $B_3$ | $B_4$ | $c_4$ |
|---|---|---|---|---|
| 2 | 2.659 | - | 3.267 | 0.798 |
| 3 | 1.954 | - | 2.568 | 0.886 |
| 4 | 1.628 | - | 2.266 | 0.921 |
| 5 | 1.427 | - | 2.089 | 0.940 |
| 6 | 1.287 | 0.030 | 1.970 | 0.952 |
| 7 | 1.182 | 0.118 | 1.882 | 0.959 |
| 8 | 1.099 | 0.185 | 1.815 | 0.965 |
| 9 | 1.032 | 0.239 | 1.761 | 0.969 |
| 10 | 0.975 | 0.284 | 1.716 | 0.973 |

**Median chart ($\widetilde{X}$ and R)**

| Sample size | $\widetilde{A}_2$ | $D_3$ | $D_4$ | $d_2$ |
|---|---|---|---|---|
| 2 | 1.880 | - | 3.267 | 1.128 |
| 3 | 1.187 | - | 2.574 | 1.693 |
| 4 | 0.796 | - | 2.282 | 2.059 |
| 5 | 0.691 | - | 2.114 | 2.326 |
| 6 | 0.548 | - | 2.004 | 2.534 |
| 7 | 0.508 | 0.076 | 1.924 | 2.704 |
| 8 | 0.433 | 0.136 | 1.864 | 2.847 |
| 9 | 0.412 | 0.184 | 1.816 | 2.970 |
| 10 | 0.362 | 0.223 | 1.777 | 3.078 |

**Individuals chart (X and MR)**

| Sample size | $E_2$ | $D_3$ | $D_4$ | $d_2$ |
|---|---|---|---|---|
| 2 | 2.660 | - | 3.267 | 1.128 |
| 3 | 1.772 | - | 2.574 | 1.693 |
| 4 | 1.457 | - | 2.282 | 2.059 |
| 5 | 1.290 | - | 2.114 | 2.326 |
| 6 | 1.184 | - | 2.004 | 2.534 |
| 7 | 1.109 | 0.076 | 1.924 | 2.704 |
| 8 | 1.054 | 0.136 | 1.864 | 2.847 |
| 9 | 1.010 | 0.184 | 1.816 | 2.970 |
| 10 | 0.975 | 0.223 | 1.777 | 3.078 |

# Appendix 3 - Sample standard SPC chart for variables ($\overline{X}$ and R)

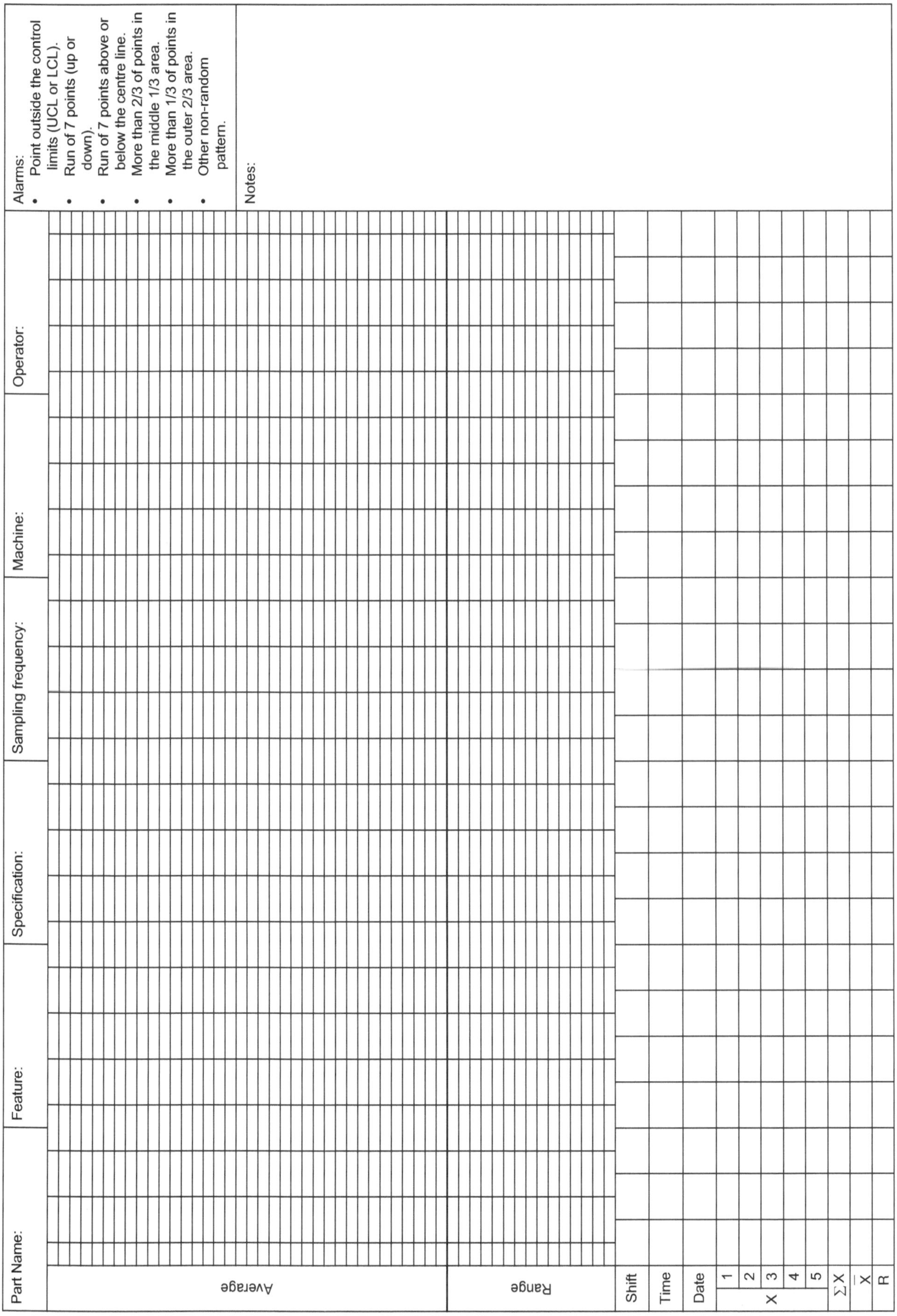

# Appendix 4 - $P_z$ table for % nonconforming

## Estimating the % nonconforming

$P_z$ = the proportion of process output beyond a single specification limit that is 'z' standard deviation units away from the process average (for a process that is in statistical control and is normally distributed).

Knowing the $P_z$ allows locating $\overline{\overline{X}}$ to minimise the out-of-specification output.

**Note:** This is only for one side of the curve.

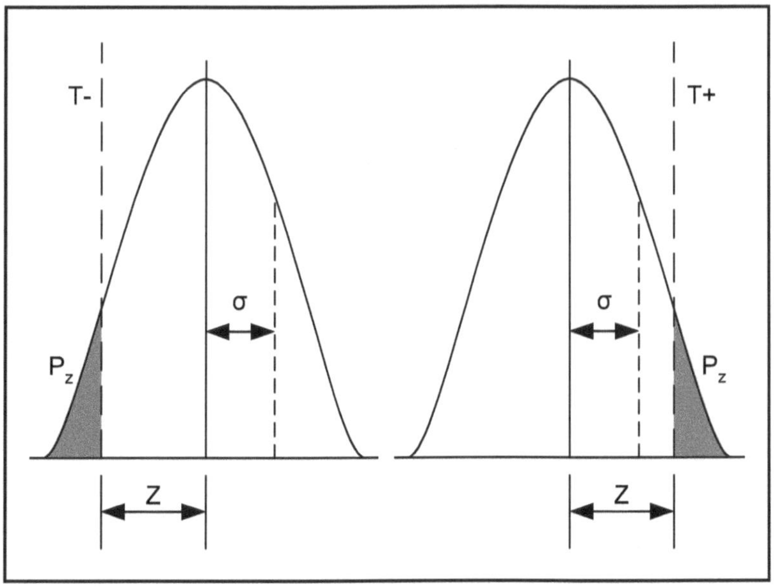


| z | 0.00 | 0.01 | 0.02 | 0.03 | 0.04 | 0.05 | 0.06 | 0.07 | 0.08 | 0.09 |
|---|---|---|---|---|---|---|---|---|---|---|
| **0.0** | 0.5000 | 0.4960 | 0.4920 | 0.4880 | 0.4840 | 0.4801 | 0.4761 | 0.4721 | 0.4681 | 0.4641 |
| **0.1** | 0.4602 | 0.4562 | 0.4522 | 0.4483 | 0.4443 | 0.4404 | 0.4364 | 0.4325 | 0.4286 | 0.4247 |
| **0.2** | 0.4207 | 0.4168 | 0.4129 | 0.4090 | 0.4052 | 0.4013 | 0.3974 | 0.3936 | 0.3897 | 0.3859 |
| **0.3** | 0.3821 | 0.3783 | 0.3745 | 0.3707 | 0.3669 | 0.3632 | 0.3594 | 0.3557 | 0.3520 | 0.3483 |
| **0.4** | 0.3446 | 0.3409 | 0.3372 | 0.3336 | 0.3300 | 0.3264 | 0.3228 | 0.3192 | 0.3156 | 0.3121 |
| **0.5** | 0.3085 | 0.3050 | 0.3015 | 0.2981 | 0.2946 | 0.2912 | 0.2877 | 0.2843 | 0.2810 | 0.2776 |
| **0.6** | 0.2743 | 0.2709 | 0.2676 | 0.2643 | 0.2611 | 0.2578 | 0.2546 | 0.2514 | 0.2483 | 0.2451 |
| **0.7** | 0.2420 | 0.2389 | 0.2358 | 0.2327 | 0.2297 | 0.2266 | 0.2236 | 0.2206 | 0.2177 | 0.2148 |
| **0.8** | 0.2119 | 0.2090 | 0.2061 | 0.2033 | 0.2005 | 0.1977 | 0.1949 | 0.1922 | 0.1894 | 0.1867 |
| **0.9** | 0.1841 | 0.1814 | 0.1788 | 0.1762 | 0.1736 | 0.1711 | 0.1685 | 0.1660 | 0.1635 | 0.1611 |
| **1.0** | 0.1587 | 0.1562 | 0.1539 | 0.1515 | 0.1492 | 0.1469 | 0.1446 | 0.1423 | 0.1401 | 0.1379 |
| **1.1** | 0.1357 | 0.1335 | 0.1314 | 0.1292 | 0.1271 | 0.1251 | 0.1230 | 0.1210 | 0.1190 | 0.1170 |
| **1.2** | 0.1151 | 0.1131 | 0.1112 | 0.1093 | 0.1075 | 0.1056 | 0.1038 | 0.1020 | 0.1003 | 0.0985 |
| **1.3** | 0.0968 | 0.0951 | 0.0934 | 0.0918 | 0.0901 | 0.0885 | 0.0869 | 0.0853 | 0.0838 | 0.0823 |
| **1.4** | 0.0808 | 0.0793 | 0.0778 | 0.0764 | 0.0749 | 0.0735 | 0.0721 | 0.0708 | 0.0694 | 0.0681 |
| **1.5** | 0.0668 | 0.0655 | 0.0643 | 0.0630 | 0.0618 | 0.0606 | 0.0594 | 0.0582 | 0.0571 | 0.0559 |
| **1.6** | 0.0548 | 0.0537 | 0.0526 | 0.0516 | 0.0505 | 0.0495 | 0.0485 | 0.0475 | 0.0465 | 0.0455 |
| **1.7** | 0.0446 | 0.0436 | 0.0427 | 0.0418 | 0.0409 | 0.0401 | 0.0392 | 0.0384 | 0.0375 | 0.0367 |
| **1.8** | 0.0359 | 0.0351 | 0.0344 | 0.0336 | 0.0329 | 0.0322 | 0.0314 | 0.0307 | 0.0301 | 0.0294 |
| **1.9** | 0.0287 | 0.0281 | 0.0274 | 0.0268 | 0.0262 | 0.0256 | 0.0250 | 0.0244 | 0.0239 | 0.0233 |
| **2.0** | 0.0228 | 0.0222 | 0.0217 | 0.0212 | 0.0207 | 0.0202 | 0.0197 | 0.0192 | 0.0188 | 0.0183 |
| **2.1** | 0.0179 | 0.0174 | 0.0170 | 0.0166 | 0.0162 | 0.0158 | 0.0154 | 0.0150 | 0.0146 | 0.0143 |
| **2.2** | 0.0139 | 0.0136 | 0.0132 | 0.0129 | 0.0125 | 0.0122 | 0.0119 | 0.0116 | 0.0113 | 0.0110 |
| **2.3** | 0.0107 | 0.0104 | 0.0102 | 0.0099 | 0.0096 | 0.0094 | 0.0091 | 0.0089 | 0.0087 | 0.0084 |
| **2.4** | 0.0082 | 0.0080 | 0.0078 | 0.0075 | 0.0073 | 0.0071 | 0.0069 | 0.0068 | 0.0066 | 0.0064 |
| **2.5** | 0.0062 | 0.0060 | 0.0059 | 0.0057 | 0.0055 | 0.0054 | 0.0052 | 0.0051 | 0.0049 | 0.0048 |
| **2.6** | 0.0047 | 0.0045 | 0.0044 | 0.0043 | 0.0041 | 0.0040 | 0.0039 | 0.0038 | 0.0037 | 0.0036 |
| **2.7** | 0.0035 | 0.0034 | 0.0033 | 0.0032 | 0.0031 | 0.0030 | 0.0029 | 0.0028 | 0.0027 | 0.0026 |
| **2.8** | 0.0026 | 0.0025 | 0.0024 | 0.0023 | 0.0023 | 0.0022 | 0.0021 | 0.0021 | 0.0020 | 0.0020 |
| **2.9** | 0.0019 | 0.0018 | 0.0018 | 0.0017 | 0.0016 | 0.0016 | 0.0015 | 0.0015 | 0.0014 | 0.0014 |
| **3.0** | 0.00135 | | | | | | | | | |
| **3.5** | 0.00023 | | | | | | | | | |
| **4.0** | 0.00003 | | | | | | | | | |

# Appendix 5 - Precision and accuracy

## They are different

Precision and accuracy are very different things as the diagram on the upper right shows. If you can imagine the diagram to be the results of a shooter then you can easily see that precision is about how close your shots are together and accuracy is about hitting the target.

- Highly precise shooters will always have good grouping but will not necessarily hit the target.
- Highly accurate shooters will always hit the target but will not necessarily hit the centre.

Precision measures the scatter of measurements or results and is affected mainly by common causes, i.e. variations that are characteristic of the limitations of the process. These cannot be removed without management action.

Accuracy measures the location of the measurements and is affected mainly by special causes, i.e. variations that can be assigned specific causes. These can be removed by the process operator.

It is possible to consider meeting specification and being in control in the same way and this is shown on the lower right.

When a process is in control then it is the equivalent of being precise and it is affected mainly by common causes that relate to the limitations of the process and can only be removed by management action on the process, e.g. better machine, better tooling, etc.

When a process is in specification then it is the equivalent of being accurate and it is affected mainly by special causes that can be removed by the process operator, e.g. material changes, setting changes etc.

It is best to be highly precise and highly accurate just as it is best to be in control and in specification.

- **Tip -** As a shooter, I first seek high precision or good grouping (remove the common causes such as poor support, good trigger control) and then adjust the sights to give high accuracy. Good grouping but poor location (high precision with low accuracy) can be adjusted by the sights to give good grouping with high accuracy.

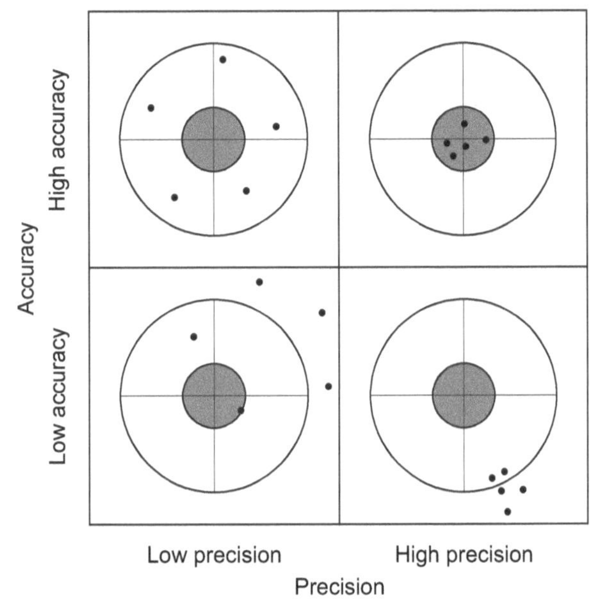


### Precision and accuracy are not the same thing

Precision is about being consistent and having low variation or bias in the results. Accuracy is about consistently achieving the desired result. The ideal is to have both high precision and high accuracy.

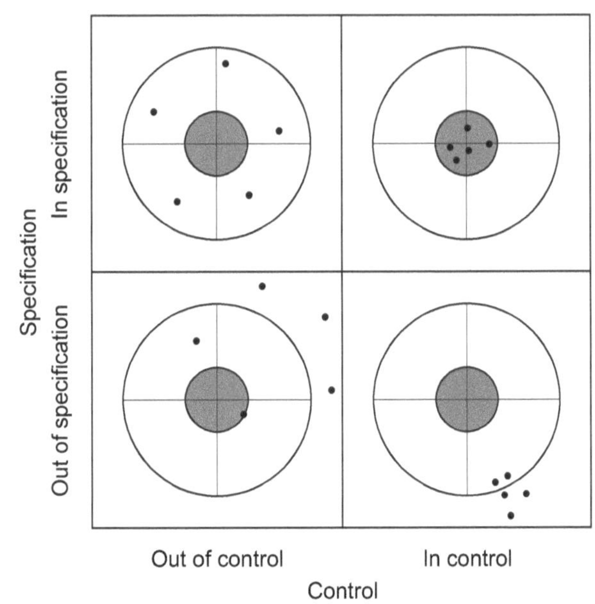


### Meeting specification and being in control are not the same thing

As with precision and accuracy, being in control and in specification are not the same thing. The ideal is to be in control and in specification.

# Appendix 6 - Measurement systems analysis

## There is no right answer

Assessing the quality of products relies on measurements but no measurement can ever be exactly correct and there is an 'uncertainty' associated with every measurement. This 'uncertainty' tells us about the overall quality of the measurement and the measurement process.

- **Tip -** Uncertainty is not the same thing as 'error' because error (wherever it comes from) can be known or estimated and corrected for.

Measurement processes are the same as any other process and they will suffer from the same type of variation in results as parts. This means that a measurement is never a single point but a distribution as shown on the upper right. When measurement and the spread of uncertainty is entirely inside or outside the desired limit then the result is either definitely good or definitely bad. The problem area is when the measurement is close to the limit and the spread of uncertainty extends across the limit. In these cases, it is impossible to absolutely decide if the result is good or bad, i.e. without a knowledge of the uncertainty then we might accept 'bad' parts or take action when it is not necessary or reject 'good' parts and take no action when it is necessary.

## The two approaches

Controlling the measurement process is a vital and developing area of quality management and there are two basic approaches:

- The 'uncertainty' approach: This is based on ISO/IEC Guide 98-3:2008 'Guide to the expression of uncertainty in measurement' (also known as GUM). This approach provides information on the sources of uncertainty and how they can be treated and combined to give an overall uncertainty for a given measurement. An excellent introduction to this approach is given in 'A Beginner's Guide to Uncertainty of Measurement' by Stephanie Bell (available free from www.npl.co.uk/publications/guides/).
- The 'measurement systems analysis' (MSA) approach: This is based largely on the AIAG publication 'Measurement Systems Analysis' (see box on opposite page). This approach provides more understanding of the actual measurement process and attempts to allocate the variations in the process to allow variation reduction and measurement improvement.

For plastics processors, we believe that the measurement systems analysis approach provides a more practical method to

Measurements tell us about the quality of the product.

Uncertainty and measurement systems analysis tells us about the quality of the measurement.

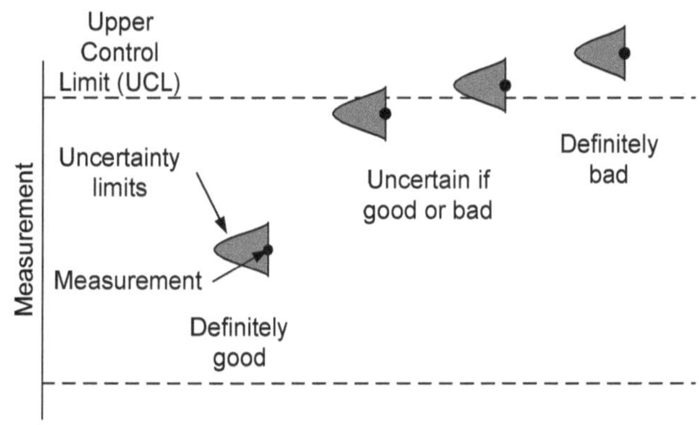


### Measurement variations can affect decisions

If we do not know the size of the measurement variation then when a measurement is close to the limit we may take action when it is not needed or, alternatively, take no action when action is actually needed.

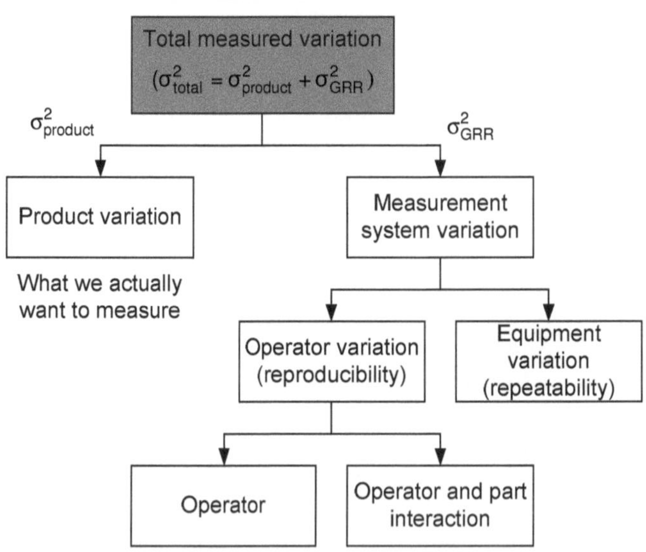


### The total measured variation

We really want to know about part variation (so we can measure and control the process) but we will always have additional variations due to the measurement system. This variation must be quantified to have confidence in the measurement.

analyse and improve the measurement process.

- **Tip** - The AIAG approach is required by most large automotive manufacturers.

## Sources of variation

Appendix 5 discussed the concepts of precision and accuracy in processes in general terms but this is equally applicable to measurement processes and the sources of variation in a measurement can be divided into those affecting accuracy and those affecting precision.

### Accuracy (location)

Accuracy is being close to the real or reference value, i.e. in a measurement system it is being 'on target'. Accuracy will be affected by:

- Bias: This is when the measured value is different from a reference value. Bias can be corrected by calibration traceable to national and international standards.
- Linearity: This is how the bias varies over the measurement range, i.e. is the bias constant over the measurement range? Linearity can be accounted for by calibration.

Accuracy is generally the result of systematic errors and can be corrected for by suitable corrections to the measured values. It is also possible to correct for low accuracy (however caused) by improving areas such as equipment quality, equipment maintenance, equipment calibration and environmental consistency.

### Precision (spread)

Precision is how close individual measurements are to each other, i.e. in a measurement system it is being 'consistent'. Precision will be affected by:

- Reproducibility (appraiser variation): This is the variation in measurements from the same measuring equipment when using different operators or appraisers, i.e. do two operators measuring the same part with the same equipment get the same result? This is the Appraiser Variation (AV).

  A high AV value will indicate a need for operator training or equipment redesign to make measuring easier.
- Repeatability (equipment variation): This is the variation between multiple measurements when using a single operator or appraiser, i.e. does the same operator measuring the same part with the same equipment get the same result? This is the Equipment Variation (EV).

  A high EV value will indicate a need for equipment maintenance or redesign.

Precision is the result of random variation and it is not possible to simply 'correct' for these variations.

The combined estimate of the measurement system reproducibility and repeatability is referred to as the Gauge Repeatability and Reproducibility (or GRR) and the variance ($\sigma^2$) of the GRR can be represented by:

$$\sigma^2_{GRR} = \sigma^2_{repeatability} + \sigma^2_{reproducibility}$$

where:

$\sigma_{repeatability}$ = the standard deviation for repeatability.

$\sigma_{reproducibility}$ = the standard deviation for reproducibility.

The resolution of measuring equipment should be at least 1/10 of the range to be measured.

It is even better if the resolution of the measuring equipment is 1/10 of the process variation.

## Total measurement system variation

As for processes, it is possible to define the short-term measurement system capability by:

$$\sigma^2_{capability} = \sigma^2_{bias/linearity} + \sigma^2_{GRR}$$

where:

$\sigma_{bias/linearity}$ = the standard deviation for any uncorrected bias or linearity.

It is also possible to define the long-term measurement system performance by:

$$\sigma^2_{performance} = \sigma^2_{capability} + \sigma^2_{stability} + \sigma^2_{consistency}$$

where:

$\sigma_{stability}$ = the standard deviation for variations in stability.

$\sigma_{consistency}$ = the standard deviation for variations in consistency.

MSA assumes a measurement system that is stable and in control. As with processes, this can be checked using control charts for stability and control.

Stability = the change in bias/ linearity with time.

Consistency = the change in repeatability with time.

### All you ever wanted to know

The AIAG Measurement Systems Analysis (MSA) Reference Manual (4th edition 2010) is a treasure trove of information on good practice and procedures in measurement systems analysis ('Measurement Systems Analysis', ISBN: 978-1-60534211-5). Get a copy from AIAG (www.aiag.org). Some of the procedures are slightly contentious (see Appendix 7) but overall it is an excellent text to begin with.

My original copy of this (2nd edition from 1995) is well worn. Be advised of changes in the $K_1$, $K_2$ and $K_3$ values between the 3rd and the 4th editions which many calculators and spreadsheets on the Internet do not seem to have updated.

# Appendix 7 - Evaluating measurement systems

## The gauge repeatability and reproducibility (GRR) study

Before conducting a gauge R&R study it is necessary to assess the accuracy elements of the measurement system. This involves assessing the bias and linearity of the measurement system. Chapter III Section B of the AIAG MSA guide (see Appendix 6) gives details of how to assess bias, linearity and stability using graphical, numerical and control chart methods.

This must be done before carrying out the gauge R&R study to remove accuracy (location) errors. The reader is referred to the AIAG guide for details of these.

## Decide on the GRR method

There are 3 different methods of carrying out a gauge study:

- ANOVA method: This is the preferred method because it is the most flexible and gives the most information but it requires good software, e.g. Minitab™, and an experienced user to interpret the output data. This method is not considered further because not all companies have both of these.
- Average and range method: This is the most common method as it can be done manually or using a simple spreadsheet. This method allows estimation of repeatability and reproducibility but not operator to part interaction. This is the method that is followed in this Workbook.
- Range method: This does not provide as much information as the average and range or ANOVA methods and it is generally used only to check that the GRR has not changed from a previous study.

## Carry out the study

The standard study involves 3 appraisers (operators) who take 3 readings on a set of 10 sample parts using the same gauge. The samples should represent the whole range of the process variation, i.e. do not choose samples that are all on specification, it is best to have some at the lower and higher limits so that the part variation is effectively represented.

1. Each sample is numbered and the sample number hidden from the appraisers so that they do not know which sample they are measuring.
2. Each appraiser (A, B and C) measures the 10 parts x 3 trials, i.e. 30 measurements, in a random order and the results are recorded by part number and appraiser.
3. For each appraiser, calculate the global average from the complete set of 30 results and record this as $\overline{X}_{appraiser}$ for the relevant appraiser.
4. Calculate $\overline{X}_{DIFF}$ from:
   $\overline{X}_{DIFF} = \text{Max } \overline{X}_{appraiser} - \text{Min } \overline{X}_{appraiser}$
5. For each set of samples and each appraiser calculate the range of the measured values and then calculate the average range ($\overline{R}_{appraiser}$) for each appraiser.
6. Calculate the global $\overline{\overline{R}}$ from:
   $$\overline{\overline{R}} = \frac{\overline{R}_{appraiser\ 1} + \overline{R}_{appraiser\ 2} + \overline{R}_{appraiser\ 3}}{\text{Number of appraisers} = 3}$$
7. Calculate $UCL_R$ from:
   $UCL_R = \overline{\overline{R}} \times D_4$ where $D_4 = 3.267$ for 2 sets of reading and 2.574 for 3 sets of readings (see Appendix 2).
8. If any readings have a range > $UCL_R$ then discard these readings and repeat using the same appraiser and sample. Recalculate the $UCL_R$ using the new readings.
9. For each sample calculate the average of all the appraisers readings ($\overline{X}_{part}$) and then calculate $R_p$ from:
   $R_p = \text{Max } \overline{X}_{part} - \text{Min } \overline{X}_{part}$.
10. You should now have 3 values, i.e. $\overline{X}_{DIFF}$, $\overline{\overline{R}}$ and $R_p$.

## Analysing the results

The conduct of the study is relatively straightforward but the analysis of the results is still contentious and there are three basic methods. These result from criticisms of the AIAG on the basis that it uses ratios which are not mathematically valid. The three methods are described below with brief comments.

### 1. AIAG method[1]

**Repeatability (EV):**

This measures the equipment variation:

$$EV = \overline{\overline{R}} \times K_1$$

where $K_1$ varies with the number of sets of readings:

| Trials | $K_1$ |
|---|---|
| 2 | 0.8862 |
| 3 | 0.5908 |

**Reproducibility (AV):**

This measures the appraiser variation:

$$AV = \sqrt{(\overline{X}_{DIFF} \times K_2)^2 - (EV^2/(nr))}$$ (default to 0 if -ve)

where:

n = number of appraisers (3 in the example).

r = number of parts (10 in the example).

$K_2$ varies with the number of appraisers:

| Appraisers | $K_2$ |
|---|---|
| 2 | 0.7071 |
| 3 | 0.5231 |

**Gauge R&R (GRR)**

$$GRR = \sqrt{EV^2 + AV^2}$$

**Part variation**

This measures the total part variation:

$$PV = R_p \times K_3$$

where $K_3$ varies with the number of parts:

| Number of parts | $K_3$ |
|---|---|
| 9 | 0.3249 |
| 10 | 0.3146 |

**Total variation**

This is the total variation of the study:

$$TV = \sqrt{GRR^2 + PV^2}$$

**ndc (no. of distinct categories)**

$$ndc = 1.41\left(PV/GRR\right)$$ should be > 5.

The AIAG method then calculates:

- %EV = (EV/TV)%
- %AV = (AV/TV)%
- %GRR = (GRR/TV)%
- %PV = (PV/TV)%

The GRR is assessed by:

- %GRR < 10% - OK.
- 10% < %GRR < 30% - Possibly OK.
- % GRR > 30% - Needs improvement.

### 2. Wheeler and Ermer methods[2,3]

Wheeler and Ermer both agree with the AIAG calculations for EV, AV, GRR and TV as estimators of the variations (Wheeler states that they are not the only ones and are not always unbiased but they do provide reasonable estimates).

However at the assessment stage, where the %'s are calculated, the AIAG simply says 'The sum of each percent consumed by each factor will not equal 100%' and no explanation is given for this.

The reason, as rightly pointed out by both Wheeler and Ermer, is that the %GRR etc. are not percentages but are trigonometric functions and should be calculated from:

$$\%EV = (EV^2/TV^2)$$

$$\%AV = (AV^2/TV^2)$$

$$\%GRR = (GRR^2/TV^2)$$

$$\%PV = (PV^2/TV^2)$$

These correct proportions will add up to 100% and then represent the actual contribution of each factor to the total variation.

### 3. Notes:

1. Considerable caution should be used with the AIAG method due to the effective use of %'s of the standard deviations (σ) rather than %'s of the variances ($σ^2$).
2. Be especially careful about condemning a gauge on the basis of the 10%, 10%-30% and >30% rules.
3. If in doubt then read the papers and form your own opinion but remember that your customers may simply want an AIAG report whether it is correct or not - this is the power of the market.
4. Be aware of changes made in AIAG MSA between the 3rd and the 4th editions - the $K_n$ values changed to reflect changes in the methods. Many of the spreadsheets available on the Internet do not reflect these changes. Check the constants before using any external spreadsheet.
5. Practitioners agree that ANOVA is best but there are still people who use average and range.


- 1. AIAG. 2010. '*Measurement Systems Analysis*'. AIAG.
- 2. Wheeler, D. 2006. '*An Honest Gauge R&R Study*'. www.spcpress.com/pdf/DJW189.pdf.
- 3. Ermer, D. 2006. '*Improved Gage R&R Measurement Studies*'. Quality Progress. March 2006.

# Postscript

As always, this book was born out of conversations with the people at the coal-face of plastics processing. Practitioners knew that they needed to do something concrete to improve perceived and actual quality levels but lacked guidance on simple ways to start. The available texts were difficult to understand and launched into high-level statistics, they provided little real guidance on the very real problems that plastics processors faced and the inevitable result was that processors continued to do what they have always done well - make products as best they could.

Hopefully this Workbook makes a complex subject simple and demystifies the statistical aspects of quality management so that it can be both understood and applied. Above all, hopefully it will enable processors to make their products better, to reduce costs and to delight the customer.

There is always an element of 'a labour of love' in the creation of any book and in this case it allowed me to get all sorts of ideas that had been rolling around in my head for nearly 40 years down on paper. There is a wealth of information on quality management obtainable from other sources but it is invariably general and not specific to plastics processing. This makes it difficult for managers in the plastics industry to access the information and to assess how relevant it is to their operations.

As with my other books, this is a Workbook and not a sacred text, it is designed for real use in real plastics processing companies. Please use and abuse it as such. All of the concepts and projects discussed in the book will work in practice but it can never be comprehensive in scope or coverage. If I have made any mistakes then they are mine alone - please do not blame any of the myriad of sources that I have used.

As ever, in the planning stage of this book I wondered if there was sufficient information specific to the plastics industry to make such a book worthwhile. Equally, in the writing stage it was more a case of what to leave out (to keep the book to a manageable size) rather than what to include - there was simply too much to fit into the available space and the book grew considerably from that originally envisioned.

The variety of the plastics processing industry means that this cannot be a definitive work - space and time do not permit this. If there are significant omissions then please let me know and I will update the text for future editions.

I have acknowledged as many of the sources as possible in the text but it is impossible to acknowledge all those people who contributed practical assistance and ideas (good and bad) during my time in the plastics industry.

"Life is an experiment without a control group."

**Joachim Krueger**

# Abbreviations and acronyms

**ABC** - Activity Based Costing
**Ac** - Acceptance Number
**ABM** - Activity Based Management
**AIAG** - Automotive Industry Action Group
**ANOM** - Analysis of Means
**ANOVA** - Analysis of Variance
**APC** - Automatic Process Control
**APQP** - Advanced Product Quality Planning
**AQL** - Acceptance Quality Level
**ASPC** - Algorithmic Statistical Process Control
**AV** - Appraiser Variation
**BOM** - Bill of Materials
**BPR** - Business Process Re-engineering
**c-chart** - Count control chart
$C_m$ - Machine Capability Index
$C_p$ - Process Capability Index, measure of process spread
$C_{pk}$ - Minimum process index, measure of process location
**CAD** - Computer Aided Design
**CEDAC** - Cause and Effect Diagram Adding Cards
**COC** - Certificate of Conformance
**CoPQ** - Cost of Poor Quality
**CPM** - Critical Path Method
**CTQ** - Critical To Quality
**CR** - Capability Ratio
**CRM** - Customer Relationship Management
**CRQ** - Consumer's Risk Quality
**CUSUM** - CUmulative SUM
**DCF** - Discounted Cash Flow
**DFA** - Design for Assembly
**DFM** - Design for Manufacture
**DFx** - Design for x (Assembly, Manufacture etc.)
**DMA** - Dynamic Mechanical Analysis
**DOE** - Design of Experiments
**DSC** - Differential Scanning Calorimetry
**DTC** - Design to Target Cost
**EBM** - Extrusion Blow Moulding
**EPC** - Engineering Process Control
**EV** - Equipment Variation
**EWMA** - Exponentially Weighted Moving Average
**FEA** - Finite Element Analysis
**FMCG** - Fast Moving Consumer Goods
**FMEA** - Failure Modes and Effects Analysis
**FMECA** - Failure Modes and Effect Criticality analysis
**FT-IR** - Fourier Transform Infrared
**GPC** - Gel Permeation Chromatography
**GRR** - Gauge Repeatability and Reproducibility
**HDD** - Heating Degree Days
**HDT** - Heat Deflection Test
**IBM** - Injection Blow Moulding
**Ishikawa Diagram** - Cause and effect diagram
**ISBM** - Injection Stretch Blow Moulding
**ISIR** - Initial Sample Inspection Report
**ISO 9001** - International Standard for quality management systems
**ISO 14001** - International Standard for environmental management systems
**ISO 19011** - International Standard for auditing management systems
**ISO 50001** - International Standard for energy management systems
**JEDI** - Just Enough Desirable Inventory
**Jidoka** - 'Human automation' of the process by standard operations to prevent errors
**JIT** - Just-In-Time
**Kaizen** - Continuous improvement by small incremental steps
**KANBAN** - Japanese word for signal
**KFS** - Key Factors for Success
**LAL** - Lower action limit
**LCL** - Lower Control Limit
**LQ** - Limiting Quality
**LSL** - Lower Set Limit or Lower Acceptable Tolerance
**LWL** - Lower warning limit
**Me-chart** - Control chart using the sample median (Me)
**MIS** - Management Information System
**MR chart** - Control chart using the moving range MR
**MSA** - Measurement Systems Analysis
**MVA** - Multi-Variate Analysis
**MTBF** - Mean Time Between Failure
**M&T** - Monitoring and targeting
**np chart** - Number of categorised units control chart
**NPV** - Net Present Value
**OC** - Operating Curve
**OEE** - Overall Equipment Effectiveness

**OHSAS 18001** - Occupation Health and Safety Assessment Series Standard

**p-chart** - proportion categorised units control chart

**P-chart** - percent categorised units control chart

**$P_m$** - machine performance capability index

**$P_{mk}$** - minimum machine performance capability

**$P_p$** - process potential index

**$P_{pk}$** - process performance index

**PAF** - Prevention, Appraisal and Failure

**Pareto Principle** - The 80:20 rule

**PCA** - Principle Component Analysis

**PDCA** - Plan-Do-Check-Act

**PDPC** - Process Decision Program Chart

**PDS** - Product Design Specification

**PERT** - Program Evaluation Review Technique

**PPAP** - Production Parts Approval Process

**Poke-yoke** - Mistake proofing or the building of fool-proof devices to ensure quality

**QA** - Quality Assurance

**QC** - Quality Control

**QFD** - Quality Function Deployment

**QI** - Quality Improvement

**QM** - Quality Management

**QMR** - Quality Management Representative

**QMS** - Quality Management System

**QTQ** - Quantified Total Quality

**R-chart** - control chart using the sample range R

**Re** - Rejection number

**RCM** - Reliability Centred Maintenance

**ROI** - Return on investment

**RPN** - Risk Priority Number

**RSM** - Response Surface Methodology

**5S** - 'seiri, seiton, seiso, seiktsu and shitsuke' or 'organisation, orderliness, cleanliness, standardised cleanup and discipline'

**σ (sigma)** - the standard deviation of a population

**6 Sigma** - quality management based on the production of parts within 6 standard deviations of the mean

**s** - the standard deviation of a sample

**s-chart** - control chart using the standard deviation, realised value

**SMED** - Single Minute Exchange of Dies

**SPC** - Statistical Process Control

**SQC** - Statistical Quality Control

**SWOT** - Strengths, Weaknesses, Opportunities and Threats

**Taguchi** - Design of experiments to allow more than one variable to be changed at a time and reduce the number of experiments

**TAC** - Total Acquisition Cost

**TGA** - Thermo-gravimetric analysis

**TMA** - Thermo-mechanical Analysis

**TPM** - Total Productive Maintenance

**TPP** - Total Product Planning

**TQM** - Total Quality Management

**u-chart** - count per unit control chart

**UAL** - Upper Action Limit

**UCL** - Upper Control Limit

**USL** - Upper Set Limit or Upper Acceptable Tolerance

**UWL** - Upper Warning Limit

**VA** - Value Analysis

**VE** - Value Engineering

**WCM** - World Class Manufacturing

**X** - individual measured value

**X (Xbar)** - subgroup average

**X (Xbar)-chart** - control chart using the sample average X